THE KING LIBRARY
KING COLLEGE
BRISTOL, TENNESSEE
WITHDRAWN

D1358378

THEORY AND OBSERVATION
OF
NORMAL STELLAR ATMOSPHERES

THEORY AND OBSERVATION
OF
NORMAL STELLAR ATMOSPHERES

PROCEEDINGS
OF THE
THIRD HARVARD-SMITHSONIAN CONFERENCE
ON
STELLAR ATMOSPHERES

Owen Gingerich, *Editor*

Owen Gingerich
20 November 06

THE M.I.T. PRESS

Cambridge, Massachusetts, and London, England

THE E. W. KING LIBRARY
KING COLLEGE
BRISTOL, TENNESSEE

523.86
H339t

Copyright © 1969 by
The Massachusetts Institute of Technology

Printed by Halliday Lithograph Corp.
Bound in the United States of America by The Colonial Press Inc.

All rights reserved. No part of this book may be reproduced in
any form or by any means, electronic or mechanical, including
photocopying, recording, or by any information storage and
retrieval system, without permission in writing from the publisher.

SBN 262 07035 9 (hardcover)

Library of Congress catalog card number: 69-19248

PREFACE

These proceedings document a conference on the theory and observations of normal stellar atmospheres. This meeting, convened in Cambridge, Massachusetts, on April 8-11, 1968, under the auspices of the Smithsonian Astrophysical Observatory and the Harvard College Observatory, was attended by nearly 100 astronomers from 19 different countries. It was the third in the series of Harvard-Smithsonian conferences on stellar atmospheres.

When the first conference took place on January 20-21, 1964, the art of computing nongray stellar atmospheres was still in its infancy. That conference was held to exchange ideas on the methods and data of such computations and to provide an early check on the mutual consistency of various programs. In the following year, a second conference assembled to facilitate discussions among the increasing number of astrophysicists who were studying the theory of spectral-line formation. The proceedings of the first two conferences were published in the Smithsonian Astrophysical Observatory's Special Reports 167 and 174, respectively.

The third conference set out to examine the extent to which quantitative spectral classification could be interpreted by "classical" model atmospheres. These were defined, for the purpose of the conference, by the assumptions of hydrostatic equilibrium, flux constancy, statistical equilibrium, and plane geometry. In a sense, the conference represented the culmination of this "classical" approach. Mathematical techniques whose absence had prevented a serious attack on this problem until about 10 years ago played an important part in the discussions of the first conference. In contrast, the third conference could deal primarily with physical problems. We hope that this analysis of the successes and limitations of classical models will pave the way for a more searching examination of the astrophysical assumptions and for the computational attack on peculiar and abnormal stars.

These proceedings include a series of papers on both the theory and the spectra of normal stars. Most of the papers were presented at the conference, but the order here has been rearranged. The transcription of the tape-recorded discussions following the papers has to some extent been modified by the participants after the conference. Their remarks as printed retain the flavor of the original discussion, yet they contain additional information, references, and in some instances entirely new contributions.

70410

Included with this volume is a standard grid of over 50 atmospheres, computed by Duane Carbon, Robert Kurucz, and myself. If the conference has achieved one of its principal goals, a grand summation of the successes of the "classical" theory, this may be the last grid of model atmospheres to be published _in extenso_. We have tried to make them useful both as a "quick-look" reference and as a comparison for other workers with their own computer programs. Especially for this latter reason, we have endeavored to produce the most realistic grid within today's state of the art, and consequently we have included our best guess at blanketing as well as a representation of convection. If further progress reveals that these features have fallen wide of the mark or if our choice of parameters and models to be tabulated seems arbitrary, the blame must rest with our cloudy crystal ball.

Unfortunately, the perfecting of the computer programs to produce the model-atmosphere grid required far more time than we had anticipated. Consequently, the release of this volume has been repeatedly postponed, much to our regret and embarrassment. We can only hope that the unprecedented scope of the standard grid compensates in part for this unhappy delay.

The conference was organized by Charles A. Whitney and myself, with the close assistance of Eugene H. Avrett and Wolfgang Kalkofen. We are very grateful to the many members of the Harvard and Smithsonian Observatory staff who helped in the organization of this meeting and the preparation of the proceedings. Special thanks are due to Mrs. Nancy Adler, Miss Carolyn Yarid, David Latham, and Jay Pasachoff. The editorial staff of the Smithsonian Astrophysical Observatory, especially Dr. Rosa Goldstein, have dealt effectively with the technical problems presented by these proceedings. Mr. Michael Connolly, of the M.I.T. Press, has served us well with his knowledge and patience.

We wish to thank the Alfred P. Sloan Foundation for a grant of $9,920 in support of travel expenses for delegates from foreign countries. The generosity of the Hilles Library of Radcliffe College in making available their conference room is appreciatively acknowledged.

We are grateful for the hospitality and support provided by our hosts, Dr. Fred L. Whipple, Director of the Smithsonian Astrophysical Observatory, and Dr. Leo Goldberg, Director of the Harvard College Observatory.

Owen Gingerich
Editor

vi

TABLE OF CONTENTS

TABLE OF CONTENTS (Cont.)

LIST OF PARTICIPANTS

L. H. Auer. Joint Institute for Laboratory Astrophysics, Boulder, Colorado.

J. R. Auman, Jr. Princeton University Observatory, Princeton, New Jersey.

E. H. Avrett. Harvard College Observatory and Smithsonian Astrophysical Observatory, Cambridge, Massachusetts.

R. C. Bless. Space Astronomy Laboratory, Washburn Observatory, University of Wisconsin, Wisconsin.

E. Böhm-Vitense. Astronomy Department, University of Washington, Seattle, Washington.

W. Buscombe. Mount Stromlo Observatory, Canberra, A. C. T., Australia.

R. Cayrel. Observatoire de Paris, Meudon, France.

G. Cayrel de Strobel. Institute d'Astrophysique, Paris, France.

G. W. Collins, II. McMillan Observatory, Ohio State University, Columbus, Ohio.

D. L. Crawford. Kitt Peak National Observatory, Tucson, Arizona.

Y. Cuny. Observatoire de Paris, Meudon, France.

P. Demarque. Department of Astronomy, University of Chicago, Chicago, Illinois.

O. J. Eggen. Mount Stromlo and Siding Spring Observatories, Institute for Advanced Studies, Australian National University, Canberra, A. C. T., Australia.

G. H. E. Elste (presented Williams' paper). Department of Astronomy, University of Michigan, Ann Arbor, Michigan.

P. Feautrier. Observatoire de Paris, Meudon, France.

C. Fehrenbach. Observatoire de Marseille, Marseilles, France.

D. Fischel. Goddard Space Flight Center, National Aeronautics and Space Administration, Greenbelt, Maryland.

I. Furenlid. Stockholm Observatory, Saltsjöbaden, Sweden.

O. Gingerich. Smithsonian Astrophysical Observatory and Harvard College Observatory, Cambridge, Massachusetts.

M. Golay. Observatoire de Genève, Switzerland.

D. Gray. The University of Western Ontario, London, Ontario, Canada.

J. R. W. Heintze. Sonnenborgh Observatory, Utrecht, The Netherlands.

L. Houziaux. Université de Liège, Liège, Belgium.

K. Hunger. Goddard Space Flight Center, National Aeronautics and Space Administration, Greenbelt, Maryland.

A. R. Hyland. Mount Wilson and Palomar Observatories, Carnegie Institute of Washington, and California Institute of Technology, Pasadena, California.

W. Iwanowska. Astronomical Observatory of N. Copernicus University, Toruń, Poland.

H. R. Johnson. Astronomy Department, Indiana University, Bloomington, Indiana.

D. H. P. Jones. Radcliffe Observatory, Pretoria, South Africa.

W. Kalkofen. Smithsonian Astrophysical Observatory, Cambridge, Massachusetts.

L. D. Kaplan (presented Connes' paper). Department of Geology, Massachusetts Institute of Technology, Cambridge, Massachusetts.

D. A. Klinglesmith. Goddard Space Flight Center, National Aeronautics and Space Administration, Greenbelt, Maryland.

H. Lamers. Sonnenborgh Observatory, Utrecht, The Netherlands.

J. Landi Dessy. Observatorio Astronomico e IMAF, Universidad Nacional de Cordoba, Argentina.

M. Lecar. Smithsonian Astrophysical Observatory and Harvard College ˣ Observatory, Cambridge, Massachusetts.

S. Matsushima. Department of Astronomy, Pennsylvania State University, University Park, Pennsylvania.

E. E. Mendoza, V. Observatorio Astronomico Nacional, University of Mexico, Mexico, and Astronomy Department, University of Chili, Santiago, Chili.

L. Mertz. Harvard College Observatory, Cambridge, Massachusetts.

D. Mihalas. Joint Institute for Laboratory Astrophysics, Boulder, Colorado.

D. C. Morton. Princeton University Observatory, Princeton, New Jersey.

V. P. Myerscough. Mathematics Department, Queen Mary College, University of London, London.

K. Nariai. Goddard Space Flight Center, National Aeronautics and Space Administration, Greenbelt, Maryland.

J. S. Neff. Department of Physics and Astronomy, University of Iowa, Iowa City, Iowa.

E. C. Olson. University of Illinois Observatory, Urbana, Illinois.

J. M. Pasachoff. Harvard College Observatory and Smithsonian Astrophysical Observatory, Cambridge, Massachusetts.

D. W. Peat. University Observatories, Cambridge, England.

D. M. Peterson. Smithsonian Astrophysical Observatory and Harvard College Observatory, Cambridge, Massachusetts.

D. M. Popper. Department of Astronomy, University of California, Los Angeles, California.

F. Praderie. Observatoire de Paris, Meudon, France.

M. J. Price. Kitt Peak National Observatory, Tucson, Arizona.

U. Sinnerstad. Stockholm Observatory, Saltsjöbaden, Sweden.

F. Spite. Observatoire de Paris, Meudon, France.

J. C. Stewart. Joint Institute for Laboratory Astrophysics, Boulder, Colorado.

D. W. N. Stibbs. University Observatory, Saint Andrews, Fife, Great Britain.

P. A. Strittmatter. Institute of Theoretical Astronomy, University of Cambridge, Cambridge, England.

S. E. Strom. Smithsonian Astrophysical Observatory and Harvard College Observatory, Cambridge, Massachusetts.

B. Strömgren. The Observatory, Copenhagen, Denmark.

T. L. Swihart. Steward Observatory, University of Arizona, Tucson, Arizona.

R. N. Thomas. Joint Institute for Laboratory Astrophysics, Boulder, Colorado.

A. B. Underhill. Sonnenborgh Observatory, Utrecht, The Netherlands.

M. S. Vardya. Tata Institute of Fundamental Research, Bombay, India.

G. Wallerstein. University of Washington, Seattle, Washington.

V. Weidemann. Institut für Theoretische Phyik und Sternwarte der Universität, Kiel, Germany.

C. A. Whitney. Smithsonian Astrophysical Observatory and Harvard College Observatory, Cambridge, Massachusetts.

REMARKS IN DEDICATION OF THIS VOLUME

April 10, 1968

Charles A. Whitney

It seems to me that, among the fields of science, those relating to spectral classification and the theory of spectral types have benefited most from the direct contributions of women.

I am, of course, thinking of the work of such women as Lady Huggins,* who held the telescope on many a cold night for her husband during the "Victorian Era" of spectral classification. And then Mrs. Henry Draper decided to memorialize her husband by supporting what she thought would be a modest program of spectrum studies under Edward C. Pickering here at Harvard. And I recall the names of Miss Maury, Mrs. Fleming, Miss Cannon, and finally Mrs. Mayall and Miss Hoffleit.

In 1920, Saha invented† an equation, and he laid the theoretical basis for the derivation of a temperature scale for stellar atmospheres. It was Miss Cecilia Payne who established this scale, and within 10 years she had published the first two monographs on the physics of stellar atmospheres. The questions posed in those books are still under discussion.

Within another few years, Albrecht Unsöld had systematically applied the quantum theory to the calculation of opacity, and had applied the theory of radiative transfer to the calculation of temperature structures. Strömgren soon initiated his fundamental studies of the Hertzsprung-Russell diagram.

There have been only three changes from those days, but these changes have drastically altered the color of our work. In the first place, we now realize that the chemical composition of stars is not internally homogeneous. In the second place, we have learned, with the help of Mrs. Böhm-Vitense, to examine the sun and its convection zone for clues to the stars; and we have learned from Mrs. Cayrel de Strobel that some stars are very peculiar, and that all stars are great fun.

In the third place, the Saha equation has now become the Saha-Boltzmann equation with the introduction of the concept of excitation. But now we have another woman, Anne Underhill, who tells us that the Saha-Boltzmann equation is passé.

Miss Payne is now Mrs. Gaposchkin, the wife of another gifted astronomer, and as an expression of our admiration for the role of women in our field, we shall dedicate to her the proceedings of this conference. May I present Mrs. Cecilia Payne-Gaposchkin.

---

*The question: "What about Caroline Herschel?" arose from the audience, and it was pointed out that she was not a spectroscopist, and we were trying to keep our classification criteria "pure."

†A member of the audience objected to the use of the word "invented," but the objection was put down with the comment that all equations are invented.

Response by Mrs. Cecilia H. Payne-Gaposchkin

Well, Chuck, I am sorry to say that I don't think you have given me my deserts. In the words of my favorite author: "Use every man after his deserts, and who shall 'scape whipping'?" Perhaps I don't recognize a whipping when I get one; in other words, perhaps I don't know when I am beaten. I remember when, as a student at Cambridge, I decided I wanted to be an astronomer and asked the advice of Colonel Stratton, he replied, "You can't expect to be anything but an amateur." I should have been discouraged, but I wasn't, so I asked Eddington the same question. He (as was his way) thought it over a very long time and finally said: "I can see no insuperable obstacle."

I had started out as a field naturalist, and I think I carried the spirit of the field naturalist into astronomy. I remember the excitement of going through the Harvard specturm plates—thousands of them. I still remember the indescribable thrill of coming on the spectrum of γ Velorum — an indescribable but not unrepeatable experience. A couple of months ago I felt the same thrill when I first saw the ultraviolet spectrum of the solar corona, but that was a deeper experience because I had had many years in which to anticipate what the corona would look like in the ultraviolet.

I suppose I am the oldest astronomer in this gathering, and as such I have accumulated a great deal of experience. In fact, I have long planned to write a book entitled How Not to Do Research, a subject on which I am a great authority. But I'm afraid it couldn't be published for a hundred years because a great deal of it would probably be actionable.

Looking back on my years of research, I don't like to dwell only on my mistakes; I am inclined to count my blessings, and two seem to me to be especially valuable. The first blessing is that the process of discovery is gradual—if we were confronted with all the facts at once we should be so bewildered that we should not know how to interpret them. The second blessing is that we are not immortal. I say this because, after all, the human mind is not pliable enough to adapt to the continual changes in scientific ideas and techniques. I suspect there are still many astronomers who are working on problems, and with equipment, that are many years out of date.

Now that I am old, I see that it is dangerous to be in too much of a hurry, to be too anxious to see the final result oneself. Our research does not belong to us, to our institution, or to our country. It belongs to mankind.

And so I say to you, the young generation of astronomers: more power to you. May you continue to expand the picture of the universe, and may you never lose the thrill it gave you when it first broke on you in all its glory.

PART I

THE EMPIRICAL BASIS OF QUANTITATIVE

CLASSIFICATION

D. L. Crawford, Chairman

QUANTITATIVE SPECTRAL CLASSIFICATION

C. A. Whitney

Smithsonian Astrophysical Observatory and
Harvard College Observatory, Cambridge, Massachusetts

1. INTRODUCTION

The first and second monographs on stellar atmospheres, both of which were written by Cecilia H. Payne-Gaposchkin during the 1920's, systematically studied the quantitative basis of spectral classification (Payne, 1925, 1930). Her aim was astrophysical, or "academic" in the sense that I defined in my opening remarks, but further development of quantitative classification has had a broader astronomical motivation. It has achieved its goals in two ways.

First, through enhanced precision, the quantitative methods have led to a hundred-fold increase in the number of discernible categories of stellar spectra. For example, if we adopt a precision of 0.005 mag for a photo-electrically determined spectrum index, the u-b index of the Strömgren-Crawford system will divide the B stars into 200 quanta of spectral type. As another example, we note that classification based on photographically determined line intensities divides the B stars into approximately 75 groups. Both these techniques have refined the spectral type by a full order of magnitude over the best modern visual systems.

From the following relationship between the absolute visual magnitude M_V and the Hβ index,

$$\frac{\Delta M_V}{\Delta \beta} = \frac{4}{0.15} = 30,$$

we estimate that an observational uncertainty of 0.005 mag in the Hβ index should correspond to 0.15 mag in M_V. Actually, the observed deviation from the calibration curve of Hβ is about 0.25 mag, and this discrepancy is attributed to the need for further parameters in classifying the B stars.

The second advantage of the quantitative techniques is the possibility of adding further parameters, or dimensions, to the classification scheme. The adopted level of sophistication is set by the astrophysical requirements and the limitations of the method of data analysis. At present, there are several systems that employ measurements in seven pass bands, and, setting aside one pair for the elimination of interstellar reddening, it would seem possible to establish systems of five dimensions. Of course, the "axes" will not be strictly perpendicular to each other, since many of the spectral features are correlated. In any case, these are more dimensions than most of us can cope with, but a clever use of electronic aids might enable us to see into these higher dimensions.

The desirable number of dimensions is determined by the number of physical parameters needed to describe a star. At the moment, it appears that we need at least the following: mass, age, initial hydrogen-helium ratio, ratio of hydrogen to selected metals (how many?), rotational velocity, and inclination of rotation axis.

With the hope of approaching the problem from a strictly objective stand-point, several authors (Deeming, 1966; Jones, 1967) have suggested applying the techniques of multivariate analysis, an n-dimensional correlation study, to the establishment and evaluation of parameters. However, since the

relative value of a particular set of parameters is judged on the basis of the
residual errors, this analysis is quite sensitive to correlations between the
measured indices and between the errors of the measurements. Both
Collins and Jones have privately reported discouraging results in attempting
to establish a meaningful statistical approach. In this regard, we should note
that Peat, at Cambridge, has established a quantitative scheme in which the
number of parameters to be determined is equal to the number of observed
indices, and in which the statistical questions are effectively set aside. Peat
describes the method elsewhere in these proceedings.

The difficulty attending the establishment of a purely objective system
emphasizes the point, made by Strömgren (1966), that virtually all work on
quantitative classification has been "on limited, scattered regions in the
H-R diagram, and it is the general framework provided by classification on
the MK system through visual inspection that ties the work together. " This
very aptly describes the central role played by the MK system as defined
in the Yerkes Atlas (Morgan, Keenan, and Kellman, 1939) and in several
subsequent addenda and modifications. Two new atlases are nearing completion,
and although these are based on the MK system, they employ criteria that were
devised for their particular spectral dispersion and resolving power, and they
represent "second-generation" classification atlases.

At the Kitt Peak National Observatory, Abt and Morgan have just completed
an atlas using a grating spectrograph with a dispersion of 128 A/mm and an
ultraviolet transmission and a resolving power that are considerably higher
than those of the prism spectrograph used for the original Yerkes Atlas
(see Abt, 1963). In South America, Landi Dessy and the Jascheks, in collaboration with Morgan, have nearly completed a new atlas at a dispersion
of 42 A/mm.   The LJJ system, described elsewhere in these proceedings
by Landi Dessy, achieves high precision by exploiting the concept of "vernier
criteria"; that is, the ratios of lines of nearly equal intensity are used to
determine precise classifications in small areas of the H-R diagram once a
rough classification has been achieved with other criteria.

To the theoretician, these new atlases are significant because they
provide a more detailed statement of classification criteria, and since the
new criteria are less subject to blending, they should be more amenable to
quantitative analysis.

2. PROPERTIES OF THE METHODS

There are a number of ways to distinguish between the various methods
of quantitative classification, and Walraven (1966) has suggested distinguishing
between the "empirical" and the "functional" techniques. The empirical
technique simply divides the spectrum into a convenient set of wavelength
bands with little regard to spectrum details, as did the original six-color
work of Stebbins and Whitford or the recent systems of Golay and of the
Walravens. For statistical studies, the empirical techniques will be very
valuable, but the theoretician will hesitate to attempt a direct interpretation
of the classification criteria. He will prefer to wait until the interesting stars
have been measured in other ways.

The functional techniques, on the other hand, measure specific attributes
of the spectrum, such as the Balmer jump or the intensity of a member of the
Balmer series. And among the functional techniques, the theoretician has
tended to distinguish between those that measure "pure" features that are
not "cosmically blended" and those that measure cosmically blended features,
such as the G band or the confluence of the high members of the Balmer series.
He has shied away from the blended features on the grounds of complexity, but
the argument no longer seems compelling, and it has become necessary to face
the task of point-by-point synthesis of detailed spectra.

The nature of the problem posed by instrumental and cosmical blending, and the technique of solving this problem, can best be discussed in terms of the absolute spectrophotometry of the so-called continuum of a star. This class of photometry is essential to the determination of photospheric temperatures and bolometric corrections, but the data are still inadequate despite several decades of hard work.

Lamla's contribution to this conference summarizes the data for α Lyr, and it indicates agreement to within about 2% among various observers for wavelengths greater than 4000 A. But at shorter wavelengths, the discrepancies reach several tens of percent, and it is quite evident that the major source of difficulty has been insufficient spectral resolution and a lack of precision in defining just <u>what</u> was being measured. Figure 1 indicates the severity of the problem, even among the early spectral types. This is Oke's (1965, Figure 7) illustration of the spectrum of three A7 stars as viewed with with a sequence of 50-A passbands; it shows how wavelength averaging drops the so-called continuum. The most practical suggestion for meeting this problem appears to be that of Oke: use a linear system, such as photoelectric photometry, with relatively broad and well-defined windows, and then make corrections on the basis of high-dispersion spectra. Where the instrumental blending depresses the spectrum by no more than 15%, high-resolution spectra can indicate the true height of the peaks. However, when the corrections are larger, they probably become meaningless, and the only answer is to synthesize spectra in detail and use only the high-dispersion spectra to verify the <u>relative</u> line intensities.

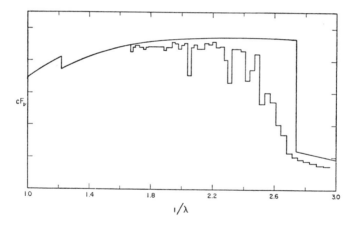

Figure 1.   The absolute energy distribution of the continuum (smooth curve) and that observed with successive 50-A bands for the average of three stars in the Hyades of spectral type A7 and θ = 0. 62 (from Oke, 1965).

Furenlid has prepared for this conference a brief review of Swedish work on spectral classification from low- and moderate-dispersion photographic spectrophotometry. This work has been valuable from the statistical standpoint, but from the astrophysical point of view, its value is limited by random errors — about 0. 05 mag in a single measurement — and by the difficulty of correcting for instrumental blending in a system that incorporates a nonlinear element such as the photographic plate. Even the excellent and numerous measurements of slit spectra obtained by Chalonge, Barbier, and their colleagues are difficult to interpret theoretically because of the influence of line blending. These classification schemes are internally consistent and are well defined from the operational standpoint, but they are "closed" systems in the sense that the measured quantities comprise a nonlinear mixture of spectral attributes and they cannot be reproduced by purely theoretical calculations; nor can they be constructed by measurements on another, external system without an elaborate calibration through the measurement of common stars.

I will confine the summary to slit spectra and photoelectric photometry.

3. SLIT SPECTRA

Adams and his colleagues at Mt. Wilson (Adams, Joy, Humason, and Brayton, 1935) and Young and Harper (1924) at the Dominion Astrophysical Observatory pioneered in the determination of spectral types and luminosity classes using semiquantitative techniques. Their classification parameters were based on estimated line intensities plotted as functions of spectral type on the Henry Draper system and of absolute magnitude.

Although subsequent work based on measured intensities has produced varying degrees of success, two points seem to have been established:

A. Errors will be intolerable unless the plate material is of a uniform and high quality.

B. Classification criteria must be carefully selected for each spectral region, spectral type, and spectral apparatus. The results of Wright and Jacobson (1966) indicate that, even with high-dispersion plates, quantitative classification will be little better than visual classification at a lower dispersion if the criteria are not rechosen for the higher dispersion.

In the first substantial effort of its kind, Hynek (1935) measured, for about 100 stars, the equivalent widths of the K line, the $H\delta$ line, and five other pairs of lines that are sensitive to absolute magnitude. He tabulated the average equivalent widths of the K line and the $H\delta$ line as functions of HD spectral class, and he studied the variation of selected pairs of luminosity discriminants that had been employed for visual classification at other observatories. Table 1 lists the lines for which Hynek measured or estimated intensities. The correlations with spectral type and luminosity found by Hynek were quite strong, and they indicated the possibility of quantitative classification to within a fraction of a subclass and to within several tenths of a magnitude.

Table 1. Luminosity criteria in Hynek's study

Identification (1)	$\lambda(1)$	:	$\lambda(2)$	Identification (2)
Sr II	4077	:	4071	Fe I
Zr II, Ti II	4161	:	4167	Mg I
Sr II	4215	:	4250	Fe I
Fe II, Fe I	4233	:	4236	Fe I
Sc II, Fe I	4246	:	4250	Fe I

At the Mount Wilson Observatory, Williams (1936a) examined absorption-line intensities in the spectra of 84 O and B stars. He tabulated total absorptions and central depths for the hydrogen lines except $H\alpha$ and for helium lines in the range 3820 to 4922 A. For virtually all other lines in the range 3820 to 4922 A, he tabulated total absorptions. Plates were available with four dispersions (10 A/mm, 27 A/mm, 39 A/mm, and 65 A/mm), and Williams concluded that, except for the weakest lines observed with the lowest of these dispersions, the central depth was independent of dispersion, while for total absorptions he found no difference even at the lowest dispersion. Intercomparison of plates on the same star indicated accidental deviations of about 6% for the hydrogen lines for all dispersions and from 20 to 35% for the lines of the metals; the larger deviations corresponded to lower dispersions.

The discrepancies Williams found between theoretical and observed hydrogen lines of the early B stars and the A stars of high luminosity have, for the most part, been eliminated by improved theories of the Stark effect.

In his next paper, Williams (1936b) established a classification of O7 to B8 stars that used line ratios for the determination of spectral type and line _intensities_ for luminosity classification. He discussed the apparent inconsistencies and ambiguities encountered in classifying B stars. He emphasized that the helium lines, as well as the hydrogen lines, are sensitive to absolute magnitude and that the B stars "are a very mixed lot" and that classification ratios should therefore be based on "lines of as many elements as possible." Much of his work was incorporated in the later work of Kopylov (1958a, b), so we need not review it in detail here except to quote his important comment on the distinction between classification methods using line ratios and those using line intensities:

> It is important to realize that line-ratio methods must be clearly separated from absolute-intensity methods. Stars in which classification ratios are identical show wide variations in the total absorption of the lines involved.

At the Yerkes Observatory, Rudnick (1936) investigated the intensities of numerous lines in the spectra of seven O stars of spectral types Oe to B9. This work was published simultaneously with that of Williams, and the two sets of papers nicely confirm each other in most respects.

The first large-scale study following the publication of the Yerkes Atlas was that of Mustel and his collaborators at the Crimean Astrophysical Observatory (Mustel, Galkin, Kumaigoradskaya, and Boyarchuk, 1958). They determined line intensities on spectrograms of moderate dispersion (72 A/mm) for 81 stars of classes F0 to K5. Table 2 lists the lines they found to be most useful for classification in this range of spectral type

Table 2.   Classification lines used by Mustel et al. (1958)

λ	Identification	λ	Identification
4046	Fe I	4272	Fe I
4064	Fe I	4325	Fe I
4072	Fe I	4384	Fe I
4144	Fe I	4405	Fe
4227	Ca I	Hβ, Hγ, Hδ	
4260	Fe I	3934	Ca II(K)

In two stages of iteration, Mustel and his collaborators derived improved spectral types, starting from the published MK classes and combining the smoothed plots of several ratios of equivalent widths. They demonstrated that a substantial reduction (typically by a factor 2) of the scatter of line ratios versus spectral class was possible, and they inferred that the quantitative technique was significantly more accurate than the visual technique.

An uncertainty of somewhat less than 0.1 spectral class was achieved. Luminosity calibration was not discussed in detail, because these stars were virtually all of low luminosity.

Oke (1957) used microphotometer tracings of spectra taken with a dispersion of 33 A/mm at Hγ to develop a two-dimensional classification for F5 to K1 stars. His criteria were the ratios of logarithms of central depths as measured on the tracings, and he based the selection of lines on the experience of Adams and of Young and Harper. Weighted means of about a half-dozen ratios were used for spectral type and for luminosity determination, and Oke found a probable error of 0.2 mag in M_v.

Kopylov (1958a, b) carried out a comprehensive quantitative study of line intensities in stars of spectral type O5 to B7. He obtained new spectra of 109 stars, with a dispersion of 75 A/mm, and, collecting virtually all other available data, he compiled line intensities for 238 stars.

A thorough investigation of the systematic differences between results of different observers convinced him that, among the early-type stars, these differences were due primarily to differences in photometric techniques, and they were only slightly, if at all, correlated with dispersion. This conclusion agrees with Wright's (1957) empirical result, but theoretical studies indicate that the influence of a finite slit-width in scanning photographic spectra should produce a systematic difference correlated with dispersion, especially in sharp-lined spectra.[*]

In two successive approximations, Kopylov derived correlations between spectral type on the MK system and the ratios of 23 lines and combinations of lines. Absolute magnitudes for 165 stars were assigned on the basis of membership in clusters or associations.

Table 3 lists the combinations of line strength used for the determination of spectral type. Graphs of these discriminants permitted the determination of spectral type with probable errors of 0.15 of a spectral subclass in the range O5 to B3 and 0.28 in the range B5 to B7. For the determination of absolute magnitude, Kopylov attempted, but failed, to find ratios of hydrogen and helium (which decrease with luminosity) to other elements (which increase with luminosity) that were independent of spectral type. Instead, he employed hydrogen by itself and several combinations of other lines, shown in Table 4. Neutral helium was not used, because the variations with spectral type and luminosity appeared to be quite complex.

By combining several criteria and adjusting his calibration curves in an iterative process, Kopylov achieved an internal consistency of several tenths of a magnitude in the determination of luminosity. There was, however, a systematic discrepancy with the absolute magnitudes obtained by Petrie. This discrepancy has since been removed by a redetermination of the absolute magnitudes (see Petrie, 1966).

The fact that the discrepancy was inherent in the calibration of the magnitudes and not in the measurements of the spectra strengthens Morgan's argument that there should be a clear operational distinction between the problems of luminosity classification and the absolute calibration of the magnitudes. In other words, investigators must publish a measure of the raw data as well as the fully processed consequences.

Sinnerstad (1961a) undertook a detailed study, at dispersions of 20 to 90 A/mm on slit spectra, of the criteria employed for luminosity determination. Line intensities were spectrophotometrically determined for spectral classes O6f to F8, and Sinnerstad's catalog provides a valuable collection of hydrogen-line intensities: from Hβ to the last measureable line of the Balmer series (but not beyond H14) in the B and A stars, and Hγ and Hδ in the F stars.

---

[*]See, for example, Cayrel (1953) and Deutsch (1954). In Cayrel's words, "the fact that the microphotometer takes the mean of transmissions and not intensities introduces a systematic error corresponding to a loss of light."

Table 3.   Lines used for two-dimensional classification (Kopylov, 1958a, b)

Number	Element	Wavelength and weight
I	He II	$0.5 \, (\lambda \, 4542 + \lambda \, 4200)$
II	O III	$\lambda \, 3962$
III	N III	$\lambda \, 4379 + \lambda \, 4097$
IV	Si IV	$\lambda \, 4116 + \lambda \, 4089$
V	C III + O II	$\lambda \, 4649$
VI	C III + O II	$(\lambda \, 4068\text{-}72) + 2(\lambda \, 4185\text{-}89)$
VII	Si III	$\lambda \, 4575 + \lambda \, 4568 + \lambda \, 4552$
VIII	N II	$\lambda \, 3995 + \lambda \, 4631$
IX	N II, O II	$\lambda \, 3954 - 56$
X	O II	$(\lambda \, 4415\text{-}17) + (\lambda \, 4317\text{-}20)$
XI	O II	$2 \, \lambda \, 4641$
XII	O II	$\lambda \, 4676 + \lambda \, 4662$
XIII	Si II	$\lambda \, 4128 + \lambda \, 4131$
XIV	C II	$\lambda \, 4267 + \lambda \, 3920$
XV	Mg II	$\lambda \, 4481$
XVI	He I	$\lambda \, 4471$
XVII	He I	$\lambda \, 4026$
XVIII	He I	$\lambda \, 3820$
XIX	He I	$\lambda \, 4388$
XX	He I	$\lambda \, 4144 + \lambda \, 4009$
XXI	He I	$\lambda \, 4713 + \lambda \, 4121$
XXII	He I	$\lambda \, 4438 + \lambda \, 4169$
XXIII	He I	$\lambda \, 3965$

Two schemes for two-dimensional classification were established for B and A stars and were calibrated against absolute magnitude. Both schemes used the equivalent widths of $H\gamma$ and $H\delta$ as luminosity indicators; for temperature, one scheme used MK spectral type and the other used the central depth of $H\gamma$ and $H\delta$. Both schemes permitted determination of absolute magnitude with a mean internal error of ± 0.25 mag. Sinnerstad (1961b) showed a very close correlation between the equivalent widths of $H\gamma$ and $H\delta$ in the B and A stars. The slight tendency for $H\gamma$ to be relatively weak in stars showing emission at $H\alpha$ is noteworthy; detailed examination of the corresponding intensities of $H\beta$ would be instructive.

Sinnerstad's line intensities showed excellent agreement with the work of Petrie, although they showed a slightly greater scatter relative to Kopylov's results.

At the I.A.U. Symposium on "Spectral Classification and Multicolour Photometry," Petrie (1966) presented a new luminosity calibration of the $H\gamma$ equivalent widths that had been determined at the Dominion Astrophysical Observatory during the preceding two decades. The new calibration was essentially in agreement with the MK calibration and with that of Sinnerstad, and it gave cluster distances that agreed with those derived photometrically, with the exception of NGC 2244. Thus, a long-standing discrepancy was removed.

Table 4.   Lines used for luminosity classification (Kopylov, 1958a, b)

Symbol	Combination of lines from Table 3	Limits (application of combination)
A	1.5(III) + (IV) + (VII) (III) + (IV) + 1.5(VII)	O8.75 - O9.75 O9.75 - B1.50
B	1.5(III) + 0.5(VII) (III) + 0.75(VII)	O8.75 - O9.75 O9.75 - B1.50
C	1.5(IV) + 0.5(VII) (IV) + 0.75(VII)	O9.25 - B0.25 B0.25 - B1.50
D	(III) + 1.5(VIII)	O9.0 - B2.0
E	(IV) + 2(VIII)	O9.5 - B2.0
F	1.5(III) + (X)	O8.75 - B1.50
G	1.5(III) + (XI)	O8.75 - B1.50
H	(IV) + (X)	O9.5 - B1.5
J	(IV) + (XI)	O9.5 - B1.5
K	(V) + (VII)	O9.5 - B1.5
L	(V) + 2.5(VIII)	O9.5 - B2.0
M	Hβ	O4.6 - B7.0
N	0.5(Hγ + Hδ)	O4.6 - B7.0
O	H7 + H8	O4.6 - B7.0
P	H9 + H10 + H11	O4.6 - B7.0

Several aspects of Petrie's Hγ measures have an important bearing on quantitative classification. He showed that the Hγ -M_v relation is insensitive to evolutionary changes among the B and A stars and that it is unchanged by stellar rotation. Equivalent widths of Hγ in stars with projected rotational velocities greater than 200 km sec^{-1} agreed to within 10% with those in stars rotating less than 85 km sec^{-1}. Petrie showed that the Hγ -M_v relation is not entirely independent of spectral type, but he provided a table of corrections to the mean relationship.

Yoss (1966) has reported the development of an automatic procedure for reducing photographic spectra, and we look forward to hearing of the further progress of this extensive program.

Jones (1967) has pursued the suggestion by Greenstein (1956) that it should be possible to establish a three-dimensional classification based on criteria that are specifically sensitive to atmospheric pressure, excitation temperature, and metal abundance. From plates with 18 A/mm dispersion on F, G, and

K stars, he derived estimates of absolute strength for the hydrogen lines and the Fe I lines λ 4045, λ 4063, λ 4071, λ4202, λ4308, and λ 4383. He also estimated 10 line ratios: 7 "excitation" ratios between lines of neutral elements with differing excitation potentials and 3 "ionization" ratios between lines of neutral and ionized atoms. He formed combined indices, using the statistical technique of multivariate analysis, which he designated as p, t, and s, corresponding to the pressure, temperature, and metal abundance. The p index was calibrated in terms of luminosity by a statistical method using radial velocities and proper motions, while the temperature index was calibrated by comparing infrared photometric gradients, obtained from spectrum scans, with predictions of model atmospheres constructed by Swihart and Fischel (1960). Absolute calibration was effected by assuming Vega to radiate as a model atmosphere with effective temperature T_{eff} = 9500° and approximate corrections for line blanketing were included. This technique gave T_{eff} = 5700°K for the Sun rather than the measured value of 5800°K, and Jones concluded that the discrepancy was too large to be acceptable, although the source was not uncovered. Jones also found a striking difference between the temperature calibrations of the t index for the dwarfs and giants. The difference was in the sense that, at a given t index, the giants had a lower infrared color temperature than the dwarfs or, at a given color temperature, the giants had a t index corresponding to a higher temperature than the dwarfs. Several explanations were discussed and rejected, and this difference remains unexplained.

The system of Peat (these proceedings) differs from that of Jones in using theoretical values of the partial derivatives of the spectral features with respect to atmospheric parameters.

Underhill (1966) has reviewed the difficult problem of establishing an orderly pattern among the early-type stars, and Morgan (1967) has pointed to the need to add a third classification parameter to the visual MK system.

4. THE OSCILLOSCOPE MICROPHOTOMETER

Hossack (1953, 1954) recognized the potential value of a method that would combine the quantitative precision of the microphotometer with the versatility and speed of the human eye; the result was his development of the "oscilloscope microphotometer," which gives a simultaneous display of the density tracings of two spectra with density scales that can be adjusted by a measurable ratio such that the depth of the classification feature appears the same in the object and in a standard star. By averaging the determinations from the four ratios, he achieved a luminosity classification with a probable error of about 0.15 of a luminosity class. Hossack concluded that the spectral type could also be determined to a precision of about 0.17 of a spectral subclass by averaging over independent determinations from five line ratios.

These impressive results encourage the exploration of techniques that thus combine the human eye and the photocell, but since Hossack's device used density tracings, the measures do not lend themselves to direct quantitative interpretation.

5. PHOTOELECTRIC PHOTOMETRY WITH NARROW (Δλ < 100 A) AND INTERMEDIATE (Δλ = 100 to 300 A) BANDS

The narrow-band techniques are, for the most part, "pure" in the sense defined above, and although the intermediate-band techniques can be pure when applied to the relatively simple spectra of hot stars, it is likely that they are "blended" for the majority of stars. But the linear response of the photoelectric systems ensures that the interpretation of blended features is straightforward, if tedious.

Three methods of wavelength selection have been employed:

A.   A spectrograph with exit slits.

B.   Filters.

C.   A scanning spectrometer whose output is selectively combined into indices.

The first technique has been employed extensively at the Cambridge Observatories, and Redman (1966) has summarized the program aimed at detecting peculiarities in the spectra of K stars. The Balmer lines have also been measured in early-type stars. Table 5 lists the published measures, based on Redman's summary and a bibliography provided by Peat.

Table 5.   Recent photoelectric determinations of line strength

Features	Band-widths (A)	Number of stars	Observers
CN 4200 A	50	712	Griffin and Redman (1960)
G band	30	212	Griffin and Redman (1960)
Mg b	29	415	Deeming (1960)
Sc I 6306 A	2	200	Griffin (1961)
Na D	15	415	Griffin (1961), Price and Deeming (unpublished)
Fe I 5250 A	11	525	Griffin (1961), Scarfe (unpublished)
Hγ	45	177	Bappu, Chandra, Sanwal, and Sinvhal (1962)
Hα	35	588	Peat (1964a)
Hα	10.5	202	Price (unpublished)
Hβ	30, 150	1217	Crawford, Barnes, Faure, Golson, and Perry (1966)
Ca I 4226 A	15	53	Fernie (1966)
Ca I triplet 6100 A	20 (3 bands)	299	Peat (1964b)
Fe I RMt 15	21.5 (6 bands)	294	Scarfe (unpublished)
Fe I RMT 1146	20.5 (6 bands)	305	Scarfe (unpublished)

Tables 6 through 12 summarize the systems using filters of narrow and intermediate bandwidth. Three points should be noted concerning these systems:

A.   Correction for interstellar reddening can be obtained either by making a suitable choice of the relative wavelenths of the filters or by determining the color-excess ratio of one pair of filters with respect to the other pairs of filters and then taking suitable difference of indices. This technique is illustrated in the discussion of the Stromgren-Crawford system below.

B.   In some cases the indices are pure, while in others they are not. An excellent example of purity is the β index of Crawford and Stromgren, which combines a pair of filters of different widths centered on Hβ. Sinnerstad has also measured several indices determining the strengths of selected helium lines.

C.   Thus far, the systems have concentrated on a relatively small fraction of the possible attributes of the spectra. For example, among the early stars attention has been focused on the hydrogen spectrum through the intensity of the lower members of the Balmer series, the so-called Balmer discontinuity, and gradients in the ultraviolet and the visual. This selection of attributes was based on the early success of the Barbier-Chalonge system.

Table 6.   The Strömgren c - l system

1. Bands (Strömgren, 1958)

	f	e	d	c	b	a
λ	3600	4030	4500	4700	4861	5000
$\Delta\lambda$	350	90	80	100	35	90

2. Indices (Chosen to be unaffected by interstellar absorption)

$l = b - 1/2 (a + c)$ H_β

$c = d + f - e$ D_B

$m = (a - d) - (d - e)$ metal lines near λ 4030

3. References

Strömgren (1963)
McNamara (1966)

Table 7.   The Crawford seven-color system

1. Bands (Crawford, 1961)

	u	v	c	b'	a	b	y
λ	3450	4090	4160	4265	4370	4700	5480
$\Delta\lambda$	380	128	44	66	62	100	220

2. Indices

$m = (y - b) - (b - v)$ metallic lines
$c = (b - v) - (v - u)$ D_B
$G = I_a / I_{b'}$ break at G band
$I = I_{b'} / I_c$ CN break at λ 4216

3. Reference

Williams (1966) (with slightly different filters)

Table 8.   The Strömgren - Crawford system

1. Bands (Crawford, 1966)

	u	v	b	y
λ	3420	4100	4700	5500
$\Delta\lambda$	400	200	200	200

2. Indices (Compensated for interstellar absorption)

$[c_1] = (u - v) - (v - b) - 0.20(b - y)$ D_B
$[m_1] = (v - b) - (b - y) + 0.18(b - y)$ metallic lines

These indices are supplemented with the H_β index (Crawford, 1966)

3. References

Strömgren (1966)
Kelsall and Strömgren (1966)(theoretical calibration)
Crawford and Strömgren (1966)(intercomparison of clusters)

Table 9.   The Borgman seven-color system

	R	Q	P	N	M	L	K
1. Bands (Borgman, 1960)							
λ	3295	3560	3750	4055	4550	5240	5880
$\Delta\lambda$	80	90	110	200	200	220	215

2. Indices (Compensated for interstellar absorption)

$\beta = P - N - 0.701(N - M)$ high Balmer lines
$\gamma = Q - N - 1.068(N - M)$ D_B
$\delta = R - Q - 0.882(N - M)$ ϕ_{uv}

3. References

Borgman (1963)(reddening)
Borgman and Blaauw (1963)

Table 10.   The Walraven five-color system

	W	U	L	B	V
1. Bands					
λ	3220	3620	3900	4260	5590
$\Delta\lambda$	150	250	290	420	850

2. Indices (Compensated for interstellar absorption)

$[B - U] = B - U - 0.66(V - B)$ D_B
$[U - W] = U - W - 0.55(V - B)$ ϕ_{uv}
$[B - L] = B - L - 0.40(V - B)$ M_v

3. Reference

Walraven and Walraven (1960)

Table 11.   The Golay seven-color (broad-band) system

	U	B_1	B	B_2	V_1	V	G
1. Bands* (Golay, 1966)							
λ	3450	4028	4270	4494	5408	5532	5850
$\Delta\lambda$	800	500	800	500	500		500

2. Indices

$\Delta = U - B_2 - 1.055(B_2 - G)$ D_B
$g = B_1 - B_2 - 1.520(V_1 - G)$ high Balmer lines

3. References

Golay (1966)
Golay and Goy (1965)

*The U, B, and V are similar to the filters of the Johnson system.

Table 12.   The Morgan and Neff Chi system

---

1. __Bands__ (Neff and Travis, 1967)

λ	3300	3700	4700	5500
$\Delta\lambda$	350	150	200	200

---

2. __Indices__

$$\chi_1 = m4700 - m5500 \qquad\qquad \phi b$$
$$\chi_3 = m3300 - m3700 \qquad\qquad \phi uv$$
$$\Delta(3700) = 2.842 \log \frac{F(4700)}{F(5500)} - \log \frac{F(3700)}{F(5500)} \qquad D_B$$

Several other combinations were also used.

---

3. __Reference__

Neff (1968)

As a specific example, we will now consider the uvby system of Strömgren and Crawford (see Table 8). Four bands of intermediate width are situated at mean wavelengths 3500, 4110, 4670, and 5470 A, respectively. The corresponding magnitudes define three indices: the color index b - y and the following color-index differences,

$$c_1 = (u - v) - (v - b), \quad \text{Balmer discontinuity index,}$$

$$m_1 = (v - b) - (b - y), \quad \text{metal-line index.}$$

The c_1 index measures the curvature of the spectrum in the neighborhood of the Balmer limit, while the m_1 index measures the curvature longward of the Balmer limit.

All three indices are influenced by interstellar reddening, but one of them can be used to compensate for reddening in the other two, if the reddening obeys an assumed law. For example, assuming the "standard" law to hold between the various color excesses,[*]

$$E(c_1) = 0.20 \, E(b - y),$$

$$E(m_1) = -0.18 \, E(b - y),$$

the following two indices, which are independent of reddening, may be constructed:

$$[c_1] = c_1 - 0.20(b - y),$$
$$[m_1] = m_1 + 0.18(b - y).$$

The indices $[m_1]$ and $[c_1]$ permit a two-dimensional classification, but the separation between luminosity classes and the detection of population differences is not unambiguous. Supplementing these indices with the narrow-band β index, which measures the strength of the Hβ line, permits accurate classification of the B, A, and F stars. The revised Hβ photoelectric system was described by Crawford and Mander (1966) and by Crawford (1966). It comprises two filters centered at 4860 A, one with a half-width of 30 A, and the other of 150 A. The transmission curves are shown in Figure 2 (see Crawford, 1966, Figure 2).

---

[*] The color excess is defined as the ratio of interstellar extinctions in the two wavelengths entering each index.

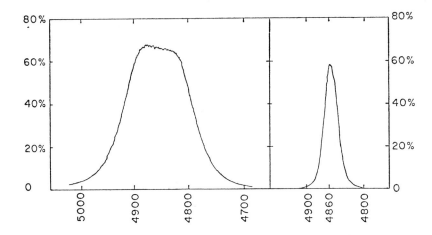

Figure 2. Transmission curves of filters for the uvby and H$_\beta$ photometry
(from Crawford, 1966).

Fernie (1965) has calibrated the measure of β against absolute visual
magnitude and obtained the curve shown in Figure 3. In constructing this
curve, he employed measures of the H$_\gamma$ intensity and exploited the fact that
the intensities of these two hydrogen lines are very closely correlated to
each other.

Among the emission-line B stars, this correlation between the strengths
of the Balmer lines is no longer unique, but Strömgren concludes that, if
Hα does not show emission on spectra of moderate dispersion, the Hβ intensity
can be assumed free of the effects of emission. He suggests that the use of
an Hδ index and an Hα index, for example, would permit simultaneous
determination of luminosity and emission intensity. Mendoza (1958) and Abt
and Golson (1966) discuss the photometry of emission-line B and A stars,
and Peterson (1967) points out that the Balmer absorption-line decrement
should be sensitive to departures from LTE in the stellar atmospheres.

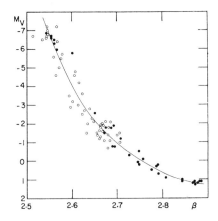

Figure 3. Relationship of the H$_\beta$ index to absolute visual magnitude
(from Fernie, 1965).

The influence of chemical composition and stellar rotation on the two-dimensional classification has been examined semiempirically in some detail. Kelsall and Strömgren (1966) have computed spectroscopic data for a grid of model atmospheres corresponding to zero-age sequences of six different compositions. They find that, for a given T_{eff}, the abundance of helium has a very small influence on the surface gravity of the stellar model, but it appreciably alters the opacity and pressure of the corresponding atmosphere. For late-type B stars, atmospheric helium influences only the molecular weight, since it is neutral and does not contribute to the opacity, so an alteration of the helium abundance produces an effect similar to an alteration of effective surface gravity. The magnitude of the effect is such that $\Delta\beta = 0.01$ mag corresponds to $\Delta\gamma = 0.15$ in that spectral range.

The apparent relationship between M_V and β shows a mean error of 0.2 to 0.3 mag, a portion of which is produced by the incidence of binary stars (a binary of identical stars will show M_V too bright by 0.75 mag for its β) and by the influence of rapid rotation. Abt and Osmer (1965) showed that among the members of an association, the slow rotators have higher values of β for equal $(U - B)_0$. The effect amounts to a few hundredths of a magnitude for the 15-A filter used by Abt and Osmer, but it is negligible for the 30-A filter used by Crawford in his $H\beta$ photometry.

For a detailed summary of the techniques of classifying the A and F stars and photometrically determining their ages, the reader is referred to Strömgren's review article (1966).

The use of b - y, c_1, and m_1 permits a three-dimensional classification on the basis of T_{eff}, g, and metal deficiency among unreddened stars; the addition of the $H\beta$ intensity allows classification of stars affected by interstellar reddening. Further theoretical analyses employing model atmospheres and data obtained from high-dispersion spectra are required for the calibration of this system.

Kraft and Wrubel (1965) investigated theoretically the influence of rotation on the relationship between (b - y) and c_1 among the A stars of the Hyades, and they found agreement with empirical results. The effect is about 0.1 mag on the M_V scale.

Among the main-sequence stars later than spectral-type G, the age effects are quite small, so a two-dimensional classification should, in principle, be adequate. On the other hand, classification of the giants is quite a complex task, because this group is replenished by stars from a wide range of the initial main sequence and because the group is luminous enough to contain members of both stellar populations.

Strömgren and Gyldenkerne (1955) based a two-dimensional classification of G and K stars on narrow-band measures of the K-line strength, the G-band discontinuity, and the cyanogen-band strength. Metal-line indices were added later, with widths of about 100 A, to isolate the influence of chemical composition. Wallerstein and Helfer (1966), on the basis of abundance analyses, demonstrated a good correlation between iron abundance and Gyldenkerne's C index.

Fernie (1966), seeking a spectrum-luminosity indicator that was monotonic rather than reaching a maximum, as do the G-band and cyanogen-band intensities, investigated the strength of Ca I 4226. He compared fluxes with a 15-A and a 150-A band centered on 4226 A, and he remarked that "because of the complexity of the region, no attempt has been made at a theoretical interpretation of the results." Measures were tabulated for 53 MK-standard stars in the range F0 to M2 and luminosity classes V to Ia, and excellent correlations were found with spectral type and luminosity.

Gyldenkerne and Helt (1966) have reported encouraging progress in the two-dimensional classification of M stars by means of narrow-band photometry between 5400 A and 4300 A.

6. SPECTROPHOTOMETRIC CLASSIFICATION

Barbier and Chalonge developed a precise two-dimensional classification based on scans of photographic spectra at a dispersion of 80 A/mm. These measures could not, of course, eliminate altogether the effects of the background of weak and moderate absorption lines, and for this reason the application and calibration of this technique become very difficult for spectral types later than about G0. The Balmer discontinuity D is measured by extrapolating the redward and blueward continua to a reciprocal wavelength of 2.70 μ^{-1}, and the second parameter λ_1 indicates the location of the last visible member of the Balmer series. (This is not the operational definition of λ_1, but it is the theoretician's description of the result.) The method and its calibration in terms of the MK system have been described by Chalonge (1956) and Berger (1962). A bibliography covering its extension to three dimensions by the inclusion of an ultraviolet gradient is given by Strömgren (1963).

Calibration curves relative to the MK system are shown in Figures 4 and 5, and they succinctly summarize the behavior of the hydrogen spectrum among the early stars. The precision of the method is indicated by the error bars of ± 0.02 in D and ± 3 A in λ_1.

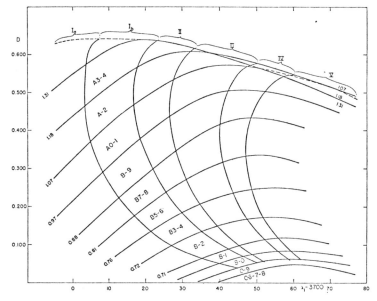

Figure 4.   Calibration curves in the λ_1-D diagram for O6 to A4 stars (from Chalonge, 1956).

7. PHOTOMETRY WITH INTERMEDIATE AND WIDE BANDWIDTH

In 1964, Morgan and Neff developed a photometric system of intermediate bandwidth, which has been described and calibrated by Neff and Travis (1967). The system, called the Chi system, is described by Neff elsewhere in these proceedings. An absolute calibration was derived from the observations of Willstrop (1960), and it was verified by comparison of the measured solar fluxes with those of similar stars. A luminosity discriminant based on the pressure broadening of the high members of the Balmer series was shown to be useful over a wide range of spectral type, and preliminary comparisons with model atmospheres were encouraging.

A seven-color system, using the U, B, and V filters supplemented with four others of similar bandwidth (500 A) and lying between 4000 A and 5900 A, has been developed by Golay (1966). The aim was the study of faint stars on

a system related to the spectrophotometry of Chalonge (see Section 8). Two-
and three-dimensional classifications were achieved for unreddened stars,
and a number of "photometrically metallic stars" were isolated on the basis
of their colors.

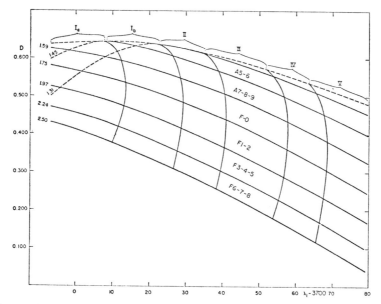

Figure 5.   Calibration curves in the λ_1-D diagram for A5 to F8 stars (from
Chalonge, 1956).

Borgman (1960, 1963) has developed an intermediate passband system
utilizing seven interference filters of widths 80 to 220 A lying between 3295 A
and 5880 A. The filters were chosen to elicit data similar to those obtained
in the French system for early-type stars, but Borgman has demonstrated the
usefulness of the system for late-type stars also.

A variant of this method has been explored by Hack (1953), who used D
with a measure of the equivalent width of Hδ. She concluded that Hδ was to
be preferred over Hγ and H8 because it is relatively unblended and because
it is less affected by emission than Hγ.

Van't Veer (1966) has initiated a program aimed at automatization of the
three-dimensional classification system of Chalonge.

Although the theoretical interpretation, or calibration, of the λ_1, D,
and ultraviolet gradient of the French spectrophotometry is afflicted with
the uncertain correction for weak absorption lines, these measures provide
a powerful tool for investigating the correlation between the metallic-line
spectrum of a star and its continuum — or more correctly, its hydrogen
spectrum.

At the David Dunlap Observatory, van den Bergh (1966) has exploited
a photoelectric spectrum scanner, designed by Oke, for the examination of
G-type main-sequence stars with moderate resolution (20 A). Of course, at
this dispersion the spectral features of G stars are, for the most part,
unresolved, so a direct theoretical interpretation of the measures is tedious.
But, as van den Bergh points out, use of this moderate resolving power
permits spectrum reconnaissance and the search for classification criteria
of more direct astrophysical interest. This concept also underlies the use
of photographic spectrum scans, with a dispersion of 42 A/mm, in the
new system of Landi Dessy and Jaschek (see the paper by Landi Dessy in
these proceedings).

8.   AN EXAMPLE OF INCONSISTENCY

Since all recent schemes of quantitative classification have been tied to the MK classification, the systematic consistency among them is, for the most part, excellent.   Also, the random discrepancies are about as large as would be expected from the observational errors.

There is, however, one example of a serious discrepancy between methods based on visual inspection of the metallic-line spectrum and the photometry of the continuum and the Balmer lines.   Walraven and Walraven (1960) displayed such a discrepancy between MK spectral types and the colors of main-sequence stars in the Scorpio-Centaurus group.   Several stars classified B7 to B9 on the MK system showed colors appropriate to classes B3 to B6.   Garrison (1967) has rediscussed the spectral classification of these stars, and he finds a similar result.   In two cases (HD 147010 and HD 148199), stars classified A2p on the basis of metal lines have colors appropriate to class B7.   Examination of Crawford's (1958) data indicates that the Hβ line strength is consistent with classification based on the continuum colors for main-sequence stars, so this is clearly not a simple luminosity effect.   The MK classification of these stars is derived from the relative strengths of the helium and silicon lines, and the silicon strength increases with advancing spectral type.   Thus, the discrepancy can be characterized by silicon lines that are too strong relative to the helium and hydrogen lines.   Whether or not this is simply an abundance effect and whether or not these stars are high-temperature analogues of the peculiar A Stars remains to be seen, but Morgan and Underhill have both remarked that the spectra often suggest a striking independence between the line-forming layers and the photospheres of the B stars.   In a private communication, Morgan draws attention to Bidelman's helium variable (HD 125823 = a Centauri), whose spectral type varies from B2 V to about B8 III while its color appears to be constant at a value corresponding to the earlier spectral type.

Quantitative spectral classification has provided the basis for a theoretical interpretation of spectral types, but it continues to raise more questions than it answers.

REFERENCES

Abt, H. A., ed. 1963, Ap. J. Suppl., 8, 99.
Abt, H. A., and Golson, J. C. 1966, Ap. J., 143, 306.
Abt, H. A., and Osmer, P. S. 1965, Ap. J., 141, 949.
Adams, W. S., Joy, A. H., Humason, M. L., and Brayton, A. M. 1935, Ap. J., 81, 187.
Bappu, M. K. V., Chandra, S., Sanwal, N. B., and Sinvhal, S. D. 1962, Mon. Not. Roy. Astron. Soc., 123, 521.
Berger, J. 1962, Ann. d'Ap., 25, 1; 25, 77.
Borgman, J. 1960, Bull. Astron. Netherlands, 15, 255.
_____. 1963, Bull. Astron. Netherlands, 17, 58.
Borgman, J., and Blaauw, A. 1963, Bull. Astron. Netherlands, 17, 358.
Cayrel, R. 1953, Ann. d'Ap., 16, 129.
Chalonge, D. 1956, Ann. d'Ap., 19, 258.
Crawford, D. L. 1958, Ap. J., 128, 185.
_____. 1961, Astron. J., 66, 281 (abstract).
_____. 1966, in IAU Symp. No. 24, ed. K. Lodén, L. O. Lodén, and U. Sinnerstad (London: Academic Press), 170.
Crawford, D. L., Barnes, J. V., Faure, B. Q., Golson, J. C., and Perry, C. L. 1966, Astron. J., 71, 709.
Crawford, D. L., and Mander, J. 1966, Astron. J., 71, 114.
Crawford, D. L., and Strömgren, B. 1966, in Vistas in Astronomy, Vol. VIII, ed. A. Beer and K. Aa. Strand (Oxford: Pergamon Press), 149.

Deeming, T. J. 1960, Mon. Not. Roy. Astron. Soc., 121, 52.

_____. 1966, in IAU Symp. No. 24, ed. K. Lodén, L. O. Lodén, and U. Sinnerstad (London: Academic Press), 188.

Deutsch, A. J. 1954, J. Opt. Soc. Amer., 44, 492.

Fernie, J. D. 1965, Astron. J., 70, 575.

_____. 1966, Mon. Not. Roy. Astron. Soc., 132, 485.

Garrison, R. F. 1967, Ap. J., 147, 1003.

Golay, M. 1966, in IAU Symp. No. 24, ed. K. Lodén, L. O. Lodén, and U. Sinerstad (London: Academic Press), 262.

Golay, M., and Goy, G. 1965, Publ. Obs. Genève, Ser. A, No. 71, 19 pp.

Greenstein, J. L. 1956, Proc. Third Berkeley Symp. on Mathematical Statistics and Probability (Berkeley: University of California Press), 3, 11.

Griffin, R. F. 1961, Mon. Not. Roy. Astron. Soc., 122, 181.

Griffin, R. F., and Redman, R. O. 1960, Mon. Not. Roy. Astron. Soc., 120, 287.

Gyldenkerne, K., and Helt, B. 1966, in IAU Symp. No. 24, ed. K. Lodén, L. O. Lodén, and U. Sinnerstad (London: Academic Press), 162.

Hack, M. 1953, Ann. d'Ap., 16, 417.

Hossack, W. R. 1953, J. Roy. Astron. Soc. Canada, 47, 195.

_____. 1954, Ap. J., 119, 613.

Hynek, J. A. 1935, Ap. J., 82, 338.

Jones, D. H. P. 1967, preprint.

Kelsall, T., and Strömgren, B. 1966, in Vistas in Astronomy, Vol. VIII, ed. A. Beer and K. Aa. Strand (Oxford: Pergamon Press), 159.

Kopylov, I. M. 1958a, Ann. Crimean Ap. Obs., 20, 123; available in English as Atronomical Papers Translated from the Russian, No. 4, Smithsonian Ap. Obs., Cambridge, Mass.

_____. 1958b, Ann. Crimean Ap. Obs., 20, 156; available in English as Astronomical Papers Translated from the Russian, No. 5, Smithsonian Ap. Obs., Cambridge, Mass.

Kraft, R. P., and Wrubel, M. H. 1965, Ap. J., 142, 703.

McNamara, D. H. 1966, in IAU Symp. No. 24, ed. K. Lodén, L. O. Lodén, and U. Sinnerstad (London: Academic Press), 190.

Mendoza, E. E., V. 1958, Ap. J., 128, 207.

Morgan, W. W. 1967, in Modern Astrophysics, ed. M. Hack (Paris: Gauthier-Villars), 83.

Morgan, W. W., Keenan, P. C., and Kellman, E. 1939, An Atlas of Stellar Spectra with an Outline of Spectral Classification (Chicago: University of Chicago Press).

Mustel, E. R., Galkin, L. S., Kumaigorodskaya, R. N., and Boyarchuk, M. E. 1958, Ann. Crimean Ap. Obs., 18, 3.

Neff, J. S. 1968, Astron. J., 73, 75.

Neff, J. S., and Travis, L. D. 1967, Astron. J., 72, 48.

Oke, J. B. 1957, Ap. J., 126, 509.

_____. 1965, Ann. Rev. Astron. Ap. 3, 23.

Payne, C. H. 1925, Stellar Atmospheres, Harvard Obs. Mono. No. 1 (Cambridge, Massachusetts: The Observatory).

_____. 1930, The Stars of High Luminosity, Harvard Obs. Mono. No. 3 (New York: McGraw-Hill Book Co.)

Peat, D. W. 1964a, Mon. Not. Roy. Astron. Soc., 128, 435.

_____. 1964b, Mon. Not. Roy. Astron. Soc., 128, 475.

Peterson, D. M. 1967, Astron. J., 72, 822 (abstract).

Petrie, R. M. 1966, in IAU Symp. No. 24, ed. K. Lodén, L. O. Lodén, and U. Sinnerstad (London: Academic Press), 304.

Redman, R. O. 1966, in IAU Symp. No. 24, ed. K. Lodén, L. O. Lodén, and U. Sinnerstad (London: Academic Press), 155.

Rudnick, P. 1936, Ap. J., 83, 439.

Sinnerstad, U. 1961a, Ann. Stockholm Obs., 21, No. 6, 64 pp.

_____. 1961b, Ann. Stockholm Obs., 22, No. 2, 57 pp.

Strömgren, B. 1958, in Stellar Populations, ed. D.J.K. O'Connell
(Amsterdam: North-Holland Publ. Co.), 245.
_____. 1963, in Stars and Stellar Systems, Vol. VIII, ed. K. Aa. Strand
(Chicago: University of Chicago Press), 123.
_____. 1966, Ann. Rev. Astron. Ap., 4, 433.
Strömgren, B., and Gyldenkerne, K. 1955, Ap. J., 121, 43.
Swihart, T. L., and Fischel, D. 1960, Ap. J. Suppl., 5, 291.
Underhill, A. B. 1966, The Early Type Stars (Dordrecht-Holland:
D. Reidel).
van den Bergh, S. 1966, in IAU Symp. No. 24, ed. K. Lodén, L. O. Lodén,
and U. Sinnerstad (London: Academic Press), 132.
van't Veer, F. 1966, Ann. d'Ap., 29, 293.
Wallerstein, G., and Helfer, H. L. 1966, Astron. J., 71, 350.
Walraven, Th. 1966, in IAU Symp. No. 24, ed. K. Lodén, L. O. Lodén, and
U. Sinnerstad (London: Academic Press) 223.
Walraven, T., and Walraven, J. H. 1960, Bull. Astron. Netherlands, 15, 67.
Williams, E. G. 1936a, Ap. J., 83, 279.
_____. 1936b, Ap. J., 83, 305.
Williams, J. A. 1966, in IAU Symp. No. 24, ed. K. Lodén, L. O. Lodén,
and U. Sinnerstad (London: Academic Press), 211.
Willstrop, R. V. 1960, Mon. Not. Roy. Astron. Soc., 121, 17.
Wright, K. O. 1957, in Transactions of the IAU, Vol. IX, ed. P. Th. Oosterhoff
(Cambridge: University Press), 527.
Wright, K. O., and Jacobson, T. V. 1966, Publ. Dominion Ap. Obs.,
Victoria, B.C., 12, 373.
Yoss, K. M. 1966, in IAU Symp. No. 24, ed. K. Lodén, L. O. Lodén, and
U. Sinnerstad (London: Academic Press), 111.
Young, R. K., and Harper, W. E. 1924, Publ. Dominion Ap. Obs.,
Victoria, B. C., 3, 3.

DISCUSSION

Crawford: Perhaps I could start the discussion by saying that we look at
these stars from several different points of view. I'm interested in these
difficulties because they will influence the kind of galactic structure I get.
If rotational velocities are affecting the spectra or affecting the absolute
magnitude calibrations, then my picture of the Galaxy is going to be wrong.
Now it is exactly these problems that yield very interesting astrophysics,
so the astrophysicists, who are the majority of the people here, are interested
in these things because it gives them a deeper insight into the astrophysics.
We discover some of these peculiarities as we go along because one looks at
the stars with a more critical eye.

In addition to the classification problems, there are the problems of
calibration. As Whitney pointed out, although we can classify some of these
stars nicely, the problem of calibrating against real life is different. There
remains a fraction of stars that we simply can't classify or calibrate in the
simple way we have been doing. It is also important to note that the stars in
that diagram are labeled peculiar not just because they don't fit the diagram,
but also because their spectra really do look peculiar, and these same stars
not only can't be classified but they don't fit the calibration, so they are the
ones I've thrown out when trying to do galactic structure.

The calibration problem enters many investigations. Sometimes the
intercomparison of different clusters indicates we have problems in calibration. My own hope in attending this conference was to find where I can have
confidence in the astronomical results in view of some of these things we
listed, such as the helium effects, the effects of rotational velocity, the
varying metal-to-hydrogen ratio, and the departures from LTE. With regard
to the B stars, which I am particularly fond of, in some cases we may not
be looking at the star at all, but rather at a very extended atmosphere. And
for the future, we should ask what wavelength regions the models are telling
us to look at, and what are the resolutions we should be trying for.

Perhaps we now have somebody who would like to get up and answer us, or just to hold forth.

Eggen: I'd like to ask a question that may be irrelevant or irreverent: At a stellar atmospheres conference I'm a little surprised to find so much emphasis on spectral type. Is the spectral type of some importance in itself?

Whitney: It represents a systematic statement of the properties of spectra.

Eggen: Does the spectral type itself mean anything?

Whitney: No, it's the systematic arrangement of spectra that we're interested in interpreting here. In trying to decide how far model atmospheres can tell us what is going on, we have chosen to discuss the so-called normal stars, and these are the ones that are encompassed by a spectral classification system.

Crawford: I think another importance of spectral classification is that in the deviations from the normal spectral types we find the interesting stars.

Underhill: From the evolutionary point of view, the basic parameters of stars are the mass and the effective temperature. The models have shown that the photometric criteria of classification line up quite well with the temperature structure of the atmosphere, and this in turn is determined by the effective temperature. The mass we can isolate by going through surface gravity using the hydrogen lines. Now the reason you find conflicts with the spectral types defined the classical way, through absorption lines, is that the absorption lines are generally the ones that can be measured on moderately low-dispersion spectra. These are therefore the strong lines, and they are sensitive not so much to temperature as to some of these peculiar and interesting physical effects that I was mentioning yesterday. I certainly agree with Crawford in saying, let's get those stars out and give them to the astrophysicists and let the galactic structure people keep the others.

Crawford: Yes, but the problem is that more and more of them keep going out so we have only a very small percentage left for galactic structure.

Cayrel: I should like to make the point that it is quite important that the number of observational criteria be greater than the number of physical parameters we should like to determine. Because if it is not so, you can always classify and classify and classify blindly, and you could include galaxies, quasars, etc. without ever finding a conflicting situation. If you have more criteria than the number of free parameters, you can detect interesting objects, and I think an important aspect of classification is detecting interesting objects.

Houziaux: The question is whether it's worthwhile to go through the usual spectral classification terminology when what we really want is to determine the physical parameters of the atmosphere and of the star. So when we characterize the star by one spectral type, we try to characterize the whole spectrum. Now that will certainly bring some confusion when we try to translate it into the physical parameters of the star.

Buscombe: Suppose we have two stars of very simple spectral class— B5 on the main sequence. Both are rapidly rotating, so we see almost no lines but the hydrogen lines. One star shows hydrogen lines with an equivalent width in the middle of the Balmer series of 6/A, and the other shows the same lines with an equivalent width of 10/A. What's the difference between the masses of the two stars?

Underhill: You put your finger on the question when you said we have two stars of class B5 on the main sequence. What does that really mean? Then you said they have the same color but they have different hydrogen lines.

Buscombe: Well, what I'm saying is that you have to describe the spectrum quantitatively.

Underhill: I think it's an in-trained prejudice in all of us that spectral type gives temperature. This is a feeling that we've all been trained to accept.

Cayrel de Strobel: Never can they be assumed to be the same thing. We always have to calibrate one against the other. Surely, I am pleased to have a spectral type for a given star, but I never know what it means until I have obtained the effective temperature from quantitative criteria. For you and for me the problems of calibrating spectral types could be different: I am not concerned with hot stars but with late spectral-type stars. You and I are not in the same stellar ring. I am cold and you are warm. Of course, I'm Italian and you are Canadian.

Eggen: I'm glad I didn't say that.

Crawford: Of course, spectral type must be reasonably correlated with temperature, because one of the original aims in setting up the spectral system was to get a smooth arrangement with color.

Underhill: This is true, but where it really begins to fall apart is in the range from B2 up to the hotter stars, as our models are now permitting us to see. On this basis one is tempted to tell the classifiers to be a little careful in thinking they have found the relationship with temperature.

Strom: I'd like to comment on the determination of temperature for the later-type stars. In my review yesterday, I tried to indicate what sort of sensitivity a narrow-band system might give with respect to effective temperature, particularly if one adopts a visual-infrared color index, a narrow-band analogue of V-I, for example. For G and K stars you are less affected by line absorption; there is another advantage in such an index in that it is a measure of flux produced outside the convection zone. This is so because in this region the atmosphere is relatively opaque and we see only the upper layers, which are less affected by convection.
   Another comment is that I don't think there has been an adequate exploitation of indices such as the β index for determining the gravities of stars. I think that such an index could be used in the A stars, but it hasn't yet been. There are a number of important problems in this area, such as the masses of the horizontal branch of stars, the masses of the metallic-line, and the peculiar A stars and the normal A stars. I think it would help to have a quantitative correlation between the indices and the equivalent widths of the relevant lines.

Underhill: I agree with you, but there is quite a severe difficulty that still remains on the theoretical side. And that is that for the stars of smaller gravity, where we deal with relatively low densities, we need a theory of line formation that can adequately account for these more complicated additional processes that can and do occur in these regions. Perhaps we can find a set of criteria that are independent of these peculiar processes and that can be described with the LTE equations.

EMPIRICAL PARAMETERS AND THE THEORY OF SPECTRAL CLASSIFICATION

C. Fehrenbach

Observatoire de Marseille, Marseilles, France

The main purpose of this conference on stellar atmospheres is to try to collate the ideas of theoreticians with the needs of observers in regard to spectral classification.

My intention is to describe briefly the needs of astronomers engaged in the study of Galaxy.

The first spectral classifications were made here at the Harvard College Observatory, and the Henry Draper Catalogue is certainly one of the most important contributions to stellar astronomy of this century. This catalog includes 225,000 stars and, with its extension, more than 300,000. Actually, we know the classification of more than 500,000 stellar spectra.

The first attempt to classify stars was made with only one parameter, which was, in fact, the surface temperature of the star. The need for a second parameter was dimly apparent in the beginning, as was proved by the introduction of suffixes, but this need was clearly recognized only 50 years later. Morgan's classification with luminosity criteria is, like the HD, an empirical classification; the second parameter is either the absolute magnitude or a datum of the same type. Theoreticians prefer to consider the surface gravity of the star.

The need for a third parameter is now evident: Astronomers engaged in the study of the Galaxy desire to know the age of the stars, whereas theoreticians prefer to know the actual chemical composition. It is interesting to bring these two points of view into confrontation.

In order for a spectrum to be described by theoretical calculations, knowledge of a certain number of physical parameters is indispensable. We can reduce this number to a minimum: The surface temperature, surface gravity, and actual chemical composition are absolutely necessary, but the turbulence in the atmosphere, the rotation of the star, and some other parameters are useful only if we wish to obtain a precise representation of the spectrum.

These parameters have a true physical significance and are useful to theoreticians, but they are not so important for astronomers who study stellar systems and, especially, the galactic structure. Their problem is the study of the position, the motion, and the age of stars; it is also the problem of the evolution of a star, the final aim being the structure and the evolution of the Galaxy.

For this purpose, the astronomer must know the position, the apparent magnitude, the apparent color, the radial velocity, and the proper motion. The spectrum, obtained with a limited resolution, will permit him to determine the absolute magnitude, the intrinsic color, and perhaps the age of the star. He can thus localize this star in space at a given moment of its evolution. These results, when they are sufficiently numerous, will supply the key to the structure of our Galaxy.

You can easily understand the difficulty that exists in reconciling these two viewpoints.

To some extent, the observers solved from empirical data the problem of the relation between spectral type and absolute magnitude (the HR diagram). The relation between spectral type and intrinsic color is also solved in practice. But the relation between spectral type and age can result only from a confrontation between observation and theory. Astronomers interested in the structure of the Galaxy are interested not necessarily in the actual chemical composition but rather in the original composition, knowledge of which is extremely important. This is a most difficult problem. Astronomers are also very interested in other physical data, although they actually find them difficult to use.

Other difficulties exist: What are the most suitable spectral characteristics for determining the parameters we need to use?

Since our spectral resolution is evidently too small for the fainter stars, we need parameters for numerous stars and do not have enough time for even the relatively brilliant stars. For this reason, Morgan's natural group classification is especially useful. It is also interesting to use new methods, such as Strömgren's for narrow-band photoelectric measurements and even wide-band photometry, or the measurement of hydrogen-line intensities.

Up to now, most of the relations between the observed parameters and the useful parameters (M_V, color) are empirical. I ask theoreticians to help us in resolving this problem with theory. In short, you, the theoreticians, have the very interesting task of relating theoretical spectra to good physical parameters: T, g, and chemical composition, among others. The choice of your parameters is good and physically significant, but we, the observers, need some other characteristics: absolute magnitude, intrinsic color, and age. A very important task for the theoreticians would be the determination of a theoretical relation between observational data and the parameters essential for a knowledge of stellar systems in particular and for the structure of the Galaxy in general.

THE EFFECTIVE TEMPERATURE – COLOR RELATIONSHIP

E. E. Mendoza, V

Observatorio Astronomico Nacional, University of Mexico, Mexico,
and Astronomy Department, University of Chile, Santiago, Chile

At present, we have only 27 stars for which direction determinations of effective temperature can be made. These are listed in Table 1, together with the B-V and V-R colors and the effective temperatures. The colors are from Johnson, Mitchell, Iriarte, and Wiśniewski (1966) and Mendoza (1966, 1967). The temperatures are taken from Hanbury Brown, Davis, Allen, and Rome (1967) and from those quoted by Johnson (1966).

An empirical procedure to determine effective temperatures of stars other than those given in Table 1 has been through broad-band photometry. During recent years most of these determinations have been based on the B-V color of the UBV system defined by Johnson and Morgan (1953). Unfortunately, for a number of stars this color is seriously affected by factors other than just interstellar extinction. For example, Be and T Tau-like stars have B-V colors that show excesses caused by line and continuum emission (Mendoza, 1958, 1966). Also, the long-period variables of spectral types M, S, and C have their B-V colors affected substantially by atmospheric chemical composition (Mendoza, 1967), in particular by the O/C ratio. Johnson (1966) obtained an I-L index that gives fair effective temperatures for a variety of stars. Unfortunately, L magnitudes (3- to 4-μ effective wavelength) are not easy to obtain, and again, Be and T Tau-like stars have a noticeable excess in this wavelength [caused most likely by the presence of shells in Be stars, and probably by circumstellar dust in T Tau-like objects (Mendoza, 1966, 1968a; Johnson, 1967; Feinstein, 1968)].

A visual comparison of red and blue spectra of Be and T Tau-like stars with the long-period variables of spectral types M, S, and C shows that the former are less affected than the latter by the various factors mentioned above. It can be shown that Be and T Tau-like objects lie slightly above the line defined by nearby stars, galactic clusters, and stellar associations in the two-color plane (V-R, B-V). Furthermore, each of the long-period variables of spectral types M, S, and C has a distinctly different position in such a diagram, and hence it is possible to separate these stars (Mendoza, 1967).

A comparison of the effective temperatures given in Table 1 with B-V and V-R shows less scatter, on the average, for log T_{eff} versus V-R than for log T_{eff} versus B-V, especially for the late-type stars. The scatter for the early-type stars does not increase by much with the use of V-R color. Therefore, we tentatively give the relationship between effective temperature and V-R in Table 2. The values in parentheses are somewhat arbitrary extrapolations to the values found in Table 1.

It should be pointed out that corrections to be applied to the V-R color, when it is affected by interstellar extinction, are equal to or smaller than those applied to the B-V color. From a practical point of view, it is possible to obtain the same accuracy in V-R as in B-V (Mendoza, 1968b). At present there are new detectors that are able to reach at least the same limiting magnitude in R as in B (Johnson, private communication, 1968).

Table 1. Effective temperatures and colors

Star	B-V	V-R	$T_{eff}(°K)$
Sun	+0.62	+0.52	5800
α Aql	+0.22	+0.14	8250
β Aur	+0.03	+0.08	10000
α Boo	+1.23	+0.97	4250
α CMa	0.00	0.00	10380
ε CMa	-0.21	-0.09	21000
α CMi	+0.42	+0.42	6450
α Car	+0.15	+0.22	7510
o Cet[*]	+1.41	+2.32	2600
β Cru	-0.23	-0.13	26600
α Eri	-0.15	-0.04	14000
YYGem	+1.49	+1.38	3720
α Gru	-0.13	-0.08	14600
α Her	+1.45	+2.10	3400
α Leo	-0.11	-0.02	13000
α Lyr	0.00	-0.04	9500
α Ori	+1.84	+1.64	3790
β Ori	-0.03	+0.01	11200
γ Ori	-0.22	-0.09	21000
ε Ori	-0.18	-0.07	21000
α Pav	-0.20	-0.09	17100
β Peg	+1.67	+1.50	3130
β Per A	-0.05	-0.04	11500
α PsA	+0.09	+0.06	9300
α Sco	+1.84	+1.55	3520
μ'Sco	-0.22	-0.12	19200
α Tau	+1.54	+0.98	3860

[*]JD 243909.66

Table 2.   The relationship of effective temperature versus V-R

V-R	T_{eff} (°K)	V-R	T_{eff} (°K)
-0.15	>26000	+0.60	5400
-0.10	19000	+0.80	4600
-0.05	12000	+1.00	4100
0.00	10400	+1.20	3750
+0.05	9500	+1.50	3500
+0.10	8800	+1.75	3000
+0.15	8250	+2.00	3100
+0.20	7800	+3.00	(2500)
+0.30	7100	+4.00	(2200)
+0.40	6450	+5.00	(2000)
+0.50	5900		

The scale presented here cannot be considered final, but it provides a basis for further comparison with interferometer measurements, which are not available for Be and T Tau-like stars, and for theoretical models of stellar atmosphere.

REFERENCES

Feinstein, A. 1968, Zs. f. Ap. , 68, 29.
Hanbury Brown, R. , Davis, J. , Allen, L. R. , and Rome, J. M. 1967, Mon. Not. Roy. Astron. Soc. , 137, 393.
Johnson, H. L. 1966, Ann. Rev. Astron. Ap. , 4, 193.
_____ . 1967, Ap. J. , 150, L39.
Johnson, H. L. , Mitchell, R. I. , Iriarte, B. , and Wiśniewski, W. Z. 1966, Comm. Lunar Planet. Lab. , 4, No. 63, 99.
Johnson, H. L. , and Morgan, W. W. 1953, Ap. J. , 117, 313.
Mendoza, E. E. , V. 1958, Ap. J. , 128, 207.
_____ . 1966, Ap. J. , 143, 1010.
_____ . 1967, Bull. Ton. y Tac. Obs. , 4, 114.
_____ . 1968a, Ap. J. , 151, 977.
_____ . 1968b, PDA No. 5, Publ. Dept. of Astron. , University of Chile, in press.

DISCUSSION

Popper: At the risk of belaboring the obvious for a moment, I would just like to point out that real effective temperatures, which are a fundamental property of real stars, can be determined by only one means. This involves determining the angular diameters and absolute fluxes at one wavelength, as well as complete spectral-energy distributions, for a pretty good set of stars. For hot stars, the last, I presume, are to be provided by the rocket ultraviolet observations, so, with the Hanbury Brown diametry, we should have real effective temperatures for the hot stars fairly soon. For the cooler stars, the prime observational desideratum is the angular diameters obtained through the use of modern Michelson interferometers. As far as I know, there are two or three groups working on this, and until results are in, we will not have effective temperatures for the cooler stars other than the sun. Spectral-energy distributions are of course another problem, which is not quite so serious for the cooler stars.

In addition, we can't properly discuss effective temperatures for the cool stars until we have reliable observations of limb darkening for stars above the main sequence. For bright stars of luminosity class III and above, a modern Michelson interferometer will, we hope, be able to provide a limb-darkening parameter directly. When these things are done, we will have a real scale of effective temperatures.

Morton: I might differ with Dr. Popper on one point. For the hottest stars my own feeling is that one still needs model atmospheres to get effective temperatures. Maybe the rocket data will be good enough to get fluxes long-ward of the Lyman limit, but the hot stars have fluxes shortward of the Lyman limit that I doubt will penetrate the interstellar hydrogen. The only way to correct for that effect is straight from model atmospheres. So, let's not give up on model atmospheres as a way of getting effective temperatures.

Popper: Well, it's just a manner of speaking. This would not be an effective temperature in the sense that an effective temperature measures the amount of flux coming out of each square centimeter of the surface of a real star. It is semantics, perhaps, but fitting a model based on theory to some observed parameter does not necessarily tell us what the integrated flux is. The numbers coming out of the model determine what you might call a "hypothetical" effective temperature. Now if the amount of the unobservable hypothetical flux is only a few percent of the total, let's not quibble about it, but if it's 50%, by my definition you are not determining an effective temperature.

Morton: Maybe we should ask why we want effective temperatures; it is for comparisons with the stellar interior results.

Popper: And the atmosphere as well.

Wallerstein: May I ask why the Michelson and not the Hanbury Brown intensity interferometer?

Popper: The Hanbury Brown interferometry in its present form is the least efficient way to use photons. It cannot be used for temperatures lower than it has been used. It's just impossible — it's a fundamental limitation having to do with the fact that, if the star is bright enough in order for you to have enough light to make the measurement with less than, say, 100 hours observing time, then the star is going to be very big angularly, and if the star is very big you can't get the mirrors close enough together to get a point on the transform and, in addition, the individual mirror will resolve the star. So you have to use a Michelson interferometer.

RESULTS FROM ECLIPSING BINARIES

D. M. Popper

Department of Astronomy, University of California, Los Angeles, California

1. ECLIPSING BINARIES AND ABSOLUTE RADIATIVE FLUXES

The contributions of eclipsing binaries to fundamental masses, radii, and surface gravities are well known. It should be emphasized once again that precise spectral types and/or colors of the components are absolutely essential if the masses and radii are to be useful, a point unfortunately often overlooked by photometric observers. The use of eclipsing binaries of known distance (β Aur, μ' Sco) in furnishing absolute radiative fluxes has not been very effective. Although there are still some possibilities in this direction, the Narrabri angular diameters provide much more trustworthy and straight-forward results. It is particularly important that current efforts to develop modern Michelson interferometers be brought to fruition.

While eclipsing binaries are of less value than one might have hoped for furnishing absolute fluxes, they do provide precise <u>ratios</u> of the absolute fluxes of the two stars in the observable spectrum. In favorable cases these ratios provide powerful checks on predictions of fluxes by other means, as well as a basis for interpolation between them. There are, of course, many eclipsing systems for which the flux ratio is known with high precision. But these are of no value for our problem unless the system is relatively uncomplicated, the flux ratio differs appreciably from unity, <u>and</u> the colors and/or spectral types of both components are known with precision. In addition, the colors for which the ratio is determined must be on a precisely calibrated system. In cases where interstellar reddening can be neglected and differential limb darkening is unimportant, the flux ratios and colors are obtained simply from five quantities, namely, the color index outside eclipse and the light within both minima in each of the two colors relative to the light outside eclipse. As applications of this method, we may consider the systems CM Lac and V477 Cyg. The relevant quantities are:

	CM Lac (Barnes, Hall, and Hardie, 1968)	V477 Cyg (O'Connell, 1968)
$(J_1/J_2)_V$ observed	2.20	3.29
$(B-V)_1$	+0.10	+0.10
$(B-V)_2$	+0.26	+0.38
Sp_1	A3	A3
(J_1/J_2) predicted	1.7	2.5

There is some uncertainty in the values of B-V because of an uncertain amount of space-reddening for each system, but the prediction of the surface-brightness ratio J_1/J_2 from models or from angular-diameter results is insensitive to small displacements along the B-V scale. The predictions, based on flux scales from models fixed to the sun, to the angular-diameter results, and to spectral-energy distributions, differ markedly from the observed values in both of the eclipsing systems. The differences are in the same direction and are of about the same amount for both cases. The explanation for the discrepancies is not clear. The most uncertain aspect of the observations is the difference in depths of secondary minima in the two colors, from which the color index of the cooler component is obtained. Photometric observations of high precision on a well-calibrated color system are required. More work is needed in this field.

2. CORRELATIONS BETWEEN PARAMETERS OF ECLIPSING BINARIES

For problems in stellar interiors in particular, as well as in atmospheres, we should like to have, for as many individual stars as possible, precise positions in the H-R diagram, with the masses of all objects known. The masses should be known over much of the diagram to better than 20% for modern problems. Visual binaries serve fairly well for the lower main sequence, although requiring a''/p'' $P^{2/3}$ to a 7% accuracy is a fairly rigid demand. But for visual binaries, one does have directly position in the H-R diagram.

For eclipsing binaries the masses are, with proper precautions, more readily tied down to better than 20% than for visual binaries, and data are available for the upper (G2-0) main sequence as well as for a few evolved systems. In contrast to visual binaries, however, we do not have M_V directly. On the other hand, the radii are available. The natural coordinates for eclipsing stars are mass, radius, and B-V or spectral type. Position in the R, B-V diagram shows the evolutionary state just as well as in the M_V, B-V diagram, except that the zero-age sequence is not directly established from other information. Rather than rely on scales of absolute visual flux versus B-V to convert to the H-R diagram, we may employ the lower envelope of observed eclipsing binaries in the R, B-V diagram as the locus of unevolved stars.

Figure 1 shows the log R, B-V plot for stars with well-determined masses and radii in the spectral range B5 to G2. The values are taken from a recent compilation (Popper, 1967) with the addition of CO Lac (Smak, 1967) and a revision of V477 Cyg (Popper, 1968). Also shown are the Narrabri stars (Hanbury Brown, Davis, Allen, and Rome, 1967) with good parallaxes. The location of a star above the lower envelope of an H-R diagram is conventionally assumed to be a consequence of evolution, the evolutionary tracks being based on computations for assumed mass and chemical composition. But since the masses of the stars shown are known (except for four of the six Narrabri stars), we can determine their unevolved loci directly if we assume that the lower envelopes of the mass-radius (Figure 2) and mass-color (Figure 3) diagrams are the loci of unevolved stars.

The stars of Figure 2 are, of course, just those for which the surface gravities, required for analyses of stellar atmospheres, are available directly from observations.

The assumption that the relationships between M, R, and B-V, obtained for eclipsing binaries with well-determined masses, are applicable to single or wide double stars is probably unobjectionable for components of detached eclipsing systems. Components filling or nearly filling their Roche limiting surfaces are not shown in the diagrams. A few primaries of semidetached systems are included. It is a moot point whether their chemical compositions and structures can be considered unaffected by their past histories.

The observations upon which many of the results reported here are based were obtained at the Mount Wilson and Lick Observatories. Financial support was provided in part by the Office of Naval Research under contract Nonr 3507(00).

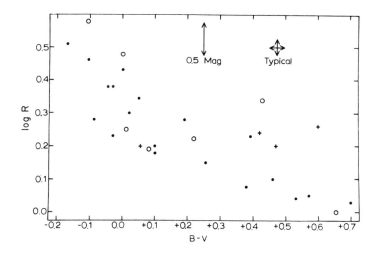

Figure 1. Colors and radii of eclipsing binaries with well-determined masses
and of Narrabri stars with parallaxes. Filled circles: eclipsing
binaries with both components on or near the main sequence;
plusses: main-sequence components of eclipsing binaries with subgiant secondaries; open circles: Narrabri stars and the sun.

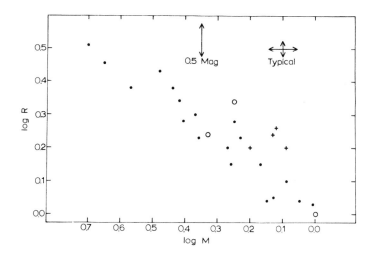

Figure 2. Masses and radii of the stars in Figure 1, omitting the Narrabri
stars without masses. Symbols as in Figure 1.

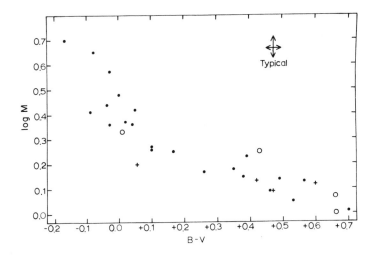

Figure 3.   Masses and colors of the stars of Figure 2 with the addition of
three eclipsing binaries and one visual binary with radii not determined. Symbols as in Figure 1.

REFERENCES

Barnes, R. C., Hall, D. S., and Hardie, R. H. 1968, <u>Publ. Astron. Soc.</u>
<u>Pacific</u>, <u>80</u>, 69.
Hanbury Brown, R., Davis, J., Allen, L. R., and Rome, J. M. 1967,
<u>Mon. Not. Roy. Astron. Soc.</u>, <u>137</u>, 393.
O'Connell, D. J. K. 1968, <u>Vistas in Astronomy</u>, in press.
Popper, D. M. 1967, <u>Ann. Rev. Astron. Ap.</u>, <u>5</u>, 85.
⸻. 1968, <u>Ap. J.</u>, <u>154</u>, 191.
Smak, J. 1967, <u>Acta Astron.</u>, <u>17</u>, 245.

DISCUSSION

<u>Neff</u>: I would like to ask about CM Lac and V477 Cygni. Are those both total
eclipses?

<u>Popper</u>: No, neither one is a total eclipse.

<u>Neff</u>: Then you are really measuring the color at the center of the star.

<u>Popper</u>: Not really. If you measure the light that is lost in two colors, you
get directly the color index of that part of a surface of a star that is being
eclipsed. There will be corrections for limb darkening, of course, but these
will be relatively small.

<u>Underhill</u>: I agree with your remarks about checking eclipsing binaries. The
problem is that theoreticians compute spectra of intrinsically infinite spectral
resolution, whereas in the eclipsing binaries you generally have two spectra
on top of each other, and the available spectra are usually of rather low dispersion because the star is never bright enough for a really powerful spectrograph.

<u>Popper</u>: Well, I think most of these are moderately bright stars, some of
them of 7th magnitude, very few of them of 5th magnitude, though. Now that's
not very encouraging if 2 A/mm are required. But if 10 A/mm will suffice,
quite a bit of material should be obtainable.

ON THE MAGNESIUM GREEN TRIPLET LINES AND THE SODIUM D LINES AS TEMPERATURE AND GRAVITY INDICATORS IN LATE-TYPE STARS

G. Cayrel de Strobel

Institute d'Astrophysique, Paris, France

ABSTRACT

The profiles of the $D_1 + D_2$ sodium lines and those of the magnesium green triplet lines were calculated by use of a grid of 42 model atmospheres. The temperature range chosen for these models was $\theta_{eff} = 0.90$ to 1.20, and the gravity range, $\log g = 4.5$ to 1.0. The models were constructed on the assumption of solar abundances everywhere. A set of equivalent widths was calculated for given θ_{eff} and $\log g$ values for the sodium and magnesium lines, and isoequivalent widths were drawn in the θ_{eff} - $\log g$ plane. Sodium lines are a better temperature indicator than magnesium lines, whereas magnesium lines are a better gravity indicator than sodium lines.

For late-type G and K stars, the Balmer discontinuity is no longer a good luminosity indicator. We must, therefore, look for other criteria if we want to determine quantitively the luminosity of late-type stars.

The magnesium triplet lines are one of the best gravity criteria for late-type stars because the wings of the lines are dominated by collisional damping with neutral hydrogen atoms, even if the gravity is low. But the intensity of the lines is temperature dependent. The sodium D lines are also good gravity and temperature indicators for late-type stars.

We have computed profiles of the magnesium and sodium lines, using a grid of 42 stellar atmosphere models. These models are rescaled solar models. The continuous absorption was calculated by a routine of Vardya's (1964), which included the absorbents H, H^-, H_2^+, and H_2, as well as Rayleigh scattering. We calculated the models assuming solar abundances everywhere. It must be emphasized that the magnesium and sodium lines are not independent of the abundances of these elements.

In our computations, we chose a grid of 42 models with θ_{eff} values of 0.90, 1.00, 1.05, 1.10, 1.15, and 1.20 and $\log g$ values of 1.0, 2.0, 2.5, 3.0, 3.5, 4.0, and 4.5. We decided to begin the computation of models with $\theta_{eff} = 0.90$, because this is the limit toward the higher temperatures at which the magnesium and sodium lines can be used as gravity criteria. For higher temperatures the broadening of the lines is due no longer to collisions but to microturbulence. In other words, the equivalent widths of these lines move down from the damping part to the flat part of the curve of growth. We measured the equivalent widths of all the magnesium and sodium multiplets resulting from the 42 models.

Figures 1 and 2 show iso-θ_{eff} lines in the $\log g - W_A$ plane for the magnesium and sodium multiplets. We see that in the length of the cooler iso-θ_{eff} lines there is a large spread in W_A and $\log g$ values. For example, in Figure 1, at $W_A = 5$, we find first a model of a cool bright giant, then a less cool giant, and finally a rather hot K dwarf. Note that we no longer have a luminosity effect for θ_{eff} values smaller than 0.90. (To emphasize this effect we also calculated the iso-θ_{eff} 0.85 line for sodium, shown in Figure 2.) Indeed, between 0.90 and 0.85, the equivalent widths change very little from the giant to the dwarf models. On the other hand, we note that for models with $\theta_{eff} = 1.20$, the range of equivalent widths is extremely large.

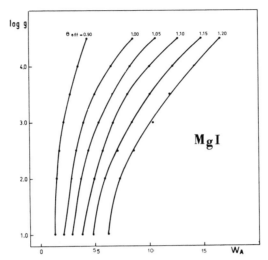

Figure 1.   Iso-θ_{eff} lines in the log g – W_A plane for the magnesium triplet lines.

From these figures we see that several temperatures and gravities are possible for a single value of equivalent widths of the magnesium and the sodium lines.   In the near future, we shall calculate the same grid of models but with nonsolar abundances.   The same equivalent widths will then satisfy a quasi infinity of models.

It is, therefore, necessary to combine our magnesium and sodium gravity and temperature criteria with two other criteria if we want to determine the three parameters of a stellar atmosphere: temperature, gravity, and metal abundance.   The temperature and metal-abundance parameters may, for example, be:

A.   Strömgren's (1963, 1966) b - y as a temperature indicator and m_1 as a metal index.

B.   Other photometric data such as Golay's (1963, 1968) indices and the six-color indices of Stebbins and Whitford (1945).

C.   Neutral and ionized curves of growth of iron.

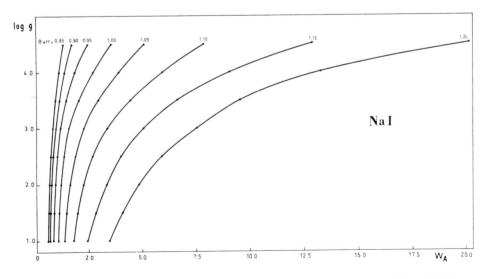

Figure 2.   Iso-θ_{eff} lines in the log g – W_A plane for the sodium D lines.

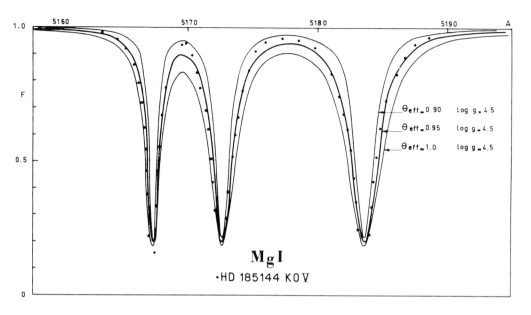

Figure 3.   Profiles of the magnesium triplet for HD 185144 K0 V.
Solid lines are computed profiles; dots are observed
profiles.

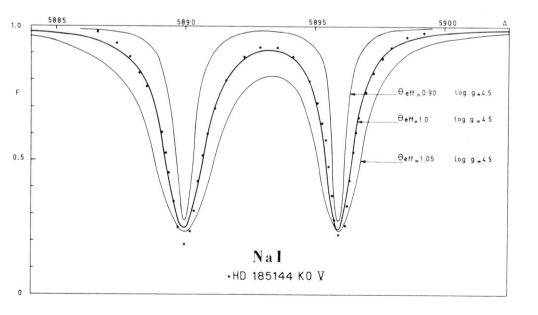

Figure 4.   Profiles of the sodium D lines for HD 185144 K0 V.
Solid lines are computed profiles; dots are observed
profiles.

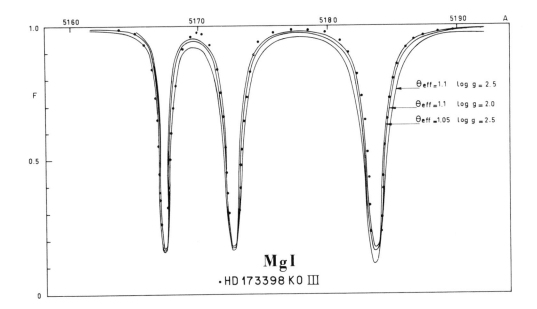

Figure 5.   Profiles of the magnesium triplet for the giant
star HD 173398 K0 III.   Solid lines are computed profiles; dots are observed profiles.

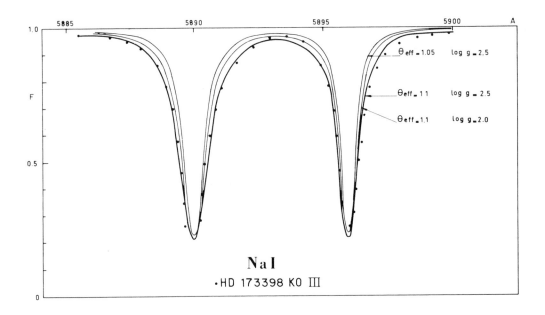

Figure 6.   Profiles of the sodium D lines for the giant
star HD 173398 K0 III.   Solid lines are computed profiles; dots are observed profiles.

We wanted to check our grid of magnesium and sodium profiles with real magnesium and sodium profiles resulting from a K0 V dwarf and a K0 III giant star. Both these stars were carefully selected in order not to show Population-II characteristics.

In Figures 3 and 4 we show one observed and three computed profiles of the magnesium and sodium lines for the dwarf star HD 185144 K0 V. We see that the best fit between the observed and the computed profiles is at θ_{eff} = 0.95 for magnesium and at θ_{eff} = 1.0 for sodium. Since this star is a dwarf, log g was taken equal to 4.5. The mean of the θ_{eff} values is 0.97, which is in agreement with the value 0.96 found by Rousseau (1968) using the six-color photometry method.

Figures 5 and 6 give the computed and observed profiles of the magnesium and sodium lines for the giant star HD 173398 K0 III. The best fit between the observed and the calculated magnesium and sodium profiles for this star is at θ_{eff} = 1.07 and log g = 2.2. We could not find a photometric θ_{eff} value for this star in the literature. We found, however, as expected because of the difference in the luminosities of the two stars, that the K0 V star is slightly hotter than the K0 III star.

REFERENCES

Golay, M. 1963, Publ. Obs. Genève, Ser. A, No. 64, 199.
_____. 1968, in Vistas in Astronomy, in press.
Rousseau, J. 1968, Thesis; Ann. d'Ap., in press.
Stebbins, J., and Whitford, A. E. 1945, Ap. J., 102, 318.
Strömgren, B. 1963, in Stars and Stellar Systems, Vol. VIII, ed. K. Aa.
    Strand (Chicago: University of Chicago Press), 123.
_____. 1966, Ann. Rev. Astron. Ap., 4, 433.
Vardya, M. S. 1964, Ap. J. Suppl., 8, 277.

DISCUSSION

Matsushima: Your comparison of equivalent widths indicates that it is difficult to distinguish between effective temperature and gravity. Does it help to compare the detailed profile rather than just the equivalent widths?

Cayrel de Strobel: Yes, but this is what I did not do.

Matsushima: So you have had to assume one of the parameters to be known when you determined the other.

Cayrel de Strobel: Yes.

Strom: I'm curious as to the effect that line blanketing would have on these results for the reason that it would change the temperature gradient in the region used in making the fit.

Cayrel de Strobel: This is what I'm calculating now.

Cayrel: Yes, and there is another effect that has to be taken into account, namely, convection. And convection will probably be more important in dwarfs than in giants.

Elste: I would like to add one word of caution. I know that in the sun, where we can measure the center-to-limb variations of the wings of the Mg b lines, there arises a problem. We know that the lower level is a metastable level that is probably excited by collisions and is then de-excited by ionizations. I find that if one takes this sort of an effect into account, the fitting of the observations is improved. And I wonder if this will affect your results.

Cayrel de Strobel: Well, I think you're right. We must take these things into account when we carry out further calculations.

SUBLUMINOUS LATE-TYPE STARS[*]

O. J. Eggen

Mount Stromlo and Siding Spring Observatories, Institute for
Advanced Studies, Australian National University, Australia

ABSTRACT

Photometric observations of 145 randomly selected proper-
motion stars yield a minimum of 19 late-type white dwarfs on the
assumption that space velocities greater than 600 km sec^{-1} do not
occur among these objects. The possible sources of large ultraviolet excesses in late-type stars are also discussed. At least four
new possible late-type white-dwarf members of the Hyades are
isolated.

DISCUSSION

Wallerstein: What is the current status of the so-called subdwarfs in the
Hyades and Praesepe that lie 1. 0 to 1. 5 mag below the main sequence?

Eggen: I was able to track down 13 such stars that had been reported as
being members of the Hyades. Seven of those have now been excluded by
better proper motions. The other half of the sample has been excluded on
the basis of radial-velocity measurements. Evidently these are subdwarfs,
all right, and they just happen to lie in the direction of the Hyades without
being members.

Peat: Does this mean that the density of subdwarfs is high in that particular
field of the sky?

Eggen: No, I don't think so.

[*] This paper has been submitted to the Astrophysical Journal.

SPECTROPHOTOMETRIC ANALYSIS OF OBJECTIVE-PRISM SPECTRA

W. Iwanowska

Astronomical Observatory of N. Copernicus University, Toruń, Poland

In view of the considerable economy in observing time and of the advantage of photographing simultaneously a wide spectral range and a vast sky region, the objective-prism technique deserves more attention. Its accuracy can be improved, and more information can be extracted. Work along these lines is being continued at the Toruń Observatory. We have undertaken a spectral sky-survey program using the 60/90-cm Zeiss Schmidt telescope with an objective prism giving 250 A/mm dispersion at $H\gamma$ with plates calibrated photometrically. I shall report briefly on two points.

A. It is well known that the main source of errors in ground-based line-depth measurements made with objective prisms is the poor and variable resolving power of spectra, depending on seeing and focusing conditions.

Mr. L. Zaleski has developed a simple method of reducing line-depth measurements to standard "best" seeing and focusing conditions. He found that atmospheric scintillation and focal setting act in a similar way on observed line depths. He established correction curves for line depths for consecutive focal settings and recommends the use of these curves for seeing corrections as well. The appropriate correction curve is selected by a visual comparison of a given plate with the standard focusing plate according to the sharpness of the spectra.

B. Mr. J. Smolinski has attempted to establish a method of estimating the microturbulent velocity, in addition to the principal classification parameters of temperature, gravity, and Fe/H abundance, from objective prism spectra of F, G, and K stars. He synthesized the solar spectrum from the Utrecht Catalogue by summing up equivalent widths of lines over 25-A wavelength intervals, taking sums for lines of the three principal curve-of-growth portions separately: 1) the linear part, reaching up to about 50 mA, 2) the flat part, reaching up to 200 mA, and 3) the damping portion.

Figure 1 shows these results and gives ϵ_λ, the percent of energy absorbed in lines, for 25 A-wavelength intervals for each of the three curve-of-growth parts, their sum, and their relative distribution. It can be seen that the distribution of ϵ_λ for the flat part (Figure 1c) of the curve of growth, which is most sensitive to microturbulent effects, is not a smooth curve. Thus, we hope to get differences in the pseudocontinuum of objective prism spectra for different microturbulent velocities. In objective prism spectra of 250 A/mm dispersions, all lines of parts 1 and 2 of the curve of growth are swallowed up by the continuum and cannot be seen individually. The lines that can be seen and whose depth or areas can be measured and converted into equivalent widths belong to the damping part 3 of the curve of growth for solar-type spectra. We assume that T_{eff}, g, and Fe/H can be determined from these lines, and then the appropriate points in the pseudocontinuum, where the part 2 lines prevail, may give the estimate of the velocity of microturbulence.

Mr. Smolinsky has calculated the change of the level in one such wavelength interval, assuming that the equivalent width of part 2 lines is proportional to $\xi_t^{0.75}$, where ξ_t is the microturbulent velocity. The results are shown in Figure 2, where we see a rather higher sensivity of this parameter for velocities exeeeding about 2 km sec^{-1}. The slope of the curve is then about 0.1 mag per 1 km sec^{-1}. Similar analyses will be performed for α Boo high dispersion spectra and, when possible, for an F-type star as reference. The method will be checked on objective-prism spectra of stars with known microturbulent velocities.

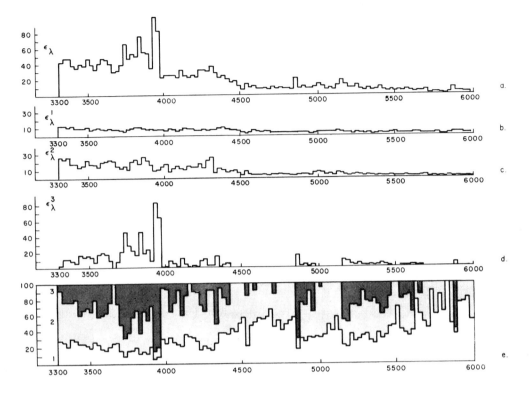

Figure 1.   The energy ϵ_λ absorbed in solar spectral lines: a) total, b) linear, c) flat, d) damping portions of the curve of growth, e) relative distribution.

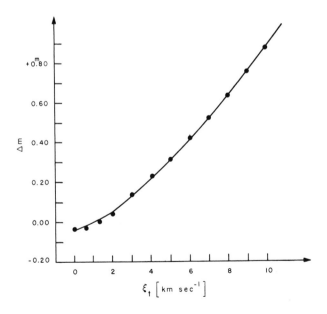

Figure 2.   The computed growth of the depression Δm at λ 4000 - λ 4025 A with increasing microturbulent velocity for solar-type spectra.

QUANTITATIVE SPECTRAL CLASSIFICATION OF MODERATE DISPERSION SPECTRA

F. Spite

Observatoire de Paris, Meudon, France

The work of Mustel, Galkin, Kumaigorodskaya, and Boyarchuk (1958) has been mentioned in Whitney's introductory paper. I cannot see how equivalent widths could be measured from spectra with a dispersion of 72 A/mm. The measurements of Mustel and his coworkers should be called indices. The physical meaning of these indices is not quite clear, since the correlation between their measurements of Hδ, for instance, and an iron or calcium line is not very good, the deviations from the mean relation appearing uncorrelated with luminosity or abundance (Lebon, 1962). Of course, it is always possible to combine indices to find a spectral type. However, reducing a set of measurements to a one-dimensional type results in a considerable loss of information.

I adopted a different technique, measuring the central depth P of the lines. This measurement is a function of the equivalent width. Empirically determined for the sun, this function appears to be more or less linear. Even the smaller lines are on the third part of the curve of growth, so that their equivalent widths are independent of microturbulence. No appreciable rotation is expected in cool stars, so that the measurements P are a function of the effective temperature T_{eff}, the gravity g, and the metal abundance (characterized by iron abundance Fe/H).

These measurements can be used in two ways. The first (and correct) way is to compute line (and blend) profiles over a small range of wavelengths from a grid of models, then to convolute these profiles with the instrumental profile, and finally to compare the resultant synthetic spectrograms with observed ones. This work, now in progress, takes full advantage of the resolution of the spectra, thus allowing us to compute line profiles over a comparatively small range of wavelengths. This method is more or less along the lines proposed by Houziaux.

A less satisfactory but quicker method is the search for empirical relations between the measurements P and the parameters T_{eff}, g (or M_v), and Fe/H of well-known stars. Since this method has been published (Spite, 1966a), I will only mention here that, with some care, the effective temperature, absolute magnitude, and iron abundance can be determined with a standard error of ±100°K, ±0.42 mag, and ±0.13 (logarithm), respectively. Applications of this method have been described in my earlier paper. In one of these applications, not only T_{eff}, M_v, and Fe/H were determined, but also the color index B-V. Thus, the _intrinsic_ color of stars can be found with a standard error of ±0.017 mag. It is possible to deduce the color of the sun; I find $(B - V)_\odot = 0.68$. Another application is to select stars with measured lines very similar to those of the sun. These stars can then be measured in any photometric system and thus provide a substitute for a direct measurement of the colors of the sun (useful for many purposes), inasmuch as two stars with the same line spectrum can be supposed to have the same colors, a point on which comments are welcome. Up to now, two stars have lines that match fairly well those of the sun: κ Cet and 51 Peg.

Other miscellaneous applications can be noted. The absolute magnitude I find for HD 175541 clearly indicates that the star is a subgiant, confirming the suggestion made by Whiteoak (1967) from energy distribution. The iron abundance I obtain for HD 145675 is higher than that of the sun. This is confirmed by the higher blanketing found by Whiteoak. I predicted (Spite, 1966b)

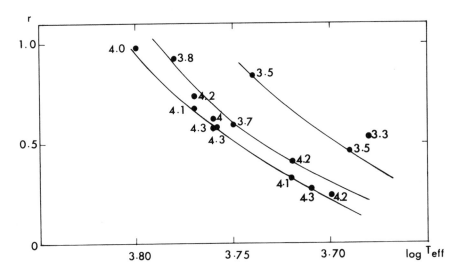

Figure 1.   Iso-g lines empirically drawn. Abscissa is log T_{eff} deduced from colors; ordinate is the measured ratio $r = \lambda\ 4077$ Sr II/$\lambda\ 4071$ Fe I.

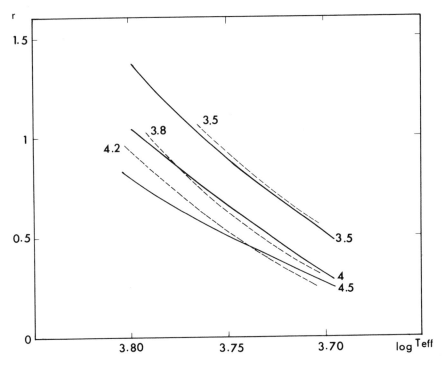

Figure 2.   Same coordinates as in Figure 1, but the ratio r is now computed (solid lines) for three values of log g (3.5, 4.0, 4.5). The dashed lines are those of Figure 1.

that HD 6755 was probably a giant, although the λ 4077 Sr II line was weak. Koelbloed (1967) found that this star is a giant or a subgiant.

A peculiar point in this work is the fact that the line λ 4077 Sr II appears not to be useful for luminosity classification. To clear up the point, the ratio r = λ 4077 Sr II/λ 4071 Fe I was computed from a grid of empirical models. Among measured stars, those with good parallaxes and the R I J K colors of Johnson, Mitchell, Iriarte, and Wiśniewski (1966) were selected. The T_{eff} can be found from calibration of these colors through stars with known radius (Johnson, 1964). Then, from M_V and evolutionary tracks of stars (Demarque and Larson, 1964; Hallgren and Demarque, 1966), the gravity g can be found. Iso-g lines are empirically drawn through the points (Figure 1); the agreement between empirical (dashed lines) and computed (solid lines) values of r is not too bad (Figure 2). Before any definite conclusion is drawn, the uncertainties involved in reducing the observed parameters of the stars to T_{eff} and g should be diminished and the influence of blends in the measurements should be taken into account.

The sensitivity of the computed ratio r to the general abundance of metals was computed and found to be relatively small. Thus, the empirical inutility noted above of the line λ 4077 Sr II as a gravity indicator remains unexplained.

I am indebted to Mrs. Spite-Lebon for computing the lines from the grid of models.

REFERENCES

Demarque, P. R., and Larson, R. B. 1964, Ap. J., 140, 544.
Hallgren, E. L., and Demarque, P. R. 1966, Ap. J., 146, 430.
Johnson, H. L. 1964, Bull. Ton. y Tac. Obs., 3, 305.
Johnson, H. L., Mitchell, R. I., Iriarte, B., and Wiśniewski, W. Z., 1966, Comm. Lunar Planet. Lab., 4, No. 63, 99.
Koelbloed, D. 1967, Ap. J., 149, 299.
Lebon, M. 1962, C. R. Acad. Sci. Paris, Ser. B, 254, 2141.
Mustel, E. R., Galkin, L. S., Kumaigorodskaya, R. N., and Boyarchuk, M. E. 1958, Ann. Crimean. Ap. Obs., 18, 3.
Spite, F. 1966a, Ann. d'Ap., 29, 601.
_____. 1966b, in IAU Symp. No. 24, ed. K. Lodén, L. O. Lodén, and U. Sinnerstad (London: Academic Press), 95.
Whiteoak, J. B. 1967, Ap. J., 150, 521.

DISCUSSION

Strom: I have some concern about your Sr-Fe index. Some recent analyses indicate that the abundance of Sr in the subdwarfs changes with respect to that of Fe. In other words, occasionally s-process elements appear to be down in abundance relative to Fe. This would presumably make you predict too high a gravity for these stars.

Underhill: This correlation of the Sr II 4077 line with the strong Fe line has been well known for a long time. This is an empirical fact, but no one yet seems to understand its theoretical basis. Spite has based his analysis on the apparent depth of these lines, and it seems to me that to interpret the variations in the strength of the Sr lines as an abundance effect is quite dangerous because there is a lot of implied theory behind it. I would just ask for caution, and point out that the Sr lines are resonant lines. Certainly they require a sophisticated theory. I have the impression that people who have made comments about the s-process abundances have not used any theory at all.

Neff: I was disturbed to hear that you consider 51 Pegasi (G4V) to have the same color as the sun. I have found it to have peculiar ultraviolet colors.

I have indirectly determined the colors of the sun by folding my sensitivity functions with the flux distribution of the sun [Neff, Astron. J., 73, 75, 1968]. However, there are inconsistencies that depend on the wavelength region. The observed color of HR 483 (G2V) agrees with the predicted color of the sun on the most sensitive color ($m_{0.37\,\mu} - m_{0.55\,\mu}$) system.

STELLAR FLUX DISTRIBUTIONS IN THE INFRARED

D. H. P. Jones

Radcliffe Observatory, Pretoria, South Africa

Johnson (e. g., 1964; Johnson, Mitchell, Iriarte, and Wiśniewski, 1966) has published photoelectric colors of stars extending to 10 μ in the infrared. Before these colors can be used to derive relative flux distributions for comparison with model atmospheres, it is necessary to attend closely to two factors: 1) the correction for extinction in the earth's atmosphere and b) the conversion from colors to flux ratios.

In modern photoelectric photometry it is usual to assume that the extinction in magnitudes is proportional to the air mass (sec z) and to extrapolate the observations to zero air mass. Johnson (1965a) has demonstrated that this procedure is preferable to the square-root law. However, there are two other phenomena that cause departures from this law. First, if the exponential extinction coefficient varies markedly within the bandpass of the photometer, a nonlinearity with air mass will result; for at small air masses the extinction will be governed by the largest extinction coefficient, and at large air masses by the smallest coefficient. Second, in the infrared the extinction is not a smooth function of wavelength but arises from many weak lines in the molecular bands. To compute their total extinction accurately would be an extremely complex calculation and would require access to data that are not readily available. To make the problem tractable, we assumed that the monochromatic extinction law was that derived by Cowling (1950) for water vapor. Water vapor is the dominant source of opacity in the infrared. Allen (1963) has tabulated the extinction coefficients of this and other gases in the format required by Cowling's law.

The response of the photometer in magnitudes is given by

$$-2.5 \log \int F_\lambda A_\lambda S_\lambda \, d\lambda \quad .$$

Considering first the J and K bands, the stellar fluxes F_λ were taken from the Stratoscope observations (Woolf, Schwarzschild, and Rose, 1964) and the S_λ from Johnson (1965b), multiplied by the reflectivity of two aluminum surfaces. We took the monochromatic extinction coefficients from Pettit and Nicholson (1928). There are more modern determinations in the literature, but theirs seem to be the values most comparable to conditions on Catalina Mountain. We plan to repeat the calculation at a later stage with the atmospheric transmissions of Gates and Harrop (1963). The variation of magnitude with air mass is shown in Figure 1 for α CMa as measured with the Stratoscope A detector and the J receiver band. The broken curve is for an exponential extinction law, and the solid curve for a Cowling law. Neither is linear with air mass, especially in the air-mass range 0 to 1. The Cowling law exaggerates the curvature; in both cases the positive curvature arises from the variation of the extinction coefficient in the bandpass. The exponential law takes into account only the macroscopic changes, while the Cowling law also includes microscopic changes. Supposing the Cowling curve to be correct, is it conceivable that this curvature has escaped detection? The extinction is usually determined over the air mass 1 to 2 (Johnson, et al./1966). If a straight line is drawn through air masses 2 and 11 the departure at 1.5 air masses is 0.006 mag. This is probably too small to be detected in practice. If the extinction is determined over the range sec z = 1 to 2 and extrapolated to 0, an error will result, equal to

$$m(0) + m(2) - 2m(1) \quad ,$$

where the extinction coefficient is m(2) - m(1). In Figure 1 this error is
-0.21 mag and the extinction coefficient is 0.10. However, the error in
itself is immaterial, provided it is the same for all stars, for it then
amounts to an arbitrary redefinition of the color system. Figure 2 shows
the errors for the J and K systems as a function of J-K. The mean of the
Stratoscope A and B detectors was used; the fluxes of Vega and the Sun are
discussed below. The total range is 0.10 mag for both J and K, and both
show a rough dependence on color.

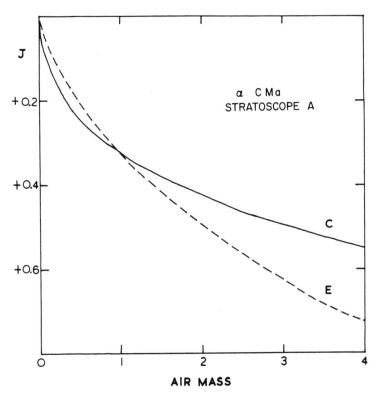

Figure 1.   Apparent magnitude as a function of air mass.
J filter band, flux distribution of α CMa from
Stratoscope A detector. Zenith extinction is
from Pettit and Nicholson (1928). E = exponential law; C = Cowling law.

    In Figure 3 the observed J-K colors of these stars, corrected by the
amounts in Figure 2, are compared with the extra-atmospheric colors for
the Stratoscope detectors A and B separately. The agreement is disappointing. It is improbable that the scatter arises from the observed J-K colors
or from the corrections in Figure 2, for both are smooth functions of spectral type. Some scatter must arise from intrinsic variability, for all but
one of the stars concerned are known or suspected variables. However,
o Ceti has a V amplitude of 5.4 mag and a J-K amplitude of only 0.15 mag,
so this can scarcely be the dominant source of scatter. Turning to the possibility of error in the Stratoscope observations, the authors give two sources
of error that may well contribute:
    A. Sensitivity of the detectors varies over the surface of the detectors
and slow drifts of tracking may have occurred.

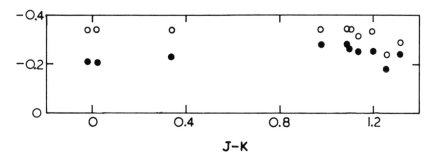

Figure 2.   Departures from linear law.   Filled circles,
J magnitudes; open circles, K magnitudes.

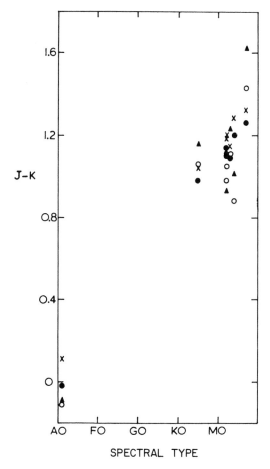

Figure 3.   J-K colors as a function of spectral type.   Filled circles,
Johnson observed; crosses, same corrected by the
amounts in Figure 2; open circles, fictitious J-K colors
from Stratoscope A detector; triangles, same from
Stratoscope B detector.

B. Detector and system noise causes an uncertainty in the zero. In short, although the agreement in Figure 3 is not so good as we would like, it does not provide conclusive evidence that the extinction model is wrong.

Johnson's L', L", M, and N photometry was examined in the same way for a Cowling law and for the extinction coefficients of Allen (1963). The allowances for extinction may well be too great, for they correspond to 1 cm of precipitable water vapor, which is probably too large for Catalina Mountain. For the Sun the departures from the linear law amount to -0.41, -0.47, -1.05, and -0.35 mag, respectively.

The second problem, that of converting colors to flux ratios, requires a simultaneous knowledge of the fluxes and colors of some star in this wave-length region—data that do not exist. However, for Vega there is good evidence that Mihalas' model, with T_e = 9600°K, log g = 4, predicts the stellar fluxes within the accuracy of the observations over the wavelength range 3000 to 10,550 A (Mihalas, 1966; Jones, 1968). Accordingly, the fluxes of this model were extended into the infrared by use of the opacity tables of Bode (1965). We made allowance for the blanketing by Paschen and Brackett lines following Kuiper (1963), and we neglected the higher series of hydrogen. The colors of Vega have been directly observed. Also, the infrared fluxes of the Sun have been directly observed; in general, these observations were made in sufficiently narrow bands for the curvature discussed above not to occur. Molecular absorption was generally specifically allowed for. The values used were taken from Allen (1963). The colors for the Sun have not been measured and must be inferred. The values adopted, taken from Table 2 of Johnson (1966), are the mean of those corresponding to the Sun's spectral type G2V and color B-V = 0.663 (Alexander and Stansfield, 1966). The calibration is presented in the form

$$2.5 \log \frac{F_{\lambda 1}}{F_{\lambda 2}} = K_{12} - \text{color } (1\text{-}2) \quad .$$

The values of K are presented in Table 1. The first three rows give the constant K with the extinction treated as outlined above; the fourth row gives the mean values of K that result if the extinction follows a strictly linear law. Johnson's values are shown for comparison. The changes introduced by the improved treatment of the extinction are surprisingly small. The overall agreement is good, especially for the first and fifth rows, which are independent.

Table 1. Values of K

	J-K	J-L'	J-L"	J-M	J-N
Vega	2.26	4.30	4.19	5.55	8.59
Sun	2.34	4.36	4.23	5.36	8.49
Mean	2.30	4.33	4.21	5.46	8.54
Mean without extinction	2.20	4.17	4.07	5.54	8.40
Johnson (1966, Table IV)	2.35	—	4.06 —	5.48	8.60

As a byproduct, the effective wavelengths for the different bands were computed from the first moments of the integrals (Table 2).

Table 2. Effective wavelengths of Johnson's photometry
at air mass 1 (unit = 1 μ)

	Vega	Sun	α Ori	Strömgren (equal F_λ)
J	1.19	1.19	1.20	1.22
K	2.16	2.17	2.16	2.20
L'	3.57	3.57		3.65
L''	3.47	3.48		3.53
M	4.73	4.73		4.76
N	9.50	9.48		10.42

REFERENCES

Alexander, J. B., and Stansfield, R. 1966, Bull. Roy. Obs., No. 119, E325.
Allen, C. W. 1963, Astrophysical Quantities (2d ed.; London: Athlone Press).
Bode, G. 1965, Die Kontinuierliche Absorption von Sternatmosphären in Abhängigkeit von Druck, Temperatur und Elementhäufigkeiten (Kiel: Inst. Theoret. Phys. und Sternw. der Univ. Kiel), 193 pp.
Cowling, T. G. 1950, Phil. Mag., 41, 109.
Gates, D. M., and Harrop, W. J. 1963, Appl. Opt., 2, 887.
Johnson, H. L. 1964, Bull. Ton. y Tac. Obs., 3, 305.
_____. 1965a, Comm. Lunar Planet. Lab., 3, No. 52 67.
_____. 1965b, Ap. J., 141, 923.
_____. 1966, Ann. Rev. Astron. Ap., 4, 193.
Johnson, H. L., Mitchell, R. I., Iriarte, B., and Wiśniewski, W. Z. 1966, Comm. Lunar Planet. Lab., 4, No. 63, 99.
Jones, D. H. P. 1968, Mon. Not. Roy. Astron. Soc., 139, 189.
Kuiper, G. P. 1963, Comm. Lunar Planet. Lab., 2, No. 25, 17.
Mihalas, D. 1966, Ap. J. Suppl., 13, 1.
Pettit, E., and Nicholson, S. B. 1928, Ap. J., 68, 279.
Woolf, N. J., Schwarzschild, M., and Rose, W. K. 1964, Ap. J., 140, 833.

A PROPOSED SYSTEM OF SPECTRAL CLASSIFICATION FOR G- AND K-TYPE STARS

D. W. Peat

University Observatories, Madingley Road, Cambridge, England

The system of spectral classification as developed by Morgan, Keenan, and Kellman (1943) (MKK system) has provided astronomers for more than two decades with a tool that is quick and easy to use and that has proved invaluable in a wide range of investigations in stellar spectroscopy and galactic structure. Nevertheless, the system has serious disadvantages. First, it is not divided into so many classes as we would like; the system does not distinguish between stars within a given luminosity class at a given spectral type, although the intrinsically brightest stars within such a group may be more than one stellar magnitude brighter than the intrinsically faintest stars. Second, "luminosity" is not a single fundamental parameter characterizing the spectrum of a star and cannot therefore be determined from a knowledge of the spectrum alone. The fundamental parameters characterizing a spectrum produced by a stellar atmosphere in LTE are, as is well known, effective temperature, surface gravity, the abundances of the elements, and one or more parameters specifying nonthermal velocity fields; these parameters should define a classification system. Third, the MKK system is one based on only two classification parameters — luminosity and spectral type; the effects of abundance variations on the strengths of absorption features are not formally taken into account, and spectra where different criteria lead to different classifications are merely designated as "peculiar."

The development of high-speed computers and modern model-atmosphere techniques has in recent years made possible the setting-up of a much improved and more refined system of spectral classification. The purpose of this brief communication is to outline such a system, applicable to stars classified on the MKK system as of spectral types G and K, based on the observational results of narrow-band photometry.

The multichannel spectrometer, described by Griffin and Redman (1960) and used subsequently in a number of Cambridge investigations, measures the total flux within a passband a few angstroms wide relative to the flux within two neighboring side bands. For a G- or K-type spectrum, the ratio R_i of these fluxes is a function of the line absorption within the passbands, and so, for a given model atmosphere of surface gravity g and effective temperature T_{eff}, it can in general be written

$$R_i \equiv R_i(T_{eff}, g, A_j \cdots, v_j \cdots) \quad , \tag{1}$$

where $A_j \cdots$ are the abundances of the elements, and $v_j \cdots$ are parameters representing nonthermal velocity fields that increase the doppler widths of the absorption lines. Hence, for any number n of observed photometric ratios within a given spectrum, we have n expressions of type (1), and if the parameters T_{eff}, g, $A_j \cdots$, $v_j \cdots$ were known, the n values of R_i could be computed. Our purpose, however, is to perform the reverse operation — that is, if for a given spectrum we have observed n ratios R_i, we wish to determine n fundamental parameters T_{eff}, g, $A_j \cdots$, $v_j \cdots$.

Now we can write the expressions (1) in differential form:

$$dR_i = \left(\frac{\partial R_i}{\partial T_e}\right)_{g, A_j} \cdots dT_{eff} + \left(\frac{\partial R_i}{\partial g}\right)_{T_e, A_j} \cdots dg + \left(\frac{\partial R_i}{\partial A_j}\right)_{g, T_{eff}} dA_j \quad ,$$

$$(2)$$

so that, to a first approximation, dR_i is the difference between the true (i.e., observed) value of R_i and the value computed for equation (1) if we assume particular values of the parameters T_{eff}, g, etc. The partial derivatives that occur in equation (2) can be determined from the values of R_i computed for a grid of model atmospheres represented by different values of the parameters T_{eff}, g, $A_j \cdots$, $v_j \cdots$. For n parameters and n observed ratios R_i in a given spectrum, we now have n linear equations of type (2), whose solutions are the first-order differentials dT_{eff}, dg, etc. Iteration of this procedure now leads to exact determination of the fundamental parameters T_{eff}, g, etc.

In general, of course, the abundances of all the elements giving rise to absorption lines within the passbands cannot be determined, because there are too many unknowns. Even if the nonthermal velocities were ignored, then, because of the unknown values of T_{eff} and g, n independent ratios could determine only (n-2) abundances. We therefore assume that the abundances of certain elements vary in the same way relative to each other; for example, our previous work (Peat and Pemberton, 1968) has suggested that, when averaged over more than approximately 50 field giant stars, the logarithmic abundance ratio (Fe/Ca) remains constant near its solar value.

CHOICE OF SPECTRAL FEATURES TO BE MEASURED

The spectral features chosen for observation must be sufficiently dependent on the fundamental parameters to allow the latter to be determined to a satisfactory degree of accuracy. Most of the features in G- and K-type spectra are markedly sensitive to temperature; they are less sensitive to abundances, however, and comparatively few are sensitive to surface gravity. The magnesium green triplet is sensitive to gravity but is needed to determine the magnesium abundances. The gravity is therefore best determined from the strengths of neutral iron lines; the group of iron lines near λ 5250 has been known for many years to be sensitive to luminosity. The dependence of neutral iron absorption in giant and supergiant atmospheres (where pressure broadening is negligible for all but the very strongest lines) arises from the approximate equality of electron pressure P_e and ionization constant ϕ for iron, where ϕ = (number of singly ionized atoms/number of neutral atoms)$\times P_e$. Since the line-absorption coefficient of a neutral element is proportional to $1/(P_e + \phi)$, lines of an element such as neutral calcium, whose ionization potential is sufficiently low for $\phi \gg P_e$, will show no dependence on P_e and hence no dependence on g.

The system briefly outlined here provides astrophysical information in addition to that arising from the classification procedure. Element abundances are, of course, of considerable interest in themselves, and in the few cases where distance moduli of the stars are available, knowledge of g and T_{eff} is sufficient to determine stellar masses and radii.

REFERENCES

Griffin, R. F., and Redman, R. O. 1960, Mon. Not. Roy. Astron. Soc., 120, 287.

Morgan, W. W., Keenan, P. C., and Kellman, E. 1943, An Atlas of Stellar Spectra with an Outline of Spectral Classification (Chicago: University of Chicago Press).

Peat, D. W., and Pemberton, A. C. 1968, Mon. Not. Roy. Astron. Soc., 140, 21.

DISCUSSION

Strom: Did you say that you could determine gravities just using the Fe I lines?

Peat: And Fe II lines. Yes, we have selected lines that are particularly sensitive to gravity because they are formed in different levels of the atmosphere where the ionization conditions are quite different.

Strom: I would expect that the turbulent velocity parameter that you determine from the curve of growth would depend on the metal abundance because the metal lines would, through the blanketing effect, influence the temperature gradient.

Peat: Yes, this effect is quite pronounced.

Stibbs: You wrote down a functional expression for R, differentiated it, and got expressions for the partial derivative of R with respect to the atmospheric parameters. But it wasn't quite clear to me what you were determining by least squares. Have you determined the partial derivatives empirically, by combining the observations of stars whose physical parameters are known?

Peat: Well, it's not a least-squares solution. The differences of the ratios, R_i, are, in fact, the difference between the observed value and the theoretical value computed on a first guess. The partial derivatives are computed theoretically from a grid of models. The models themselves are "standard models," computed with radiative equilibrium and LTE.

Stibbs: It seems to me you're assuming at the outset that you know all about the star in order to put it into the right box.

Peat: No, you only need to guess the first values, and then you proceed by iteration. The initial guesses, of course, can be based on almost any sort of data.

THE QUANTITATIVE BASIS FOR CLASSIFICATION OF LATE-TYPE STARS IN THE CORDOBA SPECTRAL ATLAS

J. Landi Dessy

Observatorio Astronomico e IMAF,
Universidad Nacional de Cordoba, Argentina

1. INTRODUCTION

The classification and atlas of stellar spectra in 42 A/mm, made with spectrograph I of the 61-inch reflector at Bosque Alegre, will be finished this year. In this paper I want to report on some quantitative tests based on Morgan-Keenan standard stars in order to define new standards in the Southern Hemisphere. We carried out the tests with the utmost possible precision, taking care, however, that the new standards coincided on an average with the MK system.

The atlas was compiled in collaboration with C. and M. Jaschek; most of the observations were made by R. Colazo, and most of the computations by A. Puch. On our transparents the spectra will be shown in a scale of 7.5 A/mm for early spectra (O to F5) and 5 A/mm for late ones (F5 to M6). The atlas will also contain microphotometric tracings of interesting spectral regions. The classification of the early spectra (O to F5) was performed by C. and M. Jaschek; that of the later spectra, by the author.

Preliminary lists of Southern standards have already been published (Jaschek and Jaschek, 1966; Landi Dessy and Keenan, 1966). Dr. Keenan's stay in Argentina in early 1966 was very helpful to the author. The system of the new spectrographic atlas was named "System LJJ" by W. W. Morgan; this abbreviation will be used throughout this paper. To make the discussion of luminosity easier, we suggest the introduction of the following table of equivalencies:

I-0	Supertitan
Ia	Large Titan
Iab	Titan
Ib	Subtitan
IIa	Bright Supergiant
IIab	Supergiant

Although we have tried to stay as close as possible to the MK system, the LJJ differs from it in the following points:

A. Use of larger dispersion (42 A/mm) and a spectrograph with diffraction grating of high separating power, which makes possible the use of new criteria.

B. Use of the ultraviolet region.

C. Introduction of quantitative parameters for the definition of types and luminosities.

D. Introduction of the parameter of metallicity with the main purpose of placing the star accurately within the H-R diagram.

E. Introduction of "vernier criteria," by means of which we can obtain a high degree of precision in short intervals of spectral type or luminosity.

F. Determination, wherever possible, of the spectral type in two ways— based on hydrogen and independent of hydrogen.

G. Mean photometric calibration of plates.

H. Introduction of Southern parallaxes (which do not modify the MK system).

First, we attempted to introduce quantitative measurements using relations of intensities between lines whose separation does not exceed 50 A. The intensity was calculated according to the depth of the line with respect to a pseudocontinuum adopted for the region; the calibration was taken into account in the calculation, which was somewhat lengthy and difficult. The essential results of this first stage have been published (Landi Dessy, 1966). The wish to increase precision and to make reduction easier led to the exchange of the King step sensitometer for a King continuous sensitometer. The original photometer was also provided with a direct intensity register. Calibration was done with a lamp fitted with a blue filter, which was considered valid for the spectral range of a IIa-O plate.

The necessity of reproducing for the atlas some microphotometric profiles, all in the same horizontal and vertical scale, showed that metric parameters could be obtained with ease and great precision if a means could be found to define the vertical scale (intensity) by lines so well represented in the interval F5 to K5 that the scale could then be used as a reference for a spectral zone; i. e., the depth of the selected line is arbitrarily taken as the constant in the spectral interval F5 to K5. The other relations are then measured in the form of differences on the ordinate. Figure 1 shows an example of how this difference is taken. It is obvious that for lines of equal intensity this difference is zero and independent of the chosen scale. If we consider only differences of intensity in the lines that are much smaller than the depth of the defining line of the vertical scale, the influence of the defining line will be small. Since the problems of photometric calibration become less critical under these conditions, quite precise results could be expected.

It can also be observed that, in a way, this method approaches the manner in which a spectrum is normally estimated by sight.

Late spectra (F5 to K5) generally show three photometric regions, so that for maximum precision a separate plate of each region should be taken, since in this way it is easier to obtain a linearized recording. Precision will not be the same in the region that is linear by itself as in the region where linearization is obtained by means of the King continuous sensitometer. Three zones were therefore chosen:

	Interval	Defining line of scale
Zone 0	UV−H and K	3820
Zone 1	H and K−4226	4045
Zone 2	4226−4600	4383

2. SEARCH FOR CRITERIA

In an attempt to obtain the greatest precision in spectral classification, some MK-standard stars were recorded and compared with one another to see, in a purely empirical way, which lines showed greater sensibility to variation in spectral type, luminosity, or metallicity. The relations of the MK system were also studied, but it must be pointed out that in our recordings some blends were resolved, such as λ 4415 to λ 4417 (see Figure 1). The goal of the investigation was to find relations in equality of intensity that would define types and luminosities. We could not always obtain criteria of equality, yet we found some criteria with small differences of intensity. It is well known that the difference between the intensity of two lines with respect to spectral type or luminosity is generally not a linear function. In some cases it may occur that the maximum of sensitivity does not correspond to the equality of intensity of two lines. Therefore, it is sometimes not easy to take into account all the factors that contribute to a higher or lesser precision of the intensity relation.

In many cases the difference in intensity between two lines is very sensitive within a small interval of spectral type or luminosity, later becoming asymptotic or losing precision, or both at the same time. We have called these kinds of relation vernier criteria, since they are valuable as interpolaters in small intervals. A good example of a vernier criterion is the relation λ 4250 - λ 4254 - λ 4260 to determine the spectral type in the K0 to K4

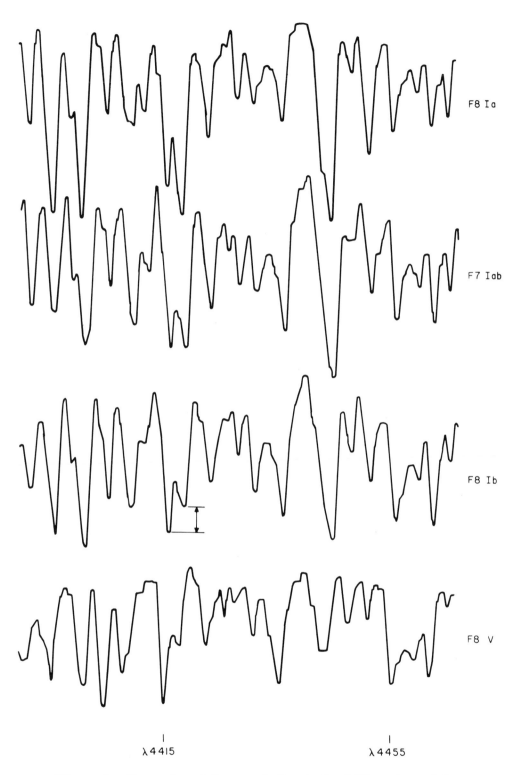

Figure 1. Examples of the resolution in the LJJ system of the
MK blend λ 4415 to λ 4417.

interval (Figure 2; see also Table 1), and λ 4415 to λ 4417 to determine the luminosities in Titans (Figure 1). The region of the validity of the vernier is marked in Figure 3, and on Figure 2 we can appreciate what we call an optical Vernier. In Figure 3 we simply see a line passing through zero, since it is difficult to transform all the details into measurable parameters.

Table 1. Vernier criteria

a) Temperature criteria

Types	λ 4250−λ 4254	λ 4254−λ 4260
G5	+2.60	-8.30
G8	+2.65	-8.10
K0	+2.45	-5.05
K1	0.00	-1.60
K2	-0.85	-0.37
K3	-1.90	+2.10
K5	-5.35	+5.30

These criteria work as a very sensitive vernier between K0 and K5 and lose sensitivity for early types.

b) Luminosity criteria (λ 4443−λ 4455)

Types	I-O	Ia	Iab	Ib	IIa	IIIa
F6				+2.2		
F7			+6.2			
F8		+12.6		+4.6		
G0				+3.4		
G2				+3.3	+1.8	
G3	+13.1					
G5				+3.6	+0.2	-3.0

In these spectral types the luminosity is determined fundamentally with the λ 4077 line of Sr II. With a larger dispersion we can apply criteria based on ionized elements Ti II, Fe II, etc. As we always tried to stay as close as possible to the MK system, the new relations were selected in such a way that they gave luminosities in agreement with the MK system, although greater precision was sometimes achieved; i. e. , the mean of a standard group of the MK system must give the same value as a mean standard of the LJJ system. For example, Keenan classifies λ Vel as K4 Ib-IIa and β Ara as K3 Ib-IIa. In our classification, λ Vel would be K4 Ib and β Ara would be K3 IIa. The relations λ 4399 (Ti II) to λ 4404 (Fe I) (Figure 4) and λ 4415 (Fe I) to λ 4417 (Ti II) permit more precise classification than λ 4045 to λ 4077 does, especially in supergiants and Titans. Table 2 lists the relations tested up to now.

3. METALLICITY

As shown in a previous publication (Landi Dessy, 1966), metallicity influences some lines that are also sensitive to luminosity, which easily leads to errors in the determination. The lines that mostly show this effect

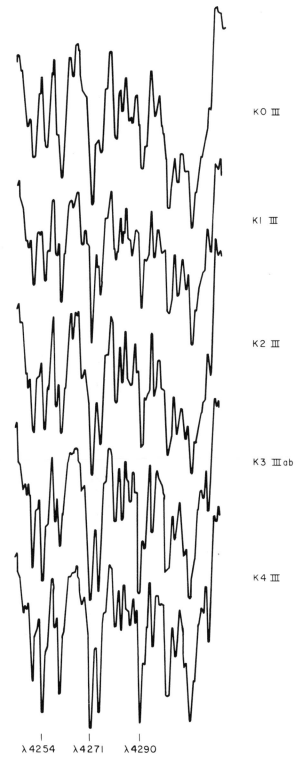

λ4254   λ4271   λ4290

Figure 2.   An example of a vernier criterion λ 4250 - λ 4254 - λ 4260
to determine the spectral type of early K giants.

Table 2.   LJJ spectral classification criteria

<hr>

Zone 0

λ 3758 - H11 (F5 - G5)	H10 - λ 3820 (F5 - G5)	λ 3895 - λ 3899 (G2 - K5)
H11 - λ 3786 (F5 - G5)	λ 3878 - H8	λ 3899 - λ 3905 (G2 - K5)

<hr>

Zone 1

λ 4045 - Hδ	λ 4077 - Hδ	λ 4167 - λ 4172
λ 4045 - λ 4077	Hδ - λ 4144 (G0 - K5)	λ 4196 - λ 4202
λ 4063 - Hδ	λ 4149 - λ 4152	λ 4202 - λ 4205
λ 4063 - λ 4077	λ 4152 - λ 4154 (G5 - K5)	λ 4215 - λ 4226
λ 4067 - λ 4071	λ 4154 - λ 4161	

<hr>

Zone 2

λ 4233 - λ 4235	λ 4404 - λ 4408 (F5 - K2)
λ 4246 - λ 4250 (F5 - K0)	λ 4408 - λ 4415 - (K0 - K5)
λ 4250 - λ 4254	λ 4415 - λ 4417 (I-II;F5 - G8)
λ 4254 - λ 4260	λ 4415 - λ 4434 (K0 - K5)
λ 4258 - λ 4260 (K0 - K5;I-II)	λ 4443 - λ 4455 (F5 - K5;I-III)
λ 4260 - λ 4271 (G2 - K5)	λ 4455 - λ 4461
λ 4290 - λ 4300	λ 4489 - λ 4494 (F5 - G8;I-III)
λ 4300 - Hγ	λ 4494 - λ 4501 (F5 - G8;I-III)
λ 4315 - λ 4325	λ 4494 - λ 4533 (F5 - G8;I-III)
λ 4325 - Hγ	λ 4554 - λ 4585 (F5 - G8;I-III)
Hγ - λ 4351	λ 4563 - λ 4585 (F5 - G8;I-III)
λ 4374 - λ 4383 (F5 - K0)	λ 4582 - λ 4585 (F5 - G8;I-III)
λ 4383 - λ 4385 (F5 - K0)	
λ 4383 - λ 4395 (I-II;F5 - K5) (III-V;F5 - G5)	
λ 4399 - λ 4404 (I-II;F5 - K5) (III-V;F5 - G5)	

are λ 4167, λ 4196, λ 4215, and λ 4233. A good example is λ 4167 with respect to λ 4172. In giant and supergiant stars, λ 4167 tends to disappear. There are, however, examples where this line is abnormally developed; e. g. , Figure 5 shows an alternate sequence of stars with and without metallicity. It is easy to imagine the anomaly that would arise in the classification if the relation λ 4215 to λ 4226 were taken to define the luminosity. The star ζ Cap G4 Ib(+) Ba II would be brighter than x Car G3 I-0, and δ TrA G2 IIa(+) brighter than α Aqr G2 Ib. The four marked lines on Figure 5 show this effect. The star ζ Cap G4 Ib(+) Ba II is Ib, as we can see by the fact that the mean intensities of the faint lines are very similar to those of α Aqr; if ζ Cap were brighter, it would show the same effect of supertitanism as x Car. On the other hand, there are relations to determine the luminosity that are practically free from this effect, e. g. , λ 4246 to λ 4250, λ 4399 to λ 4404, and λ 4415 to λ 4417.

In the LJJ system we measure the relations λ 4167 to λ 4172, λ 4196 to λ 4202, λ 4215 to λ 4226, and λ 4233 to λ 4235 to determine the effect of metallicity; other relations also show this effect, namely, λ 4063 to λ 4077 and λ 4045 to λ 4077 (see Figure 6).

So far there has been no attempt to establish degrees of metallicity, because the well-known cases are few, but nothing prevents us in the future from adding (+1) or (-2). Up to now it has been sufficient to call attention to the necessity of determining the other parameters accurately.

Table 3 lists the preliminary standards of the F5 to K5 region not contained in the previous lists (Jaschek and Jaschek, 1966; Landi Dessy, 1966).

This method allowed us to make numerical tables to calibrate the spectral types and luminosities, and with the help of these tables we are setting up the new standards.

The diameter of the circles in Figures 3, 4, 6-9 shows the mean error of the differences in intensity.

REFERENCES

Jaschek, M. , and Jaschek, C. 1966, in IAU Symp. No. 24, ed. K. Lodén, L. O. Lodén, and U. Sinnerstad (London: Academic Press), 30.
Landi Dessy, J. 1966, in IAU Symp. No. 24, ed. K. Lodén, L. O. Lodén, and U. Sinnerstad (London: Academic Press), 33.
Landi Dessy, J. , and Keenan, P. C. 1966, Ap. J. , 146, 587.

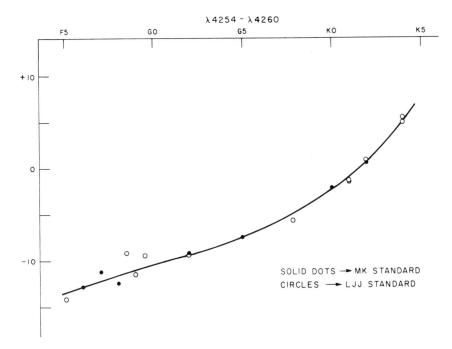

Figure 3.   The difference in central depths between λ 4254 and λ 4260
provides a spectral classification for late-type stars.

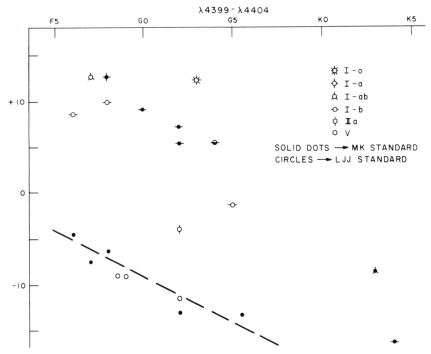

Figure 4.   The difference in central depths between λ 4399 and λ 4404
provides a luminosity criterion from F5 to G5.

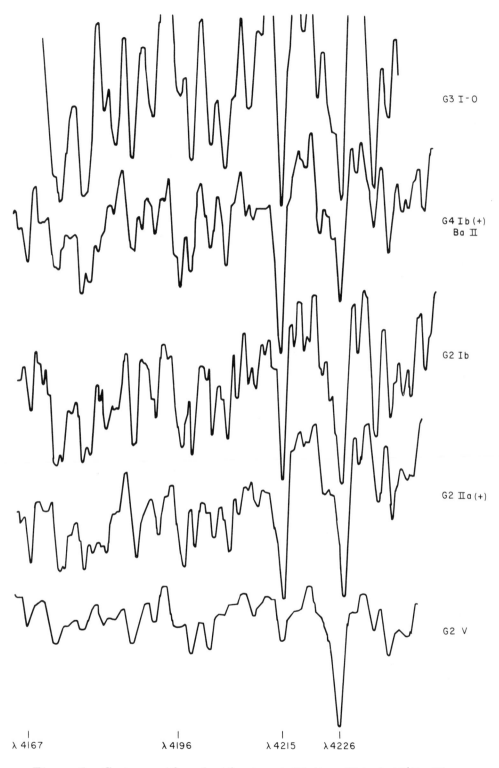

Figure 5.   G stars with and without metallicity.   Note λ 4167 with
respect to λ 4172.

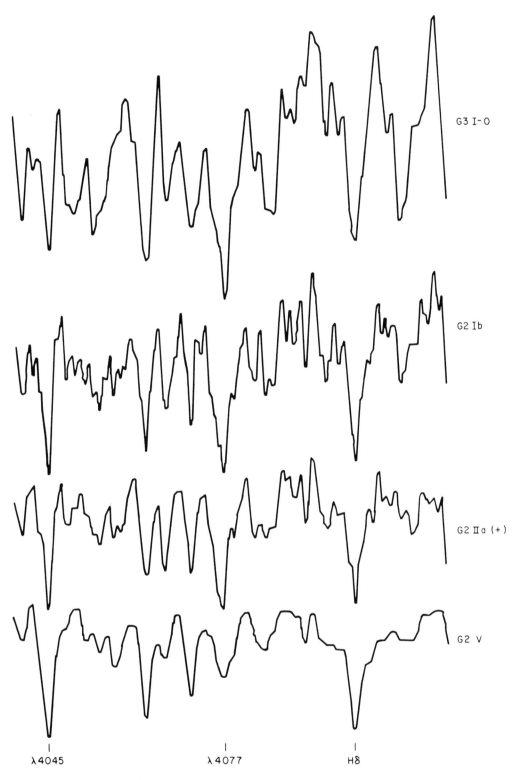

Figure 6.   Other metallicity effects in G stars.

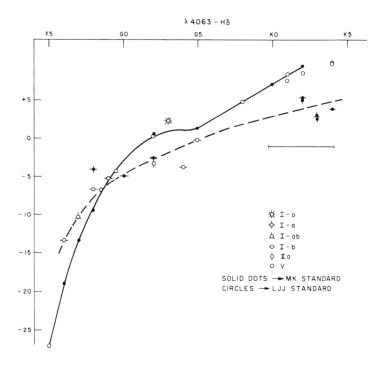

Figure 7.   The difference in central depths between λ 4063 and Hδ (or the ordinate) provides a more sensitive spectral classification criterion for late F stars in the MK system than in the LJJ system.

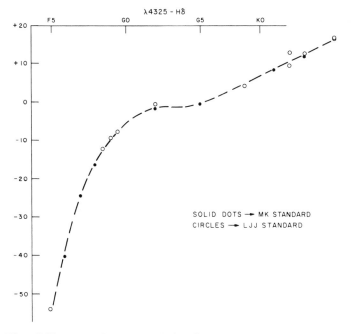

Figure 8.   The difference in central depths between λ 4325 and Hγ provides a sensitive spectral classification criterion for late F stars in both the MK and LJJ systems.

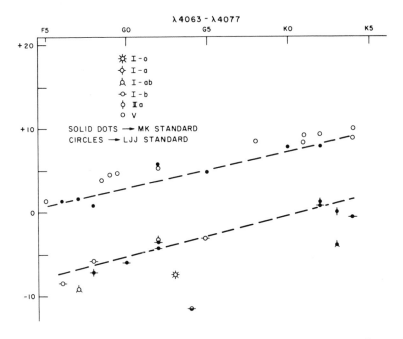

Figure 9. The difference in central depths between λ 4063 and
           λ 4077 provides a luminosity criterion from F5 to
           K5.

Table 3.   Provisional Southern standards (F5 to K5)

Name	Star BS	HD	1900	1900	V	Type
ψ Cap	7936	197692	20^h40^m	-25°38'	4.13	F5 V
ν² Cen	5260	122223	13 55	-45 07	4.34	F7 Ib
γ¹ Nor	6058	146143	16 10	-49 49	4.96	F8 Iab
δ CMa	2693	54605	7 04	-26 14	1.84	F8 Ia
δ Vol	2803	57623	7 17	-67 46	3.97	F8 Ib
ζ TrA	6098	147584	16 18	-69 52	4.90	F8,5 V
ν Phe	370	7570	1 11	-46 04	4.95	F9 V
ζ Tuc	77	1581	0 15	-65 28	4.22	F9,5 V
β Aqr	8232	204867	21 26	- 6 01	2.89	G0 Ib
ζ Mon	3188	67594	8 04	- 2 42	4.35	G2 Ib
α₁ Cen	5459	128620	14 33	-60 25	0.33H	G2 V
δ TrA	6030	145544	16 06	-63 26	3.84	G2 IIa(+)
α Aqr	8414	209750	22 01	- 0 48	2.93	G2 Ib
x Car	4337	96918	11 04	-58 26	3.90	G3 I-0
ζ Cap	8204	204075	21 51	-22 51	3.73	G4 Ib(+) Ba
ξ Pup	3045	63700	7 45	-24 37	3.34	G5 Ib
τ Cet	509	10700	1 39	-16 28	3.50	G8 Vb
o² Eri	1325	26965	4 11	- 7 49	4.42	K1 V
α₂ Cen	5460	128621	14 33	-60 25	1.70H	K1 V(+)
ε Eri	1084	22049	3 28	- 9 48	3.73	K2 V
o¹ CMa	2680	50877	6 50	-24 04	3.78	K3 Iab
	7703	191408	20 05	-36 21	5.32	K2 V
β Ara	6461	157244	17 17	-55 26	2.84	K3 IIa
λ Vel	3634	78647	9 04	-43 02	2.30?	K4 Ib
		156026	17 10	-26 23	6.34	K5 V
ε Ind	8387	209100	21 56	-57 12	4.67	K5 V

DISCUSSION

Crawford: I would like to show just one diagram. I think it will illustrate some of the problems that I get into in calibration attempts. This is a plot of the Balmer discontinuity index against the color index b-y in Strömgren's system, for stars between about F0 and G2. I have taken data from the Strömgren-Perry catalog for field stars, and there seems to be a little scatter about a mean relation. The points for the stars in the Coma cluster lie along the lower envelope of the field star points, in other words, right near the zero-age line. On the other hand, the Hyades stars are not on the zero-age line, contrary to what one would expect. Only a small part of this discrepancy can be attributed to abundance difference or a difference in the blanketing effect, because there are some stars that show a larger metal index and yet lie quite close to the zero-age line.

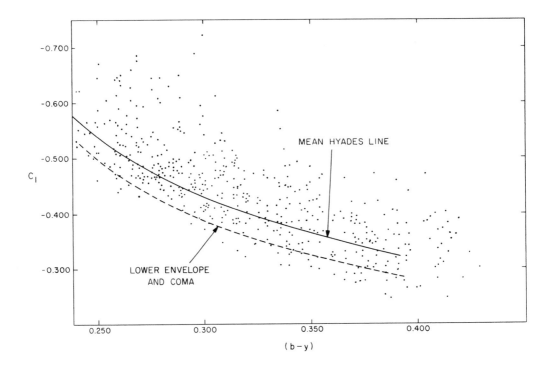

THE SPECTRAL INTENSITY DISTRIBUTION OF α LYR AS A STANDARD FOR SPECTRAL PHOTOMETRY

E. Lamla

Sternwarte, Astronomische Institute, Bonn, Germany

ABSTRACT

From a study of the results of all available absolute calibrations, it is possible to determine a mean distribution of the spectral intensities of α Lyr for the wavelength range $3000 \leq \lambda \leq 10500$. This gives the stellar continuum disturbed by all faint lines and should be used as a standard in spectrophotometric work. The mean intensity distribution at $\lambda < 4000$ A shows considerable differences among all the data used. The calculated color indices are practically in agreement with the observed ones. It will be proved which of the stellar model atmospheres yields the same intensity distribution as that observed. By a comparison of the theoretical with the observed intensity distributions, an effective temperature of $9700 \pm 200°$K is determined for α Lyr. Attention is directed to the necessity of new observational data from the earth and also from satellites, especially for the ultraviolet spectral region on the short-wavelength side of the Balmer jump.

1. INTRODUCTION

The spectral intensity distribution of the continuum of a star can be determined in two steps by photographic or photoelectric spectrophotometry. If we measure in any wavelength range the brightness differences of program stars against one or more selected standard stars, we get first the spectral intensity distributions of these program stars relative to the standard stars used ("relative calibration"). In the second step, if the flux of the standard stars is compared with the flux of a blackbody of known effective temperature, we get the spectral intensity distributions of the standard stars in absolute units ("absolute calibration").

Technical reasons make it difficult to carry out an absolute calibration (cf. Oke, 1965), and since the review by Code (1960), an absolute calibration has been made only by Bahner (1963), Charitonov (1963), Glushneva (1964), and Willstrop (1960, 1965).

Some authors compared their results with the data of others, but there is up to the present no discussion of all the results together. For the reduction of the above-mentioned relative calibrations into absolute energies, the values of the star α Lyr A0V compiled by Code (1960) were used. This compilation disagrees with the results of the absolute calibrations for the ultraviolet spectral range. There are also differences between the observed intensity distribution for α Lyr and the results fitted from theoretical stellar model atmospheres used by Oke (1964) for the calibration of the intensity values in the infrared wavelength range.

Hence, I found it desirable to continue my work (Lamla, 1959a, b; 1965) of comparing all the published results dealing with the absolute calibration of the stellar radiation in order to get a spectral intensity distribution of the main-sequence star α Lyr. This absolute distribution of energy can be used in the future to reduce all relative calibrations to an absolute base.

2. THE SPECTRAL INTENSITY DISTRIBUTION

The results of all absolute calibrations for the star α Lyr have been used in plotting Figure 1. For this, the absolute calibration of the mean flux of eight different stars of the mean spectral type A0, including α Lyr itself, could be taken only from Kienle, Wempe, and Beileke (1940); the mean flux of four different A0 stars was taken only from Willstrop (1960).

In most cases, the fluxes are not given by the authors in absolute units directly but in units with a constant term (additive in the magnitude scale). This term is determined in such a way that the spectral distribution of the intensities of a star with the spectral type A0V gives a flux according to its mean absolute brightness at the isophote wavelength of the visual magnitude system, i.e., $M_V = 1.00$ ($\lambda_i = 5460$ A).

Some calibration systems without measurements in the visual brightness range have been fitted together to give the same results in the common wavelength ranges. This has to be done only for the absolute calibration data of Hall and Williams (1942). For the gradient ϕ_1, the values of Chalonge and Divan (1952) and of Cayrel de Strobel (1961) have been taken. The mean intensity level given by this gradient for $\lambda > 4000$ A was fitted to the mean intensity distributions of the other calibration systems. Then with the size of the Balmer jump, the level of the gradient ϕ_2 is given. All the data used are collected in Table 1.

Table 1. The observed gradients and the Balmer jump for the star α Lyr

| ϕ_1 | Balmer jump | | ϕ_2 | Reference |
	D	m		
1.14	0.494	1.24	1.40	Barbier and Chalonge (1941)
0.96	0.516	1.29	1.43	Chalonge and Divan (1952)
1.0	0.53	1.33	1.36	Cayrel de Strobel (1961)
	0.54	1.35		Bahner (1963)

It is sometimes stated in the literature that the computed spectral intensity distribution of α Lyr by Code (1960) is the result of an absolute calibration. Actually, this is not the case for the intensity data in the violet wavelength range. Code and Whitford tried such an absolute calibration in this range, but it proved unsuccessful because of systematic errors (Code, 1960, p. 73). Instead, Code chose for his calculation for $\lambda < 3700$ A only the values of the ultraviolet gradients determined by Chalonge and Divan (1952) and Cayrel de Strobel (1961), and for $\lambda > 3700$ A a mean value from all measurements available up to 1960. It is unfortunate that Code and Whitford did not publish the absolute calibration mentioned previously in the longer wavelength range.

The heavy line in Figure 1 is the mean of the distribution of absolute energy of the continuum of the star α Lyr, including all faint lines, given in magnitudes for outside the earth's atmosphere. For the wavelength range $4050 \leq \lambda \leq 6600$ A, this mean is identical with the scanning data of two A0 stars given by Willstrop (1965).

From Figure 1 it can be seen that the results of the different calibration systems are in good agreement for all $\lambda > 4000$ A. Estimating a mean error, we get about 2% for that spectral region. In the wavelength range $6000 \leq \lambda \leq 7200$ A, the resulting data of several of the calibration systems

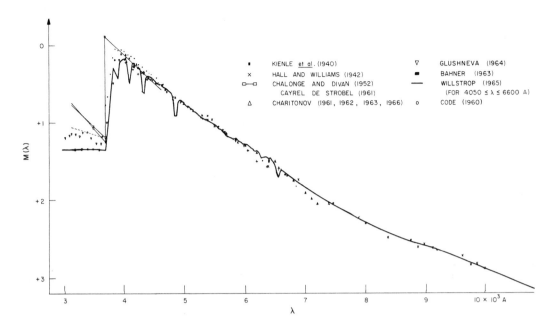

Figure 1.   The mean intensity distribution of the continuous spectrum of
α Lyr, outside the earth's atmosphere.

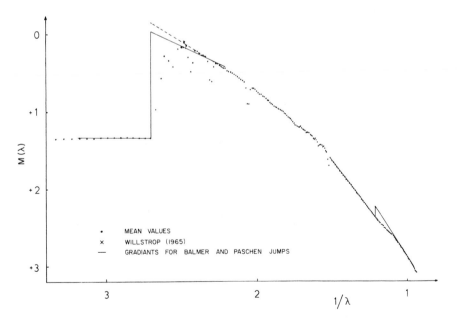

Figure 2.   The mean intensity distribution of α Lyr as a function of 1/λ.

lie at wavelengths where the sensitivity of the photographic plates and multi-
pliers is decreasing. Hence, the systematic differences from the mean flux
values are explained.

But in the region $\lambda < 4050$ A, the apparent differences are much larger.
One reason for this up-to-the-Balmer jump is the effect of the crowded
Balmer lines. For still shorter wavelengths the special measuring method
used could yield the systematically higher values.

In Figure 1 the intensity distribution for a Lyr computed by Code (1960)
is given in order to compare it with the mean flux computed here. The differences are as large as 0.30 mag. To clarify this, we must remember that the
mean intensities in this spectral region are those measured by Bahner (1963).
It was hoped that this calibration system would give better results than the
old one of Chalonge and Divan. The new values are quoted from a manuscript
(Bahner, 1962) that contains the brightness differences of all measured stars
(also a Lyr) against the flux of a blackbody. Therefore, the absolute intensity
distribution for the standard star a Lyr itself could be computed without difficulty. In addition, the intensities given in that manuscript were recalculated
in such a way as to compare them with the data in Bahner's publication (1963).
They were found to be in complete agreement.

In order to show in which wavelength range the mean intensity distribution
of a Lyr can be replaced by gradients, Figure 2 gives the brightness $M(\lambda)$ as
a function of $1/\lambda$. The gradients for calculating the Balmer and Paschen jumps
are also plotted. We get the following values, which are comparable with the
data in Table 1:

$$-2.5 \ D = 1.37 \text{ mag or } 1.48 \text{ mag} \quad ,$$

$$-2.5 \ P = 0.14 \text{ mag} \quad .$$

The highest value for the Balmer jump has been computed from the combined
data of Willstrop (1965) for $\lambda < 4500$ A and Bahner (1963) for $\lambda < 3700$ A.
Thus, the flux in the calibration system of Willstrop seems to be systematically too high in the region $4000 \leq \lambda \leq 4350$ A. It could be that this difference
is a real one, because the data of Willstrop are not valid for a Lyr but are
valid for two other A0 stars.

The absolute intensity distributions of stars related to a Lyr computed by
Code and Oke can be found in papers referred to in Table 2. For all these
energy data to be fitted to the new values for a Lyr computed in this paper,
the corrections in the table are given in units of 0.01 mag.

3. COMPARISON BETWEEN OBSERVATIONS AND THEORY

In most papers on absolute calibrations, the goal has been to connect the
temperature scale of the stars with the temperature scale normally used in
the field of physics by observation of the flux of a blackbody. The main result
of such an absolute calibration was knowledge of the color temperature of the
investigated stars in addition to the intensity distribution in the stellar spectra.
Because the stars are not blackbody emitters, the color temperature is a
function of the wavelength; therefore, in a physical sense, it is not really a
simple function of the conditions in the atmosphere of a star. It is only the
effective temperature T_{eff} that can be determined if the size of the radiant
surface of the star (radius) and the total flux (bolometrically corrected) are
well known. But both problems remain to be solved.

Knowledge of the stellar intensity distribution is used in a different way.
The absolute calibration should be applied principally to yield the intensities
in absolute units, e.g., for an arbitrary star with a visual brightness of
$M_V = 0.00$ mag (Willstrop, 1960). The importance of the energy distribution
of some stars as standard light sources for the observations of other stars
was shown above with the example of a Lyr. Several questions can be explored
if the flux in the stellar continuum from stars of different spectral types and
luminosity classes is known; see, e.g., Seitter (1962), Schmidt-Kaler (1961),
and Lamla (1959b).

Table 2. Corrections for the new calibration of the absolute intensity distribution of α Lyr

λ (A)	Oke (1960)	Oke (1964)	Whiteoak (1966)	λ (A)	Bonsack and Stock (1957)	λ (A)	Aller, Faulkner, and Norton (1966)	λ (A)	Charitonov (1963)	Charitonov (1966)	λ (A)	Code (1960)	Melbourne (1960)	Bless (1960)
3390	21	24		3475	−29	3392	29	3200		33:	3400	22	22	21
3448	21	22	13	3520	−7	3441	28	3300	29	30:	3650	17	17	16
3509	19	20	11	3570	−1	3504	24	3400	22	28	3860	15		
3571	18	20	11	3620	0	3567	21	3500	20	25				
3636	18	18	9	3675	1	3634	20	3600	19	23				
3704	16	15	8	3730	−4	3703	22	3700	18	26				
6800		6						6700		6				
7100		8						6800		7				
7530		8						7000		9				
7850		10						7200		10				
8080		9						7500		11				
8400		10						8000		13				
8805		10						8500		14				
9700		13						9000		15				
9950		10						9500		15				
10250		10						10000		18				
10400		10						10300		14				
10800		9						10800		16				

By the modern theory of stellar atmospheres it is now possible to determine the effective temperature of a star by a comparison of its observed intensity distribution with that computed from a stellar model atmosphere.

In Table 3 the results of several papers dealing with this problem are collected. Following the procedure of α Lyr, it has to be proved first which of the stellar models used in the literature mentioned in Table 3 will give intensity distributions coinciding with the observations in the wavelength range $3000 \le \lambda \le 10,500$ A. Oke (1964) has tried to calibrate his observed relative measurements of α Lyr at $\lambda > 6500$ A by a model atmosphere according to Mihalas (1965) with $T_{eff} = 9500°$K and log g = 4.44. The continuum values of this model fit the observations at $\lambda < 6500$ A very well, but not those in the red and infrared spectral region. Here the theory gives systematically brighter magnitudes (see Figure 3).

The new fitting procedure shows agreement between theory and observation for a model with $T_{eff} = 10,080°$K and log g = 4.44 in the total wavelength range but not in the ultraviolet part. The same agreement can be reached by other models. The results are given in Figure 3.

For the origin of the previously discussed divergence between the observed and theoretically determined intensity distributions, two reasons can be taken into consideration:

A.   The theory does not include an absorption effect in the ultraviolet parts of the stellar spectra.

B.   Errors are not known in the absolute calibration of Bahner (1963). The only possibility of clearing up the second point is to reobserve the intensity distribution of the star α Lyr and to try to get a completely new absolute calibration.

In this matter it is important to put more stress on the measurements of the stellar fluxes from rockets and satellites in the wavelength range from 3000 to 3600 A to fit these data with the earth-bound measurements. The results may be better as long as the wavelength regions are the same for all the measurements.

4.   THE EFFECTIVE TEMPERATURE OF α LYR

The determination of the effective temperature of the star α Lyr by theoretical stellar model atmospheres has been made possible from the interferometric measurement of its angular diameter by Hanbury Brown, Davis, Allen, and Rome (1967). By using a proper value for the limb darkening (Popper, 1959) and the known distance of that star, Hanbury Brown et al. obtained $T_{eff} = 9200 \pm 300°$K. With this value, we can compare all the results in Tables 3 and 4.

The values in Tables 3 and 4 have been taken from the literature cited. In Table 3 the results are based on different parts of the stellar spectra, but in each case with the remark "author," the different authors used, in the ultraviolet wavelengths, the computed intensity data by Code (1960). Because the mean intensity distribution determined here cannot be represented by the emergent flux from any model atmosphere, an attempt was made to calibrate an effective temperature without use of points shorter than $\lambda = 4000$ A. The results are given in Table 3 along with the others.

We see that all temperatures are higher than the one "observed" by Hanbury Brown et al. For a statistical mean value without weights, all the five temperature determinations by use of only stellar atmospheres with a normal metal abundance are taken; the result is 9654°K. Hence, it seems reasonable to adopt the following value as the best at this time:

$$T_{eff} = 9700 \pm 200°K \quad .$$

In Figure 4 the quantity defined by Avrett and Strom (1964),

$$\Delta m(1/\lambda) = -2.5 \log \left[\frac{F(\nu)_{\lambda = 5560 \text{ A}}}{F(\nu)_{\lambda = 3600 \text{ A}}} \right] \quad ,$$

Table 3.   Temperature calibration for α Lyr, the observed stellar continuum fitted by theoretical ones
X: H/He

Model	Spectral range (A)	Teff (°K)	log g	X	Remarks	Reference
LTE	4000 - 10500	10080	4.44	0.85		Mihalas (1965)
LTE blanketing of Balmer and Lyman	4000 - 9500	10000	4.0	0.83		Strom and Avrett (1965)
Non-LTE		9000	4.0	0.83	author	Strom and Kalkofen (1966)
LTE		9500	4.0	0.83	author	
Non-LTE	4000 - 11200	9000	4.0	0.83		
LTE, blanketing Normal metal	3300 - 5560	9500 - 10000	4.0	0.83	author	Strom, Gingerich, and Strom (1966)
10 × metal	3300 - 5560	9000 - 9500	4.0	0.83	author	
Normal metal	4000 - 10500	9500	3.7 + 4.0	0.83		
10 × metal	4000 - 10500	9500	4.0	0.83		
LTE, blanketing	3300 - 5500	9600	4.0	0.85	author	Mihalas (1966)
LTE, blanketing	3300 - 8000	9500	4.0	0.83	author	Avrett and Strom (1964)
Interferometer measured		9500 ± 300				Hanbury Brown et al. (1967)

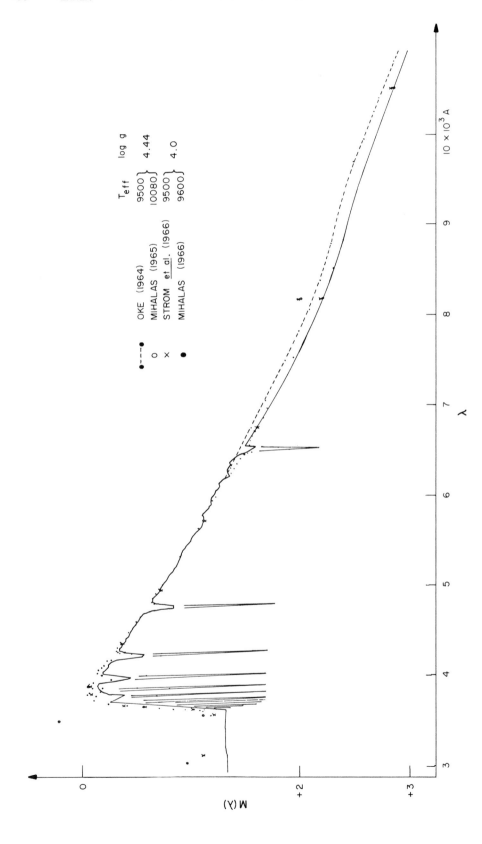

Figure 3.   The mean absolute intensity distribution of α Lyr and the results from theoretical stellar model atmospheres.

is plotted for different stellar model atmospheres as a function of the effective temperature. In the same figure the observed data (old and new) are plotted (both rectangles). This diagram also shows that no stellar model atmosphere can fit the observed stellar flux for λ < 4000 A.

Table 4. Temperature calibration for α Lyr, the observed line profile
of Hγ fitted by theoretical profiles. X: H/He

		log g	X	Reference
LTE, normal metal	9500	3.7	0.83	Strom et al. (1966)
10 × metal	9500	4.0	0.83	
LTE, blanketing of Balmer + Lyman	10080	4.0	0.85	Mihalas (1966)
LTE, blanketing of Balmer + Lyman	9500	3.8	0.83	Avrett and Strom (1964)

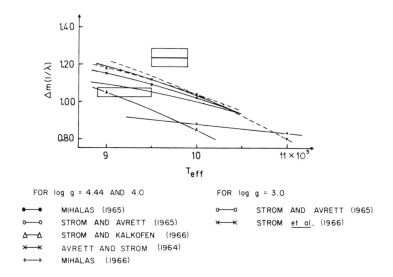

FOR log g = 4.44 AND 4.0

●—● MIHALAS (1965)
○—○ STROM AND AVRETT (1965)
△—△ STROM AND KALKOFEN (1966)
×—× AVRETT AND STROM (1964)
+—+ MIHALAS (1966)

FOR log g = 3.0

○—○ STROM AND AVRETT (1965)
×—× STROM et al. (1966)

Figure 4. The quantity Δm (1/λ) as a function of T_{eff}.

I wish to thank Dr. S. E. Strom, Smithsonian Astrophysical Observatory, Cambridge, Massachusetts, for sending me the tables of the stellar atmospheres quoted in his paper (1966). Also many thanks are due Dr. K. Bahner, Heidelberg, for a list of his absolute calibrations. Dr. Grewing, Bonn, is to be especially thanked for his program for computing color indices. I thank A. F. J. Moffat for reading the manuscript.

REFERENCES

Aller, L. H., Faulkner, D. J., and Norton, R. H. 1966, Ap. J., 144, 1073.
Avrett, E. H., and Strom, S. E. 1964, Ann. d'Ap., 27, 781.

Bahner, K. 1962, private communication.

———————. 1963, Ap. J., 138, 1314.

Barbier, D., and Chalonge, D. 1941, Ann. d'Ap., 4, 30.

Bless, R. C. 1960, Ap. J., 132, 532.

Bonsack, W. K., and Stock, J. 1957, Ap. J., 126, 99.

Cayrel de Strobel, G. 1961, Ann. d'Ap., 24, 509.

Chalonge, D., and Divan, L. 1952, Ann. d'Ap., 15, 201.

Charitonov, A. V. 1961, Izv. Ap. Inst. Alma-Ata, 12, 27.

———————. 1962, Izv. Ap. Inst. Alma-Ata, 15, 52.

———————. 1963, Sov. Astron. —AJ, 7, 258.

———————. 1966, Publ. Ap. Inst. Alma-Ata, 7, 70.

Code, A. D. 1960, in Stars and Stellar Systems, Vol. VI, Stellar
    Atmospheres, ed. J. L. Greenstein (Chicago: University of Chicago
    Press), 50.

Glushneva, I. N. 1964, Sov. Astron. —AJ, 8, 163.

Hall, J. S., and Williams, R. C. 1942, Ap. J., 95, 225.

Hanbury Brown, R., Davis, J., Allen, L. R., and Rome, J. M. 1967,
    Mon. Not. Roy. Astron. Soc., 137, 393.

Kienle, H., Wempe, J., and Beileke, F. 1940, Zs. f. Ap., 20, 91.

Lamla, E. 1959a, Astron. Nachr., 285, 12.

———————. 1959b, Astron. Nachr., 285, 33.

———————. 1965, Landolt-Börnstein, Neue Ser., Vol. I, ed. H. H. Voigt
    (Berlin: Springer-Verlag), p. 375.

Melbourne, W. G. 1960, Ap. J., 132, 101.

Mihalas, D. 1965, Ap. J. Suppl., 9, 321.

———————. 1966, Ap. J. Suppl., 13, 1.

Oke, J. B. 1960, Ap. J., 131, 358.

———————. 1964, Ap. J., 140, 689.

———————. 1965, Ann. Rev. Astron. Ap., 3, 23.

Popper, D. M. 1959, Ap. J., 129, 647.

Schmidt-Kaler, Th. 1961, Astron. Nachr., 286, 113.

Seitter, W. C. 1962, Veröffentl. Bonn, No. 64, 60 pp.

Strom, S. E., and Avrett, E. H. 1965, Ap. J. Suppl., 12, 1.

Strom, S. E., Gingerich, O., and Strom, K. M. 1966, Ap. J., 146, 880.

Strom, S. E., and Kalkofen, W. 1966, Ap. J., 144, 76.

Whiteoak, J. B. 1966, Ap. J., 144, 305.

Willstrop, R. V. 1960, Mon. Not. Roy. Astron. Soc., 121, 17.

———————. 1965, Mem. Roy. Astron. Soc., 69, 83.

SEVEN-COLOR PHOTOMETRY

M. Golay

Observatoire de Genève, Switzerland

1. DESCRIPTION OF THE PHOTOMETRIC SYSTEM

Our photometric system has been discussed in several papers published in Publications de l'Observatoire de Genève since 1963. See Golay (1963) for a description of the filters used. A network of 200 stars has been established that defines our standard photometric system. Table 1 from Rufener, Bartholdi, Maeder, and Genoud (1968) gives the electric response curves $S(\lambda)$ of the seven bands used. Their paper also gives the procedures employed to obtain $S(\lambda)$ and to find the accuracy. The average wavelengths and half-bandwidths follow.

	U	B	V	B_1	B_2	V_1	G
λ_0 (A)	3458	4248	5508	4022	4480	5408	5814
μ (A)	170	283	298	171	164	202	206

$$\mu^2 = \frac{\int (\lambda - \lambda_0)^2 \cdot S(\lambda)\, d\lambda}{\int S(\lambda)\, d\lambda}$$

We note that filters U, B, and V are very near those of the Johnson-Morgan three-color photometry. Filters B_1 and B_2 cover half of filter B, and filters V_1 and G cover half of filter V.

2. CATALOGS

Each catalog that we have published and will publish gives and will give the seven colors of all stars measured from the beginning of our work to the date of the publication. All colors published are reduced to the same photometric standard. It is in the interest of the user to refer always to the latest catalog (Rufener et al., 1968), which includes almost 1000 stars. To date, the colors of 1500 additional stars have been measured or are in the process of being measured and will be published next year.

With a view to investigating the application of our photometry to various problems, we have measured the following:

A. Stars with known MK classification for O - G8 on the main sequence and with known λ_1, D, ϕ_b.

B. Stars with known MK classification for O - G8 and with known λ_1, D, ϕ_b, for luminosity classes IV, III, II, I_a, and I_b, for which many have a spectroscopic absolute magnitude.

C. Stars with known MK classification, with known λ_1, D, ϕ_b, and with various interstellar reddening.

D. Stars with known metal-hydrogen ratio [Fe/H].

E. Stars with known rotational velocity v sin i.

F. Peculiar stars Am, Ap, subdwarf, weak lines, strong lines.

G. Stars that are members of stellar groups.

The 1968 catalog contains colors of stars that are members of the galactic clusters Pleiades, Praesepe, Hyades, and Coma Berenices and are brighter than magnitude 8.5 to 9.0; and colors of stars that are members of NGC 752, 1545, 1647, 1662, 2168, 2169, 2244, 6871, 6633, 7092, 7160, h + χ Per, IC 1805, IC 348, IC 4665, and M 103 and are brighter than magnitude 10.5 to 11.0. Also listed in the catalog are members of various associations and streams.

3. PARAMETERS

The parameters we used are:

$$B_2 - V_1 \quad ,$$

$$d = (U - B_1) - 1.6 (B_1 - B_2) \quad ,$$

$$\Delta = (U - B_2) - 1.055 (B_2 - G) \quad ,$$

$$m_2 = (B_1 - B_2) - 0.69 (B_2 - V_1) \quad ,$$

$$g = (B_1 - B_2) - 1.520 (V_1 - G) \quad .$$

A discussion of the parameters appears in Golay and Goy (1965), Hauck (1968), and Golay (1968a), and a general study of the subject, in Golay (1968b).

These parameters are affected by interstellar reddening. Correcting them we obtain

$$d_0 = d + 0.16 \, E_{B_2 - V_1} \quad , \quad (m_2)_0 = m_2 + 0.23 \, E_{B_2 - V_1} \quad ,$$

$$\Delta_0 = \Delta + 0.4 \, E_{B_2 - V_1} \quad , \quad g_0 = g + 0.06 \, E_{B_2 - V_1} \quad ,$$

$$(B_2 - V_1)_0 = (B_2 - V_1) - E_{b_2 - V_1} \quad ;$$

d_0 and Δ_0 measure the Balmer discontinuity, and $E_{B_2 - V_1}$ is the color excess of the index $B_2 - V_1$.

3.1 Effective Temperature

O to B5V: We have roughly the differential relationship

$$\log T_{eff} = 0.6 \, \delta\Delta \quad .$$

A0 to G5: $\theta_{eff} = 0.727 (B_2 - V_1) + 0.649 \quad .$

For $(B_2 - V_1) \geq 0.230$, we must correct $B_2 - V_1$ for a blanketing effect given by $(B_2 - V_1) = 1.11 (\Delta m_2 + 0.07)$.

3.2 Absolute Magnitude

O to B5: if $v_{rot} < 70$ km sec^{-1} ,

$$M_v = M^0_{v, sp} + 13 (\Delta_0 - [\Delta \, sp]) \quad ,$$

where $M^0_{v, sp}$ is the absolute magnitude on the zero-age main sequence of a star of spectral type sp. For a star of spectral type sp on the zero-age main sequence, $[\Delta_{sp}]$ is Δ_0.

Class V, IV, III-IV:

A2 to F0:
$$\frac{\delta M_v}{\delta d} = 5 \quad,$$

where $\delta d = d_0 - d_{0, (B_2 - V_1)_0}$ and $\delta M_v = M_v - M_{v, (B_2 - V_1)_0}$.

F1 to F5:
$$\frac{\delta M_v}{\delta d_0} = 1 + 43.2 \, (B_2 - V_1)_0 \quad,$$

where $d_{0, (B_2 - V_1)}$ and $M_{v, (B_2 - V_1)_0}$ are d_0 and M_v of the zero-age main-sequence star with the same $(B_2 - V_1)_0$.

F5 to G2:   same as above, with d_0 and $(B_2 - V_1)_0$ corrected for residual effects of blanketing.

3.3 Chemical Composition

$$m_2 = (m_2)_0 - (m_2)_{(B_2 - V_1)_0} \quad,$$

where $(m_2)_{(B_2 - V_1)_0}$ is the value of (m_2) defined by a reference sequence ;

$\Delta m_2 > 0$ for metallic-line stars,

$\Delta m_2 < 0$ for metal-deficient stars.

If $\Delta m_2 > -0.140$, we have

F8 to G0: $[\frac{Fe}{H}] = 6.75 \, \Delta m_2 + 0.25.$

Earlier F_5 stars: Δm_2 must be corrected for the effect of luminosity.

3.4 Rotation

B0 to B5V: $\delta d = 1.8 \times 10^{-4} v \quad,$

where v = rotational velocity in km sec^{-1} and $\delta d = d_{0(\Delta 0)} - d_0$; the term $d_{0(\Delta 0)}$ is the value of d_0 in a standard relationship for zero rotational velocity.

3.5 Binarity

The seven-color photometry is able to separate the effects of rotation and binarity in the spectral range A2 to F6. The method is explained in Maeder (1968) and uses the diagrams V, $(B_2 - V_1)$, and V, d.

REFERENCES

Golay, M. 1963, <u>Publ. Obs. Genève</u>, Ser. A, No. 64, 199.

_____. 1968a, <u>Publ. Obs. Genève</u>, Ser. A, in press.

_____. 1968b, in <u>Vistas in Astronomy</u>, in press.

Golay, M., and Goy, G. 1965, <u>Publ. Obs. Genève</u>, Ser. A, No. 71, 19.

Hauck, B. 1968, Thesis, Observatoire de Genève; <u>Publ. Obs. Genève</u>, Ser. A, in press.

Maeder, A. 1968, <u>Publ. Obs. Genève</u>, Ser. A, No. 74, in press.

Rufener, F., Bartholdi, P., Maeder, A., and Genoud, A. 1968, <u>J. des Obs.</u>, in press.

SPECTROPHOTOMETRY OF LOW-DISPERSION SPECTRA

I. Furenlid

Stockholm Observatory, Saltsjöbaden, Sweden

ABSTRACT

This report is a short account of the methods of spectral classification by means of spectra of low dispersion, as carried out in Sweden. Objective-prism spectra are used with a dispersion of 200 to 500 A/mm. The methods, developed by Lindblad (1922, 1925) and by Lindblad and Stenquist (1934), were originally based on density estimates in certain spectral regions. For early-type stars the method yielded essentially a one-dimensional luminosity classification, and for late-type stars, a two-dimensional classification. Öhman (1927, 1930) obtained similar results by measurements in microphotometer tracings. He also showed that widened as well as unwidened spectra could be used. Much work has since been done along these lines by, among others, Schalén, Malmquist, Ramberg, Elvius, Westerlund, Ljunggren, and Oja. References and a good summary of the extent of this work have been given by Elvius (1966).

1. THE EARLY-TYPE STARS

The Lindblad scheme of classification is given by Lindblad and Stenquist (1934). Figure 1 shows the measured wavelengths for early-type stars of classes B to F8. The hydrogen lines Hγ and Hδ are measured, together with adjacent regions, at λ 4400, λ 4260 and λ 4140, λ 4050, respectively. The microphotometer tracings are obtained with a wide analyzing slit, corresponding to some 20 to 40 A in wavelength. The slit thus embraces more or less completely the hydrogen lines, and the measure of their depth is actually a measure closely correlated to their equivalent widths. The hydrogen lines are measured as the difference in magnitude between the average of the adjacent regions and the depth of the line. A gradient is computed as a straight line fitted by a least-squares solution to the magnitudes at λ 4400, λ 4260, λ 4140, λ 4050, and λ 4215, with the abscissa given in $1/\lambda$ units. The gradient serves as a temperature criterion, which together with the strength of the Balmer lines yields, at least in principle, a two-dimensional classification, since the absorption in the hydrogen lines is closely correlated to luminosity. For the early-type stars the measure of the hydrogen lines gives a luminosity classification. Lindblad divided these stars into groups called τ, τ-, σ+, σ, σ-, and μ, which represent, in that order, decreasing luminosity. Öhman (1930) introduced a measure of the K line as a temperature criterion and could establish a two-dimensional classification of A stars. Westerlund (1951, 1953) recorded spectra with a rather narrow slit and measured Hγ and Hδ with a planimeter relative to a drawn continuum. He also measured the depth of these hydrogen lines, the K line and several other criteria giving independent checks between the methods. A more recent summary of the methods currently in use can be found in a paper by Ljunggren and Oja (1961). For stars earlier than A2 they obtain a classification in luminosity by measuring hydrogen absorption (with a bandwidth of 20 to 30 A). This is also the case for stars of types A2 to F2, although here the K line is added as a criterion to separate these stars from those of the previous group. For F2 to G0 stars luminosities are obtained from the ratio of a measure of the G band and hydrogen-line absorption. A one-to-one correspondence between the luminosity measures for these groups of stars and intrinsic color is assumed, giving a one-dimensional classification.

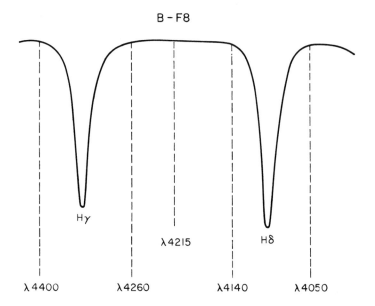

Figure 1.   Selected wavelengths for early-type stars.

In current programs for early-type stars, short spectra are obtained with Schmidt telescopes where the ultraviolet is also recorded.   These programs aim at a two-dimensional classification based on hydrogen-line absorption and the Balmer jump.   Particular interest is also devoted to the luminosities of A stars.

2.   THE LATE-TYPE STARS

The wavelengths utilized in Lindblad's scheme of classification for late-type stars (Lindblad and Stenquist, 1934) are shown in Figure 2.   The measured points are the G band and two surrounding points λ 4360 and λ 4260, the line λ 4227 of Ca I, λ 4180 and λ 4140 affected by cyanogen absorption, and λ 4095.   The following quantities, measured in magnitudes, are formed:

g = λ 4360 - λ 4260   ,

c = λ 4260 - λ 4180   .

The gradient k is defined in the same way as for early-type stars, but in this case it is fitted to λ 4360, λ 4260, λ 4180, λ 4140, and λ 4095; g is a measure of the break in the continuum at the G band and closely correlated with temperature, while c measures the cyanogen absorption, which is a sensitive luminosity criterion.   The gradient k is sensitive mainly to temperature but also includes a luminosity dependence by λ 4180 and λ 4140.   A two-dimensional classification is thus obtained from g and c, and for the later types, also k and c.   Ramberg (1941) showed that the line λ 4227 is useful as a luminosity criterion for K and M dwarfs.   Ljunggren and Oja (1961) separated giants and dwarfs by g and c and then derived intrinsic color and luminosity by g for each group separately.

Currently, work is being done on a three-dimensional classification of late-type stars by inclusion of a population-sensitive parameter.   In particular, more accurate absolute magnitudes of late-type giants are desirable.

The main importance of low-dispersion spectrophotometry lies in its application to stellar statistics, where it has proved to be an effective tool. The methods can be applied to faint stars in large numbers.   Some 40,000 stars have so far been treated, the limiting magnitude being around 14.   The limitation in ability to resolve structural features in the Galaxy is set by the accuracy of the measurements. Generally speaking, the mean error in a single determination of an index is about 0.05 mag. The errors in the absolute magnitudes range from about 0.5 to 1.0 mag, depending on the spectral class.

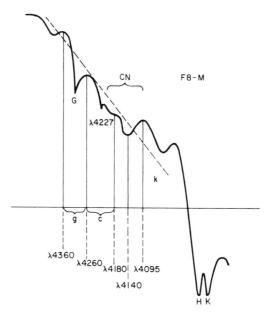

Figure 2. Selected wavelengths for late-type stars.

The measured quantities for one single star are not accurate enough to have a precise meaning in terms of astrophysical parameters. A considerable increase of accuracy is required to give astrophysically significant measures. To this purpose, equipment for digital treatment of spectra will be installed shortly in Stockholm and Uppsala; the equipment will also permit an increased speed in the reduction procedure. Other means of improving the accuracy are also being tried, and a report on a method of photographic narrow-band photometry is being prepared.

3. REFERENCES

Elvius, T. 1966, in IAU Symp. No. 24, ed. K. Lodén, L. O. Lodén, and
      U. Sinnerstad (London: Academic Press), 333.
Lindblad, B. 1922, Ap. J., 55, 85.
_____. 1925, Nova Acta Reg. Soc. Sci., Uppsala., Ser. 4, 6, No. 5,
      99 pp.
Lindblad, B., and Stenquist, E. 1934, Ann. Stockholm Obs. 11, No. 12,
      75 pp.
Ljunggren, B., and Oja, T. 1961, Ann. Uppsala Univ. Astron. Obs., 4,
   No. 10, 40 pp.
Öhman, Y. 1927, Medd. Astron. Obs. Uppsala, No. 33, 41 pp.
_____. 1930, Medd. Astron. Obs. Uppsala, No. 48, 109 pp.
Ramberg, J. M. 1941, Ann. Stockholm Obs., 13, No. 9, 168 pp.
Westerlund, B. 1951, Ann. Uppsala Univ. Astron. Obs., 3, No. 6, 57 pp.
_____. 1953, Ann. Uppsala Univ. Astron. Obs., 3, No. 10, 50 pp.

NARROW-BAND PHOTOMETRY AT THE STOCKHOLM OBSERVATORY

U. Sinnerstad

Stockholm Observatory, Saltsjöbaden, Sweden

A two-channel photometer is used in the Cassegrain focus of the 1-m reflecting telescope. In the case of line photometry, four filters (f_1, f_2, f_3, f_4) are used for each line. The line index is determined by means of simultaneous measurements through a wide (w) and a narrow (n) interference filter centered on the spectral line in question and arranged in the following way:

Direct beam	Reflected beam
f_1 (n)	f_2 (w)
f_3 (w)	f_4 (n)

Since the two filter slides take six filters each, three lines can be measured in succession. The beam splitter allows approximately the same flux to pass in the two beams, and since the wide filters are always combined with neutral filters, all measurements can be made with the same amplifier gain.

Let $m_1 \ldots m_4$ be the deflections, expressed in magnitudes, through the filters $f_1 \ldots f_4$. If the two wide filters and the two narrow filters, respectively, have about the same characteristics, the line strength can be derived as the sum of two indices as follows:

$$\ell' + \ell'' = [m_1(n) - m_2(w)] + [m_4(n) - m_3(w)]$$
$$= \{m_1(n) - m_3(w)\} + \{m_4(n) - m_2(w)\} \quad , \tag{1}$$

where simultaneous measurements are inside the square brackets.

Suppose now that the characteristics of the wide filters or the narrow filters differ from each other. In this case, we also make the following measurements:

$$\ell' - \ell'' = [m_1(n) - m_4(n)] + [m_2(w) - m_3(w)]$$
$$= \{m_1(n) - m_3(w)\} - \{m_4(n) - m_2(w)\} \quad . \tag{2}$$

When the difference is due to unequal bandwidths of the wide filters, it depends on the features in the adjacent continuum of the line; but the difference depends on the line profile when the bandwidths of the narrow filters significantly differ from each other.

The method outlined above is applied to early-type stars and is used to determine 1) line intensities of Hβ, Hγ, and He I 4471, and 2) v sin i by means of the line profile of Hβ.

The accuracy of the measured line intensities is high and corresponds to a mean error in equivalent width of about ±0.14 A for one set of measurements [two measurements according to equation (1)]. Therefore, in a plot of Hβ against Hγ, all Be stars are easily separated. A comparison between the line index of He I 4471 and the equivalent width of the same line, obtained from the Edinburgh data, is given in Figure 1 for the stars in common.

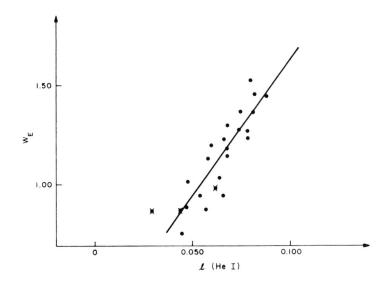

Figure 1.   Relation between the He I 4471 index ℓ(He I) and the
equivalent width W_E of He I 4471 from the Edinburgh
data.   Crosses indicate supergiants.

The index related to the projected rotational velocity v sin i is obtained
according to equation (2), where the two narrow-band filters have halfwidths
of 6 A and 28 A, respectively.   Figure 2 shows the relation between this line
index and the spectrographically determined value of v sin i for main-sequence
stars of spectral class from B0. 5 to B6.   The accuracy in the index determined corresponds to ± 45 km sec^{-1}.
    Further details of this investigation, the ultimate aim of which is to im-
prove the determination of the absolute magnitude of the B stars, are given
by Sinnerstad, Arkling, Alm, and Brattlund (1968).

REFERENCE

Sinnerstad, U. , Arkling, J. , Alm, S. H. , and Brattlund, P.   1968, Arkiv.
    f. Astron. , 5, No. 5, in press.

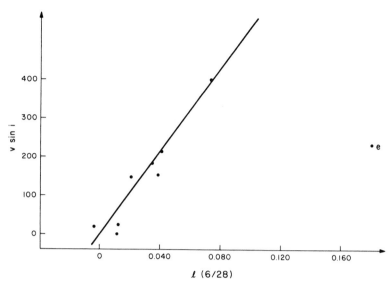

Figure 2.   Relation between the value of v sin i and the index $\ell(6/28)$ expressing the magnitude difference of the measures through two narrow filters with halfwidths of 6 A and 28 A, respectively, both centered on Hβ.

AN ANALYSIS OF FAR-ULTRAVIOLET FILTER OBSERVATIONS OF STARS[*]

R. C. Bless,[†] A. D. Code, and T. E. Houck

Space Astronomy Laboratory, Washburn Observatory,
University of Wisconsin, Wisconsin

ABSTRACT

After analyzing filter observations of stars in the spectral region λ 2800 to λ 1100, we have found the following: The best observational accuracy currently attained is about ± 0.25 mag; blanketing by strong absorption lines within a filter pass band is the most important single factor affecting the comparison of observations and model atmospheres; blanketed models together with the Morton-Adams temperature scale adequately represent most ultraviolet observations of the 35 main-sequence stars analyzed; observations of a dozen giants show equally good agreement with models, although a comparison here was possible only in the λ 2000 to λ 2800 region; a value of the interstellar extinction at λ 1115 was derived and is consistent with other determinations at longer wavelengths.

[*]This paper will appear in the Astrophysical Journal as Part III of a series entitled "Astronomical radiation measurements."

[†]Currently at NASA Goddard Space Flight Center, Greenbelt, Maryland.

PART II

PROPERTIES OF SYNTHETIC SPECTRA
AND THEIR SENSITIVITY TO
UNCERTAINTIES OF PHYSICAL THEORY

D. Mihalas, Chairman

THE USE OF MODEL ATMOSPHERES IN THE INTERPRETATION OF STELLAR SPECTRA

S. E. Strom

Smithsonian Astrophysical Observatory and
Harvard College Observatory, Cambridge, Massachusetts

The future progress of theory is a harder subject for prediction than the future progress of observation. But one thing is certain: observation must make the way for theory, and only if it does so can the science have its greatest productivity.

C. Payne (Gaposchkin), 1925

1. INTRODUCTION

The primary purpose of this review is to examine the role of stellar atmospheres in the determination of the basic parameters describing the physical state of the outer layer of a star, i. e., effective temperature, gravity, and chemical composition. The sensitivity of these parameters to the assumed physical state of the atmosphere and the possibility of making critical observational tests of the applicability of the models will also be discussed. I shall try to indicate how the evaluations of T_{eff}, g, and chemical composition from models and from observations affect the determinations of stellar mass, luminosity, chemical composition, and age.

I think it is fair to state that the problem of obtaining radiative-equilibrium plane-parallel models in hydrostatic equilibrium is pretty well understood, so that I will emphasize the relatively nonstandard problems of rotation, departures from LTE, blanketing, convection, and atmospheric motions, and their effects on stellar atmospheres.

My report will be organized as follows: I shall discuss each spectral region of the H-R diagram, starting with the B stars and working toward lower effective temperatures, and indicate how these five effects influence the determinations of effective temperature, gravity, or chemical composition from the lines and/or the continuum.

2. THE MAIN-SEQUENCE B STARS

In the range of effective temperature between 10,000 and 20,000°K, that is, between spectral types approximately B3 and B9, it has been common practice to use the Balmer discontinuity as a measure of effective temperature and to obtain the surface gravity from the equivalent width of $H\gamma$ or from the profiles of the Balmer lines. In this effective temperature range the change of the Balmer discontinuity Δm_B with temperature is given by

$$\frac{\Delta m_B}{\Delta T_3} = 0.06 \quad ,$$

where T_3 is the temperature in units of 1000°K. In principle, therefore, it is possible to determine temperatures from observations of Δm_B having an accuracy of about ±0.01 mag with an internal accuracy of about 200 to 300°K.

For gravity determinations we know that

$$\Delta[g]/\Delta[W] = 4.45 \text{ at } T_{eff} = \text{const} \quad ,$$

or

$$\Delta[W] = 0.225 \Delta[g] \quad .$$

Hence, to obtain an accuracy of ±0.1 in log g, equivalent widths must be determined to an accuracy of about 5%. We would like to have accuracies of the order of 0.1 in log g to attack such problems as mass loss for horizontal branch stars and, in general, to compare locations of stars in the T_{eff} - log g plane directly with computed evolutionary tracks. In addition to the observational errors in measuring the equivalent width, we must realize that the equivalent width is a function not only of gravity but of effective temperature as well. The change of equivalent width with temperature is given by $\Delta[W] = 0.043 \Delta T_3$, and hence we see that the effective temperature determinations must be accurate to about 500°K in order to get an accuracy of 0.1 in the determination of log g. Now let us consider the various effects that have not usually been incorporated in the models computed to date and see how they affect the determination of the equivalent width of Balmer lines and the Balmer discontinuity.

2.1 Departures from LTE

Since we are interested in the effects of departures from LTE on the determination of Balmer discontinuity and the wings of the hydrogen-line profiles, we are concerned about possible departures (primarily for atomic hydrogen) for continuum optical depths greater than about 0.05. Although it is of considerable interest, we do not need to know the detailed structure of the upper photosphere, nor do we need to solve for the behavior of the line source function at small continuum depths. At continuum optical depths $\tau > 0.05$, the monochromatic depths of the line centers of the Balmer lines are greater than 100 and even greater for the Lyman lines.

Kalkofen (1964, 1966) was the first to take advantage of this well-known situation in proposing that, owing to the presumed saturation of the lines at depths where the continuum flux and line wings are formed, the lines be treated as if they were in detailed balance. This simplification meant, of course, that the radiation field in the lines could be omitted from the problem. Therefore, the simultaneous solution of the transfer equation and the equations of statistical equilibrium is tremendously simplified, and the problem can be treated in a manner quite straightforward and consistent with the rest of the calculations made in the course of computing a model atmosphere. Kalkofen (1964) presented the results of a computation for a pure-hydrogen model of $T_{eff} = 10,000$°K, log g = 4 at the First Conference on Stellar Atmospheres. Strom and Kalkofen (1966) then solved the problem where departures were allowed in atomic hydrogen and the other sources of opacity were included in LTE. The results of these investigations suggested that the second level was underpopulated and the third level overpopulated in the B-star spectral range. For main-sequence stars, Strom and Kalkofen found that, as a result of these departures, the Balmer discontinuity was decreased by an amount requiring a change in the effective temperature scale of between 1000 and 2000° for stars in the B3 to B9 range. However, their models included only bound-free radiative and collisional terms and failed to take proper account of bound-bound collisions.

Later, Kalkofen and Strom (1966) published models in which the bound-bound collisions were included, and they found that the departures were significantly reduced. For main-sequence stars the required change in the effective temperature scale was reduced to a maximum of 500°K. They also noted, however, that for stars more luminous than main-sequence stars, where the surface gravity is lower, the effects of the departures would be more pronounced, approaching the values suggested in their original work. Mihalas (1967a) then computed the expected departures in B-star atmospheres, using a 10-level H atom and more recent laboratory data for the bound-bound collision rates. These rates were approximately five times those used by Kalkofen and Strom and, as a result, the departures for main-sequence stars were reduced even more. Work reported by Kalkofen at this conference suggests that these higher collision rates, rather than the addition of more levels, are primarily responsible for the decrease in the departure coefficients noted by Mihalas. However, it should be emphasized that the departures are always in the sense that the second level is underpopulated and the third level is overpopulated for B stars.

Strom and Kalkofen (1967) suggested that, since $b_2 < 1$ and $b_3 > 1$, the Balmer discontinuity would be decreased and the Paschen discontinuity would be increased. They therefore proposed an observational test in which the ratio of the Paschen to the Balmer discontinuity measured for late B stars could be used to determine the magnitude of the departures from LTE; the larger the departures from LTE, the larger should be the ratio of the Paschen to Balmer discontinuity. Moreover, this ratio should increase as the luminosity of the star increases. Using the photographic data of Bloch and Tcheng (1956), they found a preliminary configuration of the estimated effects of departures from LTE on this ratio. However, recent photoelectric work by Hayes (1967) would seem to suggest that at least for main-sequence stars departures from LTE have a negligible effect on this ratio. Considerably more work, particularly directed toward observations of higher luminosity stars, would provide a very useful test for the non-LTE calculations.

The contours of the Balmer lines can also be used to provide a test for departures from LTE. Deane Peterson of the Harvard-Smithsonian group made the following perceptive suggestion: since $b_2 < 1$ and $b_3 > 1$, the source function for Hα, which can be written as

$$S_{H\alpha} = 2h\nu^3/c^2[\exp(h\nu/kT)b_2/b_3 - 1]^{-1} \quad ,$$

has a larger value than in the case of LTE; in fact, if we neglect stimulated emission, the source function for Hα is larger than its LTE counterpart by the ratio b_3/b_2. Since Hα is formed rather high in the atmospheres of B stars and therefore in a relatively low-density region, b_3/b_2 is on the order of 1.2 to 1.4. Therefore, the source function for Hα is increased considerably and, as a result, the line looks weakened in the wings; Hβ, Hγ, and the higher Balmer lines will be progressively less affected by departures from LTE since the upper levels are forced to LTE via collisions with the continuum. Thus the ratio of the equivalent widths of Hα and Hγ is a good measure of departures from LTE. Peterson will report later in this conference the results of an observational test for the departures from LTE and their effects on the hydrogen lines. His work suggests good observational agreement with the predicted effects. Peterson also made some estimates of the validity of the saturation approximation and found that its application was quite satisfactory over a wide range of physical conditions, essentially from solar-type stars up through the B stars.

As I have already stated, the calculations for the line wings and the continuum are all based on the detailed balance assumption. Work is currently under way by Peterson, Kalkofen, and Avrett using integral equation techniques, and by Mihalas, Hummer, and Rybicki and the French group using differential equation techniques to develop numerical methods to handle the detailed line problem. We may expect in the next year or so a treatment of the full line-plus-continuum problem, which will considerably extend our understanding of the atmosphere in the region of the stellar boundary.

Mihalas and his coworkers (Mihalas and Stone, 1968) extended the ideas proposed by Kalkofen to explore the problem of departures for neutral and singly ionized helium in B stars. They solved the statistical equilibrium equations under the detailed balance assumption, which, as they admit, may be considerably weaker for helium than for hydrogen. They found underpopulation for both the singlets and the triplets for stars in the spectral range B3 to B9. No interpretation of existing data has been attempted on the basis of their results, and, in view of the simplifying assumptions, the work published to date should be considered only as an exploratory venture. Peterson and others have begun to develop the techniques required to handle the full line problem. This merits a good deal of attention because of its obvious significance in allowing us to discuss more accurately the state of B-star helium abundances. I should comment in passing that we probably should not ignore the effect of metal-line blanketing on the ultraviolet radiation field, since the latter controls the helium-line populations. In fact, ignoring this effect may lead to errors, particularly in the evaluation of the spectra of objects that have enhanced metal-to-hydrogen ratios.

For the early B and late O stars departures from LTE in the Lyman continuum should be important. Mihalas (1967b) and his coworkers have already begun an analysis of the effects of the Lyman continuum on the non-LTE problem. For the early B stars one expects qualitatively that b_1 will be less than unity and that b_2, at least in the continuum-forming regions, will be tied very closely to b_1 through the Lyman-α transition.

The calculations made to date for main-sequence stars suggest that the effects of departures from LTE should be minimal, at most an effect of 200 to 300° on the choice of effective temperature. The problem for B stars of higher luminosity is, of course, more severe and will be discussed later. The effect of these departures on log g for main-sequence stars is at most 0.1. However, we should note that the present uncertainties in the Stark-broadening theory also give rise to errors of this size. In fact, the difference between the current Griem (1967) theory and the semiempirical broadening mechanism proposed by Edmonds, Schlüter, and Wells (1967; hereafter referred to as ESW) amounts to about 0.2 in log g. One way of testing the validity of each of these theories is to use the ratio $W(H\alpha)/W(H\gamma)$. The calculated values of this ratio are 0.72 for the Griem theory and 0.62 for the ESW mechanism for LTE models. However, when one takes into account departures from LTE, the Griem theory would seem to be slightly favored by the observational evidence available at present. More observations and further work on the non-LTE problem are required before we can make a definite choice between the two available broadening theories.

2.2 Rotation

As is well known, B stars have a mean rotational velocity of the order of 200 km sec^{-1}, which is approximately one-third of the so-called critical velocity (for which the centripetal and gravitational forces balance at the equator). Rotation affects the shape of the star, both in the surface layers and in the interior. Through the temperature and density gradients set up as a result of rotational distortion of the star, circulation currents may be set in motion, and these in turn must be examined to see what effect, if any, they have on the structure of the atmosphere both as a mechanism of support and as a mechanism for energy transport. Mestel (1965), Roxburgh (1964), and others have suggested that stars may adjust to rotational distortion through an increase in angular velocity toward the center, instead of through circulation currents. In either case, the effects of rotation will be felt both in the interior and in the atmosphere, with the result that the luminosity, mean effective temperature, and gravity for a star of a given mass will be considerably affected. Moreover, we expect these effects to be a function both of rotational velocity and of the angle between the rotation axis and the line of sight (the aspect).

The results of interior model calculations have been used to determine the shapes of the equipotential surfaces as well as the luminosity for rotating stars. Estimates by Roxburgh and his coworkers (Roxburgh and Strittmatter, 1965; Roxburgh, Griffith, and Sweet, 1965) showed that the equipotential surfaces are close to those of a Roche model. Collins (1963) and his collaborators and Faulkner, Roxburgh, and Strittmatter (1968) have used the resulting variations of local gravity to compute the variations of temperature over the surface by applying von Zeipel's theorem. Hardorp and Strittmatter (1968a) found that the flux carried by circulation currents and the effects on the hydrostatic equilibrium equation are negligible. In addition, except at the equator, the thickness of the photospheric layers is small compared with the local radius of curvature, and therefore it is legitimate to fit a plane-parallel atmosphere parallel to the surface at each point.   These authors used the computed variations of temperature and gravity over the surface and, dependent on the local conditions, fitted model atmospheres of the appropriate parameters and computed the resulting emergent fluxes and line profiles.

In the color-magnitude array the effects for B stars are 1) for a given B-V color the luminosity for a rotating star is greater than for a nonrotating star; 2) for a given equatorial rotational velocity the main sequence for B stars viewed equator-on is brighter than for stars seen pole-on, although the situation is reversed in the A-star region.   For a star of a given mass the direction away from the zero-rotation main sequence and increasing rotational velocity, v, is as shown in the accompanying figure (see Figure 1). Hardorp and Strittmatter (1968a) give the following extreme case as an example of the effects of rotation on the choice of stellar age.   For a star rotating at 99% of the critical velocity viewed at an inclination of 90°, i. e., equator-on, they find that a star of 2 solar masses has about the same colors as those of the star of 1. 5 solar masses with an equatorial rotational velocity of 0.   If the location of the turnoff point in the color-magnitude diagram is used to estimate cluster ages, then an error of about a factor of 3 is made in estimating the age of the 2-solar-mass star, in the sense that its age would be overestimated by this amount.   Hardorp and Strittmatter suggest only sharp-line stars be used in age determinations, in which case the difference between the ages deduced for pole-on rapid rotators and intrinsically slow rotators is about 25%.

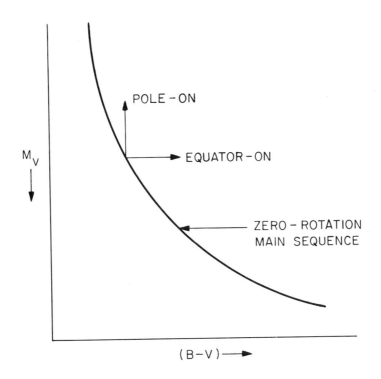

Figure 1.

In general, despite the aspect, the effect of rotation is to decrease the mean effective temperature and to lower the mean surface gravity (although for pole-on rapid rotators we get an almost negligible <u>increase</u> in the mean g owing to the slight flattening of the star at the pole and the limb-darkened contributions from the low-gravity equatorial regions). As a result, the Balmer discontinuities for rapidly rotating B stars are increased and the equivalent width of Hγ is decreased. Figure 2 shows qualitatively the effects of rotation on the equivalent width of Hγ versus U-B. Strom and Peterson have made use of this behavior to estimate B-star surface gravities from observed equivalent widths of Hγ and observationally deduced values for $(U-B)_0$ (the unreddened U-B color) for a variety of late B stars. From the available data they plotted the equivalent width of Hγ against U-B and chose the upper envelope of this curve as representative of the zero-rotation relation as suggested by the calculation of rotating star models. Then, using the calibration of $(U-B)_0$ in terms of effective temperature, from the work of Morton and Adams (1968), they chose a mean surface gravity of log g = 4.25 for B stars (ESW broadening theory) and 4.0 (Griem theory).

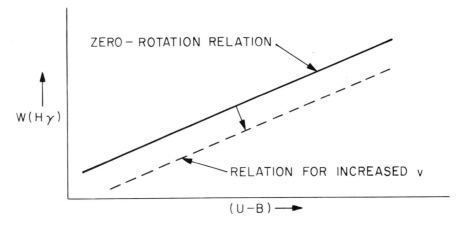

ZERO – ROTATION RELATION

$W(H\gamma)$

RELATION FOR INCREASED v

$(U-B)$ →

Figure 2.

In the $W(H\gamma)-(B-V)$ plane, an effect analogous to that on the $W(H\gamma)-(U-B)_0$ diagram occurs, in the sense that for a given B-V the rotating star has a smaller value of the equivalent width of Hγ. Furthermore, the effects of rotation on the mean gravity and on the luminosity are such as to introduce a scatter of about 0.5 mag in the $W(H\gamma)-M_V$ relation. Computations of the effects of evolution off the main sequence suggest that a scatter of about 0.3 to 0.4 mag is introduced in the course of the star's evolution, so that Hardorp and Strittmatter are not completely correct in attributing all of the scatter in the $W(H\gamma)-M_V$ relation to rotation.

Hardorp and Strittmatter (1968b) have also suggested that the effects of rotation on the spectrum can be used to estimate the absolute value of the rotational velocity. It is then possible to use the observed v sin i to obtain the inclination of the star. Their method depends upon the predicted increase in the absolute visual magnitude at a given B-V in the color-magnitude array. Their result suggests that the spread δM_V, due to aspect for a given rotational velocity, is small compared with the increase, ΔM_V, in the absolute visual magnitude for a given B-V; ΔM_V is related to the rotational velocity by $\Delta M_V = cv^2$. The value of c depends critically on the nature of the interior model adopted. Thus far the best observational work available (Dickens, Kraft, and Krzeminski, 1968) on the Coma cluster suggests a value of c smaller than that predicted from the models adopted by Hardorp and

Strittmatter.   Tentative results would seem to favor the models proposed by
Roxburgh and his associates (Roxburgh, Griffith, and Sweet, 1965).

Rotation is expected to have a large effect on the ultraviolet spectrum of
B stars.   This region is quite sensitive to the effective temperature because
the Planck function peaks in the near ultraviolet.   As a consequence, fluxes
observed shortward of the peak of the Planck function, which corresponds to
the effective temperature of the star, will be very sensitive to temperature
variations over the surface of the star.   Both the absolute value of the rotational velocity and the aspect strongly affect the observed emergent spectrum
in the ultraviolet, but unfortunately no combination of photometric indices
has yet been proposed to separate these effects and test the theoretical results.
It is not entirely clear that this in fact can be done, but it would be worthwhile
to pursue this further.

It also appears from examination of the calculations that the slope of the
accessible Balmer-continuum region is sensitive, although weakly, to the
effects of rotation, in the sense that it is steeper for the higher rotational
velocities.   The observational accuracy required to detect the effect would be
somewhat demanding.   Hardorp and Strittmatter (1968b) and Collins and
Harrington (1966) have also examined the effects of rotation on spectral lines.

Hardorp and Strittmatter (1968b) suggest that 1) the previous estimates
of the maximum rotational velocity are too low because of gravity darkening,
and there are, in fact, stars that exist with velocities approaching the critical
velocities; 2) the weak-line helium stars, such as 20 Tau, cannot be explained
by invoking the hypothesis that they are pole-on rotators as proposed first by
Huang and Struve (1956), and later by Guthrie (1965); 3) to get the velocity
values of rapid rotators, lines that increase in strength as the temperature
and gravity are decreased should be used, i.e., lines that become enhanced
in the equatorial regions of the star.

From analysis of idealized spectral lines, Collins and Harrington (1966)
suggest that line ratios used to obtain spectral types should not be affected
much by rotation if the velocities are less than or equal to two-thirds the
critical velocity.

Finally, I should make the following comments: Departures from
LTE will enter in a way to counteract the effects of rotation on the Balmer
discontinuity and to enhance the effect on the equivalent width of Hγ.   In the
low surface-gravity regions, the Balmer discontinuity will be decreased
owing to departures from LTE; in LTE these regions would be expected to
have a larger Balmer discontinuity because of their lower temperature.
Moreover, the effect of departures from LTE on the hydrogen lines is to
decrease their equivalent width because of the increase of the ratio of the
departure coefficients b_5/b_2.

Although the trend of the effects of rotation is well established, more
attention needs to be paid to the interior models and to obtaining more observational checks.   It should also be possible to compute more sophisticated
rotating model atmospheres that include more detailed effects, particularly
those due to line blanketing in the hotter models.

2.3 Blanketing

The problem of estimating the effects of line blanketing on the spectra of
B stars was first attacked in a systematic way by Gaustad and Spitzer (1961),
who predicted line strengths for the strongest lines expected in the ultraviolet
spectra of early-type stars.   No attempt was made to include the lines in a
self-consistent calculation of a model atmosphere.   Avrett and Strom (1964)
used the published equivalent widths predicted by Gaustad and Spitzer to compute a model in which the lines were allowed to block the radiation proportional to their equivalent widths.   The detailed variation of line opacity with
depth was not taken into account.

Mihalas and Morton (1965) computed a model of a B1 V star in which the
strong lines suggested by Gaustad and Spitzer were included explicitly in the
calculation.   Guillaume, Van Rensbergen, and Underhill (1965) and, most

recently and prolifically, Morton and his collaborators at Princeton (Adams and Morton, 1968; Hickok and Morton, 1968) have computed additional line-blanketed B-star models. For stars hotter than B5, more than half the total emergent flux arises from the Balmer continuum. At these high temperatures many elements, including the abundant light elements, are twice or three times ionized, and, as a result, strong lines of many of these elements fall in the ultraviolet, specifically in a region where the majority of the flux emerges. Thus, the effects of blanketing on the temperature structure can be severe, and neglect of the blanketing can lead to serious error in the determination of temperature and gravity. Qualitatively, the effects of blanketing in the continuum-forming region of the model, that is, where the lines are saturated, can be understood when we realize that at these depths the lines effectively block emerging radiation. Since for a model of a given effective temperature we demand a given fixed value of the flux constant throughout the atmosphere (the radiative-equilibrium condition), and since the flux cannot "escape" at the line frequencies, the temperature must be increased to compensate for the effect of the lines. Therefore, in two models of the same effective temperature, one blanketed and the other not, we expect in the optical-depth regions where the lines are saturated that the temperature will rise. In the case of pure absorption the surface temperature will drop in the region where the line radiation can act as an efficient cooling mechanism.

The above qualitative comments apply to lines that are formed in LTE; however, for a more realistic model of line formation the behavior of the temperature structure is not easy to compute, although it is fair to state that qualitatively the surface-temperature drop should not be so severe as predicted from the LTE calculation. The primary effect of blanketing, in most cases, will be to make the blanketed atmosphere appear hotter than an unblanketed atmosphere at the same effective temperature. Therefore, if unblanketed atmospheres are used to estimate the effective temperature, too high a value of T_{eff} will be chosen. The results of Morton and his coworkers suggest that the effective temperature is overestimated by as much as 2000° at spectral type B2, and about 500° at spectral type B5. Although Morton and his team have included the strongest lines in their calculations, our knowledge of f values, broadening parameters, etc. for the weaker lines is quite limited, so that the completeness of these models at the moment is not known. However, Morton's T_{eff} scale agrees well with that determined by Hanbury Brown and his associates.

For the late B stars, Mihalas (1966) computed models that are hydrogen-line blanketed by including explicitly many frequency points. The Smithsonian group has a similar set of models (unpublished) extending to higher effective temperature. However, in the late B stars, blanketing in the region between 2000 and 3000 A may be quite important. Very little has been done on this problem, mainly because of the paucity of atomic data, but also somewhat because of the difficulty of handling very large numbers of lines. For stars having unusually high metal abundances, the effects of line blanketing may be quite severe, and since accurate knowledge of the location of these peculiar stars in the H-R diagram is crucial to the discussion of their origin and age, the problem of line blanketing warrants more careful scrutiny. Strom and Kurucz (1966) have suggested a method that permits the detailed treatment of large numbers of lines. This method is also applicable to stars of later spectral type, where the detailed method of the Princeton group would fail owing to the numbers of lines that must be considered.

2.4 Convection and Atmospheric Motions

All the work on convection reported here is based on the essentially dimensional mixing-length theory and variations thereon. I shall not attempt to discuss in detail the problem of convective theory. However, it is clear that, in order to understand the behavior of the upper regions of stellar photospheres and the source of atmospheric motions, considerable advances must be made in applying hydrodynamics to stellar-atmosphere problems.

For main-sequence B stars, convection does not carry much flux, although the velocities computed in the region of convection instability are on the order of several tens of meters per second. Although convection does not carry much flux in main-sequence or higher luminosity B stars, atmospheric motions in the upper parts of the photosphere of these stars may be of considerable significance in terms of affecting the temperature-pressure structure as well as abundance determinations. To test the effects of turbulent pressure as a mechanism of support for the upper photospheres of B stars, I included in a crude way the effects of turbulent motion as a support term in the hydrostatic equilibrium equation. The support term was set to $1/2(\rho v^2)$, with v equal to the sound velocity throughout. The effect on the structure of the atmosphere depends more on the gradient of velocity than on its absolute value, but it was felt that this case represented, in a rough sense, a good estimate of the maximal effects to be expected from turbulent element support. For B stars the predicted emergent spectrum (that is, the hydrogen lines and continuum) looks like that of a star with a gravity lower by the mean value of the derivative of ρv^2 with depth. Certainly, from our estimates, it is not possible to remove any of the abundance anomalies associated with the peculiar late B stars by invoking turbulent pressure as a mechanism for extending the atmospheres.

2.4.1 Diffusion

In view of the recent discussion of the role of element separation by diffusion in the interpretation of the Population II halo B-star spectra, a few words on the subject are in order. Greenstein, Truran, and Cameron (1967) suggest that for slowly rotating B stars, where the effects of meridional circulation currents can be neglected, diffusion is possible during the estimated lifetime of a halo B star. They assume either that there are no convection zones in the atmospheres of these stars or that these zones are relatively thin and have no effect on the downward diffusion. From some of my calculations it seems clear that, while convection does not carry much flux, convective instability due to He I and He II ionization sets in at optical depths between 1 and 10 (depending upon the temperature under consideration), and extends to several hundred optical depths in the photosphere. Convective elements near the top of these zones have velocities on the order of several meters to several tens of meters per second, much in excess of the 10^{-5} cm sec^{-1} estimated by Greenstein et al. (1967). With velocities in the convection zone of this order, it is difficult to see how the diffusion mechanism can operate successfully if convective overshoot is at all important. However, in some cases, the conditions at the base of the convection zone are still favorable for gravitational settling.

2.4.2 Other problems of atmospheric motions

Underhill and her coworkers (de Groot and Underhill, 1964; Underhill, 1963; Underhill and de Groot, 1964) have computed a number of line profiles for early B and late O stars and find that, in order to match observed profiles, very high values of the turbulent velocity, sometimes exceeding the sound velocity, must be assumed. However, Hummer and Mihalas (1967) have suggested, with inspiration from Chandrasekhar (1948, 1950) and Münch (1948), that scattering due to free electrons is an important, and heretofore neglected, source of line broadening. Whereas Münch and Chandrasekhar assumed as a model a line-forming photosphere surrounded by a scattering layer, these authors correctly argue that this picture is a significant over-simplification since electron scattering occurs along with continuous and line absorptions and line scattering. They have recently discussed this problem qualitatively, and we can hope for a more thorough reexamination of this problem in the context of the interpretation of B-star line profiles in the near future.

2.5 Sources of Opacity

It has been commonly assumed that the only important sources of continuous opacity in most B stars arise from hydrogen, neutral and singly ionized helium, and electron scattering. Recent work suggests, however, that, for the silicon-rich Ap stars, a vital source of opacity has been omitted, namely, the bound-free opacity arising from neutral silicon. Gingerich (1964), Rich (1967), and Strom, Gingerich, and Strom (1966) have explored the role of silicon opacity in stars of cooler effective temperature and found its effects to be important. The values of the silicon opacity are based on the recent laboratory work of Rich (1966), which agrees reasonably well with the quantum-defect calculations of Miss Peach. From recent model atmospheres calculated at the Smithsonian Astrophysical Observatory for the silicon-rich B stars, we find:

A. The opacity decreases the Balmer discontinuity through increasing the temperature and decreasing the T gradients in the regions in which the Balmer discontinuity is formed. This results in an overestimate of the effective temperature as deduced from the observed Balmer discontinuity for late B stars with high silicon abundances.

B. There is an increase in the equivalent width of Hγ for a given Balmer discontinuity for stars cooler than 11,000°K. This results in an overestimate of up to 0.2 in log g for late B and early A stars with high silicon abundances.

C. An absorption edge is readily observable at 1527 A for stars with effective temperatures as high as 14,000°K and with silicon overabundances of a factor of 30. Changing the silicon abundance does not affect 1) the surface gravities deduced from Balmer-line profiles or equivalent widths for stars hotter than 11,000°K; 2) the gravities deduced by locating stars in the Balmer discontinuity—equivalent width of the Hγ plane. It would seem prudent in future estimates of the stellar effective temperature and surface gravities to take into account the role played by silicon opacity.

Summary

I summarize below the changes in effective temperature and gravity for main-sequence B stars brought about by the effects discussed in the previous paragraphs.

A. Non-LTE: temperature changes at most between 200 to 300°K.

B. Rotation: temperature changes of the order of 5 to 10% of the effective temperature.

C. Blanketing: changes of about 2000°K at B2 and lower at lower effective temperatures. There is some incompleteness in the description of blanketing owing to the fact that the near-ultraviolet lines have yet to be included in the calculations.

D. Convection: not important.

E. Turbulent pressure: in simple cases the effect is essentially a change of gravity that can amount to as much as 0.2 in the log.

F. Opacity: silicon overabundances of a factor of 30 result in temperature changes of about 1000 to 1500°K.

3. THE BLUE HORIZONTAL BRANCH STARS

The stars under consideration here have the effective temperatures of B stars, with surface gravities on the order of log g = 5 to log g = 6. I have already discussed the diffusion problem with respect to the blue horizontal branch stars, and raised my doubts concerning the validity of the diffusion hypothesis in view of the computed convective velocities.

I shall now summarize the effects of perturbations on the usual model calculations.

3.1 Departures from LTE

At these surface gravities, departures will be unimportant because of the high collision rates.

3.2 Blanketing

No blanketing calculations have yet been made; however, we can expect that the effects of line blanketing will be even more significant for stars of higher surface gravities because of the increased electron and atom densities and the resulting consequences on the line broadening. For Population II stars, however, where the metal-to-hydrogen ratio is significantly reduced, the blanketing effect will probably not be too important. For Population I O and B subdwarfs, considerable care should be taken in assigning effective temperatures and gravities for models computed without the effects of absorption lines.

3.3 Convection

Computations show that even at these relatively high surface gravities and particle densities, convection is not an important mechanism of energy transport for the stars.

3.4 Mass Determination

Perhaps the most interesting problem regarding these stars concerns the determination of their masses. The mass follows from determination of luminosity, effective temperature, and gravity. The determinations of effective temperatures from models are not likely to be worse than about ±500°, which leads to an error of less than 20% in the mass. The determination of luminosity, of course, is a problem that depends on the interpretation of cluster color-magnitude diagrams and can lead to significant absolute errors in the mass determinations, although the relative determinations of luminosity for a cluster are likely to be good. The uncertainty in the gravity determination from hydrogen-line profiles due to the uncertainty of the broadening theory is the biggest source of error. There is a difference of 0.2 in the logarithm between the gravities predicted from the Griem theory and those from the ESW broadening theory proposed on the basis of laboratory experiments. However, because of the high surface gravities of these stars, there are no non-LTE effects and these objects are ideal for testing the validity of the ESW versus the Griem theory by observing the Hα to Hγ equivalent-width ratio.

The results of Strom and Peterson (1968) and Olson (1968) for main-sequence B stars would suggest that errors in the gravity determinations from hydrogen-line profiles are at present on the order of ±0.1 in log g. Considerably more progress, particularly in the refinement of the broadening theory, would allow us to attack such problems as the determination of mass loss from the main sequence to the horizontal branch, and the problem of mass determinations of such stars as the blue stragglers. I should mention in passing that the masses derived from Mihalas' published hydrogen-line profiles for B stars must certainly be increased by a factor of 1.5 to 2, in view of more recent calculations based on the Griem and ESW theories.

3.5 Rotation

For Population II stars on the horizontal branch, rotation does not play an important role in determining the atmospheric structure.

4. WHITE DWARFS

The basic problem is to determine the gravities and effective temperatures that, combined with the luminosity, permit a determination of the mass and radius for these stars. These values can then be compared with computed mass-radius relations from which mean molecular weights and checks on the validity of the interior theory can be obtained.

The perturbations on normal models are the following.

4.1 Departures from LTE

It is probably fair to say that owing to the very high densities in these stars the departures play an unimportant role.

4.2 Rotation

Ostriker and Bodenheimer (1968) have suggested that stars higher than the Chandrasekhar limit can exist if they are rotating very rapidly, that is, several thousand kilometers per second. Their computations suggest, as expected, that in some cases the stars would be highly flattened at the poles and extended at the equators. They suggest that the DC white dwarfs are possible candidates for the rapidly rotating class of white dwarfs. It would seem quite straightforward to construct rotating white dwarf models by means of techniques already developed. E. Wright of the Harvard College Observatory is carrying out such an investigation using white dwarf models recently computed here. Scans of DC stars would be most useful in allowing a check of this hypothesis.

4.3 Convection

The role of convection in white-dwarf atmospheres has been neglected to date; however, near the cool end of the B-star white-dwarf region, convection may yet prove to be quite important. The importance of convection in these models is due to the very high density, and to the consequent fact that the ionization region extends over a wide region. Some preliminary calculations indicate that models near 10,000° K in which convection has been neglected result in the underestimation of effective temperature by a few hundred degrees. The effects for stars cooler than 10,000° K are no doubt considerably larger. Qualitatively, we expect that if the effective temperatures obtained from radiative models are too small, the predicted radius will be too large; therefore, the mass is also predicted too large. This is not likely to be significant for late B stars, but possibly for the A-type white dwarfs this effect will be of somewhat greater importance.

4.4 Line blanketing

The role of line blanketing is very important but has not yet been included in a self-consistent manner. Terashita and Matsushima (1966) have published models computed without lines, for which they used the resulting temperature-pressure structures to predict the line spectrum. Of course, owing to the high densities, Stark widths for the hydrogen lines in white dwarfs are very large. As a consequence, the lines have an important effect on the temperature structure. Some very recent unpublished calculations of Matsushima and Terashita (1968) would suggest fairly significant changes in their original models when lines are included. Estimates based on models constructed at SAO as compared with the original Matsushima and Terashita models suggest temperature changes on the order of 1000° K, and surface gravity changes of about 0.5 when the effects of lines on the models are computed properly. In addition to including hydrogen lines, it is no doubt of considerable significance to include other sources of blanketing in the ultraviolet for white dwarfs in the effective temperature range 10,000 to 20,000° K.

5.   B STARS OF LUMINOSITY CLASS III AND IV

5.1 Departures from LTE

Departures from LTE become important for surface gravities less than or equal to about log g = 3.5.   The effective temperature is chosen too high from LTE models and the gravity from hydrogen-line profiles is chosen too low.   The ratio of the Hα to Hγ equivalent widths should begin to show a distinctly different value from that predicted from LTE models.   For surface gravities of about log g = 2 to 2.5, departures from LTE cause the temperature to be overestimated by about 1000°.

5.2 Rotation

Owing to the larger radius, the rotation velocity is decreased for the B stars that have evolved off the main sequence.   For the Be stars, needless to say, the effects of rotation should be important.   Departures from LTE should also affect the line and continuum spectra for these stars.   From the low-gravity equatorial regions it is possible to obtain Hα and Hβ in emission, since the ratio of departure coefficients is $b_3/b_2 \gg 1$.   Of course, all observed features of the complex Be spectra cannot be attributed to non-LTE effects, but it may be of some interest to bear these effects in mind in their analysis.   Deane Peterson of Harvard is currently considering this problem in the context of his survey of the effects of departures from LTE on hydrogen lines in B stars.

According to Christy (1967), the β Canis Majoris stars may in reality be rapid rotators.   If this is indeed the case, a comparison between rotating models and observations should be made in order to test this hypothesis.

5.3 Blanketing

The effects of hydrogen-line blanketing and metal-line blanketing in the ultraviolet have not as yet been included in luminosity class III and IV B-star calculations.

5.4 Convection and Atmospheric Motion

Owing to the low densities, the effects of convection are unimportant in these stars.

5.4.1 Diffusion

Diffusion may be of some importance if the convective velocities are low enough.   If the rotational velocity is small, the diffusion velocity is larger than the circulation velocity and, therefore, the time for mixing through circulation is larger than the diffusion time.   However, the time for downward diffusion is on the order of the lifetime of the luminosity class III star evolving off the main sequence.   Therefore, one might expect separation of elements as a function of charge-to-mass ratio.

6.   HIGH-LUMINOSITY B STARS

Very little has been done in computing models for these stars, but a much more detailed study is possible, and should be made, to determine accurate temperatures and gravities for these stars.

6.1 Departures from LTE

Departures from LTE are of major importance, and effective-temperature changes of the order of 1000 to 2000°K are to be expected in these low-density objects.   The hydrogen lines are also affected by departures from LTE, and we would expect emission to result from very high b_3/b_2 ratios.

6.2 Atmospheric Motion

Previous work in the literature has indicated that these stars exhibit very high macro- and microturbulent velocities, and it is important to discover how much of the support in these atmospheres comes from turbulent motions.

6.3 Blanketing

The effects of blanketing are in general less important owing to the low densities, although turbulent broadening does increase the line width somewhat. These effects will be quite difficult to estimate.

Incidentally, it is now possible to handle cases such as the high-luminosity B stars in which scattering is an important fraction of the total opacity. A rapidly convergent procedure for determining directly the source function in a scattering atmosphere has been developed by Robert Kurucz of Harvard, and these results are being prepared for publication.

7. THE HELIUM-RICH B STARS

Among the stars in the B-star temperature range there is a wide range of helium abundances. Several stars observed to date appear to be essentially pure helium (with hydrogen down by a factor greater than 100). Böhm-Vitense (1967), Hunger and Van Blerkom (1967), and Klinglesmith (1967) have computed models for these stars. These investigators predict essentially the same photometric properties as for normal B stars but much steeper temperature drops at the surface. Hunger and Klinglesmith (these proceedings) have shown that it is possible, and necessary, to perform a self-consistent analysis of these stars from the results of helium-rich model computations.

To my knowledge, the problem of helium-rich white dwarfs has not yet been attacked with models and appropriate opacity sources, although such an investigation is well within the realm of current capability.

8. THE EARLY A STARS

Stars in the effective temperature range 8000 to 10,000°K should be kept separate from the discussion of other A stars since the problems of spectral classification and temperature and gravity determinations are so complex for stars in this range. The complexity is mainly due to the onset of the importance of H^- as an opacity source. This opacity makes the Balmer discontinuity sensitive to gravity as well as effective temperature. Owing to the contribution of H^- to the total opacity and the effects of decreasing T_{eff} on the level population, the hydrogen lines are both gravity and temperature sensitive. Moreover, the slope of the Paschen continuum is still not very sensitive to effective-temperature changes. Blanketing in the near ultraviolet at higher T_{eff} and then in the accessible regions of the spectrum at lower T_{eff} is of increasing importance, since the atoms are essentially now all neutral and singly ionized at A-star effective temperatures. Furthermore, a wide range of metal abundances has been observed among the early A stars. The problem of determining accurate effective temperatures and gravities from Ap and early Am stars is much more difficult owing to the effects of the lines as well as the metal opacities on the atmospheric structure.

First, let us examine the sensitivity of the commonly used observational quantities to temperature and gravity changes. Using B-V as a measure of the slope of the Paschen continuum, we find

$$\Delta(B-V)/\Delta T_2 \sim 0.016 \quad , \quad 9500°K \gtrsim T \gtrsim 8000°K \quad .$$

For a known metal abundance and for surface gravities in the range $3 < \log g < 4.5$,

$$\Delta D_B / \Delta[g] = 0.22 \quad .$$

Thus, in order to obtain surface-gravity accuracies to 0. 1 in the log, Balmer discontinuities must be obtained to an accuracy of 0. 02 mag. It is possible, in principle, to achieve such accuracy, but the results are quite sensitive to the correction for line blocking, particularly in the region directly below 3650 A. In addition to the difficulty in making accurate blanketing corrections shortward of the Balmer discontinuity, we still have the difficulty (largest in the UV) of absolute calibration of the spectrophotometric standards. Progress has been made recently by Wolff, Kuhi, and Hayes (1968), who find good agreement between existing models and spectrophotometric observations on the basis of Hayes' calibration. Further work in this area is definitely needed. Another difficulty in using the Balmer discontinuity as a measure of the surface gravity is the fact that this quantity is also sensitive to the metal-to-hydrogen ratio. For the early A stars, we find that

$$\Delta D_B / \Delta [Z] = 0.10 \quad .$$

Hence a factor of 2 increase in the metal-to-hydrogen ratio, or more specifically the silicon-to-hydrogen ratio, results in a change of the Balmer discontinuity by almost 0. 03, or effects of more than a tenth in log g. Therefore, surface gravities cannot be obtained without first determining, at least roughly, the metal abundance.

Let us now review the effects of perturbations on normal models.

8.1 Departures from LTE

Departures from LTE are still relatively small for main-sequence A stars. However, it is of some academic concern to note an interesting effect of the metal opacities on the departures from LTE. The departure coefficient b_n in the low-density approximation is given by

$$b_n - 1 \sim \frac{\int_\nu a_\nu (B_\nu - J_\nu) \, d\nu}{\int_\nu a_\nu J_\nu \, d\nu} \quad .$$

The value of the numerator of this equation essentially controls the magnitude and direction of the departures. For cases in which the metal opacities are negligible, the numerator is positive longward of 1600 A and negative shortward of 1600 A. The departures resulting in $b_2 < 1$ are brought about by the negative contribution outweighing the positive contribution. However, when the metal opacities are included, the negative contribution shortward of 1600 A is much reduced because J_ν is brought close to B_ν, where the silicon opacity predominates. In fact, detailed computations show that the departures where the metal abundances are high are such that the second level is over-populated, although by a very small amount. Line blanketing would have an analogous effect if it contributed more significantly below 1600 A, or an opposite effect if its importance were greater at wavelengths longer than 1600 A. One should note the delicacy of the determination of the departure coefficients and their sensitivity to such apparently small effects.

8.2 Opacity

The silicon-to-hydrogen ratio is clearly important in fixing the observed spectrum in the visible and in the ultraviolet. The silicon discontinuity at λ 1527 ranges from 0. 5 to 1. 5 mag even for normal silicon-to-hydrogen ratios in this temperature range. Near A0, the high silicon-to-hydrogen ratio has the effect of causing one to choose too high a gravity from the equivalent width of Hγ when T_{eff} is fixed by either the slope of the Paschen continuum or the Balmer discontinuity. This can amount to 0. 2 in the log for an increase by a factor of 10 of the silicon-to-hydrogen ratio.

In addition, ultraviolet blanketing in the 2000 to 3000 A range is doubtless important and is yet to be considered in detail. This problem is of considerable importance in discussions of the location of such stars as the Am, Ap, and slowly rotating A stars in the H-R diagram. It would seem premature to conclude definitely that these stars have essentially the same effective temperatures and surface gravities as normal A stars without detailed account being taken of the effects of line blanketing and metal opacities.

8.3 Convection and Atmospheric Motions

The effect of convection for main-sequence A stars is not important for temperatures greater than 8500° K.

8.3.1 Turbulent pressure

Since high values of turbulent velocity have been determined for some early A stars and, in particular, for the early Am analogues, it might be speculated that turbulent pressure plays an important role in supporting the atmospheres of these stars. By setting the turbulent velocity equal to the velocity of sound, we find that a model with turbulent pressure imitates the behavior of a model having normal surface gravity lower by about 0.2 in the log. Therefore, when we assign the true gravity from, say, the equivalent width of Hγ, the value of surface gravity deduced may be affected by the magnitude of the turbulent velocity. We should thus be cautious when interpreting the values of gravity deduced for peculiar and Am stars. From this crude experiment it appears again as if the spectral abnormalities of the early Am analogues cannot be explained by a change in the atmospheric structure caused by turbulent pressure. Praderie (1967, 1968a, 1968b) had reached this conclusion in her already published work on the Am stars, although she did not include the turbulent pressure in her atmosphere calculations in a self-consistent way, as was the case in the investigation reported here.

8.4 Rotation

From the crude arguments presented previously, we expect rotation to result in a decrease in the mean surface gravity and the mean effective temperature for most aspect angles. Decreasing the mean surface gravity will increase the Balmer discontinuity since H$^-$ is relatively less important as an opacity source, and this will in turn increase U-B. Decreasing the effective temperature will increase B-V. Unfortunately, the relative sensitivity of U-B (or analogously c_1 in the Strömgren system) to small changes in surface gravity makes tests of the effects of rotation difficult for the early A stars, although, as we shall see, it is possible to comment on the changes for slightly cooler stars where the Balmer discontinuity is almost a factor of 2 more sensitive to a given change in surface gravity. As yet very little work has been done in fitting models for A through F stars, owing in part to the fact that no extensive grids of models are available for these spectral regions. It should be possible to supply blanketed non-LTE models in the near future, and we should therefore expect progress in the area of computing realistic models for rotating stars in this spectral region. However, because of the lack of ultraviolet line blanketing in the models computed to date, the possibility of higher silicon-to-hydrogen ratio, and the higher turbulence in the Am-star atmospheres, we should be wary of using the metallic-line A stars to establish the zero-rotation main sequence.

9. A STARS WITH LOG GRAVITIES IN THE RANGE 2 TO 4

We include in this category a mixed bag of stars ranging from stars still in the pre-main-sequence contraction phase, as in young clusters like NGC 2264 to old horizontal branch stars, and luminosity class III stars evolving

off the main sequence. Little published material is available for stars with
atmospheric parameters in this temperature and gravity range. It would be
important if we were able to assign accurate parameters to the pre-main-
sequence stars and the horizontal branch stars. For the horizontal branch
stars, we have the problem of determining quantitatively any mass loss.
For the pre-main-sequence stars, we would like to be able to investigate
more critically the type of problem posed by Iben, namely, the determination
from comparison of observed and computed positions of pre-main-sequence
stars in the M_{bol} - T_{eff} plane of the rate of star production as a function of
mass. Unfortunately, these will be difficult problems because of the following effects.

9.1 Departures from LTE

The departures are in a sense such that $b_2 < 1$ or the second level is
underpopulated, and hence the Balmer discontinuity is reduced. Since the
Balmer discontinuity is a measure of the surface gravity, this leads to a
choice of too large a surface gravity by about 0.2 to 0.3 in the log if normal
models are used in the analysis. Moreover, the equivalent width of $H\gamma$ will
be affected, and for the early A stars this will reduce the estimated surface
gravity for a given spectral type.

9.2 Blanketing

The effects of line blanketing have not yet been included in models of
this type. Again, the problem will be most serious in the ultraviolet,
although the hydrogen-line blanketing is less important because of the lower
surface gravity.

9.3 Convection

Convection is not important.

9.4 Opacity

Metals are still important, particularly in the case of young stars in the
pre-main-sequence contraction phase, where the Balmer discontinuity may
be reduced because of the possible enhancement of the metal-to-hydrogen
ratio. Again, the surface gravities will be significantly affected by such an
increase.

10. WHITE DWARFS

Very little work has been done on white dwarfs with effective temperatures
less than 10,000°K. Hydrogen-line blanketed models could be computed, but
as yet have not. Moreover, other sources of opacity such as H^-, C, N, and
O should also be included for the effects of composition changes to be explored.
Miss Cuny has suggested recently that resonance broadening for the far wings
of Lyman α is of considerable importance in solar-type stars. Some preliminary investigations suggest that this source of opacity could be quite important in the case of the high particle densities that obtain in these low-T_{eff}
white dwarfs.

In the A-star region at the densities appropriate to white dwarfs the role
of convection is quite important. Work on this problem is being done by
Böhm (1968). Preliminary calculations carried out at SAO suggest that the
effects result in a change of temperature of nearly 200°K. Too small a
value is chosen if convection is neglected. The effect should become increas-
ingly more significant as the effective temperature is decreased.

11.   THE MIDDLE A TO EARLY F STARS

For stars in this temperature range, H⁻ grows in importance as an opacity source; as a result, the Balmer discontinuity becomes very sensitive to gravity. The peak of the Planck function falls in the visible and the slope of the Paschen continuum is therefore sensitive to effective temperature. Moreover, the hydrogen lines with H⁻ now the background opacity are essentially independent of surface gravity and, because of the rapid decrease of the population of level 2 relative to level 3, are sensitive to effective temperature. For temperatures less than $8000°K$ we find the following variation of observable parameters with temperature:

$$\Delta(B-V)/\Delta T_2 = 0.03 \quad ,$$

$$\Delta[W]/\Delta T_2 = 0.028 \quad .$$

To obtain accuracies of $\pm 100°$ in the effective temperature, we need accuracies of ± 0.03 in B-V and about 6 or 7% in the equivalent-width measurements for hydrogen lines. It should be noted that B-V is about twice as sensitive to changes in the temperature as in the early A-star region. To obtain surface gravities, variation of Balmer discontinuity with gravity is given by

$$\Delta D_B/\Delta[g] = 0.40 \quad ,$$

which suggests that, in order to obtain gravity accurate to 0.1 in the log, we need to obtain Balmer discontinuities accurate to ± 0.04 mag. We ask again what affects the observed B-V, Balmer discontinuity, and equivalent width of hydrogen lines in this region.

11.1 Departures from LTE

With H⁻ now the dominant source of opacity, we should consider the effects of departures from LTE for H⁻. Qualitatively, we expect from the work of Cayrel (1966) and others that 1) the departure coefficient b_-, expressing the ratio of the true H⁻ number density to the LTE value, will be greater than unity — i.e., there will be more H⁻ than we expect on the basis of LTE calculations; and 2) the boundary temperature will be greater than that computed from LTE models. The cross section C_- for the associative detachment reaction for H⁻,

$$H + H \underset{\longleftarrow}{\overset{\longrightarrow}{\rule{1cm}{0pt}}} H_2 + e \quad ,$$

controls the size of the departures. By using the value of $C_- \sim 10^{-9}$ computed by Dalgarno and obtained experimentally by Ferguson's group at the Joint Institute for Laboratory Astrophysics, we find for main-sequence stars with $T_{eff} < 8000°K$ that the effects of departures from LTE in H⁻ will be negligible. However, for neutral hydrogen, because of the rapid increase of flux in the Balmer continuum with increasing depth, the second level is considerably underpopulated even for main-sequence stars. With $b_2 < 1$, the Balmer discontinuity is smaller, and hence gravities will be overestimated by about 0.2 in the log. Moreover, the hydrogen lines are weakened since $b_n/b_2 > 1$, for $n \geq 3$. Therefore the effective temperature is chosen too small from the equivalent width of the Balmer lines.

11.2 Blanketing

For temperatures lower than $8000°$K, the effects of spectral lines are now becoming of crucial importance, not only in the determination of the temperature structure, but also because of the effects of blocking. Particularly difficult is the problem of determining the Balmer discontinuity, since in the regions directly below λ 3650 the blanketing is severe. It seems clear that errors greater than 4% in determining blocking coefficients will be common, and consequently lead to errors of greater than 0.1 in the determination of log g. One might mention in this regard that Danziger and others (Danziger and Oke, 1967; Danziger and Dickens, 1967; Kuhi and Danziger, 1967) found it difficult to obtain reasonable values for the gravities for such stars as the δ Scu variables, although in these cases it is not at all clear that all or most of the trouble arises from inappropriate blanketing coefficients. Careful study of double-star systems in which mass and luminosity are known could lead to a test of model atmospheres in this temperature range, and we hope that some observational effort in this direction will be made in the near future.

11.3 Atmospheric Motions and Convection

11.3.1 Turbulent pressure

No calculations analogous to those previously reported for the B and A stars have yet been made, but clearly the effects of microturbulence are of some importance in interpreting the spectra of late Am stars. Moreover, recent unpublished work by Chaffee (1968) at Kitt Peak National Observatory would also indicate that the microturbulence reaches a maximum in its effect on the line spectra near F0, decreases toward F5, and increases again near G0, so that the influence of the turbulence on the atmospheric structure may be important and should be investigated more closely. Praderie (1967) has looked at this problem but, again, not in a completely self-consistent way.

11.3.2 Convection

For temperatures lower than $8000°$K, the effects of convection are of some importance in determining both the temperature structure and, consequently, the emergent spectrum. The existing models in the literature are those of Mihalas (1965), which go down only to temperatures around $7700°$K.

It should be pointed out, however, that if one uses colors in the spectral region longward of 5000 A instead of using B-V or b-y, the influence of convection on the atmospheric structure is not significant in the region where the flux in this wavelength region is formed. This is true because convection becomes significant in terms of the amount of flux carried only for optical depths greater than 0.5 or 1. Because of the peak of the H⁻ opacity near 8000 A, we are looking systematically higher in the atmosphere in the region longward of 5000 A than is the case for, say, the B filter.

Mihalas has explored the problem of the effect of convection on the B-V colors. He finds that the effect of convection on the determination of effective temperature from observations of the blue-green spectral regions is about $100°$; the effective temperature is underestimated by radiative models. More extensive estimates for lower surface gravities have recently been made by Strom and will be described shortly.

Searle and Oke (1962) have suggested that the equivalent width of Hγ that measures the effective temperature can be used to obtain the slope of the Paschen continuum, which also depends upon the effective temperature. If this is the case, we can use the predicted slope of the Paschen continuum from a model having a given equivalent width of Hγ and then deduce from the observed spectral-energy distribution the shape of the interstellar reddening curve. For an equivalent width of Hγ known to about ±10%, the slope of the Paschen continuum can be determined to about ±0.03 despite the effects of convection on the emergent spectrum. Although convection affects the slope of the Paschen continuum in the direction to make the star look cooler, the equivalent width of Hγ is also changed by an amount such that the correct value of the slope of the Paschen continuum is obtained for a given observed equivalent width of Hγ. However, Danziger has argued that the equivalent width of Hγ predicts a higher effective temperature than the slope of the Paschen continuum for unreddened stars. This inconsistency may possibly be removed by Hayes' calibration, although it would again be useful to look at stars of known effective temperature to check out the model, for example, Procyon.

11.4 Rotation

Kraft and others (Kraft, 1965; Anderson, Stoeckly, and Kraft, 1966; McGee, Khogali, Baum, and Kraft, 1967) have attempted to test for the effects of rotation in stars in this spectral region. The mean effective temperatures and the mean surface gravities are lower in rapidly rotating stars. Hence, the Balmer discontinuity and slope of the Paschen continuum correspond to cooler, lower gravity stars. This has the effect of increasing U-B and also of increasing B-V. Kraft and Wrubel (1965) examined the color-color diagram for the Coma cluster. They found a correlation (in the direction expected from the computed rotating models) between v sin i and the displacement in the measured U-B of the star from the normal main-sequence color-color relation. Roxburgh, Sargent, and Strittmatter (1966) studied the displacements of Praesape stars in the color-magnitude diagram and deduced the velocity of rotation from the distance ΔM_V of the stars above the main sequence, as suggested by previous discussions of the effects of rotation on luminosity and effective temperature. Their results showed a good agreement between theoretical predictions of interior and atmosphere models and observations, although recent observations by Kraft and Dickens using new rotational velocities and new photometry give much less satisfying correlations. Certainly more observational work is needed here, and one might suggest that the slope of the Balmer continuum is a sensitive measure of rotational velocity. However, the observational difficulties are quite severe, primarily because of the effects of line blanketing.

11.5 Stars of Lower Gravity in the A-F Star Region

In view of Christy's (1966) detailed predictions concerning the nature of RR Lyrae variable pulsation, it would be of some interest to reexamine the accuracy of temperature and gravity determinations for the atmospheres of RR Lyrae stars. Of course, one questions the mechanism of support, but certainly a simple approach to reexamine some questions about these stars seems in order. I cannot go into a detailed comparison with observations, but shall mention some trends indicated by recent model calculations.

The determination of the high effective-temperature end of the RR Lyrae instability strip provides a measure of the helium content of these stars: the greater the effective temperature at the high-temperature boundary of the instability strip, the greater is the helium content. We might expect that because of their low surface gravities, the RR Lyrae stars would have convection zones that carry very little flux and hence would have a negligible effect on the emergent spectrum. However, recent computations indicate the following.

A. Using convective models, as compared with radiative models, we find that the effective temperatures to date have been underestimated by at least 100 to 200° K. The densities in the convection zones are certainly lower in RR Lyrae stars than in main-sequence stars of the same effective temperature. However, the onset of convection occurs higher in the atmosphere because of the lower gravity and the consequent onset of hydrogen ionization further up in the atmosphere. As a result, convection is important and carries enough flux in the regions of the temperature-depth relation that determine the shape of the emergent continuum flux.

B. Despite the error of 200° in assigning effective temperature, the surface-gravity determinations for the Balmer discontinuities have been unaffected.   The change in temperature gradient brought about by convection is just the right amount to cause the relation between the Balmer discontinuity and the slope of the Paschen continuum to be essentially identical for convective and radiative models.   Thus, although the temperatures assigned heretofore are wrong, the surface gravities are unaffected by convection.

C. Departures from LTE result in $b_2 < 1$, and, at these gravities, $b_- > 1$.   Hence the Balmer discontinuity is smaller than the value one would predict from LTE models.   The value of surface gravity is, therefore, overestimated by perhaps as much as a factor of 2.

D. If we use the relation between equivalent width and slope of the Paschen continuum to determine the Paschen continuum slope from models and observed widths, we find that for surface gravities log g $\lesssim$ 3, the equivalent width of Hγ is not very sensitive to temperature.   Although the population of level 2 relative to level 3 decreases with decreasing T_{eff} at these low surface gravities, the relative importance of H¯ as an opacity source is much decreased, since much of the H¯ is dissociated.   Thus line formation in these low-gravity stars is analogous to line formation in B stars, where the equivalent width of Hγ essentially measures the surface gravity rather than the temperature.   This result has a serious consequence on the reddening determinations, based on the equivalent width of Hγ, and the consequent slopes of the Paschen continuum.   Again, I might mention the importance of the absolute calibration in fixing the effective temperatures and gravities.   Small errors in the absolute calibration at these temperatures can lead to rather large errors in the resulting temperatures and gravities.   An error in the calibration leading to a change of 200° in effective temperature (as suggested by Hayes' work) can lead to a gravity that is incorrect by as much as 0. 5 in the log.

For metal-to-hydrogen ratios less than unity, we find that the slope of the Paschen continuum, the equivalent width of Hγ, and the Balmer discontinuity are unaffected by the metal content except through the influence of metal lines.   This was, of course, anticipated as a result of previous calculations.

These general remarks made for the specific case of RR Lyrae stars also hold true for other stars with log g less than about 3. 5.   Therefore, these results should be kept in mind in discussions of the problems of δ Scu stars and also stars in the pre-main-sequence contraction phase.

Synthesis of the spectrum in the region of the Balmer jump would be of considerable importance in improving our estimates of the surface gravity based on the Balmer discontinuity, since blanketing is quite crucial in the accurate determination of the Balmer jump.   Also, we would recommend that, because of the blocking effect of the lines and the problem of convection, we use red colors analogous to V-R or V-I in order to determine the effect of temperature for stars later than A5.

12. LATE F AND G STARS

For stars with effective temperatures lower than about 6000 to 6500°K, we are seeing unevolved or slightly evolved objects with compositions ranging from those representative of the early composition of the galaxy to those of stars formed as recently as 10^7 to 10^8 years ago.   In this region we must, therefore, concern ourselves with stars having compositions ranging from 0. 01 times the solar metal content to a few times the solar metal content. This heterogeneity in composition influences 1) the blanketing effect, 2) the structure in the outer parts of the convection zone that affects the temperature and gravity choice, 3) the location of stars in the luminosity−effective-temperature plane because of the effects on the interior opacity, and 4) our impression of the long-term secular changes in atmospheric motions as measured by turbulent velocities and possibly by the calcium emission-line strength.

For stars of these spectral types, H⁻ is still the primary opacity source, and at temperatures lower than 6000°K is so important that for main-sequence gravities the Balmer discontinuity is rendered so small that it is no longer capable of allowing an accurate measurement of log g for stars near the main sequence.

The determination of effective temperature follows from the slope of the Paschen continuum. However, to avoid the effects of convection and blanketing on the emergent flux, we are almost forced to go to the red to measure the slope of the Paschen continuum. If I define my measure of the slope of the Paschen continuum as

$$\Delta m = m(5500) - m(7500) \quad ,$$

the sensitivity of this index to temperature is

$$\Delta m / \Delta T_2 = 0.02 \quad .$$

Rotation does not play an important role for middle F and later stars, so that at least we do not have to concern ourselves with this problem. The effects of convection are important but do not influence the red color index substantially. The importance of convection increases as the metal-to-hydrogen ratio decreases, since for a given optical depth the gas pressure and thus the density at a given optical depth increase and a larger fraction of the flux is carried by convection.

Dennis (1968) has computed convective models in the G-star effective-temperature range having a variety of temperatures and metal contents. His major conclusion is that convection may play a significant role in locating the subdwarfs in the color-color diagram. Dennis suggests that the temperatures of subdwarfs have been underestimated by about 250 or 300°K because the convective models look cooler in B-V than radiative models. By analyzing the method by which the blanketing corrections for subdwarfs have been made and the temperature scale has been calibrated, he concludes that subdwarfs have been located too far to the right in the luminosity-effective temperature plane. As a result of his work it appears as if the subdwarfs should be moved about 200 to 300°K to the left. The subdwarf sequence is thus forced below the Population I main sequence; this implies a high helium content. However, this conclusion is in no way based on a self-consistent picture for the subdwarfs of the combined effects of blanketing and convection on the model structure, although it is an indication that an important effect in determining the location of subdwarfs in the $M_{bol} - T_{eff}$ plane may have been overlooked.

At this stage we should consider the possible importance, heretofore overlooked, of Lyman-α resonance broadening as a significant opacity source in the near-ultraviolet region of subdwarf spectra. Miss Cuny has suggested that resonance broadening is a significant source of opacity in the solar ultraviolet and, in fact, has been able to improve considerably the agreement between the observed and the computed solar spectra in the region longward of 1500 A by including the effects of the Lyman-α wing opacities. She will report her results later at this meeting. Since this effect is significant for the sun where opacity due to metal continua and metal lines are important, it should be even more important for the subdwarfs where the metal-to-hydrogen ratio can be decreased by as much as a factor of 100. Moreover, since the densities at a given optical depth for a subdwarf are higher, the effects of the Lyman-α wing opacity are relatively more important in subdwarfs than in main-sequence Population I stars. I have computed subdwarf models and main-sequence models including the Lyman-α wing opacity. The results show that the subdwarf effective temperatures heretofore determined have been underestimated by between 100 and 200°K. Therefore, we must shift them to the left in the $M_{bol} - T_{eff}$ plane, which drops them even further below the main sequence. Even if we accept the Eggen and Sandage (1962) observation that the subdwarfs coincide with the Population I main sequence in the $M_{bol} - T_{eff}$ plane, the subdwarfs, because of the effect of Lyman α, must now fall below the Population I main sequence by between 0.15 and 0.25 mag. In my opinion this is additional confirmation of the already suspected high (i.e., near solar) helium content of these objects. I should also mention here that Vardya (1967) suggested the importance of the quasi-H_2 molecules in opacity source in the ultraviolet spectrum of subdwarfs. The role of quasi-H_2 should be further explored in some detail in model calculations, although it appears at the moment that its effect would be much the same as that for Lyman-α resonance broadening.

12.1 Blanketing

I have already indicated the nature of the effects of blanketing on the temperature structure. Proper treatment of the depth dependence and a more detailed examination of the formation mechanism are still needed, as demonstrated by some of the numerical experiments of Athay and Skumanich (1968) based on a two-level atom.

I should like to point out here the indirect effects of line blanketing on curve-of-growth analyses. Let us consider the atmospheric structure of a low metal-content star versus one having a high metal-to-hydrogen ratio. Furthermore, let us require that these stars have the same emergent spectrum in the red so that at first glance they would be assigned the same temperature. However, owing to the effects of line blanketing, the boundary temperature of the low metal-to-hydrogen ratio star is greater than that of the high metal-to-hydrogen ratio star. As a consequence, the weak lines in the case of the low metal-to-hydrogen ratio star are stronger in comparison with the strong lines of the neutral elements such as iron, since the strong lines are formed nearer to the boundary where the temperature is higher and there is more Fe II relative to Fe I. A comparison of the curves of growth indicates that, when we match up the linear portions, the low metal-to-hydrogen ratio curve falls below the high metal-to-hydrogen ratio curve, which leads us to believe that the star with low metal-to-hydrogen ratio has lower turbulence. However, this result is caused almost entirely by the relatively higher boundary temperature in the low metal-to-hydrogen star. Hence, when we perform a differential curve-of-growth analysis, we choose too low a value of the turbulent velocity for the low metal-to-hydrogen ratio star. Also, for exactly the same reason, the excitation temperature is found to vary with line strength for the low metal-to-hydrogen ratio star if the high-metal star is used as the standard or, in specific terms, if subdwarfs are used as program stars measured relative to the sun. Therefore, comments such as those suggesting that the turbulent velocities of stars decrease with age or with composition would seem to be invalidated if the results are based on the differential curve-of-growth techniques. Model-atmosphere analyses of subdwarfs performed at SAO would suggest that the turbulent velocities for these stars are near the solar value, if not slightly higher.

12.2 Departures from LTE

For main-sequence stars, the departures from LTE are not important, because of the high value of the reaction rate for associative detachment. However, the second level for hydrogen is significantly underpopulated, which affects the Balmer discontinuity as well as the hydrogen-line strengths.

There is clearly a great need for the development of a spectroscopic criterion for determining gravity because of the weakness of the Balmer discontinuity. Several people have attempted to use empirical criteria based on the strengths of singly ionized to neutral metal lines. Jones (1966) has developed such an empirical classification scheme by using multivariate analysis techniques. However, calibration of such empirical classification schemes has thus far been thwarted because of the extreme difficulties in synthesizing the regions of the spectrum containing the lines that are gravity sensitive. This is not so much a computational problem as one arising from the paucity of accurate f values and damping constants.

12. 3 F and G Stars of Higher Luminosities

For surface gravities lower than log g = 2, particle densities have been so reduced in the atmospheres of these stars that departures from LTE in H^- become significant. We recall that $b_- > 1$ and the boundary temperature is higher than the LTE value. The consequences of departures from LTE in H^- on G and K giants have been discussed by Strom (1967). Basically, the flux in the free-free continuum of H^- is raised relative to that in the bound-free continuum. This may provide a method for observationally testing for departures from LTE. With $b_- > 1$ and increasing as we approach the boundary, we can have important effects on the regions where the stronger lines are formed. For example, the ratio of line-to-continuum opacity is smaller for lines formed in the H^- bound-free continuum as we approach the boundary compared to the value computed in LTE. Hence, we find that departures from LTE affect the Fe I to Fe II ionization equilibrium. On the average, the Fe I lines are formed closer to the boundary than are the Fe II lines. They therefore look weaker compared with the Fe II lines that are formed further in, where b_- is closer to unity. We therefore suggest that ignoring the effects of departures from LTE will result in too low a value of the electron pressure and thus of the surface gravity being chosen. This would mean that for a given luminosity and effective temperature the mass for a G or K giant would be chosen too low. Departures are likely to be more important in the relatively rare G giants than is the case for the cooler giants.

Convection does not have an important consequence on the flux in the generally accessible regions of the spectrum.

Rotation of course is still unimportant.

13. COOL STARS

The problem of computing atmospheres for cool stars is rapidly becoming viable and, moreover, an important topic in astrophysics. The problems of computing the detailed structure of those stellar envelopes are formidable. Molecular opacity, and the resulting strong dependence of the structure on composition, the problem of the source of electrons, the importance of convection, the role of Rayleigh scattering, and so on are all very difficult problems, but ones that are slowly becoming controllable. However, the gain from any effort in this field may be great. For example, accurate determinations of compositions, particularly of the light elements, could lead to a discussion of the detailed mechanism of grain formation. We could also explore the effects of convective mixing on envelope compositions.

Pioneering work in understanding cool stars was begun at Harvard by Gingerich (Gingerich and Kumar, 1964) and continued more recently (Gingerich, Latham, Linsky, and Kumar, 1967).

In the range 2500 to 4000° K the peak of the blackbody curve falls between 8000 A and 1.65 μ, a minimum in the continuum opacity curve for the H^- bound-free opacity. Gingerich and coworkers (Carbon, Gingerich, and Latham, 1968) suggested that, as a result, the emergent fluxes for stars of this effective temperature would peak sharply at 16,500 A and, furthermore, that the peak would be sensitive to luminosity because of the increased dissociation of H^- as the surface gravity is decreased. The peak and its luminosity dependence were observed in the Stratoscope observations. However, recent work by Auman (1967) shows that both effects arise primarily not from the minimum in the H^- opacity but from a minimum in the water-vapor opacity. The increased association of water vapor with decreasing surface gravity produces the luminosity dependence of the peak.

In their exploration of this problem, Gingerich and coworkers found for very cool stars that the electron number density depends sensitively on the abundances of a few contributors having low ionization potentials, some of which were cosmologically rare, e. g. , rubidium.   They took account of the molecular equilibrium problem in their computations, but at first did not include opacity due to the molecules.   They demonstrated the importance of Rayleigh scattering due to H_2, and, although the problem of treating an essentially pure scattering atmosphere was difficult at that time, it is now possible to cope with it.   The most important <u>continuum</u> opacity sources in addition to Rayleigh scattering were found to arise from H^-, H_2^-, and, in some regions, He^-.   When water-vapor opacity is included, the importance of Rayleigh scattering decreases, at least at longer wavelengths.   As the flux removed by water vapor increases, the temperature at a given optical depth increases and results in dissociation of H_2.   The ambitious and extensive work of Auman has allowed him and others to include the effects of water-vapor opacity, which is crucial in determining the temperature structure of these late-type stars.

Another difficult problem in computing cool star models is that convectively unstable regions extend to rather shallow depths; the treatment of this problem in a realistic way would seem to be formidable and sensitive to the assumptions implicit in the mixing-length theory.   However, recent results suggest that the temperature-pressure relation may very quickly reach a point where dT/dp is adiabatic and, consequently, the details of the treatment of convection may prove to be unimportant.   Certainly more work must be done, particularly directed toward including molecular opacity sources such as carbon monoxide, before it is possible to discuss in a meaningful way the determination of temperature and gravity criteria.   However, we can be optimistic since Gingerich and his group and Auman, within a relatively short time, have models with temperature-pressure relations well enough defined to permit abundance analyses.

I should like to acknowledge the generous cooperation in the preparation of this manuscript of my colleagues   Drs. E. H. Avrett, O. J. Gingerich, W. Kalkofen, Messrs. R. L. Kurucz, D. M. Peterson, T. Simon, Miss R. Peterson, and my wife Karen.   Part of this work was supported by NASA grant NGR 22-024-001.

REFERENCES

Adams, T. F. , and Morton, D. C. 1968, Ap. J. , 152, 195.

Anderson, C. M. , Stoeckly, R. , and Kraft, R. P. 1966, Ap. J. , 143, 299.

Athay, R. G. , and Skumanich, A. 1968, Ap. J. , in press.

Auman, J. , Jr. 1967, Ap. J. Suppl. , 14, 171.

Avrett, E. H. , and Strom, S. E. 1964, Ann. d'Ap. , 27, 781.

Bloch, M. , and Tcheng, M. -L. 1956, Publ. Obs. Lyon, 3, No. 24, 6 pp.

Böhm, K. H. 1968, presented at Low-Luminosity Stars Conf. , Charlottesville, Virginia, proc. to be published.

Böhm-Vitense, E. 1967, Ap. J. , 150, 483.

Carbon, D. , Gingerich, O. , and Latham, D. 1968, presented at Low-Luminosity Stars Conf. , Charlottesville, Virginia, proc. to be published.

Cayrel, R. 1966, J.Q.S.R.T. , 6, 621.

Chaffee, F. 1968, Ph. D. Thesis, University of Arizona.

Chandrasekhar, S. 1948, Proc. Roy. Soc. , Ser. A, 192, 508.

_____. 1950, Radiative Transfer (Oxford: Clarendon Press).

Christy, R. F. 1966, Ap. J. , 144, 108.

_____. 1967, Astron. J. , 72, 293 (abstract).

Collins, G. W. , II. 1963, Ap. J. , 138, 1134.

Collins, G. W. , II, and Harrington, J. P. 1966, Ap. J. , 146, 152.

Danziger, I. J. , and Dickens, R. J. 1967, Ap. J. , 149, 55.

Danziger, I. J. , and Oke, J. B. 1967, Ap. J. , 147, 151.

de Groot, M., and Underhill, A. B. 1964, Bull. Astron. Netherlands, 17, 280.

Dennis, T. R. 1968, Ap. J., 151, L47.

Dickens, R. J., Kraft, R. P., and Krzeminski, W. 1968, Astron. J., 73, 6.

Edmonds, F. N., Jr., Schlüter, H., and Wells, D. C., III. 1967, Mem. Roy. Astron. Soc., 71, 271.

Eggen, O. J., and Sandage, A. R. 1962, Ap. J., 136, 735.

Faulkner, J., Roxburgh, I. W., and Strittmatter, P. A. 1968, Ap. J., 151, 203.

Gaustad, J. E., and Spitzer, L., Jr. 1961, Ap. J., 134, 771.

Gingerich, O. 1964, Proc. First Harvard-Smithsonian Conf. on Stellar Atmospheres, Smithsonian Ap. Obs. Spec. Rep. No. 167, 17.

Gingerich, O., and Kumar, S. S. 1964, Astron. J., 69, 139 (abstract).

Gingerich, O. J., Latham, D. W., Linsky, J. L., and Kumar, S. S. 1967, Smithsonian Ap. Obs. Spec. Rep. No. 240, 42 pp.

Greenstein, G. S., Truran, J. W., and Cameron, A. G. W. 1967, Nature, 213, 871.

Griem, H. R. 1967, Ap. J., 147, 1092.

Guillaume, C., Van Rensbergen, W., and Underhill, A. B. 1965, Bull. Astron. Netherlands, 18, 106.

Guthrie, B. N. G. 1965, Publ. Roy. Obs. Edinburgh, 3, 261.

Hardorp, J., and Strittmatter, P. A. 1968a, Ap. J., 151, 1057.

——————. 1968b, Ap. J., 153, 465.

Hayes, D. S. 1967, Ph.D. Thesis, University of California at Los Angeles.

Hickok, F. R., and Morton, D. C. 1968, Ap. J., 152, 203.

Huang, S.-S., and Struve, O. 1956, Ap. J., 123, 231.

Hummer, D. G., and Mihalas, D. G. 1967, Ap. J., 150, L57.

Hunger, K., and Van Blerkom, D. 1967, Zs. f. Ap., 66, 185.

Jones, D. H. P. 1966, Bull. Roy. Obs., No. 126, E219.

Kalkofen, W. 1964, Proc. First Harvard-Smithsonian Conf. on Stellar Atmospheres, Smithsonian Ap. Obs. Spec. Rep. No. 167, 175.

——————. 1966, J.Q.S.R.T., 6, 633.

Kalkofen, W., and Strom, S. E. 1966, J.Q.S.R.T., 6, 653.

Klinglesmith, D. A. 1967, Astron. J., 72, 808 (abstract).

Kraft, R. P. 1965, Ap. J., 142, 681.

Kraft, R. P., and Wrubel, M. H. 1965, Ap. J., 142, 703.

Kuhi, L. V., and Danziger, I. J. 1967, Ap. J., 149, 47.

McGee, J. D., Khogali, A., Baum, W. A., and Kraft, R. P. 1967, Mon. Not. Roy. Astron. Soc., 137, 303.

Matsushima, S., and Terashita, Y. 1968, presented at Low-Luminosity Stars Conf., Charlottesville, Virginia, proc. to be published.

Mestel, L. 1965, in Stars and Stellar Systems, Vol. VIII, ed. L. H. Aller and D. B. McLaughlin (Chicago: University of Chicago Press), 297.

Mihalas, D. 1965, Ap. J., 141, 564.

——————. 1966, Ap. J. Suppl., 13, 1.

——————. 1967a, Ap. J., 149, 169.

——————. 1967b, Ap. J., 150, 909.

Mihalas, D. M., and Morton, D. C. 1965, Ap. J., 142, 253.

Mihalas, D., and Stone, M. E. 1968, Ap. J., 151, 293.

Morton, D. C., and Adams, T. F. 1968, Ap. J., 151, 611.

Münch, G. 1948, Ap. J., 108, 116.

Olson, E. C. 1968, Ap. J., 153, 187.

Ostriker, J. P., and Bodenheimer, P. 1968, Ap. J., 151, 1089.

Praderie, F. 1967, Ann. d'Ap., 30, 773.

——————. 1968a, Ann. d'Ap., 31, 15.

——————. 1968b, Ann. d'Ap., in press

Rich, J. C. 1966, Ph.D. Thesis, Harvard University.

——————. 1967, Ap. J., 148, 275.

Roxburgh, I. W. 1964, Mon. Not. Roy. Astron. Soc., 128, 157.
Roxburgh, I. W., Griffith, J. S., and Sweet, P. A. 1965, Zs.f.Ap., 61, 203.
Roxburgh, I. W., Sargent, W. L. W., and Strittmatter, P. A. 1966, Observatory, 86, 118.
Roxburgh, I. W., and Strittmatter, P. A. 1965, Zs.f.Ap., 63, 15.
Searle, L., and Oke, J. B. 1962, Ap.J., 135, 790.
Strom, S. E. 1967, Ap.J., 150, 637.
Strom, S. E., Gingerich, O., and Strom, K. M. 1966, Ap.J., 146, 880.
Strom, S. E., and Kalkofen, W. 1966, Ap.J., 144, 76.
_____. 1967, Ap.J., 149, 191.
Strom, S. E., and Kurucz, R. L. 1966, J.Q.S.R.T., 6, 591.
Strom, S. E., and Peterson, D. M. 1968, Ap.J., 152, 859.
Terashita, Y., and Matsushima, S. 1966, Ap.J. Suppl., 13, 461.
Underhill, A. B. 1963, Bull. Astron. Netherlands, 17, 161.
Underhill, A. B., and de Groot, M. 1964, Bull. Astron. Netherlands, 17, 453.
Wolff, S. C., Kuhi, L. V., and Hayes, D. 1968, Ap.J., 152, 871.

DISCUSSION

<u>Morton</u>: As I understand the latest information from Griem, it would appear that the new Griem theory and the ESW results are not in disagreement at all. According to Griem there is no difference between his theory and Schlüter's experiments.

<u>Strom</u>: That would not appear to be confirmed by my calculations.

<u>Morton</u>: But that's not the point. Your calculations, made on the basis of Griem's theory, have not been made for the temperatures and densities appropriate to Schlüter's experiments. When that is done, Griem claims that the experimental and theoretical results are in agreement. It doesn't appear that there is any strong astrophysical evidence that Griem is wrong.

<u>Strom</u>: From a practical point of view, I cannot overemphasize the importance of reducing the margin of error in our estimates of surface gravities from the hydrogen lines. The difference between the Griem and the ESW theories amounts to 0.2 in the log in the sense that Griem predicts that the surface gravity is smaller by this amount. When, for example, one is interested in obtaining masses of horizontal branch stars to determine whether mass loss has taken place, an error or indeterminacy in log g of this amount is quite serious.

<u>Underhill</u>: You have to be very careful that you have the continuous opacity as well as the line-absorption coefficient correct before you can make accurate estimates of surface gravity. Another thing you have to keep in mind is that not until far out in the wings, either in Hα or Hγ, do you get to the point where you can use simple line-formation theory. And you also have to worry about atmospheric motions as well.

<u>Strom</u>: I don't quite see the relevance of your remark. The Doppler width corresponding to the thermal velocity of the hydrogen atoms is on the order of 200 mA. The thermal velocity is on the order of 10 km sec^{-1}. To have any significant effects on line wings, the atmospheric motions you referred to would have to be considerably in excess of 10 km sec^{-1} or considerably in excess of the sound velocity.

Insofar as the mechanism of line formation is concerned, at least for the late B stars as Peterson will discuss later, an LTE line-formation theory is quite applicable at distances about an angstrom and greater from line center, so that looking at the wings beyond 1 A, LTE seems quite valid.

Mihalas: In reference to Strom's remark concerning the lack of discussion of observational results compared with the calculations by Stone and me, I should like to emphasize that the calculations were of an <u>exploratory</u> nature only. As Strom pointed out, we assumed detailed balance in the lines, which of course is dubious for the conditions under consideration. The singlet-triplet problem, incidentally, cannot even be discussed on the basis of the calculations by Stone and me. My final point is that the helium departures are strongly affected by those of hydrogen. Therefore, one must construct a model including departures in <u>both</u> hydrogen and helium.

Johnson: A graduate student and I at Indiana are currently working on the He I non-LTE problem. I had hoped to present some grand results and some sweeping conclusions regarding interpretation of the singlet-triplet ratios, but unfortunately the calculations are not yet complete. However, I can say that the departures that we estimate on the basis of the Kalkofen approximation are smaller than we had hoped for, and it looks as if it will be very difficult to get an explanation of the singlet-triplet problem.

Mihalas: But I would caution you again that a complete solution of this problem requires a careful calculation not only for helium but for hydrogen as well, since hydrogen, the primary opacity source, controls the model structure.

Underhill: I looked at this problem some time ago, and I decided that it was extremely hard. I am glad to see that you are working on it. I think you ought to realize, however, that when you try to predict the line profiles, the cores are important, for the classifications are based on the cores of the line, which are formed in the outer parts of the models.

Johnson: I agree with you that it is a difficult problem.

Elste: Wouldn't the addition of more levels in the treatment of the non-LTE problem reduce the departures even further?

Mihalas: No, I don't think so, certainly not for helium.

Strom: I might also point out that the difference between Mihalas' results and mine is not, as first thought, due to the inclusion of more levels but rather to an increase in the collision rates in the sense that Mihalas used larger values.

Elste: Won't the ultraviolet lines have an effect on the departures?

Strom: Yes, certainly. It depends on where the lines fall, in wavelength, with respect to the helium continuum edges. The lines drive B_ν closer to J_ν and thus tend to reduce the contribution to the integral over the continuum at the line wavelengths.

Cayrel: Do you still consider as valid your former statement [Strom and Kalkofen, <u>Ap. J.</u>, <u>149</u>, 191, 1967] that the ratio of the Paschen jump to the Balmer jump is a sensitive test for LTE in view of the fact that you used rather low values for the bound-bound collision rates at that time?

Strom: Yes, if you restrict your attention to stars of lower surface gravity than, say, about log g = 3. For main-sequence stars I no longer regard this as a sensitive test. However, I believe that the test involving hydrogen-line profiles, to be discussed later by Peterson, represents a more sensitive measure of departures.

Mihalas: My computations indicate that the Paschen-to-Balmer ratio is virtually unchanged from its LTE values for log g = 4, that is, for main-sequence stars. For lower log g's I do find a small but measurable effect. The fact that the Paschen-to-Balmer ratio does not depart from its LTE value for main-sequence stars is consistent with Hayes' photoelectric results. I would conclude that, unless the influence of the lines on the problem is much greater than we anticipate at present, these results will be virtually unchanged.

Houziaux: In this regard ($b_3 > 1$) I would like to comment on the enhanced strength that is found when one looks at the Paschen lines in early Be stars.

Following Strom's Presentation of Rotation, Section 2.2

Eggen: What are the effects of rotation on the location of stars in the color-color plot?

Strom: According to Collins' calculations the effects are very small in the sense that the directions of the vectors in the color-color diagram are along the standard nonrotating sequence. It was my impression, however, that Strittmatter and Hardorp got a slightly different result. They feel the ratio of total to selective absorption that one obtains for the B stars will be slightly affected by the presence of rotating stars; that is, the direction of the rotation vectors is not quite along the main sequence. Would Collins like to comment on this?

Collins: I should like to emphasize that my remarks on the location of stars in the color-color plot refer only to the B stars, and the results may be quite different for the A stars. I also feel that there is no predicted effect of rotation on the colors that is larger than the observational scatter. I think we have to admit the possibility of observational errors.

Strom: Of course, I think it's useful to point out that, although the models calculated by Strittmatter and Hardorp and by Collins and his group have been most valuable in pointing out the direction and approximate magnitude of the results, they are by no means the final word. For example, several refinements such as the inclusion of ultraviolet line blanketing are necessary before we can make a more detailed comment in regard to the comparison of observations with models.

Collins: I couldn't agree more. I'd be the last one to take such models very seriously. However, I am basically interested in predicting differential effects, and I think any systematic effects such as blanketing may be second order since the differential effects are quite small to begin with. I would tend not to worry about these effects too much. Also in regard to the departures from LTE, as you correctly point out this should only be important in the equatorial regions, but of course owing to the low luminosity at the equatorial regions, these areas on the star surface are not contributing too much to the total flux.

Strom: I'd agree with that, but certainly for a star viewed equator-on it might be of some importance.

Collins: I'd also like to correct one remark that you made concerning my estimates of the effects of rotation on spectral type. Anybody who ventures into the area of quantitative treatment of spectral classification is opening a Pandora's box. The original investigation I embarked on was based on scattering lines. However, if one then investigates what happens to lines formed by an absorption mechanism, the situation is not quite so pleasant. There is at least one line ratio, the helium-to-silicon line ratio, that I'm told is useful for classifying early B stars, that is adversely affected by rotation. Helium lines look stronger, silicon weaker, and therefore one might expect a change of a tenth of a spectral class, perhaps in the case of extreme rotation. Possibly some of the helium-weak B stars could be explained as follows: The helium lines may appear weak for a given spectral type, but in fact one chooses too early a spectral type (owing to the effects of rotation on the line spectrum) and therefore anticipates too strong a helium line.

Strittmatter: Dr. Strom has given a most comprehensive summary of the work by Collins and by Hardorp and myself on synthetic spectra for rapidly rotating stars. There is little new that I can add, though perhaps I might clarify one or two points raised by Dr. Strom.

The interior models of rapidly rotating stars, in which the circulation (Eddington) currents are set equal to zero and in which the nonuniform rotation field is then uniquely determined, have been shown by Goldreich and Schubert and, independently, by Fricke to be unstable. Equally, in the absence of special magnetic fields (or some other effective viscosity), the rotational velocity distribution will vary with time in any model in which circulation currents do not vanish. Thus, we do not have an entirely self-consistent time-independent set of interior models, although we hope the uniformly rotating models will give us a meaningful average description.

The estimates given by Hardorp and myself for the effects of atmospheric circulation currents (e.g., contributions to energy flux, effect on hydrostatic support approximation, etc.) were obtained by our using values of the circulation velocity applicable to a region in which the atmosphere is just ceasing to be opaque. The precise details of the flow in the atmosphere remain an unsolved problem.

The question of rotational effects in the two-color diagram has been raised. I think Collins and I agree that, for B stars, the maximum spread due to rotation is ~ 0.09 m in U-B at a given value of B-V. As Becker has shown, any such spread must increase the derived ratio of total to selective absorption. I should also like to point out that any attempt to correct colors for reddening by conventional methods will simultaneously reduce the rotational effects. One must be very careful about looking for the influence of rotation among stars in clusters in which differential reddening is suspected. In such cases, Balmer-line indices combined with Balmer-discontinuity measures will be more appropriate.

Houziaux: You mentioned that by using sharp-line stars you could reduce the error in the age as judged from the turnoff point, but how do you tell which ones of the sharp-line stars are pole-on and which are intrinsically slow rotators?

Strom: It doesn't make any difference. The whole point is that, whether a star is rapidly rotating seen pole-on or slowly rotating, the change in location in the color−magnitude diagram is only large enough to produce a change of about 25% in the estimated age.

Strittmatter: The age of clusters is usually determined by comparing the colors of stars at the evolutionary turnoff point with those obtained from studies of stellar interiors. For equator-on stars the reddening in B-V due to rotation could easily cause cluster ages to be overestimated by as much as a factor 3. For pole-on stars that remain at essentially the same color as nonrotating stars of the same mass, the cluster age could be underestimated by up to 25%. This arises from the reduction in total luminosity due to rotation, which thus extends the star's lifetime on the main sequence.

I should like to show a slide of the Coma cluster main sequence, which I think illustrates the point fairly well [Strittmatter and Sargent, Ap. J., 145, 135, Fig. 5, 1966]. The stars at the upper end of the main sequence seem

to provide a well-determined turnoff point.   However, their position in the
color-magnitude diagram is entirely consistent with their being more massive
stars in rapid rotation, which from their spectra they certainly are.   (This
is even true of the F0 shell star 14 Comae, which has a rotational velocity
v sin i ~ 280 km sec[-1].) On the other hand, the two stars 17 Comae and 21
Comae, which are usually referred to as "blue stragglers," may well provide
a better cluster age since, being Ap stars, they are presumably nonrotating.
The cluster could, therefore, be younger by a factor of 2 to 3 than currently
estimated from the turnoff point.   Such effects should not, of course, be
important in globular clusters since the stars at turnoff point are of later
spectral type and, hence, presumably slow rotators.

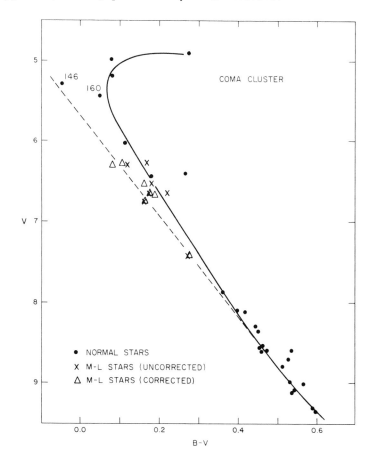

Color—magnitude diagram for the Coma cluster showing the effect of
correcting the colors of the Am stars.   The solid line is the sequence of
normal stars and the dashed line is the estimated zero-rotation main sequence.
Stars 146 and 160 are Ap stars with small v sin i.

Strömgren: From an observational point of view, of course, one would like
to be assured that the age determination from the narrow-band color observations are not affected by rotation.   It would appear at this point that one
has to be careful to eliminate the very rapid rotators but that one could live
with rotational velocities on the order of 100 or 200 km without seriously
affecting the age results.   The important question is really to make sure that

this is indeed the case and that at the end we have large-enough observational samples with which to work. If we have to restrict ourselves to the very slowly rotating stars, it will be a very difficult problem indeed.

Following Strom's Presentation of Sources of Opacity, Section 2.5

Morton: Does the high silicon abundance of Sirius explain why Hanbury Brown finds Sirius hotter by about 900° than Vega?

Strom: I don't know. I haven't worked through the effects of silicon opacity on the absolute fluxes in the visible that are used by Hanbury Brown and his collaborators. [Note added: Assuming Sirius has a normal silicon content, one deduces from Hanbury Brown's absolute flux that $T_{eff} \approx 10,270°K$; if silicon is enhanced by a factor of ten, $T_{eff} \approx 10,000°K$].

Underhill: Of course, when you talk about silicon-rich stars you're basing your deduction of high silicon abundances on rather strong lines. You have to be careful about interpretations based on these lines because such lines are formed in regions outside your model atmospheres and are affected by nonequilibrium processes.

Strom: I'd be very hard pressed to believe that departures from LTE could account for departures from 30 to 100 in the deduced silicon abundances.

Fischel: We have looked into the problem of Sirius from the point of view of interpreting the silicon discontinuity in the ultraviolet, and we believe that we find an overabundance of silicon. However, this result is based on the assumption that the Hanbury Brown temperature is the correct one. I should like to discuss this problem in detail later in the session. [Editor's note: The paper in question has been withdrawn from publication.]

Strom: Yes, I think that the result you get is consistent with the result that Gingerich and I obtained from an LTE analysis of the line spectrum. Also, as you will probably point out, the silicon discontinuities in the ultraviolet provide a mechanism for detecting silicon-rich stars.

Underhill: In reference to the weak-helium B-star problem, I think it's a problem of the line-formation mechanism. I just don't know whether you can say that these stars are truly underabundant by factors of 100.

Strom: It seems extremely difficult to see why stars that otherwise appear to have normal atmospheric properties (at least at large depths where the continuum is formed) should show such extreme departures from LTE in the helium lines. I don't think you can explain a factor of 100 this way. Moreover, the Population II halo B stars under discussion here have surface gravities 30 times those of normal main-sequence stars, and it would be very hard to explain a factor of 100 as a non-LTE effect in these high-density objects.

Underhill: Have you ever stopped to consider that the weak-helium stars might be the normal ones and the strong-helium stars might be the abnormal ones? It seems quite difficult to avoid the effects of departures from LTE.

Nariai: For white dwarfs for about 15,000° you state that the convection is not very important in affecting the temperature structure. You mean of course hydrogen-rich white dwarfs, but in the helium-rich white dwarfs couldn't there be an important helium ionization zone?

Strom: I hadn't considered that. It seems reasonable that that might very well be the case. It certainly is a point to be investigated.

Buscombe: Do we know any rotational velocities for white dwarfs?

Strom: I'm sorry, I can't answer that. Certainly the discrimination level would be on the order of 50 to 100 km sec^{-1}.

    Do the sharp temperature drops found for the surface layers of the helium-rich stars contain a hint as to the nature of the Garrison stars, that is, stars with B3 V colors but spectral type seemingly appropriate to B8 II-III that he discusses in the context of his study of the Sco associates?

Eggen: As I recall, they are probably just like any peculiar stars.

Crawford: Yes, it's a matter of the colors indicating B2 V or B3 V, but the spectral type is B 8p.

Following Strom's Discussion of the Middle A and Early F Stars, Section 11

Auman: I think it's of some relevance to point out that if one uses the mixing-length theory to estimate the temperature fluctuations at the top of the hydrogen ionization zone, one finds that they are quite large. I believe that this was first pointed out by Mrs. Böhm-Vitense [Zs.f.Ap., 46, 108, 1958]. In the late A dwarfs, $\Delta T/T$ at the top of the ionization zone is of the order of 0.1 or 0.2, while it may be as large as 0.4 or 0.5 in the giants and supergiants. The question is, how much do you want to believe these models when you have temperature fluctuations that large?

Strom: I think I'd agree with your remarks. I certainly feel that, given the current status of the convective theory, we can regard all examinations of the role of convection as purely exploratory to determine the approximate magnitude and direction of the effect.

Auman: But I want to emphasize that these temperature fluctuations are serious and the meaning of an average temperature for A and F giants is just not known. I consider this a serious defect of the currently accepted calculations. I don't know how we can do any better, but we should certainly be aware of these difficulties.

Praderie: In his paper, Strom mentioned the problem of the diffusion of atoms in an atmosphere. The possibility of modifying the chemical composition of a stellar atmosphere through a diffusion process was already considered by Eddington and by Wildt before 1940. More recently, Aller and Chapman [Ap. J., 132, 461, 1960] studied the influence of thermal diffusion and gravitational separation on the distribution of the elements at the base of the solar convection zone. In this region, indeed, the temperature and pressure gradients act so that they produce a differential motion of the heavy atoms relative to the bulk of material (i.e., hydrogen), and the atoms sink through the base of the convection zone. The work by Aller and Chapman has been extended by Delcroix [Mémoire de License, Univ. de Liège, 1967].

    On the other hand, Delache [Ann. d'Ap., 30, 827, 1967] has solved the problem of the stationary state in the transition zone between the solar chromosphere and corona, where the transport phenomena of the ions are superimposed on the presence of the solar wind; as one result, he accounts for the observed abundances in the solar corona.

    Considering the downward diffusion, whose effect should be to empty continuously the subphotospheric convective zone of its heavy elements, I want to point out that the physical conditions (temperature, density) computed in the framework of the mixing-length theory are drastically different at the bottom of the convection zone in the sun and in an A star. In the sun, ρ is of the order of 10^{-2} g cm^{-3}; in an A star ($T_{eff} = 8000°$K), $\rho \simeq 10^{-7}$ g cm^{-3}. Diffusion is therefore very rapid in the A star as compared with the sun, and the computed characteristic times of diffusion for all atoms except H and He are shorter than the lifetime of the star; this is not the case in the sun, as was first shown by Aller and Chapman. Although these authors mentioned the situation of very rapid diffusion in stars hotter than the sun, they object that one should find in such stars a dependence of the abundances on the atomic weight of the elements. In fact, one does find a dependence in Am stars, not on the atomic weight, but on the atomic number. Moreover, it is interesting to note that the elements Cu, Ni, Zn, Ga, Kr, Xe, and rare earths, which are overabundant in Am stars, or in Ap stars, or in both, are precisely among those that should have disappeared from the atmosphere in the shortest time. One is therefore tempted to conjecture that in the peculiar and metallic-line stars, either diffusion is prohibited or it is "washed" by some mixing mechanism that prevents the sinking of the heavier elements toward the interior; in the "normal" stars, on the contrary, the transport of ions acts to diminish continuously the abundances of the elements. As pointed out by Delache, a quasi-stationary theory of diffusion may not be at all valid in A stars.

Böhm-Vitense: I'd like to emphasize that it's not the absolute value of the turbulence that counts in the support term but rather the gradient of the turbulent velocity.

Strom: I agree completely.

Gray: I agree that it's possible and important to get good surface gravities from binaries, and I intend on Wednesday to discuss this question for Sirius, γ Virginis, and so on. I think we can get accuracies of 15 to 20% or 0.07 to 0.08 in the log.

Houziaux: You talk about Balmer and Paschen discontinuities. Would you care to define what you mean in terms of directly observable quantities?

Strom: No. [Laughter.] But I will anyway. Merely as a matter of convenience (although I realize that these quantities are not directly observable), I have presented the Balmer and Paschen jumps as predicted from line-free models in most of the cases discussed today. However, with no loss of generality these arguments could easily be made by replacing my D_B by a ratio of a flux measured at some point in the Balmer continuum to a flux at some point in the Paschen continuum. So while I used a very convenient definition from the point of view of the simple models, a useful observational parameter could equally well be employed with the same qualitative results.

Following Strom's Discussion of the Helium Content of the Subdwarfs, Section 12

Gray: If you calculate the radius for Groombridge 1830 using my method, you cannot avoid the conclusion that this star lies below the main sequence by a considerable amount.

Strom: This follows also from Cayrel's careful analysis.

TEMPERATURE CORRECTION PROCEDURES: AN INTEGRAL EQUATION METHOD

D. M. Peterson

Smithsonian Astrophysical Observatory and
Harvard College Observatory, Cambridge, Massachusetts[*]

ABSTRACT

An integral equation for the temperature correction has been derived. The equation, based on the Φ operator, is analogous to the integral equation, based on the Λ operator, derived by Böhm-Vitense, but avoids the difficulties of applying the large-depth boundary conditions.

The method has been found to be both powerful and practical in the calculation of model atmospheres.

1. INTRODUCTION

The computation of a model stellar atmosphere normally proceeds in an iterative manner. An initial guess is made of a temperature − optical depth $(T - \tau)$ relation. The pressure − optical depth relation is then obtained by integrating the equation of hydrostatic equilibrium. The integrated flux and depth derivative of the flux may then be calculated in order to determine whether flux constancy has been satisfied.

Since, in general, the provisional temperature structure will not yield the flux desired, an algorithm must be provided to relate errors in the flux at each depth to corrections to the initial T-τ relation. The model is considered converged when the relative errors in both the flux and flux derivative are sufficiently small (normally $\sim 10^{-3}$) at all depths.

The construction of the necessary algorithm has received considerable attention over the past two decades. Methods suggested by Avrett and Krook (1963), Lucy (1962), and Lecar (1963) have proved to be adequate for modern high-speed computers.

Recently, interest in the problem has been revived both from the desire to speed convergence and thereby reduce the cost of producing large grids of models, and from the increasing concern that the atmospheres have not been adequately driven to flux constancy at small depths. The latter point is of particular importance in the calculation of spectral lines.

In this paper I will discuss a temperature-correction procedure that is quite powerful, and at the same time meets the requirements of limited core memory and computation time.

2. THE Λ OPERATOR

Böhm-Vitense (1964) considered the equation defining the flux derivative:

$$\int_0^\infty d\nu \; k_\nu(\tau)[J_\nu(\tau) - B_\nu(\tau)] = \int_0^\infty d\nu \; k_\nu(\tau)[\Lambda - I]B_\nu(t) = \frac{dH(\tau)}{d\tau} \quad , \quad (1)$$

where Λ is the lambda operator, and I the identity operator. A necessary and sufficient condition for flux constancy is that this integral be zero.

---
[*]Now at Mount Wilson and Palomar Observatories, Pasadena, California.

A temperature structure will not, in general, satisfy this requirement. To obtain the correction to the temperature, Böhm-Vitense considered a first-order expansion of the source function:

$$B_\nu [T(\tau)] = B_\nu [T^0(\tau)] + \dot{B_\nu}[T^0(\tau)] \Delta T(\tau) \quad . \tag{2}$$

The temperature dependence of the opacities and the Λ operator was explicitly ignored.

Equation (2) is then substituted into the expression for the flux derivative to obtain the desired integral equation for the temperature correction $\Delta T(t)$:

$$\int_0^\infty d\nu \ k_\nu (\tau)[\Lambda - I] \dot{B_\nu} (t) \Delta T(t) = - \frac{d H^0(\tau)}{d\tau} \quad , \tag{3}$$

where the flux derivative is computed from the provisional temperature distribution.

The integral equation can be solved for the temperature corrections. However, this scheme has a major drawback in that it is difficult to apply the boundary conditions at large depths. The method, as formulated above, will force the flux to an asymptotic value as defined by the current large-depth T-τ relation. Since, in general, the asymptotic behavior of the temperature does not define a unique flux, the effective asymptotic flux will change as a function of iteration. This seriously impairs the convergence properties of the method.

3.   THE Φ OPERATOR

In order to overcome the difficulties and yet maintain the power and simplicity of the method suggested by Böhm-Vitense, we will consider the temperature correction procedure based on the Φ operator.

We write the condition for the flux constancy in the form

$$\int_0^\infty d\nu \ H_\nu = \int_0^\infty d\nu \ [\Phi] B_\nu \ [T(t)] = \cancel{H} \equiv \frac{\sigma}{4\pi} T^4_{eff} \quad . \tag{4}$$

Expanding the source function as in equation (2), we obtain

$$\int_0^\infty d\nu \ [\Phi] \dot{B_\nu} (t) \Delta T(t) = \cancel{H} - H(\tau) \equiv \Delta H(\tau) \quad . \tag{5}$$

This expression may be solved for the required temperature corrections.

4.   THE SOLUTION OF THE Φ EQUATION

While there are perhaps many methods that might be used to solve equation (5), I will describe below what seems to be the most straightforward technique.   This method depends on having the Φ operator pretabulated as a matrix operator on some previously chosen τ set.   Kurucz (1968) has discussed the problem of accurately pretabulating the Φ and Λ operators and their uses in the construction of model atmospheres, and has kindly made the matrices available for this investigation.

Given the matrix operator on the standard τ set, we must transform it to the appropriate monochromatic τ set at each frequency point.   This is accomplished (Kurucz, 1968) with the matrices that take functions on monochromatic set and map them onto the standard set, and vice versa. For example, let the superscripts (ex), (in) stand for functions on the monochromatic (external) and standard (internal) τ sets, respectively.   Then, to obtain the monochromatic flux, the source function is mapped onto the standard set by the matrix a:

$$B^{in} = a \ B^{ex} \ .\tag{6}$$

The flux on the standard set is then obtained by applying the pretabulated Φ operator:

$$H^{in} = \Phi^{in} B^{in} \ .\tag{7}$$

Finally, the flux is mapped onto the monochromatic set by the matrix β:

$$H^{ex} = \beta \ H^{in} \ .\tag{8}$$

Combining these operations,

$$H^{ex} = \beta \ \Phi^{in} a B^{ex} \ ,\tag{9}$$

we identify the appropriate monochromatic flux operator Φ^{ex} as

$$\Phi^{ex} \equiv \beta \ \Phi^{in} a \ .\tag{10}$$

With the Φ operator suitably mapped at each frequency point, the integration can be carried out as a simple summation with appropriate weights,

$$\int_0^\infty d\nu \rightarrow \sum_i W_i \ ,\tag{11}$$

and the matrix

$$\Gamma(\tau, t) \equiv \sum_i W_i [\Phi]_{\nu_i} \dot{B}_{\nu_i} (t) \tag{12}$$

(where $\dot{B}_{\nu_i} (t)$ is treated as a diagonal matrix) may be accumulated.

Equation (5) may now be written as a matrix equation,

$$\Gamma(\tau, t) \Delta T(t) = \Delta H(\tau) \quad , \tag{13}$$

and solved directly for ΔT:

$$\Delta T(t) = \Gamma^{-1}(\tau, t) \Delta H(\tau) \quad . \tag{14}$$

The same procedure can be used for solving the Λ operator equation.

5. RESULTS AND DISCUSSION

This technique has been tested on a variety of standard model atmospheres, including the gray model, a $10,000°(T_{eff})$ atmosphere with neutral hydrogen as the only source of opacity, and a $10,000°$, $\log g = 2$ atmosphere with neutral hydrogen and electron scattering opacities. In all cases the integral equation procedure corrected the models to a prescribed accuracy (3×10^{-3} error in both flux and flux derivative) in 30% fewer iterations than the Avrett-Krook procedure. A typical example, starting with 600% flux errors, converged in four iterations for the Φ method and in seven iterations for the Avrett-Krook method.

It should be mentioned that in the case of severe scattering it is necessary to account for the decoupling of the radiation field from the local temperature. This is accomplished by modifying the matrix operator

$$\Gamma(\tau, t) = \int_0^\infty d\nu [\Phi] [1 - \gamma] [1 - \gamma\Lambda]^{-1} \dot{B}_\nu \quad , \tag{15}$$

where

$$\gamma = \frac{\sigma}{k + \sigma} \tag{16}$$

is the fraction of the opacity due to monochromatic scattering. I have learned that Auman (1968) has independently developed essentially the same approach, and has used it successfully to construct model atmospheres for cool stars.

The author acknowledges extensive discussions with W. Kalkofen, R. L. Kurucz, and G. B. Rybicki. This research was completed during my tenure of Harvard University and Smithsonian Research Foundation fellowships.

REFERENCES

Auman, J. R. 1968, private communication.
Avrett, E. H., and Krook, M. 1963, Ap. J., 137, 874.
Böhm-Vitnese, E. 1964, Proc. First Harvard-Smithsonian Conf. on
    Stellar Atmospheres, Smithsonian Ap. Obs. Spec. Rep. No. 167, 99.
Kurucz, R. L. 1968, Ap. J., in press.
Lecar, M. 1963, Unpublished Ph.D. Thesis, Yale University.
Lucy, L. 1962, Unpublished Ph.D. Thesis, Manchester University.

DISCUSSION

Klinglesmith: Are these convergence criteria that you are talking about valid only at the surface or throughout the entire atmosphere?

Peterson: Throughout—that is, the maximum error throughout, with the exception of the last point.

A RELAXATION METHOD FOR THE COMPUTATION OF MODEL STELLAR ATMOSPHERES

M. J. Price

Kitt Peak National Observatory, Tucson, Arizona

A Monte Carlo relaxation method has been developed for the computation of model stellar atmospheres. The equations of statistical equilibrium for the atomic-level populations are solved simultaneously with the radiative-transfer and electron-temperature equations by use of the constraints of hydrostatic and radiative equilibrium. Selected LTE boundary conditions are applied deep in the atmosphere.

In the solution of the statistical-equilibrium equations, all possible electron collisional and radiative processes are considered. No assumptions are made concerning detailed balance for the bound-bound radiative processes. In formulating the source function, all atomic processes contributing to the absorption and emission coefficients are considered in microscopic detail. The local electron temperature is obtained by solving the kinetic-energy balance equation for the electron gas. All electron collisional and radiative processes whereby kinetic energy is gained and lost by the electron gas are considered. Thomson scattering is taken into account in formulating the equation of radiative transfer. The free-bound and bound-bound radiation is transferred throughout the atmosphere by use of the Monte Carlo technique.

The equilibrium physical state of the atmosphere is obtained, for fixed boundary conditions, by a relaxation technique. Boundary conditions of temperature T_0 and density ρ_0 are adopted at the lower base of the atmosphere. The atmosphere is divided into a number of homogeneous plane-parallel zones. A flux of radiation characteristic of a blackbody at temperature T_0 is fed into the atmosphere across the lower boundary. The atmosphere is initially (and arbitrarily) put into a stretched nonequilibrium physical condition, isothermal at the base boundary temperature and in LTE throughout. All atomic- and ionic-level populations are given by the Boltzmann and Saha equations at this stage in the relaxation. The radiant energy emitted per second from every part of the atmosphere is transferred by Monte Carlo source particles, which have the properties of photons. Escape of these particles from the top of the atmosphere causes the upper layers to cool and show departures from LTE in their atomic- and ionic-level populations. After the transfer is completed, the energy absorbed in each layer is used to redetermine the physical state of the atmosphere. The density and temperature profiles, and the local values for the atomic- and ionic-level populations and the electron density, are redetermined. The relaxation is continued until convergence of the temperature profile and of the local values of the atomic- and ionic-level populations and the electron density. The relaxation is extremely rapid. The reason for the stability of the method is that the physical state of the atmosphere is arbitrarily held constant during each iteration while the radiation is being transferred. No attempt is made to solve the time-dependent radiation-transfer problem. Convergence is obtained within three iterations, and the final solution suffers only from small statistical fluctuations due to the nature of the Monte Carlo process.

The computational method has been used successfully to compute a pure-hydrogen, high-temperature model atmosphere. A hydrogen atom with five bound levels plus the continuum was considered; T_0 was taken as 50,000°K for ρ_0 of 4.0×10^{-9} g cm^{-3}. Substantial departures of the atomic-level populations from their LTE values occur in the outermost layers of the atmosphere. Even at an optical depth $\tau_{4000} \simeq 1$, the lowest bound states show non-negligible departures from LTE. For that optical depth we have $b_1 \simeq 0.1$,

$b_2 \simeq 0.7$, $b_{3,4,5} \simeq 0.8$, and $b_{continuum} \simeq 1$. For small optical depths the lower bound levels become even more depopulated relative to the corresponding LTE values.

Details of the computational method and a discussion of the results of the model-atmosphere calculation will be published elsewhere.

DISCUSSION

<u>Whitney</u>: What is your boundary condition at great optical depths? Is it not an isothermal atmosphere with no flux?

<u>Price</u>: I have a lower base boundary in the atmosphere, above which there are 18 zones. Across the lower boundary, which I selected in the atmosphere at a density of 5×10^{-8}, I put in a flux of blackbody radiation characteristic of a temperature of 50,000°K.

<u>Whitney</u>: That is, you assume a black opaque surface at that temperature?

<u>Price</u>: Yes.

<u>Nariai</u>: In choosing the boundary condition, you must be careful to choose the base density sufficiently high.

<u>Collins</u>: How many photons do you run through?

<u>Price</u>: About 2 million photons. I transfer 36 continuum points and 10 lines. For each line I transfer one species of photon, or source particle, by defining a mean opacity, which depends upon the place where the source particle started and the place where it is absorbed.

<u>Auman</u>: I would worry a little about the use of such a small number of optical depths. This may create errors in the atmosphere, because using only one or two points each decade is probably too coarse a grid of optical depth points.

<u>Price</u>: I want to emphasize that this Monte Carlo method is very stable. I had to have a compromise between the number of zones and the homogeneity in each zone. I chose the run of pressure by assuming $\log (Pi/Pi+1) = 0.2$ between adjacent zones i and i + 1 and uniform density in each zone.

<u>Avrett</u>: Your solution satisfies the statistical-equilibrium and radiative-transfer equations and preserves constant flux.

<u>Price</u>: To within a few percent.

<u>Avrett</u>: Have you also solved the problem of imposing LTE?

<u>Price</u>: Yes, I put a lid on the top of my atmosphere, not allowing the photons to get out, and sure enough it stayed in LTE and all the level populations remained at their LTE values.

<u>Avrett</u>: You use a Monte Carlo method to solve simultaneously the equations of statistical equilibrium, radiative transfer, and radiative equilibrium. Investigations using standard methods have shown that this is a problem of enormous complexity with many difficulties and pitfalls. If the solutions you obtain are valid, then this approach is of very great importance. It must be established that your solutions would not qualitatively change if, for example, you used a greater number of depth points and more than a single frequency for each line. Direct comparisons could be made in limited cases with solutions obtained by other methods, e.g., with radiative-equilibrium solutions assuming LTE, or with non-LTE solutions assuming a given atmospheric structure. Do you plan to make such comparisons?

Price: You could do that, but I did not see why it was necessary. The program determines the structure.

Nariai: The following comments are based on an unpublished work made by Dr. Osaki, who is now at Columbia University, and by me a few years ago at Tokyo.

We thought that nongray model atmospheres can be obtained with the heat-flow equation

$$C_p \, \rho \, \frac{\delta T}{\delta t} = \int_0^\infty \kappa_\upsilon (J_\upsilon - B_\upsilon) \, d\upsilon \quad .$$

To understand how the temperature distribution changes, we calculated for a simple picket-fence model starting with a gray distribution. Then the equation becomes

$$\frac{\delta T}{\delta t} = \sum_{i=1}^{Z} \frac{\kappa_i}{C_p \, \rho} \, (J_i - B_i) \, a_i \quad ,$$

where the suffix i denotes the region, and a_i the fraction of the region.

The problem of this method is that, if we stick to small time intervals, the calculation would take a lot of computer time, but if we are too quick in lengthening the time interval, we may overcorrect the temperature.

It is clear that the estimation of time interval is an essential problem in this method. Therefore, it is hard for me to understand why this problem does not come into Price's formulation.

Price: Unlike Nariai, I have not set up the stellar-atmosphere problem as time-dependent. Consequently, no time-dependent terms appear in my formulation. The relaxation technique I have used is merely a mathematical method for obtaining a solution to the steady-state atmosphere problem. While I start from an initial nonequilibrium physical state, I do not follow the relaxation of the atmosphere in time. During each iteration the physical state is arbitrarily held constant while the radiation is flowing through the atmosphere. This arbitrarily imposed constraint is the essential feature of the relaxation technique. The relaxation is completed within three iterations from the initial arbitrarily selected isothermal LTE state. The solution is stable thereafter.

OPACITY CROSS SECTIONS FOR He, C, N, O, and H⁻

S. P. Tarafdar and M. S. Vardya

Tata Institute of Fundamental Research, Bombay, India

1. INTRODUCTION

Peculiar stars frequently show enrichment in the abundance of He, C, N, and O, as well as of other elements, along with a deficiency of H, as compared to the solar photosphere. Most of these stars are giants or supergiants. In the atmosphere of these stars, Rayleigh scattering may be an important source of opacity. We have calculated Rayleigh-scattering cross sections for these elements and compared them with the corresponding absorption coefficients.

We also discuss some preliminary results on the effect of pressure ionization on H⁻ opacity and on the forbidden continuum produced by H⁻.

2. RAYLEIGH-SCATTERING CROSS SECTIONS OF He, C, N, AND O

The Rayleigh-scattering cross section σ_λ for wavelength λ (in centimeters) can be written (e.g., see Griem, 1964, p. 35)

$$\sigma_\lambda = \frac{128\,\pi^5}{3} \frac{a^2}{\lambda^4} \,. \tag{1}$$

Here the polarizability a (in cubic centimeters) is the sum of a_d, the contribution from the discrete excited levels, and a_c, the contribution from the continuum states. In the limiting case where the wavelength of the scattered light is longer than the longest wavelength emitted by the atom from a fixed state, we can write (Shore and Menzel, 1965; Bethe and Salpeter, 1957; Condon and Shortley, 1935)

$$a_d = \frac{4}{3} a_0^3 \sum_i \frac{R_i^2 f_i^2}{(2L+1)\,E_i} \left(1 + \frac{x^2}{\lambda^2 E_i^2} + \frac{x^4}{\lambda^4 E_i^4} + \cdots\right) \tag{2}$$

and

$$a_c = \frac{4 \times 10^{18}}{8.067} a_0^3 \int_{E_{ion}}^{\infty} \frac{a\,dE_i}{E_i^2} \left(1 + \frac{x^2}{\lambda^2 E_i^2} + \frac{x^4}{\lambda^2 E_i^4} + \cdots\right) \,, \tag{3}$$

where

a_0 = the radius of the first Bohr orbit in centimeters,

a = the photoionization cross section in square centimeters,

E_i = the energy in Rydbergs of the ith level with respect to the ground level,

143

R_i = the multiplet factor,

L   = total orbital angular momentum of the initial state,

E_{ion} = the energy in Rydbergs corresponding to the series limit,

x   = $4\pi a_0/a'$, where a' is the fine-structure constant,

$\mathcal{f}_i$ = the radial factor defined in the dipole length formalism by

$$\mathcal{f}_i(\ell - \ell') = \sqrt{\ell}_> \, (-1)^{\ell_> - 1} \int_0^\infty R_{0\ell} R_{i\ell'} \, r \, dr \quad , \tag{4}$$

and in the dipole velocity formalism by

$$\mathcal{f}_i(\ell - \ell') = \sqrt{\ell}_> \, (-1)^{\ell_> - 1} \frac{2}{E_i} \int_0^\infty R_{i\ell'} \left\{ \frac{dR_{0\ell}}{dr} + \left[(\ell - \ell')(\ell_> + \ell - \ell') - 1 \right] \frac{R_{0\ell}}{r} \right\} \, dr \quad . \tag{5}$$

Here, $\ell_>$ is the larger of the two orbital quantum numbers ℓ and ℓ' of the jumping electron, and $R_{0\ell}/r$ and $R_{i\ell'}/r$ are the radial wave functions for the initial and intermediate states, respectively.

We have assumed the atoms to be initially in the ground state. The energy values of various levels are taken from Moore (1949) for He, N, and O, and from Johansson (1966) for C. The photoionization cross sections for He are taken from Stewart and Webb (1963), for C from Praderie (1964), for N from Bates and Seaton (1949), and for O from Dalgarno, Henry, and Stewart (1964). The multiplet factors of different transitions have been calculated following the method of Shore and Menzel (1965). The radial factors have been calculated from equations (4) and (5) with wave functions of the form used by Praderie (1964) with slight modification. The wave functions used behave as $r^{\ell+1}$ near the origin.

The Rayleigh-scattering cross section can be expressed as

$$\sigma_\lambda = \frac{A}{\lambda^4} \left(1 + \frac{B}{\lambda^2} + \frac{C}{\lambda^4} + \cdots \right) \quad , \tag{6}$$

where A, B, C for different elements are as shown in Table 1. The fairly good agreement between the values obtained by length and by velocity formalisms gives us some confidence in the wave functions used and, hence, in the computed cross sections. The Rayleigh-scattering cross sections of H, He, C, N, and O are approximately in the ratio of 9 : 1 : 108 : 46 : 27 at long wavelengths. Figure 1 shows the variation of σ_λ with λ at short wavelengths.

It is worthwhile to compare the scattering cross sections of these elements with their corresponding photoionization cross sections at a few wavelengths and temperatures. The photoionization cross sections for C, N, and O have been extrapolated from the results given by Peach (1967); and for He, from Vardya (1964). The results are presented in Table 2, which shows that the Rayleigh scattering may be an important source of the opacity in the atmosphere of low-temperature peculiar stars. These cross sections should also be compared with electron scattering and with the absorption due to negative ions. Electron scattering should be negligible for low-temperature stars except at very long wavelengths and at very low pressures. The contribution from negative ions should be considered, though it is difficult to state its relative importance without the electron pressure being specified as well.

Table 1.   Coefficients of Rayleigh-scattering formula [equation (6)]

Element		A	B	C
He	Length	6.138(-46)	0.483(-10)	0.191(-20)
	Velocity	5.472(-46)	0.473(-10)	0.184(-20)
C	Length	6.653(-44)	3.086(-10)	8.515(-20)
	Velocity	6.583(-44)	3.260(-10)	9.257(-20)
N	Length	2.819(-44)	2.034(-10)	3.485(-20)
	Velocity	2.897(-44)	2.032(-10)	3.449(-20)
O	Length	1.671(-44)	1.416(-10)	1.945(-20)
	Velocity	1.433(-44)	1.352(-10)	1.852(-20)

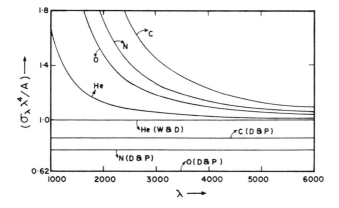

Figure 1.   Departure of the Rayleigh-scattering cross section at short wavelengths from the $1/\lambda^4$ law is plotted as a function of $\lambda(A)$. The ratios of the Rayleigh-scattering cross section for C, N, O, and He obtained from static polarizabilities to the corresponding A/λ^4 law are plotted as a comparison. D & P: computed by Dalgarno and Parkinson (1959); W & D: computed by Wikner and Das (1957).

Table 2.   Comparison of Rayleigh-scattering and photoionization cross
          sections

Elements	Temperature (°K)	Wavelength (A)	Rayleigh-scattering cross section (cm²)	Photoionization cross section (cm²)
He	20160	3953	2.590(-28)	1.423(-22)
	10080	3953	2.590(-28)	1.281(-28)
	10080	7038	2.525(-29)	6.508(-28)
	7000	3953	2.590(-28)	6.594(-34)
	7000	7038	2.525(-29)	3.591(-33)
C	7000	4100	2.856(-26)	5.166(-24)
	4000	4100	2.856(-26)	2.453(-29)
	7000	6000	5.606(-27)	5.569(-24)
	4000	6000	5.606(-27)	2.453(-29)
N	7000	4192	1.028(-26)	9.795(-26)
	4000	4192	1.028(-26)	3.125(-32)
	7000	6000	2.303(-27)	8.814(-26)
	4000	6000	2.303(-27)	9.731(-33)
O	7000	4000	7.157(-27)	1.189(-25)
	4000	4000	7.157(-27)	1.135(-31)
	7000	6000	1.341(-27)	7.525(-26)
	4000	6000	1.341(-27)	1.724(-32)

3.   PRESSURE IONIZATION AND H⁻ ABSORPTION

In the atmospheres of red dwarf stars, H^- is an important source of
opacity. All the calculations that have been done (see, e.g., Geltman, 1962;
John, 1960; Doughty and Fraser, 1964) are based on wave functions belonging
to single atoms. This simple picture is not really valid, especially in the
very cool red dwarf stars, where the densities are rather high. We have here
attempted to compute approximately the effect of pressure ionization on the
H^- opacity, making use of the strong-electrolyte approach of Debye.

The Hamiltonian, H, for a two-electron system, in atomic units
(energy = 2 Ry), can be written as

$$H = -\frac{1}{2} (\nabla_1^2 + \nabla_2^2) - \frac{Z \exp(-r_1/\lambda_D)}{r_1} - \frac{Z \exp(-r_2/\lambda_D)}{r_2} + \frac{1}{r_{12}} \quad , \quad (7)$$

where subscripts 1 and 2 refer to electrons one and two, Z is the nuclear
charge equal to unity for H^-, and λ_D is the Debye length, given by

$$\lambda_D^{-1} = \left(\frac{4\pi e^2}{KT} \Sigma N_a Z_a^2 \right)^{1/2} a_0 \quad , \quad (8)$$

where T is the temperature, K the Boltzmann constant, e the charge of an electron, a_0 the radius of the first Bohr orbit, and N_a the number density of the species of charge $Z_a e$. Note that in equation (7) we have replaced the coulomb potential, Z/r, by the Debye potential, $Z \exp(-r/\lambda_D)/r$, to take into account the effect of other particles.

Let us assume that the ground-state wave function, ψ_d, is of the form

$$\psi_d = \frac{P(r_1) \, P(r_2)}{r_1 \, r_2} \quad , \tag{9}$$

where

$$P^2(r) = (\tfrac{1}{2}) \, \beta^3 \, r^2 \, \exp(-\beta r) \quad , \tag{10}$$

so that P is normalized. The energy, E, of the system is then given by

$$E = \frac{\int \psi_d^* \, H \, \psi_d \, d\tau}{\int \psi_d^* \, \psi_d \, d\tau} = 2\mathcal{J} + F_0 \quad , \tag{11}$$

where

$$2\mathcal{J} = -\int_0^\infty P(r) \left[P''(r) + \frac{2 \, Z \exp(-r/\lambda_D)}{r} P(r) \right] dr \tag{12}$$

$$= \frac{1}{4} \beta^2 - \frac{Z \beta^3}{(\lambda_D^{-1} + \beta)^2} \quad , \tag{13}$$

and

$$F_0 = \int \frac{P^2(r_1)}{4 \pi \, r_1^2} \left[\int \frac{1}{r_{12}} \frac{P^2(r_2)}{4 \pi \, r_2^2} \, d\tau_2 \right] d\tau_1 \tag{14}$$

$$= (\tfrac{5}{16}) \, \beta \quad . \tag{15}$$

We have made use of equation (10) in deriving equations (13) and (15). Now we can write

$$E = \frac{1}{4} \beta^2 - \frac{Z \beta^2}{(\lambda_D^{-1} + \beta)^2} + \frac{5}{16} \beta \quad . \tag{16}$$

Minimizing the energy with respect to β gives

$$\beta^4 + \beta^3 (3 \lambda_D^{-1} + \frac{5}{8} - 2 Z) + \beta^2 (3 \lambda_D^{-1} + \frac{15}{8} - 6 Z) \lambda_D^{-1}$$

$$+ \beta (\lambda_D^{-1} + \frac{15}{8}) \lambda_D^{-2} + \frac{5}{8} \lambda_D^{-3} = 0 \quad . \tag{17}$$

Equation (17) has no positive real root if

$$\lambda_D \leq \frac{1}{2 Z - 5/8} \quad . \tag{18}$$

This means that for the condition (18), the ground state merges with the continuum.

For $\lambda_D \gg 1$, equation (17) can be approximated to

$$(\beta + 3 \lambda_D^{-1})(\beta + \frac{5}{8} - 2 Z) = 0 \quad . \tag{19}$$

This gives the positive root for β as

$$\beta = 2 Z - \frac{5}{8} \quad . \tag{20}$$

This is independent of λ_D . Thus, the ground-state wave function is not disturbed in the first-order perturbation. The shift, ΔE, of the ground-state energy due to pressure can be written as

$$\Delta E = E - E_{\lambda_D = \infty} = - \frac{Z \beta^3}{(\lambda_D^{-1} + \beta)^2} + \frac{Z \beta^3}{\beta^2} \simeq 2 Z \lambda_D^{-1} \quad , \tag{21}$$

if higher powers are neglected.
The shift, ΔI, in the ionization energy in Rydbergs is then given by

$$\Delta I = - 2 \left(E - E_{\lambda_D = \infty} \right) \simeq - 4 Z \lambda_D^{-1} \quad . \tag{22}$$

If we let (N_{H^-}) and $(N_{H^-})_0$ be the number density of H^- ions when λ_D is finite and tends to infinity, respectively, then

$$\frac{N_{H^-}}{(N_{H^-})_0} = \exp\left(\frac{\Delta I}{KT}\right) = \exp\left(\frac{-4Z}{\lambda_D KT}\right) \quad . \tag{23}$$

The photoionization cross section, a_ν, in the dipole length formalism can be written as

$$a_\lambda = 6.812 \times 10^{-20} \, k(k^2 + I) \left| \int \psi_d(Z_1 + Z_2) \, \psi_c \, d\tau \right|^2 \, cm^2 \quad . \tag{24}$$

Here, ψ_d as given by equation (9) and

$$\psi_c = \frac{1}{\sqrt{2\pi}} \left[\exp(-r_1 + ikZ_2) + \exp(-r_2 + ikZ_1) \right] \tag{25}$$

defines the wavefunctions for the ground state of H$^-$ and for the continuum, respectively; k is the momentum (atomic units) of the ejected electron, and I is the electron affinity in Rydbergs.

With the help of equations (9) and (25), equation (24) reduces to

$$a_\lambda = 6.812 \times 10^{-20} \, k \, (k^2 + I) \frac{(4\beta)^2 \, (8\pi\beta^2)^3 \, k^2}{\left(1 + \frac{\beta}{2}\right)^6 \left(k^2 + \frac{\beta^2}{4}\right)^6} \quad . \tag{26}$$

If we let a_λ and $(a_\lambda)_0$ be the cross sections when λ_D is finite and tends to infinity, respectively, then,

$$\frac{a_\lambda}{(a_\lambda)_0} = \frac{k^3(k^2 + I)}{\left(k^2 + \frac{\beta^2}{4}\right)^6} \cdot \frac{\left(k_0^2 + \frac{\beta^2}{4}\right)^6}{k_0^3 \left(k_0^2 + I_0\right)} \quad , \tag{27}$$

where k and k_0 are the momenta of the ejected electron, and I and I_0 the electron affinities when λ_D is finite and tends to infinity, respectively. As the wavelength, $\lambda(A)$, of the absorbed photon is the same in the two cases, we have

$$k^2 + I = k_0^2 + I_0 = \frac{911.3}{\lambda} \quad . \tag{28}$$

This gives, with the help of equation (22),

$$k = (k_0^2 + 4Z\lambda_D^{-1})^{1/2} \simeq k_0 \left(1 + \frac{2Z\lambda_D^{-1}}{k_0^2}\right) \quad , \text{ if } k^2 \gg 4Z\lambda_D^{-1} \quad . \tag{29}$$

Equation (27) can now be approximated to

$$\frac{a_\lambda}{(a_\lambda)_0} \simeq 1 + 6\, Z\, \lambda_D^{-1} \left[\frac{1}{k_0^2} - \frac{4}{k_0^2 + (\beta^2/4)} \right] \; . \tag{30}$$

The ratio of absorption coefficients, $K_\lambda = a_\lambda N_{H^-}$ with and without pressure effect, can now be written as

$$\frac{K_\lambda}{(K_\lambda)_0} = \left\{ 1 + 2\, Z\, \lambda_D^{-1} \left[\frac{3}{k_0^2} - \frac{12}{k_0^2 + (\beta^2/4)} - \frac{2}{KT} \right] \right\} \; . \tag{31}$$

The behavior of the ratio $a_\lambda/(a_\lambda)_0$ as a function of k_0^2 (and λ) is shown in Figure 2, and that of $K_\lambda/(K_\lambda)_0$ in Figure 3, for three values of λ_D at $T = 4737\,^\circ K$.

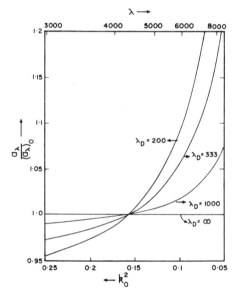

Figure 2. Photodetachment cross section of H^- at various values of λ_D, relative to that at $\lambda_D = \infty$, as a function of $\lambda(A)$.

The horizontal line, for which these ratios are unity, corresponds to $\lambda_D = \infty$, i.e., no pressure effect. The pressure ionization decreases the absorption coefficient of H^- at short wavelengths and increases it at long wavelengths. The critical value of the wavelength at which the ratio becomes unity is independent of pressure in the first approximation but shifts to shorter wavelengths with increasing temperatures.

The results obtained here are rather crude, and attempts are being made to improve them by using better wavefunctions for the ground state of H^-, though we do not expect any drastic qualitative change in the results.

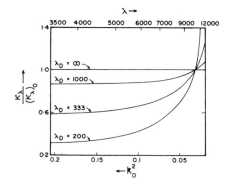

Figure 3. Absorption coefficient of H⁻ at various values of λ_D, relative to that at $\lambda_D = \infty$, as a function of $\lambda(A)$.

4. FORBIDDEN CONTINUUM PRODUCED BY H⁻

Recently, Weinberg and Berry (1966) considered the photodetachment of electrons from H⁻ due to local electric fields in an ionized plasma and computed the cross section for detachment, using a binary collision model. This is essentially equivalent to considering the effect of pressure on the photodetachment cross sections. We have looked into this interesting investigation but find that we disagree with their results. With the help of Dr. Weinberg, we are investigating this discrepancy.

REFERENCES

Bates, D. R. , and Seaton, M. J. 1949, Mon. Not. Roy. Astron. Soc. , 109, 698.
Bethe, H. A. , and Salpeter, E. E. 1957, Handbuch der Phys. , 35, 88.
Condon, E. U. , and Shortley, G. H. 1935, The Theory of Atomic Spectra (Cambridge: Cambridge University Press).
Dalgarno, A. , Henry, R. J. W. , and Stewart, A. L. 1964, Planet. Space Sci. , 12, 235.
Dalgarno, A. , and Parkinson, D. 1959, Proc. Roy. Soc. , Ser. A, 250, 422.
Doughty, N. A. , and Fraser, P. A. 1964, in Atomic Collision Processes, Proc. Third Intern. Conf. on Phys. of Electronics and Atomic Collisions, ed. M. R. C. McDowell (Amsterdam: North-Holland Publ. Co.), 527.
Geltman, S. 1962, Ap. J. , 136, 935.
Griem, H. R. 1964, Plasma Spectroscopy (New York: McGraw-Hill Book Co.).
Johansson, L. 1966, Arkiv. f. Fysik, 31, 201.
John, T. L. 1960, Ap. J. , 131, 743; Mon. Not. Roy. Astron. Soc. , 121, 41.
Moore, C. E. 1949, Atomic Energy Levels, Vol. I, NBS Circular 467 (Washington, D. C. : U. S. Government Printing Office).
Peach, G. 1967, Mem. Roy. Astron. Soc. , 71, 1.
Praderie, F. 1964, Ann. d'Ap. , 27, 129.
Shore, B. W. , and Menzel, D. H. 1965, Ap. J. Suppl. , 12, 187.
Stewart, A. L. , and Webb, T. G. 1963, Proc. Phys. Soc. , 82, 532.
Vardya, M. S. 1964, Ap. J. Suppl. , 8, 277.
Weinberg, M. , and Berry, R. S. 1966, Phys. Rev. , 144, 75.
Wikner, E. G. , and Das, T. P. 1957, Phys. Rev. , 107, 497.

DISCUSSION

<u>Auman</u>: How high a pressure do you have to have before the influence of pressure ionization on the H⁻ opacity appears to be important? Do you have numerical values?

<u>Vardya</u>: At about 6000 A and with an electron number density of about 2×10^{17} cm^{-3} and a temperature of about 4700°K, the ratio of absorption coefficient with and without pressure ionization effect is about 0.4.

A SERIES OF MODEL ATMOSPHERES WITH EFFECTIVE
TEMPERATURE 6000°K

V. P. Myerscough

Mathematics Department, Queen Mary College,
University of London, Mile End Road, London

ABSTRACT

A grid of model atmospheres for an effective temperature of
6000°K has been computed to permit consideration of the general
effects of various abundances and surface gravities on the opacity
and stratification.

A series of model atmospheres has been computed for an effective temperature of 6000°K and varying abundances and surface gravities, under the
approximations of LTE and purely continuous absorption and scattering. The
initial temperature law used in the radiative region was the approximation of
Henyey (1967); radiation pressure was neglected in the hydrostatic equation;
and the convective regions were taken into account by solution of the appropriate equation for the temperature in the superadiabatic layers with the use
of mixing-length theory (Hofmeister, Kippenhahn, and Weigert, 1964). The
aim of the models was to consider the general effects of various abundances
and gravities on the opacity and stratification, and also on the surface flux in
the red and infrared that depends little on line blanketing. The total integrated fluxes for all the models were within a few percent of the flux constant
$\sigma_R T_e^4$, and hence the models were not iterated for this general analysis.

Four series of models were computed; the atmospheric parameters are
shown in Table 1. The model al with $g = 10^4$ and solar abundances is in
quite good agreement in stratification with the essentially exact model of
Gingerich (distributed to participants before the conference opened) for
$\tau < 3.0$, below which convection becomes important. The surface flux for
this model is also within a few percent of that derived by Gingerich for
$\lambda > 5000$ A, and these results indicate that the model is fairly reliable.

The relevant opacities are shown in Figure 1; the new source of opacity
in the deeper layers is the forbidden continuum of H^-, which arises from the
distortion of the H^- ion in the field of a positive ion. The cross section for
this forbidden continuum has been calculated by Weinberg and Berry (1966);
it is roughly a constant function of temperature between 6000°K and 8000°K
for fixed ion density, is linear in positive ion density, and has about the same
value as the allowed cross section at λ 8201 A for an ion density of 5×10^{14}
cm^{-3}. However, since the forbidden continuum does not contribute to the
opacity till fairly deep in the model, it has no significant effect on the output
flux (Myerscough, 1968).

In the models with lower values of g, hydrogen becomes relatively more
important as an opacity source as the pressure decreases and the convection
begins progressively deeper in the atmosphere. Figure 2 shows the contributions to the opacity at optical depth unity for the model a3 with $g = 10^2$ and
solar abundances. For lower values of g, the hydrostatic equation begins to
break down in the superadiabatic layers, radiation pressure becomes significant, and mixing-length theory appears inadequate since the convective
velocities tend toward supersonic values. The surface flux in the red and
infrared remains very much the same for all four models in series (a), except
for slight enhancements of the magnitude decrements at the various hydrogen

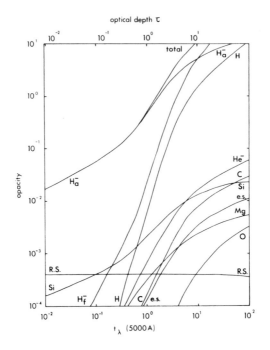

Figure 1. Contributions to the opacity (cm^2 g^{-1}) as a function of monochromatic optical depth t_λ at λ 5000 A (lower scale) and mean optical depth τ (upper scale) for the model a1 with g = 10^4 and solar abundances.   R.S. = Rayleigh scattering; e.s. = electron scattering; H_a^- = allowed continuum; H_f^- = forbidden continuum.

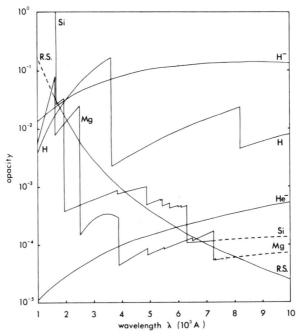

Figure 2. Contributions to the opacity (cm^2 g^{-1}) at optical depth τ = 1.0 as functions of wavelength for the model a3 with g = 10^2 and solar abundances.

ionization thresholds. The results for the models in series (b) are qualitatively similar to those for series (a), and again the surface flux in the red and infrared is fairly constant for the various values of g.

Table 1. Models with effective temperature 6000°K

Series (a). Solar abundances

$$\frac{n(He)}{n(H)} = 0.1 \qquad \frac{n(Fe)}{n(H)} = 3.72 \times 10^{-6}$$

Series (b). Reduced hydrogen and metal content

$$\frac{n(He)}{n(H)} = 1.0 \qquad \frac{n(Fe)}{n(H)} = 3.72 \times 10^{-6}$$

Series (c). Helium stars

$$\frac{n(He)}{n(H)} = 100.0 \qquad \frac{n(Fe)}{n(He)} = 3.72 \times 10^{-5}$$

Series (d). Helium-carbon stars (RCrB type)

$$X = 5.00 \times 10^{-4} \qquad Y = 9.10 \times 10^{-1}$$

$$Z = 0.09 , \qquad Z_c = 0.75 Z \quad (Searle, 1961)$$

Tabular values of g

g:	10^4	10^3	10^2	10^1	10^0
model:	a1	a2	a3	a4	
	b1	b2	b3	b4	
		c1	c2	c3	c4
		d1	d2	d3	d4

The theoretical removal of most of the hydrogen from the model atmospheres causes lower opacities and higher temperature gradients in the superadiabatic regions and, consequently, the effective ionization of H, C, N, and O nearer the surface. The opacities for the model c1 with $g = 10^3$ and a helium-type atmosphere are shown in Figure 3; on the basis of this model a small percentage of the flux is transported by convection in the H-C-O ionization region before helium ionization sets in. For lower values of g, this effect disappears completely, and the models for helium supergiants remain completely radiative to optical depths of about 20.0 because of the low opacities. Figure 4 shows the contributions to the opacity at optical depth unity for the model c2 with $g = 10^2$ and a helium-type atmosphere.

The effects discussed above are even more enhanced in the d series of helium-carbon atmospheres, as shown in Figure 5; in the models with higher values of g, there is a small amount of flux transported by convection in the carbon ionization region, but the models remain essentially completely radiative for lower values of g. The behavior of the stratification parameters p_g, p_e, κ as functions of optical depth τ is shown in Figures 6, 7, and 8 for the four models in series (d). The surface flux in the red and infrared is again constant for both series (c) and (d); however, as can be seen from Table 2, the behavior of the surface flux for these models is somewhat different from that for series (a) and (b), and this presents an interesting result from the point of view of comparison with scanner data.

Table 2. Output flux (erg cm^{-2} sec^{-1} Hz^{-1}) for models with $T_{eff} = 6000°K$, g = 10^2

| Wavelength | Model | | | |
	a3	b3	c2	d2
32806 R	3.041-5	2.986-5	2.722-5	2.689-5
32806 V	3.036-5	2.983-5	2.722-5	2.689-5
22782 R	5.684-5	5.559-5	4.930-5	4.835-5
22782 V	5.646-5	5.536-5	4.930-5	4.835-5
16419	9.468-5	9.253-5	8.040-5	7.816-5
14580 R	1.021-4	1.004-4	9.405-5	9.154-5
14580 V	1.005-4	9.924-5	9.401-5	9.154-5
9893	1.109-4	1.106-4	1.414-4	1.424-4
8201 R	1.114-4	1.113-4	1.557-4	1.467-4
8201 V	1.097-4	1.099-4	1.549-4	1.466-4
7961 R	1.093-4	1.095-4	1.529-4	1.432-4
7961 V	1.093-4	1.095-4	1.495-4	1.361-4
5917 R	9.726-5	9.742-5	9.760-5	1.010-4
5917 V	9.726-5	9.742-5	9.753-5	1.008-4
4967 R	8.336-5	8.350-5	7.635-5	7.913-5
4967 V	8.336-5	8.350-5	7.616-5	7.858-5
3646 R	5.199-5	5.216-5	2.078-5	2.572-5
3646 V	2.835-5	3.014-5	2.021-5	2.556-5
3461 R	2.437-5	2.604-5	1.691-5	2.166-5
3461 V	2.437-5	2.604-5	1.678-5	2.078-5

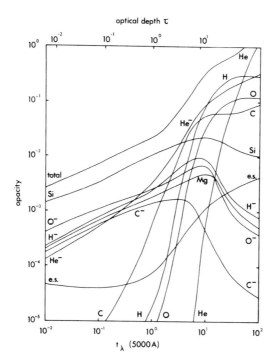

Figure 3.   Contributions to the opacity for the model c1 with
g = 10^3 and helium-type atmosphere.

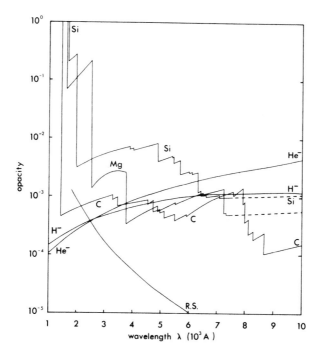

Figure 4.   Contributions to the opacity for the model c2 with
g = 10^2 and helium-type atmosphere.

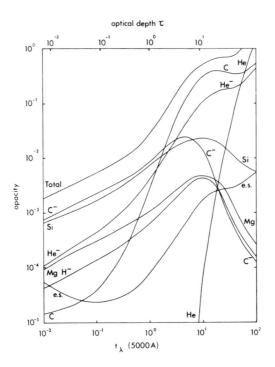

Figure 5.   Contributions to the opacity for the model d1 with
$g = 10^3$ and helium-carbon atmosphere.

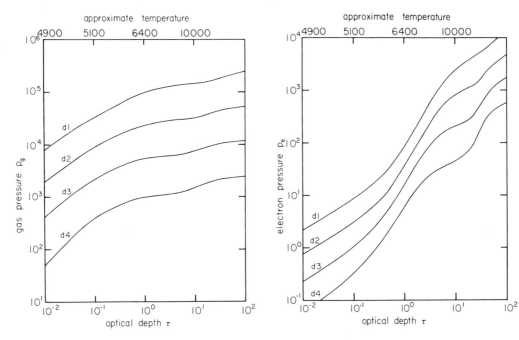

Figure 6.   Gas pressure p_g (dynes
cm^{-2}) for the series (d) of
helium-carbon atmospheres.

Figure 7.   Electron pressure p_e
(dynes cm^{-2}) for the
series (d) of helium-
carbon atmospheres.

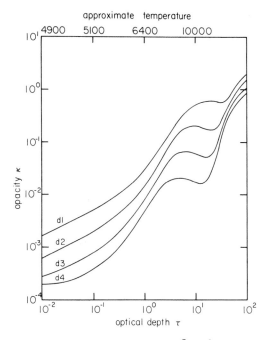

Figure 8.   Opacity κ (cm^2 g^{-1}) for
the series (d) of helium-
carbon atmospheres.

REFERENCES

Henyey, L. 1967, Ap. J. , 148, 207.
Hofmeister, E. , Kippenhahn, R. , and Weigert, A. 1964, Zs. f. Ap. , 59, 215.
Myerscough, V. P. 1968, Ap. J. , 152, 1115.
Searle, L. 1961, Ap. J. , 133, 531.
Weinberg, M. , and Berry, R. S. 1966, Phys. Rev. , 144, 75.

DISCUSSION

Auman: Do you have a reference for that paper describing the forbidden
transition in H$^-$?

Myerscough: Yes, it is Phys. Rev. , 144, 75, 1966.

THE EFFECTS OF THE CARBON-TO-HELIUM AND THE HYDROGEN-TO-HELIUM RATIOS IN HELIUM ATMOSPHERES

D. A. Klinglesmith and K. Hunger[*]

Goddard Space Flight Center,
National Aeronautics and Space Administration, Greenbelt, Maryland

ABSTRACT

Model stellar atmospheres of helium stars have been computed with several values of the C/He and H/He ratios in the range of $16,000 \leq T_{eff} \leq 20,000$ and $2.5 \leq \log g \leq 4.0$. The C/He ratios (by number) used were 0.012, 0.06, and 0.36, and the H/He ratios (by number) used were 0.0 and 0.004. For each model, He I profiles were computed by the use of Stark broadening (Griem, 1964) and Doppler broadening. A typical example of the effects of the variation of the ratios is the equivalent width of the He I profiles. A profile computed from a model with $T_{eff} = 18,000$, H/He = 0.004, and a model with $T_{eff} = 20,000$ and H/He = 0.0 are identical for all $\log g$. Varying the carbon content by a factor of 2 (from 0.012 to 0.06) would have a similar effect.

In other words, when a poor absorber like helium is being dealt with as the dominant source of opacity, any trace element that is a good absorber can alter the temperature-pressure distribution of the model in a manner such that the detailed fine analysis requires exact knowledge of these trace elements. As a result, the abundance analysis of the helium stars is complicated by the need to know not only T_{eff} and $\log g$, but also the ratios C/He and H/He.

These results will be reported in greater detail in a forthcoming paper.

REFERENCE

Griem, H. R. 1964, Plasma Spectroscopy (New York: McGraw-Hill Book Co.).

DISCUSSION

Elste: If helium is the main contributor to opacity in the visible, I suppose this means you can derive the carbon-helium abundance ratio without too much difficulty.

Klinglesmith: Yes.

Popper: What about the models with a hydrogen abundance of 0.001? Do you see any trace of the hydrogen lines at all?

Klinglesmith: We saw only Hγ and Hδ in this particular star BD 10°2179, and the equivalent widths were less than 0.1 for these two lines.

Buscombe: Is the line of C II λ4267 something you can predict?

Hunger: Yes, and it's unexpectedly strong. From the intensity of that line we derive a carbon abundance of about 100 percent, which cannot be true.

Underhill: Were the helium populations computed with the Saha-Boltzmann equation?

Klinglesmith: Yes, you have to start somewhere.

[*]NAS-NRC Senior Research Associate, 1968.

ON THE TEMPERATURE DISTRIBUTION IN THE SOLAR ATMOSPHERE

J. R. W. Heintze

Sonnenborgh Observatory, Utrecht, The Netherlands

ABSTRACT

Two recent solar models are discussed: the Bilderberg model and one newly constructed by the author. These two models can be assumed to be extreme cases. Neither of them explains all the observations. In comparison with observations, both models suggest that there is an extra absorber, certainly at 2000 A and possibly already at 4600 A. At 2000 A this extra absorber should be about 10 times the normal adopted one; at 4000 A it could be 0.5 times the normal adopted one.

1. INTRODUCTION

A solar model published by the author in 1965 (Heintze, 1965; hereafter referred to as Paper I) has been criticized several times. Noyes, Gingerich, and Goldberg (1966) show that the predicted center-to-limb variation (CLV) in the infrared does not fit the observations at all; the model shows a remarkable limb brightening. Withbroe (1967) points out that the model does not explain the observed CLV of the CH lines, whereas other models do.

An attempt is made to modify the 1965 model so that (1) the CLV will not show a limb brightening as far as possible in the infrared, and (2) the predicted CLV of the CH lines will be in agreement with observations. Detailed information on this model will be found in Heintze (1968). The solar temperature distributions for the Bilderberg (BCA) model (Gingerich and de Jager, 1968) and my 1965 and 1968 models are shown in Figure 1.

In the wavelength region 5 to 15 μ, the BCA tends to give too low relative intensities. The new model starts to show a limb brightening for $\lambda > 9$ μ. This limb brightening is small, and Lena (1968) has shown that it will not show up in measurements because of the diffraction of the instrument.

2. CLV OF THE CONTINUUM IN THE REGION $1.0 \leq \cos \theta \leq 0.15$

Figure 2 shows that even at $\cos \theta = 0.8$ and 0.7 the BCA does not describe the CLV in the visible; these measurements have to be regarded as rather accurate. The deviations between predicted and observed CLV could be removed by a steeper temperature gradient. My 1968 model is constructed by successive approximations, essentially in the same way as described in Paper I. The CLV are calculated for the wavelengths 4605, 5000, 5700, 6430 A; 1.07, 1.6481, 3.282, 3.5, 8.3, and 10.2 μ. For the visible, the absorption coefficients calculated by Bode (1965) for his standard mixture are used, and for the infrared ($\lambda \geq 1.6481$ μ), John's (1964) values are used. For every set of calculations, the deviations between calculated and observed relative intensities that showed up at certain values of $\cos \theta$ and at certain wavelengths were traced back to positions in the $T(\tau_0)$ relation where changes in temperature most effectively influence the wrong value of the relative intensity at that value of $\cos \theta$ and at that wavelength. The methods by which to find those positions in the $T(\tau_0)$ relation and the direction in which to change the temperature are given in Section 6.6 of Paper I.

As Figure 1 shows, the expected result is a temperature gradient steeper than that of the BCA. However, it seemed to be impossible to obtain a correct CLV for all wavelengths mentioned above. It is obvious that more weight is given to the observed CLV in the visible.

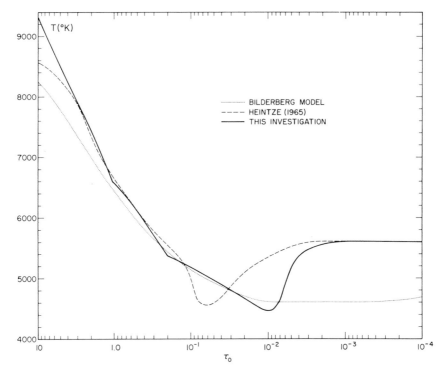

Figure 1.   Solar temperature distribution.

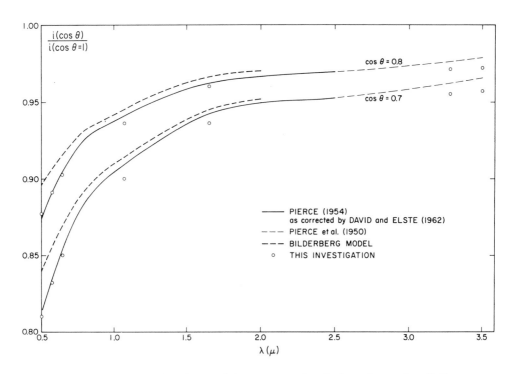

Figure 2.   Relative intensities at cos θ = 0.8 and cos θ = 0.7
from 5000 A to 3.5 μ.

3. INTENSITY PROFILE AT THE EXTREME LIMB

3.1 The Profile in the Region -6000 to -1000 km from the Limb

In Figure 3 the results of the balloon observations of Gaustad and Rogerson (1961) are shown. In Section 5.4 of Paper I, it was noted that this profile at λ 5490 agrees remarkably well with the measurements of ten Bruggencate, Gollnow, and Jäger (1950) at λ 6400 and of Dunn (1959) at λ 6563. Zirin (1966, p. 293) also emphasizes the good agreement and concludes that these measurements have to be accepted as correct.

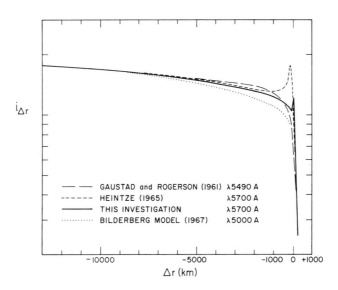

Figure 3.   Comparison of observed and calculated intensity profiles at the very limb.

In our computer program the curvature is taken into account at the extreme limb in order to obtain reliable predicted intensities at the limb. This program is also used to calculate the CLV for the BCA. From $\cos \theta = 1.0$ to 0.1 the results agree within several tenths of a percent with those obtained by Gingerich (see Gingerich and de Jager, 1968). The results at the extreme limb are shown in Figure 3. The maximum differences between the observations and the BCA are of the order of 20%. We find from our experience that it is very difficult to reduce this difference. The better agreement with the observations shown by our calculations is due to the increase of the temperature for $\tau_0 < 10^{-2}$ (see Figure 1).

The displacement of the temperature minimum from $\tau_0 = 0.075$ to $\tau_0 = 0.01$ (Figure 1) gives not only a slightly smaller intensity in the region -6000 to -2000 km from the limb with respect to our 1954 results, but also reduces considerably the predicted limb brightening (Figure 3). It is even possible to choose a temperature distribution in the region $10^{-2} \geq \tau_0 \geq 10^{-3}$ such that the limb brightening disappears completely.

3.2 The Intensity Profile at and just beyond the Limb

Another objection to the BCA is that it does not provide the correct scale height for the exponential dropoff at the extreme limb. At λ ≈ 5000 A our calculations give a scale height of ∼ 80 km for the BCA and ∼ 110 km for my new model, whereas it should be ∼ 120 km according to the observations (see Table 16 of Paper I). For a single-stream model it will be nearly impossible to obtain a scale height of ~ 100 km with a constant temperature of $\sim 4600°$K in the region $10^{-2} \geq \tau_0 \geq 10^{-4}$. The scale height could be slightly increased by introducing excessively high microturbulence.

4. THE ABSOLUTE CENTRAL INTENSITIES

The steeper temperature gradient in my model (compared with that of the BCA and needed to obtain the correct CLV of the relative intensity in the visible) causes, especially in the visible, higher absolute central intensities than in the case of the BCA (see Figure 4). The absolute central intensities in the visible, according to the BCA, lie very close to the observed values by Labs and Neckel (1962), which are regarded as the best measurements at this time. However, my model gives central intensities very close to the mean intensities as derived from the published values of the Russian astronomers Murasheva and Sitnik (1963) and Makarova (1963). The mean values used in Figure 4 are taken from Section 4.5 of Paper I.

4.1   The Absolute Intensities for $\lambda > 6000$ A

At wavelengths longer than 6000 A the difference in intensity between the BCA and my 1968 model becomes smaller (see Figure 4). At $\lambda \approx 6000$ A the difference is about 10%. At $\lambda \approx 10$ μ the difference between Saiedy's (1960) measurements and our calculations is practically zero (see lower part of Figure 5). At $\lambda = 1$ mm there is also agreement between the radio observations and our calculations. Unfortunately, no observations exist in the critical region between 10 μ and 1 mm.

4.2   The Absolute Intensities for $\lambda < 6000$ A

The absolute intensities calculated from the BCA agree much better with the observations near 4000 A of Houtgast (1968) (see Figure 4), who has carefully derived brightness temperatures for the highest intensity points. We must pay special attention to the fact that, according to Houtgast, the brightness temperature from 3100 to 3400 A is constant, namely, 6300°K. There seems to be no doubt about this result, because the density of lines in this wavelength region is much smaller than above the Balmer jump. The models definitely do not give a constant brightness temperature in that wavelength region, which suggests that an extra absorber must be working there. This suggestion is also given by the observations of Sandlin and Widing (1967), who find much lower brightness temperatures than the models predict in the wavelength region 1700 to 2070 A (see Figure 5). In the middle part of Figure 5 the temperature distributions according to the BCA and my 1968 model are given. In the upper and lower parts of this figure calculated brightness temperatures are given as a function of λ. We have attempted to link the brightness temperature at various wavelengths as well as possible to the "effective" optical depth at that wavelength, which is defined as the optical depth (τ_0) for which the local temperature of the model is equal to the brightness temperature at that wavelength. Note that the points observed by Sandlin and Widing (1967) fall far below the brightness temperatures predicted by either model. If the real absorption coefficient at 2000 A were roughly 10 times the adopted one, the effective mean optical depth of formation would be ~ 0.5 that obtained in the calculations. This would shift the region of formation out toward the temperature minimum, and the predicted intensity would drop to a value closer to that actually observed.

If Saiedy's (1960) measurements of the absolute central intensity around $\lambda = 10$ μ are correct, and if the adopted absorption coefficient at these wavelengths is correct, then according to Figure 5 our temperature distribution in the layers around $\tau_0 = 0.05$ could be quite realistic because the observed and the calculated brightness temperatures at $\lambda = 10$ μ agree so well. Fortunately, the agreement between the BCA and my 1968 model in temperature distribution in the region $\tau_0 = 0.02$ to 0.13 is very good, so that we can have

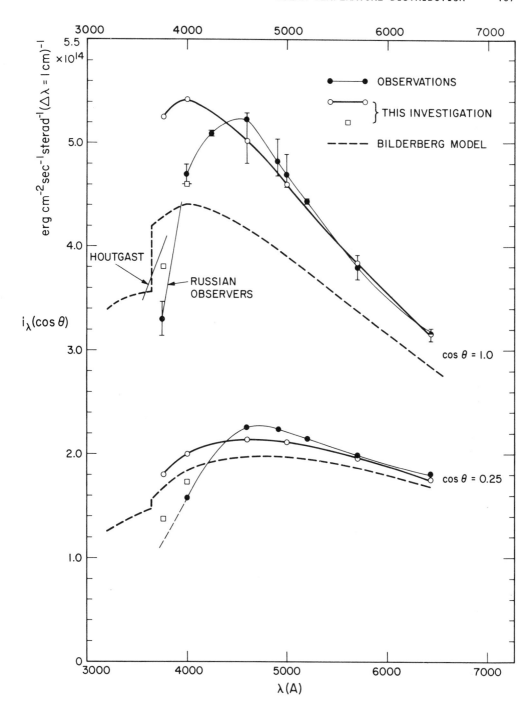

Figure 4.   Comparison of observed and calculated absolute intensities
at the center and at cos θ = 0.25.   The observed values are
taken from Murasheva and Sitnik (1963) and Makarova
(1963).   The open squares refer to calculations at 4000 and
3700 A, where the absorption coefficient is taken as 1.25
and 1.45 times the absorption coefficient after Bode (1965).

even more confidence in this region of the temperature distribution.   The
temperature differences between both models from τ_0 = 0. 13 to 1. 0 are
rather small.   These considerations make an increase of the absorption
coefficient around 2000 A rather probable.

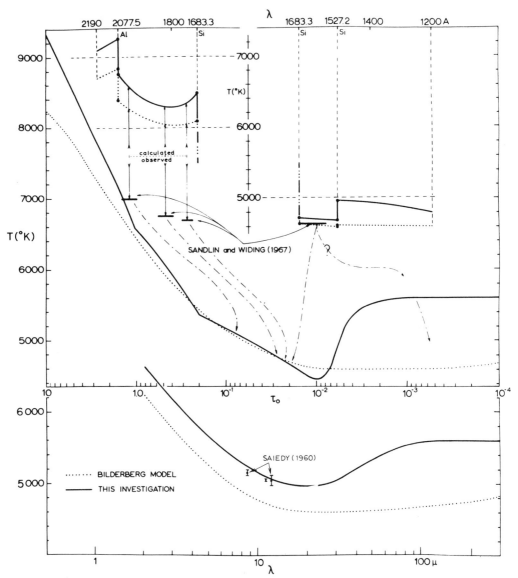

Figure 5.   Correlation between wavelength and "effective" optical depth
where the radiation at that wavelength mainly originates.   The
two lower horizontal axes give the correlation between the
wavelengths in the infrared and "effective" optical depths.
The upper horizontal axis and the middle axis give the correlation between the wavelengths in the far ultraviolet and
the effective optical depths according to the model calculations.   The arrows point to the "effective" optical depths
required by the measurements of Sandlin and Widing (1967).

We did not calculate a possible increase of the absorption coefficient around 3000 A. However, we made some calculations at 4000 and 3700 A. For $\lambda \le 4600$ A the absolute central intensity drops rapidly according to Russian observers. The calculations do not show such a drop. However, the absolute central intensities at 4000 and 3700 A according to my 1968 model can be brought into agreement with the observations by increasing the adopted absorption coefficient by a factor 1.25 and 1.45, respectively.

5. CONCLUSIONS

A rather sharp increase of the temperature for $\tau_0 \le 10^{-2}$ seems to be necessary in order to explain the intensity profile in the continuum for -6000 to -1000 km at the limb.

It is very probable that an extra absorber is working in the solar atmosphere around 2000 A, where it should be about 10 times the adopted absorption coefficient. Possibly this extra absorber is already working at $\lambda < 4600$ A. There is some evidence that the extra absorber at 4000 A should be approximately 0.5 times the adopted one.

REFERENCES

Bode, G. 1965, Die Kontinuierliche Absorption von Sternatmosphären in Abhängigkeit von Druck, Temperatur und Elementhäufigkeiten (Kiel: Inst. Theoret. Phys. und Sternw. der Univ. Kiel), 193 pp.

David, K.-H., and Elste, G. 1962, Zs.f.Ap., 54, 12.

Dunn, R. B. 1959, Ap. J., 130, 972.

Gaustad, J. E., and Rogerson, J. B., Jr. 1961, Ap. J., 134, 323; Ap. J., 135, 660.

Gingerich, O., and de Jager, C. 1968, Solar Phys., 3, 5.

Heintze, J. R. W. 1965, Rech. Astron. Obs. Utrecht, 17, Part 2, 88 pp.
_____. 1968, Bull. Astron. Netherlands, 20, 1.

Houtgast, J. 1968, Solar Phys., 3, 47.

John, T. L. 1964, Mon. Not. Roy. Astron. Soc., 128, 93.

Labs, D., and Neckel, H. 1962, Zs.f.Ap., 55, 269.

Lena, P. 1968, private communication.

Makarova, E. A. 1963, Observatory, 83, 183.

Noyes, R. W., Gingerich, O., and Goldberg, L. 1966, Ap. J., 145, 344.

Pierce, A. K. 1954, Ap. J., 120, 233.

Pierce, A. K., McMath, R. R., Goldberg, L., and Mohler, O. C. 1950, Ap. J., 112, 289.

Saiedy, F. 1960, Mon. Not. Roy. Astron. Soc., 121, 483.

Sandlin, G. D., and Widing, K. G. 1967, Ap. J., 149, L129.

ten Bruggencate, P., Gollnow, H., and Jäger, F. W. 1950, Zs.f.Ap., 27, p. 223.

Withbroe, G. L. 1967, Ap. J., 147, 1117.

Zirin, H. 1966, The Solar Atmosphere (Waltham, Mass.: Blaisdell Publ. Co.).

DISCUSSION

Thomas: Could we see your first slide again? I like the direction of change of $T_e[\tau]$ in your model relative to the Bilderberg model, namely, higher temperatures in the lower chromosphere. I may be prejudiced, because we published such a model in 1961 [Thomas and Athay, Physics of the Solar Chromosphere, Interscience Publ.], and each new model I see in this series of models sponsored by the Utrecht group seems to move closer to our 1961 model. I do think that you make T_e rise too rapidly at the lower heights; I think this will give you trouble with the sodium D lines. If you would push your rise in T_e out another decade in τ, I think things would be better.

Cayrel: I'm afraid that this model gives too strong an emission in the H and K lines.

Heintze: That could be, but we don't know how the non-LTE effects would be taken into account in this model. The same is true with the strong metal lines.

Cayrel: I am referring to correct non-LTE computations.

Heintze: Well, then the conclusion is that it seems to be impossible to find a one-stream model that can explain the line profiles as well as the intensity profile in the continuum at the limb.

Böhm-Vitense: I am worried about the deep layers. How does the steep temperature increase compare with theoretical expectations?

Heintze: The derived temperature distribution in the deeper layers is a purely empirical one. I have only tried to find a one-stream model that, especially in the visible, could explain the relative center-to-limb variations of the continuum as well as possible. The derived temperature distribution explains the center-to-limb variations for wavelengths greater than λ 4600 rather well. Below λ 4600 difficulties arise, which possibly could be removed by introducing an extra absorber. Of course, the temperature distribution can change if the extra absorber extends beyond λ 4600 and/or if more streams are introduced.

Böhm-Vitense: Which opacities did you use?

Heintze: I used Bode's (1965) tables of the absorption coefficients in the stellar atmospheres for $\lambda < 1.65$ μ for what he calls the standard mixture. For $\lambda > 1.65$, I used John's (1964) results.

Gingerich: As an author of the Bilderberg Model, I would be the first to admit its imperfections, and we must thank Heintze for clearly pointing to several of them. The one point, that the temperature and the temperature gradient in the deeper layers should be higher in order to produce the desired intensity in the visible spectrum, was immediately recognized and is stated in the original paper. The temperature distribution was bent down to allow for convection, and lies between the purely radiative and adiabatic cases according to calculations we have carried out recently, but the deep temperatures in the Bilderberg Model are too low.

On the other hand, it is necessary to select among various conflicting observational data when an empirical solar model is being constructed. Although Heintze has succeeded in representing certain observations, I believe that in two very critical areas his model gives false predictions, and either one of these is serious enough to invalidate his model. First, Labs and Neckel [Zs. f. Ap., 69, 1, 1968] have thoroughly discussed all determinations of the solar photospheric intensity, and in particular they demonstrate the complete untrustworthiness of the Russian observations. Hence, we must conclude that the agreement of Heintze's model with the continuum determination of Makarova, Sitnik, and Murasheva carries little weight and, in fact, his predictions in the visual region must be regarded as 15 to 20% too high.

Second, in the ultraviolet region Heintze has graphed his results as if each portion of the spectrum arose from a comparatively limited zone in the solar photosphere, a simplistic assumption immediately disproved by his own graph, for example, where his model and the BCA predict rather different intensities in the infrared in the region where the concordance shows that both models have the same temperature distribution. Because contributions come to the emergent spectrum from an extensive range in depth, particularly in the ultraviolet, the abrupt temperature rise to the chromospheric plateau in Heintze's model will have the following effects: (a) the model will predict substantial limb brightening at 1600 A, whereas this region shows neither limb darkening nor limb brightening; (b) the silicon ground-state emission edge at 1525 A will be predicted as much too strong; and (c) the model will predict observably large emission cores for spectrum lines of moderate strength, as our calculations have shown. For these reasons I believe that the BCA represents the actual sun much more accurately in the region $\tau_{5000} = 0.01 - 0.001$ than does Heintze's model.

Heintze: With respect to your first point, it is true that the predicted absolute central intensities according to the proposed model do not agree with Labs' values. On the other hand, the Bilderberg model does not explain the limb darkening in the visible spectral region, of which rather accurate measurements exist. My model is constructed so that it does predict the observed darkening from 4600 to 8000 A very well. A direct consequence is that the predicted central intensities are higher than Labs and Neckel find. The conclusion that has to be drawn is: if the adopted absorption coefficients are correct and if Labs and Neckel's values are correct, no one-stream model exists that explains the observed central intensities as well as the observed limb darkening in the visible spectrum region at the same time.

With respect to your second point, I first want to say that Figure 5 does not imply an exact one-to-one correspondence between wavelength in the calculated continuous spectrum and the optical depth of the model. It only tries to give an impression from which zones (not too limited) in the solar photosphere the several positions of the continuous spectrum mainly arise. Second, the results drawn in the upper part of Figure 5 cannot be compared with the observations. This figure enables us only to estimate the influence of the extra absorber around 1900 A. The predictions in the ultraviolet cannot be compared with the observations, because the extra unknown absorber is not taken into account in the calculations. Therefore, your points (a) and (b) are irrelevant.

With respect to your point (c), I know that emission cores show up in spectrum lines of moderate strength. However, Grevesse (private communication) has shown that even according to the Bilderberg Model the Ca II $\lambda 8498.02$ line at $\cos \theta = 1$ and the C I $\lambda 10691.24$ line at $\cos \theta = 0.3$ show emission peaks, which are not observed. It is quite possible that the applied theory of line formation (LTE) is too simple. On the other hand, I have some evidence that the position of the temperature minimum has to be moved somewhere to $\tau_0 = 0.004$. This will decrease the emission peaks but not eliminate them according to the LTE theory. The position of the temperature minimum will be discussed in a paper to be published in the Bull. Astron. Netherlands.

Elste: Concerning the behavior of the very deep layers, David's investigation [Zs.f.Ap., 53, 37, 1961], in trying to explain the center-to-limb variations of the wings of the Balmer lines, found a flattening in the temperature distribution versus log τ_0, as predicted by the mixing-length theory. This result was somewhat different from the Bilderberg Model, but definitely not as steep as Heintze's result.

Heintze: My model is definitely an upper limit.

INFLUENCE OF THE LYMAN-α WING ON THE SOLAR SPECTRUM IN THE 1500 A TO 8000 A WAVELENGTH RANGE

Y. Cuny

Observatoire de Paris, Meudon, France

1. INTRODUCTION

The Bilderberg continuum atmosphere (BCA), a model established at an international week in Holland in April 1967, succeeds in predicting correctly many features of the observed solar continuum. Nevertheless, it fails in several respects, the most notable being the prediction of far too great a flux in the ultraviolet between 1700 and 3600 A. Some of these discrepancies were already noted in the Proceedings of the "Bilderberg Conference" in the January 1968 issue of Solar Physics, but no fully satisfactory explanation was offered.

The results reported here will show that at least some of these disagreements between the predicted spectrum and the observations may be resolved by considering the resonance broadening opacity from the Lyman-α wing at great distances from the line center.

2. COMPARISON OF THE SPECTRUM PREDICTED BY THE BCA MODEL WITH OBSERVATIONS

Between 1683 and 8000 A the theoretical central intensities and the center-to-limb variations are not satisfactorily predicted. The discrepancies between the observed and the predicted central intensities throughout the wavelength range 3000 to 5000 A are reported by Carbon, Gingerich, and Kurucz (1968). Redward of the Balmer jump the predicted intensities are smaller than the observed ones. The discrepancies may be partially ascribed to the model, as is suggested by the results given by the Utrecht model (Heintze, Hubenet, and de Jager, 1964) and the Mutschlecner model. Violetward of the Balmer jump, all the models give intensities and fluxes larger than those observed. The observations of Bonnet and Blamont (1968) in the range 1980 to 2885 A show the same systematic discrepancies. At shorter wavelengths a comparison of the observations of Detwiler, Garrett, Purcell, and Tousey (1961) with the theoretical results, for the flux, is reported by Gingerich and Rich (1968). The discrepancies between the observations and the theoretical results decrease at 1683 A, where the absorption due to the second level of silicon is very large. The different results are shown in Figures 1 and 2. The differences between our results and those of Gingerich will be explained later.

In this region the center-to-limb variations are not well predicted by the BCA model. Although the discrepancies between the observed and the theoretical intensities might be ascribed to the model at long wavelengths, it seems this is not the case for the center-to-limb variations. The model specially computed by Elste (1968) to predict correctly the center-to-limb variations in fact fails in the 3000 to 6000 A range: the computed center-to-limb variation is always smaller than the observed one. The same effect is confirmed by the observations of Bonnet between 2514 and 2885 A (Figure 3). Beyond 2514 A the discrepancy is in the opposite direction; at 1980 A it is very large (Figure 4). The spectral region between 1525 and 1683 A shows neither darkening nor brightening. The BCA model gives a slight limb brightening.

We shall now explain the origin of the differences between our results and those of Gingerich.

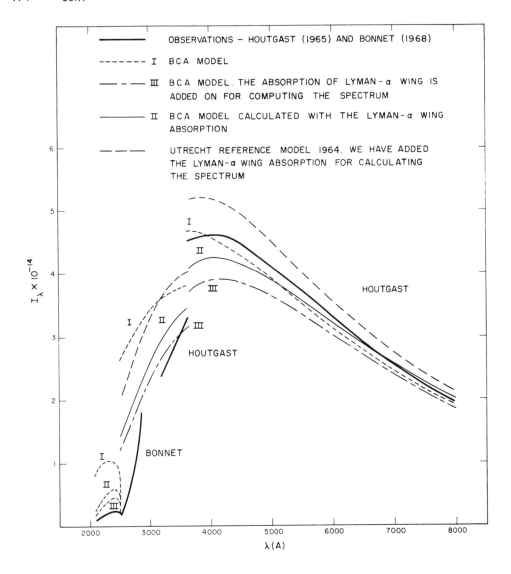

Figure 1. Central intensity.

We have computed the absorption of the quasi molecule H_2 with the results of Doyle (1968), which give an absorption two times greater than the values found by Solomon (1964); at 1700 A the absorption of the quasi molecule H_2 is not negligible. We have extrapolated the experimental photoionization cross section of Bötticher (1958) for the second level of magnesium, using the theoretical photoionization cross section of Peach (1962). The absorption by magnesium at 2000 A is thus five times greater than the absorption calculated by Gingerich. Furthermore, we have computed the absorption by silicon and magnesium with a limited number of levels, nine for silicon and eight for magnesium, and we have ignored the contributions of the upper levels. By calculating the absorption by a 64-level silicon atom, with the hydrogenic approximation, we have checked that the influence of the upper levels is, in fact, negligible. Restricting the level number to nine and eight for silicon and magnesium, respectively, appears to be sufficient. At 3700 A the sum used with the hydrogenic approximation for the upper levels gives a greater

absorption. Using our absorption subroutine we have calculated the model
starting from the (T - τ) relation of the BCA model at 5000 A.

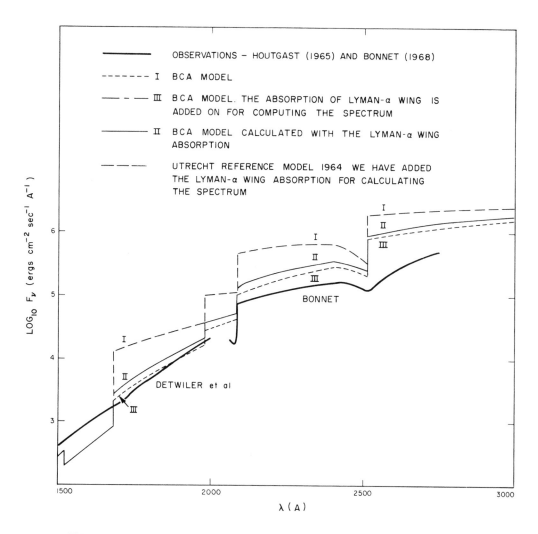

Figure 2.   Ultraviolet flux (same explanation as in Figure 1).

Although at long wavelengths the values of the intensity and flux may be
partially improved by changing the model, it is obvious that the ultraviolet
spectrum requires a larger absorption in order to be predicted correctly.
We must therefore seek an additional absorber.

3.   THE LYMAN-α WING

To evaluate the unknown absorption we have empirically added absorption
to obtain intensities close to the observed ones. The jumps due to the metals
make possible a rough determination of the variation of this absorption with
the optical depth; this variation must be similar to that of the absorption by
a neutral atom. This absorption increases with the frequency: the absorption
required to remove the jump of silicon at 1683 A is larger than that needed to
decrease the flux at longer wavelengths. Beyond 1683 A the absorption by
silicon is very large, and the extra absorber negligible in comparison.

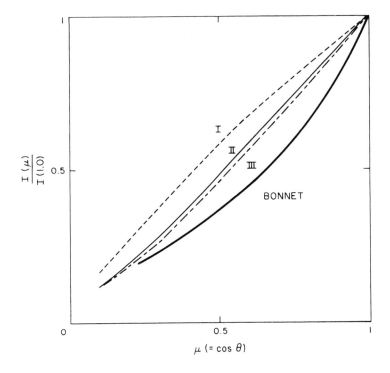

Figure 3.   Center-to-limb variation at 2665 A.

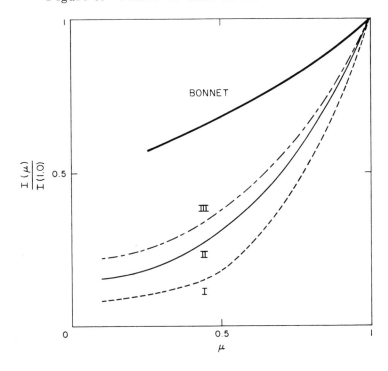

Figure 4.   Center-to-limb variation at 1985 A.

Various absorbers have been suggested by several authors. The wing of Lyman α has been proposed by Menzel and Doherty (1960), who have computed the influences of the natural broadening and of the Stark effect. In the photosphere, however, the resonance broadening is always greater than the two former effects (Cuny, 1968). It is possible that absorption from the Lyman-α wing may play a significant role in the interpretation of the ultra-violet and visible spectrum.

We calculate the resonance broadening by Griem's formula (1964):

$$\gamma_c = \frac{3e^2}{mc} \left(\frac{g_1}{g_2}\right)^{1/2} f\lambda N \quad , \tag{1}$$

$$\gamma_c = 0.6412 \times 10^{-7} N \quad . \tag{2}$$

The natural damping usually taken into account in the Rayleigh scattering is

$$\gamma_R = 0.4675 \times 10^9 \quad . \tag{3}$$

The smooth heavy line in Figure 5 gives the curve for log (γ_c/γ_R); the abscissa is log τ_{5000} computed by Gingerich and de Jager (1968). In fact, the graph shows the variation of the hydrogen neutral atom density N. At large optical depths, $1 < \tau < 25$, the absorption by the second and third levels of neutral hydrogen increases very rapidly with temperature; furthermore, the ionization of hydrogen becomes important. In this region the neutral hydrogen density, and consequently the Lyman-α absorption, are practically constant.

Near the center of the line the Lorentz profile is given by the classical theory of oscillators:

$$s_\nu = f\frac{e^2}{mc} \frac{\gamma}{4\pi} \frac{1}{(\nu - \nu_0)^2 + \left(\frac{\gamma}{4\pi}\right)^2} \quad , \tag{4}$$

which assumes as valid the approximation

$$\nu_0^2 - \nu^2 \simeq 2\nu(\nu - \nu_0) \quad .$$

Without this approximation the profile is

$$s_\nu = f\frac{e^2}{mc} \frac{\gamma}{4\pi} \left(\frac{2\nu}{\nu + \nu_0}\right)^2 \frac{1}{(\nu - \nu_0)^2 + \left(\frac{\gamma}{4\pi} \frac{2\nu}{\nu + \nu_0}\right)^2} \quad , \tag{5}$$

which far from the center of the line yields

$$s_\nu = f\frac{e^2}{mc} \frac{\gamma}{4\pi} \left(\frac{2\nu}{\nu + \nu_0}\right)^2 \frac{1}{(\nu - \nu_0)^2} \quad . \tag{6}$$

We have used formula (6) in the hope that the discrepancy from the real profile is not too large.

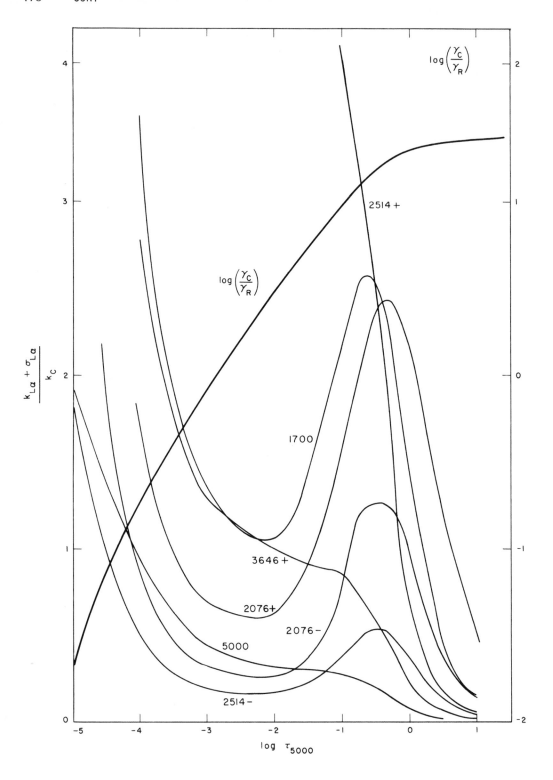

Figure 5. Lyman-α resonance broadening. Lyman-α wing absorption at various wavelengths.

The absorption of Lyman α is

$$k_{L\alpha} = 2.798 \times 10^{-4} \; \gamma_c \left(\frac{2\nu}{\nu + \nu_0} \right)^2 \frac{1}{(\nu - \nu_0)^2} \quad . \tag{7}$$

Figure 5 also shows the ratio of $k_{L\alpha} + \sigma_{L\alpha}$ (where $\sigma_{L\alpha}$ is the Rayleigh scattering of Lyman α) to the absorption k_c of the other absorbers at various wavelengths. For this graph $\sigma_{L\alpha}$ is calculated from expression (7), with γ_R used instead of γ_c.

The absorption of Lyman α is especially important redward of the magnesium jump at 2514 A. For wavelengths shorter than 2514 A, the maximum of the graph separates the region where the absorption of the second level of neutral hydrogen becomes very large, toward large optical depth, from the region where the relative importance of the absorption due to the metals increases, toward small optical depths.

We note also the Lyman-β absorption is about 4% of that of Lyman α; the absorption of the helium resonance line is negligible. These preliminary computations include only the Lyman-α wing.

4. RESULTS

We shall now discuss the results obtained assuming pure absorption for the resonance broadening. We have calculated the spectrum in two ways. First, starting with the BCA model, we have added on the Lyman-α wing absorption; the absorption of the Lyman-α wing being not negligible at 5000 A (Figure 5), this calculation gives, at this wavelength, a new T - τ relation. Then we have computed the spectrum with a model in hydrostatic equilibrium, using the T - τ relation of the BCA model at 5000 A and taking into account the absorption by the Lyman-α wing.

As seen in Figures 1, 2, and 6, the Lyman-α resonance broadening has an influence on the predicted spectrum throughout the wavelength region 1683 to 8000 A and gives generally a better agreement between observed and theoretical intensities or fluxes. The agreement with the observations is especially good at wavelengths shorter than 1900 A. Redward of 1683 A, the Lyman-α wing decreases the flux by a factor of about 6. The discrepancies between the observed and theoretical intensities are always less than a factor of 2, except near the magnesium jump at 2514 A, where the flux is depressed by many lines, as shown by Bonnet (1968).

We shall now analyze more accurately the spectrum predicted by the BCA model when the Lyman-α resonance broadening is taken into account; we shall try to interpret the discrepancies with the observed spectrum in terms of either model or absorption.

The BCA model, without Lyman α, predicts at long wavelengths too small a central intensity. Redward of the Balmer jump the Lyman-α wing increases the discrepancy with the observations (Figure 1). Nevertheless, the center-to-limb variations in this wavelength range are better predicted when Lyman α is taken into account (Figure 6). The BCA models, calculated with the same (T - τ) relation, with or without the Lyman-α wing, give for the Balmer jump $\log_{10} (I_{4000}/I_{3600})$ the same value 0.08, which is smaller than the observed value. In the whole of this long-wavelength range, the predicted spectrum may probably be improved by changing the model.

Violetward of the Balmer jump the agreement with the observations of Houtgast (1965) and Bonnet is satisfactory, but this is not the case toward the short wavelengths (Figures 1 and 2). Between 2665 and 2070 A, except near 2514 A where the flux is depressed by many lines, the calculated central intensities are too large by a factor of 2 compared with the observations of Bonnet (1968). The predicted jump at 2076 A is too small by a factor of 2 or 3. The calculated jump at 1985 A is very small; in Bonnet's observations it disappears almost completely. The center-to-limb variation in this same wavelength range is not satisfactorily predicted; the discrepancy increases

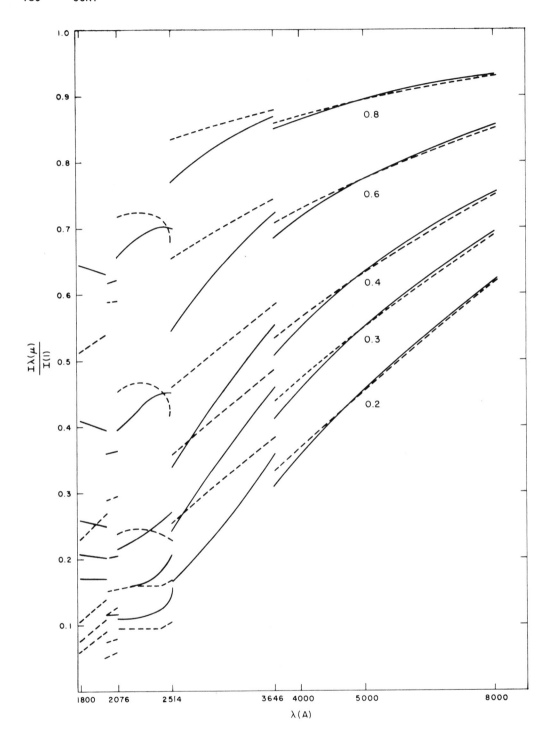

Figure 6. Center-to-limb variations between 1800 and 8000 A.

toward the short wavelength, with a reversing of direction at 2514 A.   The predicted center-to-limb variation is too small redward of the magnesium 2514 A jump and too large to the violet side (Figures 3 and 4).   For very small values of μ (μ ≤ 0. 1) the center-to-limb variations decrease from the red to the violet side of the magnesium 2514 A jump (Figure 6).   Bonnet (1968) has observed this fact for all the values of μ at this wavelength.

The center-to-limb variation at 1980 A is very large compared with the observed one.   For μ = 0. 1 the relative intensity is too small by a factor of approximately 2. 5 or 3.   The predicted central intensity is too large by a factor of 2 and the intensity for μ = 0. 1 is too small by a factor of 1. 25 or 1. 5.   Now for μ = 0. 1, the theoretical intensity is very close to the value of the source function of the temperature minimum.   An increase of the temperature minimum of 100°K gives a factor of 1.4.   This seems to corroborate that the temperature minimum is close to 4600°K, perhaps equal to 4700°K, as Bonnet has already deduced from his observations.

The discrepancies in the 1900 to 2800 A range between the observations and the theoretical results may be ascribed partially, perhaps, to an unknown absorption.   In this spectral region there are the edges of the continuous absorption of many levels of iron.   The iron absorption calculated with the hydrogenic approximation decreases the flux about 5%.   We have also added absorption by increasing the absorption of Lyman α; with a Lorentz profile (formula 4) the central intensity at 1980 A is correct, but the center-to-limb variation is not well fitted.   Certainly the theoretical results may be partially improved by changing the model.

Beyond 1980 A the agreement with the observations is better (Figure 2). For wavelengths shorter than 1683 A the results are the same with or without Lyman α, since the absorption of silicon is very important, but in this spectral region the non-LTE effects are perhaps not negligible; we will discuss this point in another paper.

I wish to thank Professor F. L. Whipple and Professor C. A. Whitney for their hospitality at the Smithsonian Astrophysical Observatory, Dr. R. M. Bonnet for providing me with his observations of the solar ultraviolet spectrum, and Professor O. Gingerich for stimulating discussions.

REFERENCES

Bonnet, R. M.   1968, private communication.
Bonnet, R. M. , and Blamont, J. E.   1968, Solar Phys. , 3, 64.
Bötticher, W.   1958, Zs. Phys. , 150, 336.
Carbon, D. , Gingerich, O. , and Kurucz, R.   1968, Solar Phys. , 3, 55.
Cuny, Y.   1968, Solar Phys. , 3, 204.
Detwiler, C. R. , Garrett, D. L. , Purcell, J. D. , and Tousey, R.   1961,
    Ann. Géophys. , 17, 263.
Doyle, R. O.   1968, Ph. D. Thesis, Harvard University.
Elste, G.   1968, Solar Phys. , 3, 106.
Gingerich, O. , and de Jager, C.   1968, Solar Phys. , 3, 5.
Gingerich, O. , and Rich, J. C   1968, Solar Phys. , 3, 82.
Griem, H. R.   1964, Plasma Spectroscopy (New York: McGraw-Hill Book
    Co.).
Heintze, J. R. W. , Hubenet, H. , and de Jager, C.   1964, Bull. Astron.
    Netherlands, 17, 442.
Houtgast, J.   1965, Proc. Roy. Netherlands Acad. Sci. , Ser. B, 68, 306.
Menzel, D. H. , and Doherty, L. R.   1960, Mem. Soc. Roy. Sci. Liège,
    Ser. 5, 4, 295.
Peach, G.   1962, Mon. Not. Roy. Astron. Soc. , 124, 371.
Solomon, P. M.   1964, Ap. J. , 139, 999.

NON-LTE EFFECTS ON THE SOLAR SPECTRUM
BETWEEN 1500 A and 1680 A

Y. Cuny

Observatoire de Paris, Meudon, France

1. INTRODUCTION

Gingerich and Rich have pointed out that the spectral region between 1500 and 1680 A is very important for the study of the temperature minimum of the solar atmosphere. The evaluation of the optical depth shows that the continuous spectrum in this spectral range arises from the region of the temperature minimum. The lack of center-to-limb variation at 1600 A bears out this fact and makes possible an estimate of the plateau extent. The BCA model (Gingerich and de Jager, 1968) has been constructed to interpret these facts with the assumption of LTE: a temperature plateau gives a source-function constant. Nevertheless, in the transition region between the photosphere and the chromosphere the non-LTE effects may not be negligible.

At the temperature minimum the non-LTE departure coefficients of the silicon atom, which is the most important absorber in this spectral range, are less than unity and increase toward the photosphere and the chromosphere. For a given model the flux calculated with the LTE assumption must be smaller than the flux calculated without this assumption; furthermore, in the low chromosphere, the center-to-limb variation calculated without the assumption of LTE must be smaller than with LTE. We shall discuss these two points, after giving some details of the calculation of the non-LTE coefficients of silicon.

2. SILICON ATOM CALCULATION OF THE NON-LTE DEPARTURE COEFFICIENTS

The analysis for the hydrogen atom (Cuny, 1968) and its various transitions is applicable for the silicon atom. We use an analogous method of calculation.

The radiation field that determines the photoionization rates of the upper levels above the second can be calculated with the LTE assumption. In the region of the temperature minimum, these photoionization rates are greater than the recombination rates, and the non-LTE departure coefficients are less than unity, which means that these levels are underpopulated. Toward the chromosphere and the photosphere the difference between the kinetic and radiation temperatures decreases, and the departure coefficients increase. We may remark that the usual models of the solar atmosphere do not predict correctly the radiation field at 1900 A; for the third level we have assigned in our calculation a photoionization rate close to the observed one.

The radiation that determines the photoionization rates of the two ground levels of the silicon atom depends very strongly on the population of these levels. The radiation field is calculated without the LTE assumption by the same method as for the Lyman continuum (Cuny, 1968).

The triplet and singlet ground levels are coupled to upper levels by collisions. In the region of the temperature minimum the collision rates are generally great enough to assure an almost Boltzmann equilibrium between the upper levels and the ground levels; hence, because the upper levels are underpopulated, the ground levels must also be underpopulated.

The problem of the determination of collisional cross sections is very difficult. For the allowed transitions with the upper levels, we have used the impact parameter method of Seaton (1962). We do not know the uncertainty of the cross section given by this method when used for an atom like silicon; but we note that the non-LTE departure coefficients do not depend very strongly on the value of the collisional rate.

For calculating the collisional rates between the three ground levels, which are coupled by forbidden transitions, we have used the cross section determined by Smith, Henry, and Burke (1967) for the carbon atom. Transitions $3p^2 \, ^3P$ - $3p^2 \, ^1D$ and $3p^2 \, ^1D$ - $3p^2 \, ^1D$ - $3p^2 \, ^1S$ are in the infrared near the Maxwell function maximum and introduce a strong coupling between the three ground levels and, more generally, between the triplet and singlet levels.

We have calculated the non-LTE departure coefficients for a six-level atom, assuming detailed balance in the lines. The levels are the following:

Triplet	Singlet
$3p^2 \, ^3P$	$3p^2 \, ^1D$
$4s \, ^3P^0$	$3p^2 \, ^1S$
	$4s \, ^1P^0$
	$3d \, ^1D^0$

3. RESULTS

Given all the simplifications (limited level number, detailed balance in the lines) and the uncertainty of the atomic data, we consider these results mainly as a qualitative indication of the non-LTE influence.

3.1 Flux

We have calculated the flux with and without the assumption of LTE for four models with broad plateaus of various temperatures: 4000°K, 4300°K, 4600°K. The results, compared with the observations of Detwiler, Garrett, Purcell, and Tousey (1961), are presented in Figure 1.

The non-LTE correction to the flux increases when the plateau temperature decreases, the ground levels of the silicon atom being more underpopulated. For the BCA model (Gingerich and de Jager, 1968) the correction is very small.

The flux value at 1600 A depends on the extent of the temperature plateau. The model with a plateau shorter than that of the BCA model (Figure 2) predicts higher values of flux.

All the models with the assumption of LTE give at 1525 A a jump in the same direction as the observed one (Tousey, 1964): there is an increase of the flux toward the short wavelengths. If the model of temperature 4000°K is excepted, the calculation of the flux with non-LTE gives an effect in the opposite direction. The decrease of the flux obtained with the model with a short temperature plateau equal to 4600°K gives a smaller decrease of the flux than the BCA model.

3.2 Center-to-Limb Variations

We have studied the influence of the plateau extent on the center-to-limb variations with the two models of temperature minimum equal to 4600°K.

Table 1 summarizes qualitatively the limb darkening and limb brightening predicted by the two models at 1600 A (Figure 3) and 1500 A (Figure 4). The observations are those of Tousey (1964).

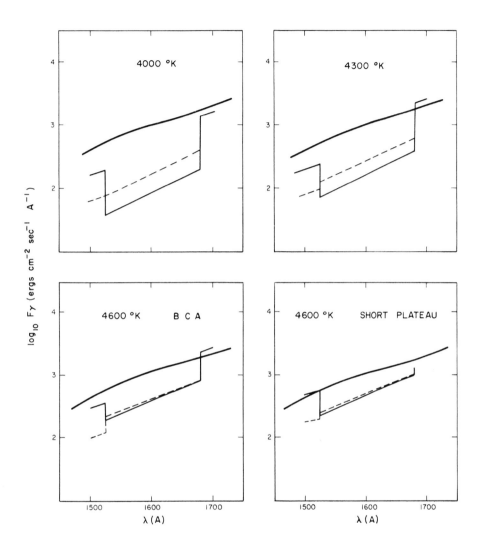

Figure 1.   Influence of the temperature minimum and of the length of the temperature plateau.

⎯⎯⎯   Detwiler et al. observations (1961)

⎯⎯⎯   LTE flux

- - - - - -   Non-LTE flux

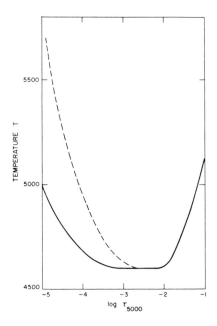

Figure 2. Models with temperature plateau equal to 4600°K: T - τ relations.

——————— BCA model

- - - - - - Model with a shorter temperature plateau

Table 1. Theoretical ultraviolet limb-darkening and limb-brightening predictions (out to $\mu = 0.1$)

		λ 1600 (A)	λ 1500 (A)
Observed		apparently neutral	pronounced brightening
BCA	LTE	brightening	pronounced brightening
	non-LTE	darkening	darkening
Short plateau	LTE	brightening	pronounced brightening
	non-LTE	darkening	brightening

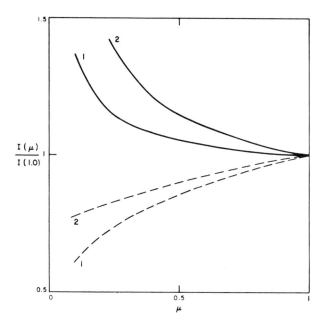

Figure 3.   Center-to-limb variation at 1600 A.
                    _____   LTE center-to-limb variation
          ------   Non-LTE center-to-limb variation
      1.   BCA Model
      2.   Model with a shorter plateau at 4600°K

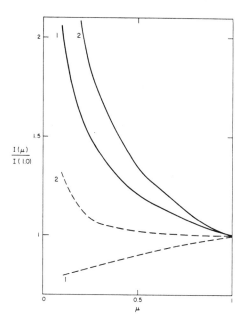

Figure 4.   Center-to-limb variation at 1500 A.   Same explanation as for
            Figure 3.

4. CONCLUSION

The non-LTE influence on the flux is not very important for models with broad plateaus of temperature close to 4600°K. This work corroborates the fact that the plateau temperature is not very different from 4600°K, perhaps slightly greater (if the plateau really exists). On the other hand, the influence of the non-LTE on the silicon jump at 1527 A and on the center-to-limb variations in all the wavelength range 1500 to 1683 A is considerable. An accurate determination of the temperature minimum and a more general determination of the transition region between the chromosphere and the photosphere seem to be possible even when the LTE assumption is removed. We intend to continue this work. It should be very interesting to have a calibrated spectrum in the range 1500 to 1700 **A**.

I wish to thank Professor F. L. Whipple and Professor C. A. Whitney for their hospitality at the Smithsonian Astrophysical Observatory, and Professor O. Gingerich for helpful discussions.

REFERENCES

Cuny, Y. 1968, Solar Phys., 3, 204.
Detwiler, C. R., Garrett, D. L., Purcell, J. D., and Tousey, R. 1961, Ann. Géophys., 17, 263.
Gingerich, O., and de Jager, C. 1968, Solar Phys., 3, 5.
Seaton, M. J. 1962, Proc. Phys. Soc., 79, 1105.
Smith, K., Henry, R. J. W., and Burke, P. G. 1967, Phys. Rev., 157, 51.
Tousey, R. 1964, Quart. J. Roy. Astron. Soc., 5, 123.

DISCUSSION

Thomas: You say you have reduced the length of the temperature plateau. Could you tell us to what length and where?

Cuny: The length is one decade in τ_{5000}; the rise begins at 2×10^{-3}. Furthermore, the gradient of the temperature of this model is greater than that of the BCA. Possibly the gradient is more important than the length of the plateau.

Gingerich: Dr. Cuny's results are an obvious example of the "Cayrel phenomenon." Those of you who attended the previous conference will remember how he explained that the right theory (i.e., non-LTE) gives the wrong results, and this is a similar situation. Dr. Cuny's results from the Bilderberg model do not predict the observed limb brightening. However, the Bilderberg model is implicitly based on LTE. So things have gotten mixed, and you can't expect to get a consistent result without going also to a temperature distribution derived from a non-LTE interpretation.

CONTINUOUS OPACITY BETWEEN λ 2000 AND λ 4000 IN THE SOLAR PHOTOSPHERE

S. Matsushima

Department of Astronomy, Pennsylvania State University,
University Park, Pennsylvania

ABSTRACT

The principal cause of the continuing disagreement between
theoretical predictions and observations of the visual spectrum of
the Sun does not seem to be in the inaccuracy of the basic theory,
but rather in the errors involved in the ultraviolet opacity. It is
found that satisfactory agreement can be obtained between theory
and observations of the intensity distributions with wavelength, as
well as of limb darkening throughout the entire spectral region, if
one assumes an additional opacity larger than metal absorption by
a factor of ~ 10 to 100 in the wavelength region between λ 2500 and
λ 3500. The effect of a possibly large error in the iron abundance
and its photoionization cross section is considered.

Because of persisting disagreement between observations and theoretical
predictions of the solar continuous spectrum, the conventional model-atmosphere approach has often been criticized. A typical example is a recent paper
by Swihart (1966, and also a paper presented at this conference), in which he
concludes that it is not possible to obtain a satisfactory agreement between
the theory and observations unless the inhomogeneities of the photospheric
structure are taken into account. Although the observations clearly indicate
the inhomogeneous structure of the chromosphere or the coronal region, the
question is whether such an inhomogeneity is sufficiently serious to make the
set of basic assumptions invalid even in the deep continuum-forming layers.
Furthermore, Swihart's conclusion is based on the difficulty in obtaining a
simultaneous agreement with observations for both intensity and limb darkening for the spectral region longer than 4000 A. However, the most serious
disagreement between the theory and observations appears in the near ultraviolet beyond the Balmer discontinuity, i. e., the region where the comparison was not made by Swihart or most of the previous workers.

Because of the basic condition of flux constancy imposed on the present-
day model-atmosphere calculations, errors in the flux computation in one
region of the spectrum would produce correspondingly large errors in the
other spectral regions. A comparison with recent observations indicates that
the emergent flux computed for a model is much larger than the observed
values in the ultraviolet region beyond the Balmer discontinuity. This seems
to suggest strongly the existence of a still unknown source of opacity much
larger than the total opacity known to us at present.

In view of the above discrepancy, we have constructed a large number of
flux-constant models in which a varying amount of additional opacity is
assumed so as to give the best possible agreement with the observed intensity
distribution with wavelength over the entire region of the spectrum. The comparison was also made with recent observations of the ultraviolet limb
darkening, so that the variation of the assumed opacity with depth may be
established.

As a result, we found that in order to obtain satisfactory agreement
between the theory and observations we have to assume an additional opacity

larger by a factor of 10 to 100 than the metal absorption in the region beyond the Balmer discontinuity. Not only does the effect of the additional opacity decrease the emergent flux and deepen the limb darkening, but the change in the T ~ τ relation affects the visual and infrared regions in such a way that the flux in these regions is increased so that better agreement with observations is seen. Thus, the origin of the disagreement in the visual region longer than 4000 A, as pointed out by Swihart, is not necessarily in the errors in the basic theory but in the errors in the ultraviolet opacity assumed in the previous calculations. The details of this study will be published elsewhere (Matsushima, 1968), and in the present report we summarize only the principal results of the calculations.

The changes in the T-P relation due to increases in the ultraviolet opacity are shown in Figure 1, in which a comparison is made for the three representative models: (1) model including only hydrogen and negative hydrogen ion as the source of opacity (shown by the dotted line); (2) model in which the metal absorption is added to H and H⁻ (broken line); and (3) model to which is further added the assumed source of unknown opacity (solid line). The values assigned to the dots specify the optical depth at λ 5000. We note that the change from model (1) to (3) is larger than that from (1) to (2), which indicates the importance of the effect of the assumed opacity on the photospheric radiation field.

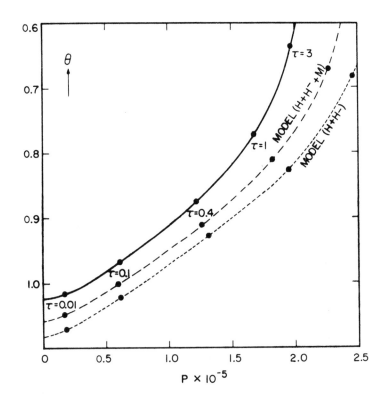

Figure 1. Comparisons of the temperature stratification between the final models obtained for the three different choices of opacity as indicated on each curve. The solid curve is for the model including the assumed unknown opacity in addition to hydrogen and the metals.

The adopted photoionization cross sections for the metals are those recently computed by Travis and Matsushima (1968), using the quantum defect method of Burgess and Seaton (1960), except for Cr, Fe, and Ni; for these, detailed hydrogenic approximations are used because of the inapplicability of the quantum-defect method. The metals that make a significant contribution to the continuous opacity at least at some region under consideration are C, Mg, Al, Si, Ca, Cr, Fe, and Ni. Since many of the absorption edges arising in the combined metal absorption have no practical importance in the result of a model-atmosphere calculation, only the large peaks are included for the shorter wavelength region, where the flux is very small in the solar atmosphere. Thus, the discontinuities at λ 1100 (C $2p^2\ {}^3P$), λ 1518 (Si $3p^2\ {}^3P$), λ 1674 (Si $3p^2\ {}^1D$), λ 2515 (Mg $3p\ {}^3P$), λ 2937 (Ca $4p\ {}^3P$), and λ 3642 (Ca $3d\ {}^1D$) are included explicitly, but other peaks such as the one at λ 2075 due to the ground state ($3p\ {}^2P$) of aluminum are smoothed out in order to reduce the amount of iterative computations.

The intensities of the emergent radiation at the center of the disk obtained for the above three models are compared with various observations in Figure 2, where I_λ vs λ^{-1} curves obtained for the three basic models are distinguished by the same choice of lines as in Figure 1. In order to exemplify the comparisons in the near infrared region, the scales of both axes are increased by a factor of 2 in this region. The effect of the increase in the ultraviolet opacity upon the visual and infrared regions is clearly seen in this figure. If we assume larger opacity in the ultraviolet, the radiation in that region emerges from more superficial regions, and the energy absorbed in the deeper layers has to be reradiated in the longer wavelength regions in order to maintain the constant flow of energy. The $T \sim \tau$ relation in the photosphere will accordingly be changed. Thus, we see the condition of flux constancy in the equality of the areas under the three intensity curves, noting that I_λ, energy per unit wavelength, is plotted against the λ^{-1} scaling.

Another striking feature is seen in the large discontinuity at λ 2515 due to the photoionization from the ^3P state of magnesium. This discontinuity is greatly reduced by adding the assumed opacity and is probably not detectable in observations. The fact that such a large discontinuity has not been observed may be considered further evidence to support the existence of a still-unknown source of opacity. In this connection, it should be remembered that the well-observed discontinuity near λ 2100 is presumably due to the photoionization from the ground state ($3p\ {}^2P$) of aluminum (Tousey, 1967; Bonnet et al., 1967). Unfortunately, as mentioned before, we have smoothed out all the metal discontinuities between λ 1674 and λ 2515 for the sake of simplifying the present computations. However, we note that the aluminum absorption peak appears to be less than the magnesium peak above H⁻ absorption at λ 2515 (see Figure 4). Hence, had there been no additional opacity in the wavelength region longer than λ 2500, a drop in intensity would be observed at λ 2515 at least as sharp as that at λ 2100.

It should be mentioned at this point that all the observations plotted in this figure refer to the continuum between the lines but not to the line-smoothed values. Hence, the total flux assigned to each model must be larger than that given by the solar constant by as much as the fraction covered by the Fraunhofer lines. In fact, we found that the best agreement was obtained by increasing the effective temperature from 5800°K to 5900°K. This corresponds to the mean blanketing coefficient of 0.071, in good agreement with the value of 0.083 directly determined from high-dispersion solar spectra by Mulders (1935).

The limb-darkening curves computed for the three basic models are compared with observed limb darkening in Figure 3. Again, the theoretical curves for the three basic models are distinguished by the choice of the same type of lines as in Figures 1 and 2. At each wavelength, the limb darkening depends on the variation of the opacity with depth, and thus the dependency of assumed opacity on T and P_e is confirmed. The differences between different models in the visual and infrared regions are insignificantly small, and, in fact, all the models give good agreement. The difference increases toward the ultraviolet and reaches maximum at λ 2190. The curves obtained for the model including the metals but no additional opacity coincide with those for which

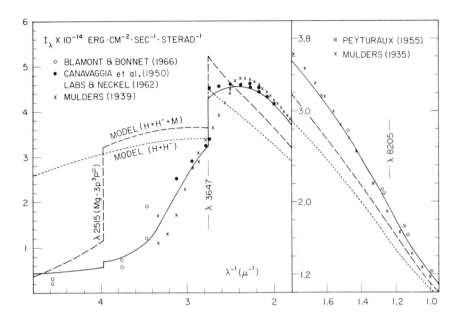

Figure 2.   The intensity variations with wavelength at the center of the Sun's disk compared with various observed values.

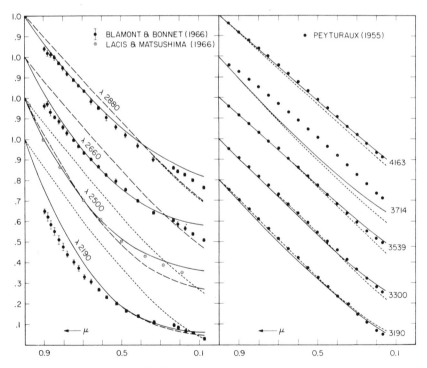

Figure 3.   Theoretical limb-darkening curves in the violet and ultraviolet regions compared with various observations.

only H is considered at $\lambda\, 2800$ and $\lambda\, 2660$. At the same time, they show little difference from the model with unknown opacity at $\lambda\, 2500$ and $\lambda\, 2190$. This clearly indicates the importance of the metal absorption relative to that of H and the assumed opacity source.

As a result of successive approximations reached by comparison of the predicted intensity and limb darkening with observations, it is found that the following expressions give the best fit to both intensity and limb-darkening observations, as shown in Figure 2 and 3:

$$\log \kappa_\lambda^u\ (\lambda \geq 0.2817\ \mu) = -4.200 - 0.30\lambda^{-1} + 1.20\lambda^{-1}\,\theta + \log P_e\ , \qquad (1)$$

$$\log \kappa_\lambda^u\ (\lambda < 0.2817\ \mu;\ \theta \geq 0.9) = 1.571 + 0.90\lambda^{-1}\,\theta - 1.48\lambda^{-1} + 0.369\log P_e\ , \qquad (2)$$

$$\log \kappa_\lambda^u\ (\lambda < 0.2817\ \mu;\ \theta < 0.9) = 4.829 - 3.62\,\theta + 0.90\lambda^{-1}\,\theta - 1.48\lambda^{-1} + 0.369\log P_e \qquad (3)$$

The relative contribution of the assumed opacity is shown in Figure 4, where the logarithms of the mass absorption coefficients are plotted for various values of T and P_e, approximately representing each optical depth in the solar photosphere as indicated in the figure. We see that, in the region between $\lambda\, 2500$ and $\lambda\, 3500$, the assumed absorption is at least an order of magnitude larger than the metal absorption and becomes as much as 100 times larger at maximum near $\lambda\, 2800$.

The foregoing calculations involve some uncertainties regarding the effect of Fraunhofer lines on the continuum. The line-blanketing effect (both back-warming and line-blocking) is partially accounted for by raising the effective temperature by an amount corresponding to the energy removed by the lines, as was discussed earlier in this paper. In fact, since line absorption affects the emergent flux in such a way that it raises the continuum between the lines (Mihalas and Morton, 1965), the line-blanketing effect does not necessarily lower the continuum as defined by the flux between the lines. However, the above procedure may not be fully appropriate for a possibly complicated depression of continuum due to overlapping of the wings of absorption lines. This effect should appear most seriously near the Balmer discontinuity. The peculiar deviation of the theoretical limb-darkening curves from observations at $\lambda\, 3714$ may be closely related to this effect (Figure 3). In order to examine this point, therefore, we have computed the emergent intensity including hydrogen lines up to H_{18} for the final model. We found that satisfactory agreement with observations can be obtained for the intensity distribution near the discontinuity if we include both the assumed source of unknown opacity and the Balmer-line overlapping. However, the line-overlapping effect alone does not seem to be sufficient to account for the depression of the continuum shown by careful observations recently made by Houtgast (1965).

Thus, it is most unlikely that the line blanketing is entirely responsible for the discrepancy between theory and observations considered in the present paper, although a more detailed treatment of absorption lines would undoubtedly require some modifications on the predicted form of unknown opacity as given by equations (1) to (3). Instead, perhaps a larger uncertainty in the present results might be caused by possibly large errors involved in the metal-absorption coefficients. Comparisons with experimentally determined values indicate that the photoionization cross sections computed on the basis of the quantum-defect method are generally accurate to within a factor of 2 or 3. We may also expect a comparable accuracy in the abundance determination (Müller, 1966). Hence, the error in the metal absorption does not seem to exceed the factor required for the additional opacity. However, the case for Fe (and possibly for Cr and Ni, too) is an exception. We have shown that the

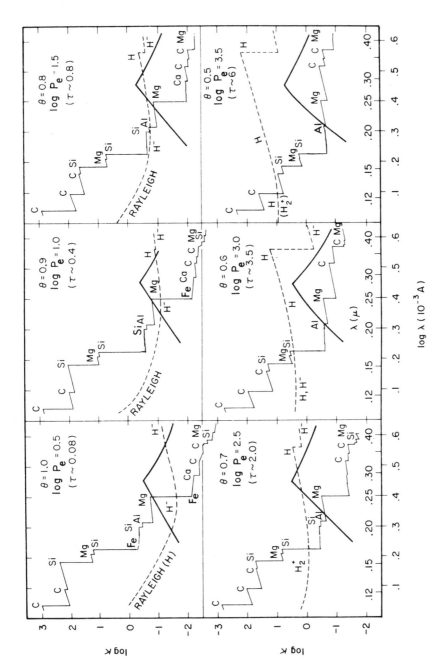

Figure 4. Comparisons of the variations of the mass absorption coefficients with wavelength and depth for the un-known source assumed (heaviest lines), for the metals (lighter solid lines), and for H, H⁻, and H₂⁺ com-bined (dotted lines). The important peaks are labeled to indicate the metal responsible for that discon-tinuity, and the dominant state of hydrogen is marked on each portion of the dotted lines. The value of τ indicates the optical depth at λ 5000 of the layer roughly corresponding to each selected pair of θ and log Pₑ in the solar atmosphere.

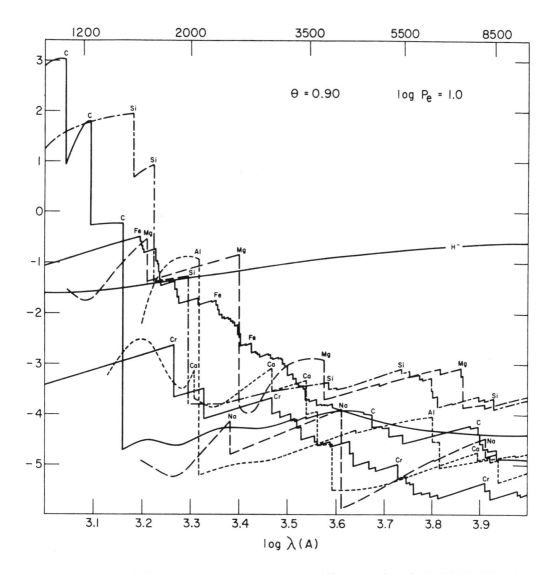

Figure 5.   Logarithms of mass absorption coefficients for the individual metals
for a solar abundance (Goldberg et al., 1960).   The conditions,
$\theta = 0.9$ and log $P_e = 1.0$, are chosen to correspond roughly to
$\tau_{5000} \sim 0.5$ for a solar-type star.   The individual absorption peaks
are labeled to indicate the various metals, and the absorption due
to H⁻ is included for comparison.

use of the hydrogenic approximation could yield an error of an order of magnitude (Travis and Matsushima, 1968).   Moreover, the iron abundance adopted
in the present calculation (Goldberg, Müller, and Aller, 1960) might involve
an error as large as a factor of 10 or more.   In a series of papers, Pottasch
(see Pottasch, 1967, for references) reports that the iron abundances determined from coronal lines or ultraviolet emission lines are larger than those
obtained from the photospheric lines by a factor of 10 to 20.   Although there
is no reason to believe that the abundance ratios should be the same between

the corona and the photosphere, Pottasch (1966) concludes that it is still too early to say that a difference in composition actually exists between the two regions. Furthermore, it is important to note the result of laboratory measurements of f values for iron recently carried out in Kiel (Koch, Köpsel, and Richter, 1967). The new values reported are lower than the previously accepted values by a factor up to about 10. This would immediately require an increase of the photospheric iron abundance by the corresponding factor.

Thus, the combination of uncertainties in the cross sections and abundance could yield an increase in the adopted absorption coefficient for iron by a factor of 100 or more. The importance of iron absorption relative to other metals is shown in Figure 5. We see that even with the currently adopted values, iron dominates over the other metals in the spectral region between λ 2500 and λ 3200. A comparison of Figures 3 and 4 also indicates that the wavelength dependency of iron absorption is quite similar to that of the assumed opacity in the wavelength region longer than λ 2800. The difference between the two curves amounts to a factor of about 200, which may be somewhat too large to account for the uncertainties mentioned above. There might arise another difficulty in the region shorter than the magnesium discontinuity at λ 2500, since the increase in iron absorption appears to yield too large opacity in this region. However, it should be remembered that the present investigation does not include a detailed comparison in the region shorter than λ 2000, and, in effect, the comparison in Figure 2 appears to show that the predicted intensity is still larger than observations in the far ultraviolet. Thus, the possibility of iron being the source of unknown opacity can not be completely ruled out.

Summarizing the foregoing discussion, we may conclude that in order to obtain satisfactory agreement between theory and observations we must assume an additional opacity of an amount about a factor of 10 to 100 larger than metal absorption. Since the effect of errors in opacity in one spectral region on the other appears to be very large, any comparison of the theoretical flux with observations is meaningful only when made on the basis of the overall wavelengths rather than on an isolated region of the spectrum. The limb-darkening observations for the visual region do not seem to yield a clue to the validity of the model, whereas the ultraviolet limb darkening appears to be very important. Finally, it should be emphasized that the result of the present investigation is a direct consequence of calculations based on the flux-constant models, suggesting the importance of the effect of a small modification of the temperature stratification upon various observable quantities, especially those related to the continuum.

It is a pleasure to thank Yoichi Terashita for his assistance in computations, and Leo Goldberg and Volker Weidemann for calling my attention at this conference to the new experiment in Kiel.

This work was supported in part by the National Science Foundation under Grant NSF-GP 8058.

REFERENCES

Blamont, J.-E., and Bonnet, R.-M. 1966, C. R. Acad. Sci. Paris, Ser. B, 262, 152.
Bonnet, R. M., Blamont, J. E., and Gildwarg, P. 1967, Ap. J., 148, L115.
Burgess, A., and Seaton, M. J. 1960, Mon. Not. Roy. Astron. Soc., 120, 121.
Canavaggia, R., Chalonge, D., Egger-Moreau, M., and Oziol-Peltey, H. 1950, Ann. d'Ap., 13, 355.
Goldberg, L., Müller, E. A., and Aller, L. H. 1960, Ap. J. Suppl., 5, 1.
Houtgast, J. 1965, Proc. Roy. Netherlands Acad. Sci., Ser. B, 68, 306.
Koch, M., Köpsel, M., and Richter, J. 1967, preprint.
Labs, D., and Neckel, H. 1962, Zs. f. Ap., 55, 269.
Lacis, A. A., and Matsushima, S. 1966, J. Opt. Soc. Amer., 56, 1239.

Matsushima, S. 1968, Ap. J., 154, 715.
Mihalas, D. M., and Morton, D. C. 1965, Ap. J., 142, 253.
Mulders, G. F. W. 1935, Zs. f. Ap., 11, 132.
_____. 1939, Publ. Astron. Soc. Pacific, 51, 220.
Müller, E. A. 1966, Proc. IAU Symp. No. 24 (London: Academic
    Press), 171.
Peyturaux, R. 1955, Ann. d'Ap., 18, 34.
Pottasch, S. R. 1966, Bull. Astron. Netherlands, 18, 443.
_____. 1967, Bull. Astron. Netherlands, 19, 113.
Swihart, T. L. 1966, Ap. J., 143, 358.
Tousey, R. 1967, Ap. J., 149, 239.
Travis, L. D., and Matsushima, S. 1968, to be published.

DISCUSSION

Elste: You have used a radiative equilibrium model, and in order to obtain a low-enough energy flux in the region 2000 to 3000 A, you need a very large opacity. But if you reduce the temperature in deeper layers, as is predicted from the mixing-length theory, then you don't need that much opacity.

Matsushima: I do not know on what basis you can say so. According to Swihart [Ap. J., 143, 358, 1966] and the result of a more recent calculation of solar models including convection by Hans G. Groth [private communication], it seems that at least for the solar-type atmospheres the effect of convection on the temperature stratification appears only in a layer too deep to show any appreciable change in the emergent flux.

Gingerich: I think that this Lyman-α resonance wing that Miss Cuny talked about can very well be the missing opacity source that you say acts in this region.

Matsushima: You mean at 2500 A?

Gingerich: Yes, I just looked at my calculations at 2100 A, and Lyman α is running a factor of 20 above the next most important absorber. So it could very easily continue right on up.

Matsushima: It's a little annoying to me, because such an extension could produce a difficulty at longer wavelengths.

Strom: We appear to agree quite well with your results for the temperatures in white dwarfs.

EFFECTS OF INHOMOGENEITIES ON THE SOLAR PHOTOSPHERE

T. E. Margrave, Jr.

Georgetown College Observatory, Washington, D.C.

T. L. Swihart

Steward Observatory, University of Arizona, Tucson, Arizona

Previous work (Swihart, 1966) indicated that purely homogeneous models of the solar photosphere could not give a completely satisfactory fit with the observed solar continuum, i.e., with the observed wavelength distribution of intensity and limb darkening at those wavelengths that are relevant to the photosphere. We attempted to make a quantitative check on this conclusion by incorporating granulation in the models. At first it appears that the inhomogeneities simply add to the degrees of freedom of the models, making an improved fit a trivial matter. It turned out to be much more difficult than expected, since a significant improvement over the homogeneous models was obtained only under very restrictive conditions; however, we found that granulation should be taken into account in models of the photosphere if a fit within observational error in the solar continuum is desired.

There have been many previous investigations of granulation in the sun, but these seem to have used two mutually contradictory assumptions: hydrostatic equilibrium and constancy of pressure on horizontal planes. These two conditions cannot both be valid if there are temperature fluctuations on horizontal planes, and apparently previous investigators thought this built-in contradiction unimportant. We found that it is quite important and has significant effects upon the photospheric structure.

We were studying the granulation not for its own sake, but for its effects on the photospheric radiation field. The discrepancies that we were trying to explain were very small, so a very crude theory of the inhomogeneities was sufficient for our purposes. Accordingly, a simple two-column model of the granulation was used.

Because of the reasons given above, the numerical details of the final model should not be taken very seriously. There are, however, a number of results that are worth noting. First, we found that the temperature differences between the hot and cold columns at equal optical depths are considerably greater than those at equal geometric depths. This is a direct consequence of our dropping the assumption of pressure equality on horizontal planes. Second, horizontal pressure differences lead to horizontal motions, and the pressure differences that we found were such that the horizontal motions would tend to destroy the granules in times of the order of 2 min. This is not inconsistent with the observed lifetimes of the granules. Finally, we made a check with Edmonds' (1962) reductions of balloon observations of granulation contrast, and the agreement is satisfactory for such an overly simplified model.

Our conclusions are that granulation does have a small but not negligible effect on the photospheric radiation, that in future granulation investigations it is important to have a realistic treatment of blanketing, and that the artificial assumption that the pressure is constant on horizontal planes should be dropped.

REFERENCES

Edmonds, F. N., Jr. 1962, Ap. J. Suppl., 6, 357.
Swihart, T. L. 1966, Ap. J., 143, 358.

DISCUSSION

Matsushima: I have a rather strong argument against your very first conclusion — that precise agreement cannot be reached without assuming inhomogeneities. In your published note [Ap. J. 143, 358, 1966] you did not show
the intensity or the limb darkening to the ultraviolet side of 4000 A, because
the region was too messy. My comment is that if you did not get the precise
agreement in that ultraviolet region, then you cannot say much more than
that, because the effect of an increase in the opacity just beyond the Balmer
discontinuity would change the temperature stratification so as to change the
observable features in the visual and infrared regions.

Swihart: I have what I consider to be excellent reasons for believing you are
wrong. In brief, I take the ultraviolet into account in a manner that is consistent with the observational accuracy in the visible. That is, I do not completely ignore it; on the other hand, an error of 50% in what's going on below
the Balmer jump will have only a minor effect in the visual region where the
criteria I am after are much more stringent and much more important. An
error of that magnitude will not affect our results.

Elste: How do you calculate the limb darkening?

Swihart: We calculated it for the cool columns in the line and in the continuum,
and for the hot columns in the line and in the continuum; and we averaged over
both line and continuum, that is, combined them to get a net result. Naturally,
when we compare with observations that are in the continuum only, we leave
the lines out.

Elste: I must criticize this strongly. I investigated this procedure last year,
and I tried to reproduce Edmonds'[Ap. J. Suppl., 6, 357, 1962] results. I
found that it is really necessary to integrate through these regions, changing
the parameters as you go through, rather than averaging. I have based my
calculations on equal pressure at equal geometrical depths, because I have
estimated the effects of a 2 km^{-1} sec streaming velocity in the hot regions as
compared with the cold regions and it had a negligible effect on the pressure.
This will not necessarily be true for pressure-sensitive lines.

Swihart: That is carrying the granulation calculations to a sophistication
beyond what we intended. Also, without pressure differences there can be
no temperature differences and, therefore, no granulation.

FORMATION OF THE SOLAR LYMAN CONTINUUM

L. H. Auer[*] and D. Van Blerkom

Joint Institute for Laboratory Astrophysics, Boulder, Colorado

There is probably only one solar chromosphere on a given day; however, as a glance at any book on the subject will show, there are far more than one model of the chromosphere. Apart from questions of shock fronts, spicules, and inhomogeneities, some of these models should be wrong. We decided it would be useful to investigate the sensitivity of the flux emitted in the Lyman continuum to the assumed model. Since this datum was not used in the construction of any of the models, such a calculation provides an independent check on their validity.

In what follows, we mean by a chromospheric model an assumed run of temperature with physical height, and one assumed value of the electron density at the base of the chromosphere—usually 500 km above the photosphere. For a given model we can construct a self-consistent model by integrating the hydrostatic equilibrium equation upward and simultaneously solving the statistical equilibrium equations and equations of radiative transfer.

Although we solved the transfer equation by varying the relevant parameters with height, we took a restricted model of the hydrogen atom. We assumed that there are only two bound levels, that Lyman α was in detailed balance, and that the Balmer continuum was formed so much deeper that the photoionization rate from the second level could be regarded as fixed. Under these assumptions we need to solve only for the radiation field in the Lyman continuum. The assumptions are probably justified inasmuch as we are mainly interested in investigating the sensitivity of the Lyman-continuum flux to parameters of the model. It is difficult to believe that changes in the atomic model could cause very large differential changes between models.

We may write the equation of transfer for this simple model neglecting stimulated emission as

$$\mu \frac{\partial I_\nu}{\partial Z} = n_1 a_\nu \left(I_\nu - \frac{n_1^*}{n_1} B_\nu \right) \, , \tag{1}$$

where n_1^* and n_1 are, respectively, the LTE and non-LTE populations of the first level, B_ν is the Wien function, and a_ν is the photoionization cross section. By using the equation of statistical equilibrium, we can eliminate the quantity n_1^*/n_1 from equation (1) to get an integro-differential equation in the Lyman-continuum radiation field.

In finding a self-consistent radiation field, we used the following iterative scheme: Given an estimate of the radiation field, the hydrostatic equation was integrated outward. At each point, values of n_e, n_H, n_1, and n_2 were found that were consistent with the total particle density. Contributions of electrons by other elements were included in the ionization balance, assuming the abundances in Allen (1963). Given the run of populations, the transfer equation is solved to find a new estimate of the radiation field.

If we are to avoid a nonlinear transfer equation, the n_1 in the opacity term of equation (1) cannot be eliminated in terms of the Lyman-continuum

---

[*]Now at Yale University Observatory.

radiation field, but rather must be taken from the solutions to the statistical equilibrium equations based on the last iteration. This causes severe convergence difficulties. The temperature is not known on an optical-depth scale, but only on a physical-depth scale. Further, the temperature decreases rapidly as we go toward the photosphere. If we underestimate the ground-state population, we will have too small an opacity and will see too deep into the chromosphere; and, since the emergent intensity is characteristic of the source function at optical depth 1, we will get much too little radiation. On the next iteration the level of ionization will be too small, and the reverse will occur. The oscillations may be effectively damped by linearizing the opacity term in equation (1); i.e.,

$$n_1 (I - S) \rightarrow n_{1\,0}(I - S) + (I_0 - S_0) \frac{dn_1}{dB_{1K}} \Delta B_{1K} \quad , \tag{2}$$

where

$$B_{1K} = \int_{\nu_{1K}}^{\infty} \frac{4\pi}{h_\nu} a_\nu J_\nu \, d\nu \quad \text{and} \quad \Delta B_{1K} = B_{1K} - B_{1K_0} \quad .$$

Quantities with subscript 0 are carried from the previous iteration, and the variables without subscripts contain the unknown radiation field. With the modification indicated in equation (2), the problem is still conveniently treated by means of difference equations. The inclusion of the linearized opacity terms improves convergence by a factor of 4 or more.

Our major result is that, while the shape and magnitude of the emitted flux in the Lyman continuum are not directly dependent upon the temperature structure, the intensity is roughly proportional to the square of the electron density in the region of continuum formation. If we change our lower boundary condition on n_e, essentially any magnitude can be found. Thus, if we increase the electron density at the base of the Bilderberg chromosphere tenfold, an emergent flux comparable to the value of $F_\lambda \approx 40$ ergs sec^{-1} cm^{-2} found by Hinteregger (1965) may be achieved (see Figure 1). This, of course, is really only parameter juggling. If we assume the structure of the atmosphere below the chromosphere to be well determined, there is no reason to believe such a large electron density exists at that depth. We can only conclude that the Bilderberg model has much too slow a temperature rise and is thus unable to support the necessary densities. The temperatures and density used by Linsky (1968) and Thomas and Athay (1961) seem to be in better agreement, though the electron density in both of these models seems too high. If the electron density is scaled down by a factor of 3, it can be brought into close agreement with the observation.

Besides the simple ad hoc variation of the electron density, there is another very important physical parameter that controls the Lyman-continuum flux, i.e., the assumed abundance of He. When the calculations were first made, this effect did not seem to be important; however, it now seems obvious. The He abundance is present in the density scale height. With a high He the atmosphere falls off more quickly, and in the region of continuum formation—the location of which is controlled by the temperature structure—n_e^2 will be smaller and thus the emission will be less. The effect is shown in Figure 2. The upper curve is the emitted flux for the Linsky temperature distribution with $n_{He}/n_H = 0.10$; the lower is for $n_{He}/n_H = 0.15$. This exactly reflects the differing electron structures shown in Figure 3. In the region of continuum formation, the two cases differ by a factor of ~ 1.8 in n_e. Squaring this factor gives the difference in the emitted fluxes. The second hump in the electron density is real. It is also found in independent calculations by Linsky.

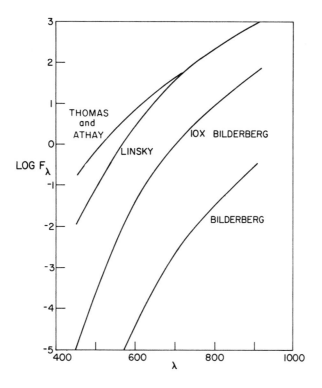

Figure 1.   Lyman-continuum fluxes from the various models.

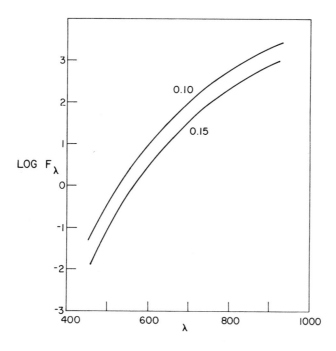

Figure 2.   Flux for different He abundances.

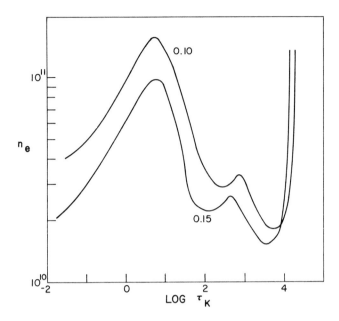

Figure 3.   Run of n_e for different He abundances.

In conclusion, we have found that the emission in the Lyman continuum is a sensitive measure of the density in the region of continuum formation. This density is primarily controlled by the length of the temperature plateau in the model, for the continuum is formed where hydrogen starts to ionize— where the temperature turns up sharply.   Thus, comparison of theoretical studies of the Lyman continuum with observation should provide good information about the height of the plateau.

We would like to thank Drs. R. N. Thomas and D. Mihalas for many useful talks during our investigation of this problem.   We would also like to acknowledge the support of NSF, contracts GP 8052 and GP 7761, and the Advanced Research Projects Agency (Project DEFENDER), monitored by the U. S. Army Research Office—Durham, under contract DA-31-124-ARO-D-139.

REFERENCES

Allen, C. W.   1963, <u>Astrophysical Quantities</u> (2d ed. ; London: Athlone Press).
Hinteregger, H. E.   1965, <u>Space Sci. Rev.</u> , <u>4</u>, 461.
Linsky, J. L.   1968, private communication.
Thomas, R. N. , and Athay, R. G.   1961, <u>Physics of the Solar Chromosphere</u> (New York: Interscience Publishers, Inc.), p. 217.

DISCUSSION

<u>Pasachoff</u>: It should be pointed out that the intensities and height scales that are being advanced as data, and are being fitted by these programs, result from a superposition of various kinds of chromospheric fine structure — spicules and interspicular regions.   These regions will be very different, and the fluxes will finally be some kind of weighted mean.   It is not at all clear that this mean will be linearly weighted.

Auer: I agree, and probably the one with a higher electron density would be weighted more strongly.

Kalkofen: Have you estimated the influence of Hα in these calculations?

Auer: No, I have not in this preliminary work.

Thomas: Pottasch and I did this calculation a long time ago [Ap. J., 132, 195, 1960], only we treated a three-level rather than a two-level atom, plus continuum; so we included the effect of Hα. Our arithmetic was not so elegant because we only did a one-point quadrature on the radiation field. Hence the Auer and Mihalas work, being done with more correct mathematics, and our work, being done with more correct physics, should be complementary. Pottasch and I treated an atmosphere with assumed $T_e[\tau]$, and then made a rough attempt to treat the actual solar atmosphere. Athay and I refined this approach later [Thomas and Athay, Physics of the Solar Chromosphere, Interscience Publ., 1961; chap. 4] to treat our solar model derived from eclipse data. We predicted the actually observed Lyman-continuum flux to within a factor of 2. Our results agreed with Auer and Mihalas in requiring a much higher T_e than the Bilderberg model shows; and I agree with Auer and Mihalas in noting that the Bilderberg model will miss by orders of magnitude in agreeing with observations (they will predict a too low Lyman continuum). I disagree with Auer's remarks on the relation between T_e and n_e; these things cannot be adjusted independently of one another. The fact that the Bilderberg model is completely inconsistent in its values of T_e and n_e, because it has used LTE ionization relations between n_H, n_{H^-}, n_e, and τ, is one of the reasons that I said the model is simply nonsense when I first saw it. The Auer and Mihalas results confirm this remark.

One of the things we really want to know is the relative importance of mechanical heating and the Cayrel planetary nebular mechanism for raising T_e. To answer this question, it is important to know the correct relation between n_e, n_H, n_{H^-}, T_e, because the question is settled by knowing where collisional effects give way to radiative effects in the H^- ionization equilibrium. Now that we know Ferguson's new determination of the associative detachment rate, all that we need to settle the question is a knowledge of these relative concentrations. So, if you publish misleading values for them, as the Bilderberg model does, you are kidding yourself that you can draw any conclusions from them.

Auer: One other comment. I think one of the most important regions for the Lyman continuum is out near 6200°K, and, in particular, exactly where the temperature rise occurs.

Thomas: The important region for the observed Lyman continuum is obviously that where $\tau[LyC] \sim 1$. This occurs just where T_e is rising rapidly, in the 1961 model by Athay and myself, near 1000 km. As I said, our model predicts Lyman continuum within a factor 2 of what it is observed to be; the Bilderberg model misses by several orders of magnitude. So I think you had better buy our T_e-model, which is rising rapidly near 1000 km, hitting just below 10,000° at about 1200 to 1300 km, and rising very rapidly there. This is in strong contrast to the Bilderberg model.

Strom: A couple of years ago I made a detailed estimate of what the value of the temperature rise would be. I considered the low-density case by setting the collision cross sections to zero, and the highest temperature that I got was 5300°, which was considerably different from the 6000°.

Thomas: A student of mine, Stuart Jordan, has just finished calculating this, as part of his thesis on chromospheric heating. He looked at the two alternatives of a strong heating in the region 0 to 500 km, and zero heating, and compared the two resulting T_e-distributions. We were curious as to whether we could use our 1961 empirical model, in combination with these calculations, to decide where and how much chromospheric heating was influential in the region 0 to 500 km. Both models gave T_e between 5800 and 6250 at the 500-km height, so it is hard to distinguish between them on the basis of present material and the sophistication of the theory used. Note, of course, that all these results lie well above the Bilderberg model, which gives something like 5000° in this region.

Cayrel: Feautrier has repeated the calculations with the latest cross sections. What did you get?

Feautrier: We found a rise of 500° above the minimum.

Mihalas: I would now like to turn from the sun to the stars...

Thomas: That was the stars, Dimitri.

Mihalas: ... and call on Miss Underhill.

EVIDENCE CONCERNING THE VALIDITY OF THE
SAHA AND BOLTZMANN LAWS IN O-, B-, AND A-TYPE ATMOSPHERES

A. B. Underhill

Sonnenborgh Observatory, Utrecht, The Netherlands

ABSTRACT

The observational evidence that non-LTE populations occur
in O-, B-, and A-type stellar atmospheres is described as dilution effects, optical pumping by radiation (fluorescent effects),
optical pumping by collision, and particular ionization effects
including autoionization. Spectra known to be susceptible to
dilution effects are He I, C III, N IV, O I, Si III, Ti II, Cr II,
Mn II, Fe II, and Ni II. The emission lines C III 5696 and N IV
4057 are excited as a result of optical pumping, while the degree
of ionization of P^+ is affected by special ionization processes,
and Ca, Ga, and Ba are known to be subject to autoionization.
Using classical model atmospheres and the standard LTE theory
of line formation, one can readily see that the electron densities
in the layers from which the normally observed lines in O, B,
and A spectra come vary between 10^{13} and 10^{14} while the temperatures vary between 8000° and 30,000°K. Since laboratory
experience indicates that in an optically thin plasma at temperatures
of the order of 10,000° an electron density of the order of 10^{16} is
required to ensure that the Saha and Boltzmann laws be valid, these
laws are not likely to be valid in O-, B-, and A-type atmospheres.
The greatest effects of deviations from LTE are expected in the
spectra listed above. These spectra furnish many of the lines used
for spectral classification.

If the hypothesis of LTE is valid in a stellar atmosphere, the problem of
predicting the strengths of lines in stellar spectra is greatly simplified. The
line strengths are then dependent only on general properties of the atmosphere
such as temperature, pressure, and chemical composition, and attention need
not be paid to the particular balance occurring between specific radiative and
collisional processes. Under the hypothesis of LTE, the Saha and Boltzmann
laws may be used for calculating the distribution of atoms and ions over their
energy levels, and Kirchhoff's relation may be used for the source function.
These convenient summaries of some aspects of the physics of a mixture of
gas and radiation should not be adopted without consideration of the conditions
under which they are valid and without reasonable proof that the necessary
conditions exist.

The observational evidence that many spectral lines in early-type spectra
are not formed under conditions of LTE may be summarized as four effects.

A. _Dilution effects._ By this we mean that particular absorption lines
arising from metastable levels, or from levels closely linked to metastable
levels, are stronger in comparison with the lines from normal levels than is
observed to be the case in the majority of stars. The majority of stars are
assumed to have "normal" spectra. The quality "normal" is not further
specified, although it is often implied that "normal" spectra correspond to
the state of LTE. Ionic and atomic spectra known to contain metastable levels
and to show dilution effects are He I, C III, N IV, O I, and Si III; dilution
effects are also said to occur in Ti II, Cr II, Mn II, Fe II, and Ni II. The
most conspicuous cases of dilution effects are found in the extended

atmospheres of supergiants, shell stars, Of stars, and Wolf-Rayet stars. Less conspicuous cases occur in the spectra of main-sequence stars.

B. <u>Optical pumping by radiation.</u> Optical pumping by radiation is filling a particular energy level as the result of absorption of radiation of one particular wavelength. Such an occurrence is often called fluorescence. Well-known cases in stellar atmospheres are the Bowen mechanism for using He II 303 quanta to excite emission lines of N III 4634, 4640, 4641 in planetary nebulae and in Of stars and the excitation of He II 4686 in emission as the result of absorbing Lyman α. In cool stars we have various "mutilated multiplets" of Fe I and the resonance lines of In I appearing in emission as a result of similar processes (see Merrill, 1956, for references).

C. <u>Optical pumping by collision.</u> In this case an atom or ion in the ground state collides with an abundant species in an excited state and the excitation energy is transferred during the collision. The result is that atoms or ions of the first species are selectively excited to one particular state. This type of mechanism may be invoked to account for the selective occurrence of N IV 4057 in emission in high-excitation Of stars, since the upper level of λ 4057 has very nearly the same energy as a He^+ ion in the n = 7 state. The required energy is too great to be obtained from any abundant resonance line in the manner postulated for optical pumping by radiation. Furthermore, the upper state of λ 4057 has the same parity as the ground state of N^{+3}; thus, λ 4057 cannot be selectively excited in emission by the absorption of a single quantum. Optical pumping by collision is a well-known way of exciting some laser transitions. It also accounts for the occurrence of many forbidden lines in nebulae.

D. <u>Particular ionization effects.</u> These effects are similar in principle to optical pumping by radiation and by collision. They perturb the degree of ionization of certain species from the values predicted by Saha's law. The process proposed by Underhill (1957) for the excitation of C III 5696 selectively in emission in relatively cool Of atmospheres is a case of particular ionization by radiation. As a result of absorbing a quantum of He II 303, C^+ ions in the 3^2D state (the lower level of λ 4267) are ionized and move to the 3^1D state of C^{++}. This excited ion then cascades to the ground state, emitting λ 5696 on the way. The process suggested by Gauzit (1966) for exciting C III 5696 selectively in emission involves absorption of the N IV line at 322.57 A (postulated to be in emission) followed by a cascade in which λ 5696 forms one step. Since it seems unlikely that the needed N IV line is in emission in the cool Of stars in which λ 5696 appears in emission, and since the C III absorption process involves going from configuration $2s^2$ to configuration 2p3s, this postulated fluorescence process is unlikely to be efficient.

The fact that unusually strong lines of P II in late B-type spectra are correlated with unusually weak lines of He I can be explained (Underhill,1968) as a result of ionization caused by collisions with neutral helium atoms in the metastable 2^3S level. The only ion of those expected to be abundant in late B-type atmospheres that has an ionization energy slightly less than the energy of metastable helium atoms is P^+. Thus, in normal stellar atmospheres where the abundance of helium atoms in the metastable 2^3S state appears to be large, this selective ionization process can be very effective. Phosphorus will then be driven preferentially to the doubly ionized stage. The P^{++} ions may not be observed readily because the accessible lines have rather high lower-excitation potentials (14.5 v) for the excitation temperatures believed to exist in late B-type atmospheres. In an atmosphere where the population of neutral helium atoms in the 2^3S state is small, the second ionization of phosphorus will not be preferentially advanced. Thus a greater abundance of P^+ ions will remain to give strong P II lines from readily excited levels. The deductions mentioned above about the population of the He I levels are from a preliminary study of the He I problem; they have not yet been established by rigorous numerical work.

Many of the atoms and ions known to produce lines of anomalous intensity in A-type stellar spectra possess autoionizing levels. Particular ionization effects could occur through autoionization. Some spectra known to be subject to autoionization are Ca I, Ga I, and Ba I.

Many of the lines involved in processes of the above types have been selected empirically as classification criteria. Thus, particularly among stars of types A2 and earlier, spectral types determined from the line spectrum reflect to some extent the result of non-LTE processes. These "tainted" spectral types are in many cases considered to be adequate for isolating normal stars. It is no wonder that conflicts and ambiguities exist between photometric and line-spectrum spectral types.

There is strong theoretical evidence that non-LTE effects must be considered when the absorption-line spectrum of models of main-sequence stars of type A2 and earlier is computed. The question to be considered is this: In the layers of the model most important for determining the emergent monochromatic flux in a line frequency, are the electron densities and temperatures such that the Saha and Boltzmann laws may be justifiably used as a reliable approximation?

In optically thin plasmas in which the electron temperature is greater than $10,000°$, the electron density must be greater than about 10^{16} for LTE to be a good approximation. Stellar atmospheres are not optically thin in all wavelengths, nor are they optically thick; thus it is difficult to apply laboratory experience directly. It is most informative to look at the predicted electron densities and temperatures in the line-forming layers of a set of model atmospheres that range in spectral type from about A2 to B0. These numbers form a zero-order approximation to the true case, for they have been obtained under the postulate of LTE and for the case that the opacity in the models is due to continuous sources of opacity only. A few results are given in Table 1. The model with $T_{eff} = 9164°$ was computed by Mihalas (1965), the others by Underhill (1962).

The emergent flux at any frequency is found by evaluating the integral

$$F_{\nu}(0) = 2 \int_{0}^{\infty} S_{\nu}(t_{\nu}) E_{2}(t_{\nu}) \, dt_{\nu} \quad ,$$

where t_{ν} is the monochromatic optical depth at the frequency ν.

The integral is evaluated by a quadrature formula of the sort proposed by Reiz (see Chandrasekhar, 1950). Frequently, the source function is put equal to the Planck function. Examination of the weights and points used shows that the weight of the solution falls in those layers of the model where t_{ν} has values between 0.3 and 0.4. The model is given in terms of the characteristic optical depth τ. We define τ_f as the value of τ at which $t_{\nu} \approx 0.4$; τ_f is the depth of formation of the emergent flux F_{ν}. This value isolates a layer having characteristic values of temperature and electron density for the evaluation of $F_{\nu}(0)$.

We shall now look at the values of n_e and T_e in Table 1. In the case of early-type models, the monochromatic optical depth in the continuous spectrum in the blue-violet spectral region is usually about the same as τ, the characteristic optical depth. Thus, for lines lying between 4000 and 5000 A, the part of the atmosphere most important for forming the continuous spectrum lies outside the layer $\tau = 0.4$. In the centers of the lines, the point at which $t_{\nu} \approx 0.4$ may fall very close to the schematic boundary level $\tau = 0.00$.

In all the models, the electron density in the layers important for forming the continuous spectrum is less than 5×10^{14}. Comparison with the rule of thumb for optically thin plasmas suggests that LTE will not be established in these layers. Detailed calculations, for example by Kalkofen and Strom (1966) and by Mihalas (1967), have shown, indeed, that significant departures from LTE do occur in the continuous spectrum of hydrogen in models of the temperature range examined here.

Table 1. The temperatures and electron densities in model atmospheres with log g = 4.0

T_{eff}	9164°K		15,333°K	
τ	T_e(°K)	n_e	T_e(°K)	n_e
0.00	6967	4.06 + 12	9600	1.50 + 13
0.01	7109	1.52 + 13	10008	2.66 + 13
0.02	7198	2.20 + 13	10344	3.45 + 13
0.05	7404	3.96 + 13	11040	5.43 + 13
0.10	7680	6.91 + 13	11736	8.55 + 13
0.20	8117	1.37 + 14	12540	1.35 + 14
0.30	8475	2.17 + 14	13116	1.82 + 14
0.40	8782	3.06 + 14	13680	2.26 + 14
0.50	9047	3.99 + 14	14100	2.69 + 14
T_{eff}	19,215°K		31,023°K	
0.00	12000	2.24 + 13	21294	3.15 + 13
0.01	12510	3.74 + 13	22226	5.61 + 13
0.02	12930	5.02 + 13	22822	7.75 + 13
0.05	13800	8.44 + 13	24419	1.30 + 14
0.10	14670	1.32 + 14	25956	2.00 + 14
0.20	15825	2.08 + 14	27972	3.03 + 14
0.30	16650	2.74 + 14	29358	3.89 + 14
0.40	17310	3.38 + 14	30492	4.62 + 14
0.50	17925	4.01 + 14	31424	5.28 + 14

What about the depth of formation of the most characteristic lines in these models? The data for the model with T_{eff} = 15,333° (this is Model 66 and corresponds roughly to type B6V) are characteristic for all the models. Some results for representative lines of He I, C II, Mg II, Si II, and Si III are given in Table 2. The adopted relative abundances by weight and by number are listed in Table 3. The He I results are from Model P13, an improved version of Model 66. In this case, the Stark broadening has been approximated by using 100 × γ_{cl}. These data have been taken from the output of a line-spectrum program; the line shape is given by thermal Doppler broadening and damping equivalent to 10 times the classical radiation damping for all lines except He I.

The depths of formation of various points in the profile of Hγ are as follows:

$\Delta\lambda$	Depth of formation τ_f in Hγ
0. 0 A	0. 000
1. 0	0. 017
2. 0	0. 05
4. 0	0. 13
10. 0	0. 31
40. 0	0. 42

Table 2.   The depth of formation of some lines in Model 66

Line		$\Delta\lambda$	τ_f	Line		$\Delta\lambda$	τ_f
He I	5876.6	0. 0 A	0. 045	Mg II	4481. 3	0. 0 A	0. 005
		0. 1	0. 056			0. 1	0. 24
		cont.	0. 25			cont.	0. 33
C II	6582.9	0. 0	0. 095	Si II	6347. 1	0. 0	0. 000
		0. 1	0. 13			0. 1	0. 02
		cont.	0. 20			cont.	0. 22
C II	4267. 3	0. 1	0. 21	Si II	4130. 9	0. 0	0. 002
		0. 1	0. 38			0. 1	0. 27
		cont.	0. 46			cont.	0. 50
C II		0. 0	0. 28	Si II	3856. 0	0. 0	0. 002
		0. 1	0. 49			0. 1	0. 39
		cont.	0. 55			cont.	0. 57
				Si III	4552. 6	0. 0	0. 32
						0. 1	0. 42
						cont.	0. 42

Table 3.   The adopted relative abundances

Element	By weight	By number
H	0. 68	1. 00
He	0. 32	1. 18 - 1
C	3. 25 - 3	3. 99 - 4
Mg	3. 28 - 4	1. 98 - 5
Si	1. 21 - 3	6. 33 - 5

Comparison of the data in Tables 1 and 2 and those for Hγ shows that the central part of Hγ ($\Delta\lambda < 4$ A) is formed chiefly in layers where the electron density is less than 10^{14}.   The same is true for the core of He I 5876 and presumably for the cores of the other strong He I lines.   Clearly, the intensities of the H and He I lines are affected by dilution effects even in main-sequence stars.

The central cores of the red C II lines, of Mg II 4481, and of all the Si II lines investigated are also formed high in the atmosphere, where the electron density is too low to make it certain that LTE exists. The weak C II lines and the weak high-excitation line Si III 4552 are formed at greater depths in the atmosphere, but the characteristic electron density is still only of the order of 2×10^{14}.

Information is also given in Table 2 about the depth of formation of the continuous spectrum at the selected lines. This depth of formation varies from about $\tau = 0.2$ in the red to $\tau = 0.6$ in the violet. The typical electron density ranges from 1×10^{14} to 3×10^{14}. Since the transparency of the atmosphere in the red differs considerably from that in the violet, study of spectral lines in both regions is a good consistency test of a model atmosphere of type B.

The above theoretical results demonstrate clearly that there is no secure basis for the blanket assumption that the Saha and Boltzmann equations may be used in the calculation of line profiles in model atmospheres of types A2 to O9.

One of the most urgent steps to be taken is to set up ways of calculating line profiles in the case of non-LTE. The atomic and ionic spectra known to be susceptible to dilution and other effects should certainly be studied, but it is not excluded that spectra such as Mg II and Si II, which contain no metastable levels, may not also show a significant sensitivity to non-LTE effects. The behavior of the Mg II and Si II lines as one goes up the spectral sequence is not well represented by standard LTE line-profile calculations in models that represent the continuous spectrum of B stars satisfactorily.

What is necessary as a first step is to develop an index of "untrustworthiness" that we can use as a guide in evaluating the behavior of various atomic and ionic spectra under the conditions in early-type atmospheres. The spectrum of He I should receive a very high rating, followed by Si III and O I. So far I have not investigated the behavior of spectra like Mn II, Fe II, and Ni II, which are known to react in a variable manner in the extended shells around some B-type stars. Part of the behavior may be due to autoionization of the neutral atom.

The fact that non-LTE populations may be expected to occur for many atomic and ionic spectra, which are seen in A-, B-, and O-type spectra, makes it somewhat meaningless to derive abundances as fitting factors between observed equivalent widths and equivalent widths computed by means of the Saha and Boltzmann laws. The presence of non-LTE populations may be obscured unless a variety of lines from levels over a wide range of excitation are studied. In the case of strong lines from abundantly populated levels, care must be taken to represent the process of line formation in a physically correct way, including noncoherent scattering. It is quite conceivable that the representation of the line-formation process that is valid for one group of lines will not be valid for another.

If we wish to use a method such as that proposed by Rybicki and Hummer (1967) to solve the transfer problem in lines, we must have a formulation for the depth dependence of the parameter ϵ. Considerations such as those discussed here may be helpful in the attempt to use the observed line strengths to indicate an empirical form for the function $\epsilon(\tau)$, which is the fractional probability that a line photon is destroyed by collision or by making another transition rather than returning in the line in which it was absorbed.

If LTE is not a good approximation in the line-forming layers of a stellar atmosphere, then the emissivity term in the equation of transfer will differ from $\ell_\nu \rho B_\nu$. It is pertinent to ask what changes will be produced in the line profile. Two extreme representations of line formation have been explored for the case of Si III 4552 in model B13, which is a line-blanketed model corresponding roughly to type B1.5 or B2. The effective temperature of this model is 23,255° K and log g is 4.0. The results are shown in Figure 1.

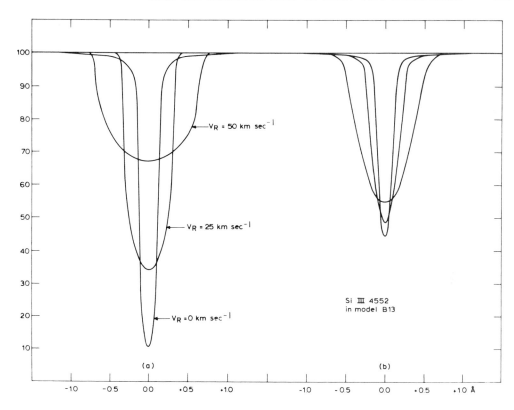

Figure 1.   A comparison of (a) line profiles formed by coherent isotropic
scattering when the projected rotation is 0, 25, and 50 km sec[-1]
with (b) profiles formed in LTE in the case of atom velocities
equivalent to 1, 2, and 4 times the thermal motions

Figure 1(a) displays the profiles resulting from line formation by
coherent isotropic scattering when the projected velocity of rotation $v_R = 0$,
25, and 50 km sec[-1]. Figure 1(b) shows the profiles resulting when the
emissivity is given by Kirchhoff's relation and the velocities of the atoms
correspond to 1, 2, and 4 times the thermal distribution. In the case of line
formation by scattering, and with $v_R = 0$ km sec[-1], the equivalent width is
376 mA and the central intensity is 0.106. In the case of line formation in
LTE, the equivalent widths are 128, 206, and 335 mA and the central intensities are 0.445, 0.486, and 0.547. Line formation by coherent isotropic
scattering gives a much deeper and stronger line, for the same number of
atoms, than does line formation in LTE. If one were certain that the star
did not rotate at all, one could reject the hypothesis of line formation by
scattering, since the lines in B2 stars are not observed to be so deep as
predicted. However, observed profiles rarely, if ever, are as narrow as
thermal broadening predicts. It is obvious from the diagram that line shape
and equivalent width alone will not enable one to decide between the representation of line formation by coherent isotropic scattering with a rotation near
35 km sec[-1] and formation in LTE with a "microturbulence" about as large
as 3 times the thermal motion. The observed shapes and depths of line in B-
type atmospheres do not automatically exclude the hypothesis that the lines are
formed in non-LTE, i.e., by a scattering process.

REFERENCES

Chandrasekhar, S. 1950, Radiative Transfer (Oxford: Clarendon Press),
    p. 67.
Gauzit, J. 1966, C. R. Acad. Sci. Paris, Ser. B, 262, 1309.
Kalkofen, W., and Strom, S. E. 1966, J. Q. S. R. T., 6, 653.
Merrill, P. W. 1956, Carnegie Inst. of Washington Publ. 610, 167 pp.
Mihalas, D. 1965, Ap. J. Suppl., 9, 321.
_____ . 1967, Ap. J., 149, 169.
Rybicki, G. B., and Hummer, D. G. 1967, Ap. J., 150, 607.
Underhill, A. B. 1957, Ap. J., 126, 28; Contr. Dominion Ap. Obs., Victoria, B. C.
    No. 55, 2 pp.
_____ . 1962, Publ. Dominion Ap. Obs., Victoria, B. C., 11, 433.
_____ . 1968, Bull. Astron. Netherlands, 19, 526.

DISCUSSION

Mihalas: If one introduces scattering in the line, say with a non-LTE source function, the line core becomes very deep. Then, accounting for stellar rotation, one can obtain profiles very similar to those observed. Auer and I have performed such calculations.

Strom: I was wondering whether you might discuss more completely the He — P II anomaly. Is weak He — strong P II favored by high or low densities?

Underhill: It's favored by the higher densities, weak helium because you get LTE population, I think. This is based on schematic calculations. I have no comment on the helium problem at the moment, but in the normal average B star most of the helium lines are formed outside the models in regions of strong non-LTE. I have the feeling that the weak-helium atmospheres are small and compact, as much like the models as they can be.

Lamers: If you accept the scattering, wouldn't you expect to find any pole-on star with very deep lines?

Underhill: Well, 30 km sec^{-1} is a small velocity; this is a projected value. We have very few stars to observe, but it is interesting that the B stars with the sharpest lines are the B peculiar stars.

THEORETICAL ULTRAVIOLET FLUX IN Am AND NORMAL A STARS

F. Praderie

Observatoire de Paris, Meudon, France

Interest in observations of the space ultraviolet spectrum of stars is manifold and well known (see, e. g. , Underhill and Morton, 1967). I will focus my attention on the predicted continuous spectrum in the spectral-type range of A stars. Until now, very few A stars have been spectrophotometrically observed outside the terrestrial atmosphere (Henize and Wackerling, 1966; Stecher, 1967; Carruthers, 1968), principally because their ultraviolet spectra are not so bright as the visible spectra, so that observational conditions (exposure time) are more difficult than for B stars.

Given a model atmosphere, the computation of the emergent flux is an easy task, provided we assume that the populations of the levels from which continuous absorption originates are governed by the LTE laws. In what follows, this explicit assumption of LTE is made.

It was shown by Gingerich (1964) that neutral silicon and magnesium are very efficient ultraviolet absorbers. The importance of these two abundant elements as absorbers extends from cool atmospheres to rather warm atmospheres, the condition being that the fraction of neutral silicon and magnesium is significant.

The consideration of these absorbers in the calculation of the ultraviolet opacity of A stars has several implications, the first of which is only a remark on a result obtained by Mihalas.

A. As was pointed out by Mihalas (1965), the information on the thermal stratification of A stars may be improved by the observation of the ultraviolet flux below 1800 A. According to Mihalas, the flux radiated by a constant-flux convective model atmosphere is smaller than that radiated by a pure radiative equilibrium model, and the relative depression increases with ℓ/h, the ratio of the mixing length to the pressure scale height. The effect appears at $\lambda \leq 1600$ A for $\theta_{eff} = 0.65$, $\log g = 4$. Mihalas' opacity includes only hydrogen as the main absorbent. If other continuous absorbers are considered (Mg I, Si I, C I), their contribution dominates over that of hydrogen for $\lambda < 1700$ A, for the same (θ_{eff}, $\log g$) values. The opacity is much higher than was assumed by Mihalas; the ultraviolet flux is then formed in regions much more superficial than the visible one, regions that are stable against convection. Therefore, the spectral distribution of the flux for $\lambda < 1700$ A reflects the structure of these superficial layers, rather than the structure of the convectively unstable ones.

B. Abundance effect on the continuous ultraviolet flux in Am, Ap, and normal A stars.

A stars comprise about 10% of Am stars, while the named "Ap" stars constitute about 11% of the B8 to A5 stars (Jaschek and Jaschek, 1962). The superficial chemical composition of Ap and Am stars differs from the solar one; among Ap stars, the silicon stars are characterized by a silicon abundance as high as 25 times the solar value (Sargent, 1964). Am stars are silicon enriched, but the factor of overabundance is less (1.15 to 15). Let us compare the ultraviolet flux radiated by a normal A star and by an Am-star model atmosphere, supposed to have the same effective temperature and the same gravity. For the first one, the abundances of the elements are those given by Goldberg, Müller, and Allen (1960; hereafter referred to as GMA). For the second, the abundances are those of a typical Am star, 63 Tau, derived by Conti (1965); the relative abundances of Conti are put into absolute values, with the assumption that the comparison star has a solar composition (GMA, revised by Müller, 1966). Values of the abundances for Si and Mg are given in Table 1.

Table 1.   Partial abundances of the elements ($a_i = N_i/N_H$)

Object	Source	Si	Mg	Remarks
Sun	GMA (1960)	3.16×10^{-5}	2.51×10^{-5}	
	Müller (1966)	6.17×10^{-5}	2.51×10^{-5}	
63 Tau	Conti (1965)	7.42×10^{-5}	1.41×10^{-5}	$[N]^*_{Si} = 0.08$ (Conti)
				$[N]_{Si} = 0.70$ (Van't Veer)

As usual, $[M]^*$ means $\log a_* - \log a_\odot$.

Both models have been computed with the same temperature distribution, derived from models in radiative equilibrium with Balmer-line blanketing (Mihalas, 1966); the effective temperature $T_{eff} = 8000°K$, and $\log g = 3.9$. The abundance of helium by number is 0.125. The Am model is that adopted in an earlier work for the Am star ζ Lyr A (Praderie, 1968). The metallic opacity is treated according to the procedure of Gingerich (1964) and Gingerich and Rich (1966), and the hydrogen opacity according to that of Vardya (1964).

Because of the difference in silicon abundances between the normal and the Am star, the relative values of the two first discontinuities due to the silicon continuous spectrum at $\lambda = 1520 A$ (D_1) and $\lambda = 1680 A$ (D_2) are such that D_1/D_2 is greater than 1 for the normal A star and smaller than 1 for the Am star. This effect is shown in Figures 1 and 2, where the flux F_ν (ergs cm^{-2} sec^{-1} H^{-1}) is given versus $1/\lambda$ ($F_\nu = \frac{1}{\pi} \times$ physical flux).

The effect is still appreciable for a model with $T_{eff} = 10,000°$, $\log g = 4$, but computation shows that the magnitude of the discontinuities is much smaller than for $T_{eff} = 8000°$; the reason is that the ionization of silicon is greater in the former model, which corresponds to an A0 star, than for models corresponding to A7-F0 stars.

C.   Influence of some strong lines on the emergent flux between 1300 and 1800 A.

Not only is the theoretical ratio of the silicon discontinuities, D_1/D_2, different in an A star and in an Am star, but the absolute values of the discontinuities are large; they exceed 1 mag, and in the case of the A star with $T_{eff} = 8000°$, D_1 reaches 4 mag.

It is necessary to discuss, in the case of A stars, the effects that may influence the computed discontinuities of Si I. Let us begin with the lines.

The presence of strong lines depresses the observed flux (blocking effect); moreover, lines affect the distribution of temperature in the model itself. Published papers dealing with the influence of the metallic lines on the ultraviolet flux refer only to B stars. A broader study has been carried out by Mihalas and Morton (1965) and by Adams and Morton (1968).

In the range of A stars, no such models exist. As a first step, I have computed some of the lines likely to appear between 1300 and 1800 A, with the above-mentioned Am model atmosphere. Because of the large continuous opacity, the expected lines are formed very superficially. Taking account of the surface temperature ($\simeq 6000°K$ for an A7-star model) and of the abundances of the elements, we expect to see lines of C I, N I, O I, Mg I, Mg II, Al I, Al II, Si I, Si II, S I, Ca II, V II, Cr II, Mn II, Fe II, Co II, Ni II, Cu I, Cu II, Zn II.

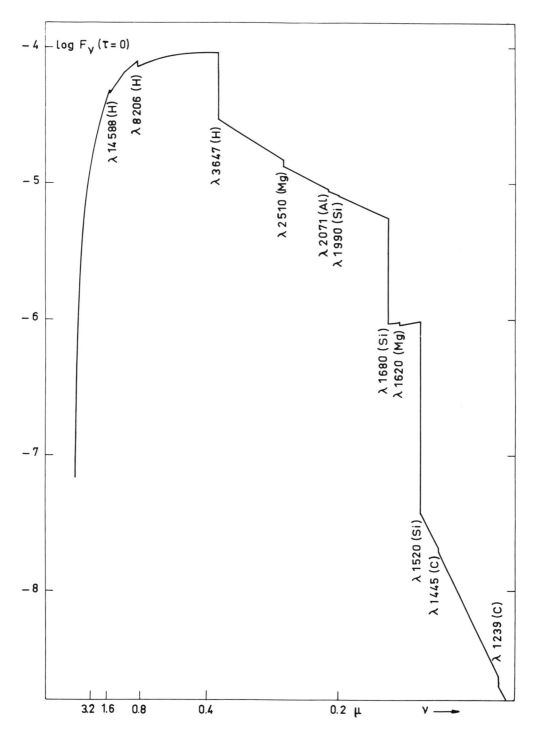

Figure 1.   Spectral distribution of the flux in an A star
(T_{eff} = 8000° K, log g = 3. 9).   Abundances of
the elements: GMA.

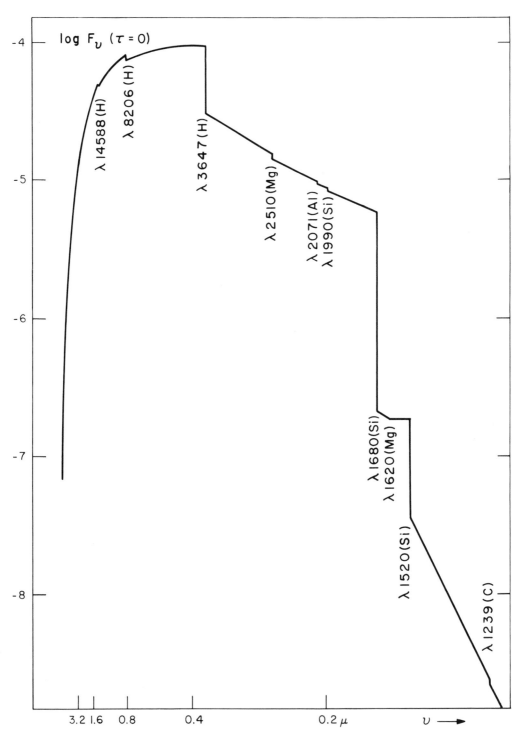

Figure 2.   Spectral distribution of the flux in an Am star
(T_{eff} = 8000° K, log g = 3. 9).   Abundances of
the elements: 63 Tau.

Only the lines given in Table 2, for which gf values can be found in the literature, have been computed, with the help of the code written by Spite (1967) and adopted by Bonnet. The broadening of the lines includes: 1) the radiative width, assumed to have the classical value; 2) the width due to electron collisions, assumed to be 10 times the preceding one; 3) the pressure width due to Van der Waals interaction of the atoms with neutral hydrogen. The adopted microturbulence velocity is that derived for the Am star, ζ Lyr A, 6 km sec^{-1}.

Figure 3 shows, as an example, the very strong lines that overlap between 1610 and 1830 A.

We must stress the very small number of elements for which lines can be computed, because of the scarcity of gf values. The appearance of the Al II line at $\lambda = 1670.81$ suggests that the line spectrum of once-ionized metals may be important. For longer wavelengths, Henize has pointed out the importance of the Fe II line spectrum in ultraviolet spectra of Sirius.

Between 1520 and 1610 A, the lines given in Table 2 are also very strong, and saturated as well. But below 1520 A, lines computed with a radiative equilibrium model originate in regions where the continuous opacity is so high that the involved regions of the model are those where the gradient of T is small. The depth of the lines is therefore shallow. We cannot conclude from this result that, below 1520 A, lines have no importance on the observed continuous flux, because the hypothesis of LTE is poor in the very superficial layers and even worse if the stars concerned have a chromosphere. Nevertheless, if they have a chromosphere, no observed emission line has as yet been reported (Stecher, 1967).

From this rough and preliminary calculation, we can infer that the discontinuities due to the silicon will be affected ("washed") by the lines; but if the abundance effect in the continuous spectrum is as important as predicted, they should still permit us to distinguish between normal A stars and Am stars (and, a fortiori, Ap silicon stars).

D. The ratio of the silicon discontinuities, D_1/D_2, depends not only on the silicon abundance, but also on the temperature profile in the very superficial layers of the star and on other possible sources of opacity for $\lambda \leq 2000$ A.

On the one hand, departures from LTE may affect in a different way the populations of the atomic levels in the A star and in the Am star. On the other hand, observations of the solar ultraviolet spectrum by Bonnet and Blamont (1968) cannot be interpreted shortward of the discontinuity observed at 2080 A: The computed flux on the violet side of the discontinuity disagrees with the observations in the sense that the computed opacity, attributed to Al I, is too small. The resonance wing of Lyman α as extra-absorber has been considered by Cuny (these proceedings); its contribution to the opacity should be included for A stars as well.

Table 2.   Ultraviolet line with gf values

λ (A)	Element	E_1 (ev)	E_2 (ev)	gf		Reference for gf
1302.174	O I	0.00	9.48	0.182	E 00	Glennon and Wiese, 1966
1302.320	S I	0.05	9.53	0.170	E 00	Lawrence, 1967
1302.860	S I	0.05	9.52	0.851	E-01	Lawrence, 1967
1303.120	S I	0.05	9.52	0.118	E 00	Lawrence, 1967
1304.370	Si II	0.00	9.51	0.295	E 00	Garstang and Shamey, 1967
1304.858	O I	0.02	9.48	0.955	E-01	Glennon and Wiese, 1966
1305.890	S I	0.07	9.52	0.129	E 00	Lawrence, 1967
1306.023	O I	0.03	9.48	0.316	E-01	Glennon and Wiese, 1966
1309.270	Si II	0.04	9.51	0.501	E 00	Garstang and Shamey, 1967
1310.569	N I	3.56	12.98	0.200	E 00	Glennon and Wiese, 1966
1310.967	N I	3.56	12.98	0.141	E 00	Glennon and Wiese, 1966
1316.290	N I	3.58	12.99	0.380	E-02	Glennon and Wiese, 1966
1319.039	N I	3.56	12.92	0.661	E-01	Glennon and Wiese, 1966
1319.717	N I	3.56	12.91	0.229	E-01	Glennon and Wiese, 1966
1326.629	N I	3.56	12.87	0.190	E-01	Glennon and Wiese, 1966
1327.960	N I	3.56	12.86	0.135	E-01	Glennon and Wiese, 1966
1328.820	C I	0.00	9.29	0.389	E-01	Glennon and Wiese, 1966
1329.099	C I	0.00	9.29	0.118	E 00	Glennon and Wiese, 1966
1329.580	C I	0.01	9.29	0.191	E 00	Glennon and Wiese, 1966
1355.605	O I	0.00	9.11	0.182	E-05	Glennon and Wiese, 1966
1358.524	O I	0.02	9.11	0.537	E-06	Glennon and Wiese, 1966
1364.140	C I	1.26	10.31	0.646	E-02	Glennon and Wiese, 1966
1381.550	S I	0.00	8.94	0.113	E 01	Varsavsky, 1961
1385.510	S I	0.05	8.96	0.900	E 00	Varsavsky, 1961
1388.390	S I	0.00	8.89	0.360	E 01	Varsavsky, 1961
1389.160	S I	0.05	8.94	0.670	E 00	Varsavsky, 1961
1392.590	S I	0.07	8.94	0.890	E 00	Varsavsky, 1961
1396.100	S I	0.05	8.89	0.111	E 01	Varsavsky, 1961
1411.937	N I	3.56	12.30	0.155	E 00	Glennon and Wiese, 1966
1425.100	S I	0.00	8.66	0.219	E 01	Savage and Lawrence, 1966
1431.595	C I	4.16	12.79	0.324	E 00	Glennon and Wiese, 1966
1432.115	C I	4.16	12.78	0.214	E 00	Glennon and Wiese, 1966
1432.538	C I	4.16	12.78	0.120	E 00	Glennon and Wiese, 1966
1433.280	S I	0.05	8.66	0.724	E 00	Savage and Lawrence, 1966
1436.940	S I	0.07	8.66	0.240	E 00	Savage and Lawrence, 1966
1448.200	S I	1.14	9.66	0.525	E 00	Lawrence, 1967
1459.054	C I	1.26	9.72	0.347	E-01	Glennon and Wiese, 1966
1463.328	C I	1.26	9.69	0.468	E 00	Glennon and Wiese, 1966
1467.450	C I	1.26	9.67	0.447	E-01	Glennon and Wiese, 1966
1469.000	C I	1.26	9.71	0.182	E-02	Glennon and Wiese, 1966
1470.200	C I	1.26	9.66	0.200	E-02	Glennon and Wiese, 1966
1473.980	S I	0.00	8.38	0.724	E 00	Lawrence, 1967

Table 2 (Cont.)

λ (A)	Element		E_1 (ev)	E_2 (ev)	gf		Reference for gf
1474.370	S	I	0.00	8.37	0.151	E 00	Lawrence, 1967
1474.540	S	I	0.00	8.54	0.107	E-01	Lawrence, 1967
1481.771	C	I	1.26	9.59	0.550	E-01	Glennon and Wiese, 1966
1483.050	S	I	0.05	8.37	0.363	E 00	Lawrence, 1967
1483.240	S	I	0.05	8.37	0.135	E 00	Lawrence, 1967
1487.120	S	I	0.07	8.37	0.162	E 00	Lawrence, 1967
1492.630	N	I	2.37	10.64	0.776	E 00	Glennon and Wiese, 1966
1492.670	N	I	2.38	10.69	0.776	E-01	Glennon and Wiese, 1966
1494.669	N	I	2.37	10.63	0.331	E 00	Glennon and Wiese, 1966
1526.720	Si	II	0.00	8.12	0.257	E 00	Wiese and Smith, 1968
1533.445	Si	II	0.04	8.12	0.525	E 00	Wiese and Smith, 1968
1560.313	C	I	0.00	7.91	0.912	E-01	Glennon and Wiese, 1966
1560.702	C	I	0.00	7.91	0.195	E 00	Glennon and Wiese, 1966
1561.292	C	I	0.01	7.91	0.214	E-01	Glennon and Wiese, 1966
1561.400	C	I	0.01	7.91	0.389	E 00	Glennon and Wiese, 1966
1649.520	Si	II	5.32	12.84	0.159	E-04	Garstang and Shamey, 1967
1650.702	C	I	0.00	7.91	0.214	E-01	Glennon and Wiese, 1966
1654.300	Si	II	5.34	12.84	0.724	E-05	Garstang and Shamey, 1967
1654.310	Si	II	5.34	12.84	0.151	E-03	Garstang and Shamey, 1967
1656.255	C	I	0.00	7.46	0.219	E 00	Glennon and Wiese, 1966
1656.918	C	I	0.00	7.45	0.170	E 00	Glennon and Wiese, 1966
1656.998	C	I	0.01	7.46	0.646	E 00	Glennon and Wiese, 1966
1657.368	C	I	0.00	7.45	0.126	E 00	Glennon and Wiese, 1966
1657.891	C	I	0.00	7.45	0.170	E 00	Glennon and Wiese, 1966
1658.113	C	I	0.01	7.45	0.219	E 00	Glennon and Wiese, 1966
1666.680	S	I	1.14	8.55	0.537	E 00	Lawrence, 1967
1670.81	Aℓ	II	0.00	7.39	0.329	E+01	Varsavsky, 1961
1706.380	S	I	1.14	8.38	0.550	E-03	Lawrence, 1967
1707.130	S	I	1.14	8.37	0.646	E-03	Lawrence, 1967
1742.734	N	I	3.56	10.64	0.398	E 00	Glennon and Wiese, 1966
1745.246	N	I	3.56	10.63	0.178	E 00	Glennon and Wiese, 1966
1751.900	C	I	2.68	9.76	0.120	E 00	Glennon and Wiese, 1966
1764.000	C	I	2.68	7.12	0.309	E-02	Glennon and Wiese, 1966
1765.000	C	I	2.68	9.71	0.100	E-02	Glennon and Wiese, 1966
1782.250	S	I	2.74	9.66	0.100	E 01	Lawrence, 1967
1807.310	S	I	0.00	6.86	0.562	E 00	Lawrence, 1967
1808.003	Si	II	0.00	6.86	0.178	E-01	Garstang and Shamey, 1967
1816.921	Si	II	0.04	6.86	0.977	E-01	Garstang and Shamey, 1967
1817.450	Si	II	0.04	6.86	0.389	E-01	Garstang and Shamey, 1967
1820.370	S	I	0.05	6.86	0.331	E 00	Lawrence, 1967
1826.250	S	I	0.07	6.86	0.110	E 00	Lawrence, 1967

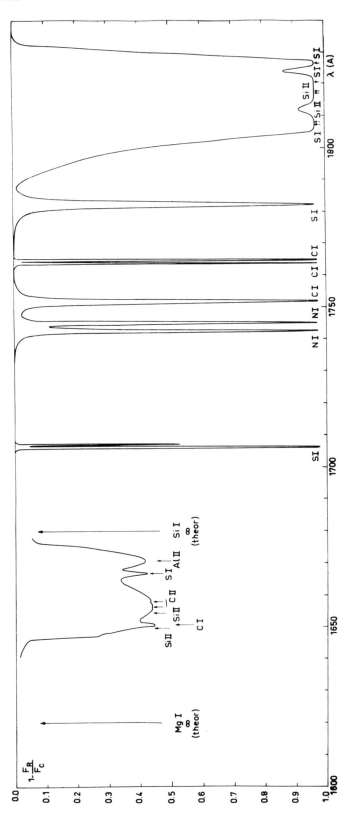

Figure 3.   Strong absorption lines predicted in the ultraviolet spectrum of an Am star.

REFERENCES

Adams, T. F., and Morton, D. C. 1968, Ap. J., 152, 195.

Bonnet, R. M., and Blamont, J. E. 1968, Solar Phys., 3, 64.

Carruthers, G. R. 1968, Ap. J., 151, 269.

Conti, P. S. 1965, Ap. J. Suppl., 11, 47.

Garstang, R. H., and Shamey, L. J. 1967, in The Magnetic and Related Stars, ed. R. C. Cameron (Baltimore: Mono Book Co.), p. 387

Gingerich, O. 1964, Proc. Frist Harvard-Smithsonian Conf. on Stellar Atmospheres, Smithsonian Ap. Obs. Spec. Rep. No. 167, 17.

Gingerich, O., and Rich, J. C. 1966, Astron. J., 71, 161.

Glennon, B. M., and Wiese, W. L. 1966, NBS Misc. Publ. 278, 92 pp.

Goldberg, L., Müller, E. A., and Aller, L. H. 1960, Ap. J. Suppl., 5, 1.

Henize, K. G., and Wackerling, L. R. 1966, Sky and Tel., 32, 204.

Jaschek, M., and Jaschek, C. 1962, in Symp. on Stellar Evolution (La Plata, Argentina: Astron Obs., University of La Plata), 137.

Lawrence, G. M. 1967, Ap. J., 148, 261.

Mihalas, D. 1965, Ap. J., 141, 564.

————. 1966, Ap. J. Suppl., 13, 1.

Mihalas, D. M., and Morton, D. C. 1965, Ap. J., 142, 253.

Müller, E. A. 1966, in IAU Symp. No. 26, ed. H. Hubenet (London: Academic Press), 171.

Praderie, F. 1968, Ann. d'Ap., 31, 15.

Sargent, W. L. W. 1964, Ann. Rev. Astron. Ap., 2, 297.

Savage, B. D., and Lawrence, G. M. 1966, Ap. J., 146, 940.

Spite, M. 1967, Ann d'Ap., 30, 211.

Stecher, T. P. 1967, quoted by the United States Space Sci. Program Rep. to COSPAR, in COSPAR Transactions No. 5 (Washington, D. C.: NAS-NRC), 271.

Underhill, A. B., and Morton, D. C. 1967, Science, 158, 1273.

Van't Veer-Menneret, C. 1963, Ann. d'Ap., 26, 289.

Vardya, M. S. 1964, Ap. J. Suppl., 8, 277.

Varsavsky, C. M. 1961, Ap. J. Suppl., 6, 75.

Wiese, W. L., and Smith, M. W. 1968, private communication.

DISCUSSION

Underhill: The problem of Si II is very difficult. Silicon does not behave very well in stellar spectra, and I suspect that when you have such strong lines you would find that the lines are formed very high in the atmosphere at electron densities of the order 10^{12} to 10^{11}. This indicates that you must use a non-LTE formulation, which can be equivalent to scattering in some ways, and consequently I am rather doubtful of the meaning of abundances calculated in this way, but I quite agree that your analysis is basically self-consistent.

Praderie: The main effect will be on the continuous spectrum, where I think there are no such difficulties. Effects of the abundance on the continuous spectrum are more interesting to test because first of all there is no effect of microturbulence. Non-LTE effects should be computed.

Strom: I think that in your remarks was embodied the thought that if the contribution of lines in the ultraviolet is important, then the relative effect of changing the silicon-to-hydrogen ratio is reduced compared to what I spoke about this morning. Second of all, I would like to point out to Miss Underhill that Hyland used fairly weak silicon lines in his determination of the silicon abundance ratios and these lines are formed very far down and should not be affected by departures from LTE.

Hyland: Also, if you have Si III lines, which may be very weak in some of these stars, they still show the overabundance.

Underhill: Silicon III is very definitely affected by dilution, which becomes important in an extended atmosphere. Furthermore, with Ap's and Am's you have a magnetic field. In addition, you can argue generally that a magnetic field must exist only in regions of rather low density, and if you are observing lines formed at such low densities, dilution effects may be important.

THE DEPENDENCE OF DEVIATIONS FROM LTE ON
SURFACE GRAVITY AND EFFECTIVE TEMPERATURE

W. Kalkofen

Smithsonian Astrophysical Observatory, Cambridge, Massachusetts

ABSTRACT

The ratio of the rate of collisional excitation of hydrogen atoms from the second to the third level to the rate of photoionization from the second level is used to predict deviations from LTE. It is estimated that early-type stars with surface gravities smaller than log g = 3.5 can show the effects of deviations from LTE in their continuous spectra.

We consider early-type stars and we pose the questions: For what values of surface gravity and effective temperature are LTE models adequate for interpreting continuous stellar spectra, and when should deviations from LTE be taken into account? In order to answer the questions, we must decide which departure coefficient most influences a continuous spectrum, and we must determine the physical process that measures the approach to LTE of the corresponding population.

In early-type stars, one of the most pronounced features of the emergent continuous spectrum is the Balmer discontinuity. This discontinuity depends on departures from LTE mainly of the population in the second level of hydrogen. We must therefore consider the departure coefficient d_2 of the second level. We determine it from the equations of statistical equilibrium,

$$(I - \omega)d = r\rho \quad , \tag{1}$$

where the d's are defined such that they are zero in LTE,

$$d_n = b_n - 1 \quad , \tag{2}$$

and I is the unit matrix.

For an atom in which the populations of the N lowest levels may depart from LTE while those of higher levels are in LTE, the elements $\omega_{n\ell}$ are given by

$$\omega_{n\ell} = \begin{cases} \dfrac{\Omega_{n\ell}}{J_n} \quad , & n \neq \ell \\ \\ 0 \quad , & n = \ell \end{cases} \tag{3}$$

$$J_n = R_{n\kappa} + \left(\Omega_{n\kappa} + \sum_{k=N+1}^{\infty} \Omega_{nk} \right) + \sum_{\substack{k=1 \\ k \neq n}}^{N} \Omega_{nk} \quad , \tag{4}$$

where $\Omega_{n\ell}$ is the collision frequency for excitation from level n to level k, $\Omega_{n\kappa}$ is the frequency of collisional ionization, and $R_{n\kappa}$ is the frequency of photoionization. The quantities r_n and ρ_n are given by

$$r_n = \frac{R_{nk}}{J_n} \quad , \tag{5}$$

$$\rho_n = \frac{\int\limits_{\nu_n}^{\infty} \frac{d\nu}{\nu} a_n(\nu) \left(1 - e^{-\frac{h\nu}{kT}}\right) (B_\nu - J_\nu)}{\int\limits_{\nu_n}^{\infty} \frac{d\nu}{\nu} a_n(\nu) J_\nu} \quad . \tag{6}$$

We see that ρ_n is essentially the normalized derivative of the flux in the nth continuum.

It follows from definitions (3) and (4) that the elements of the matrix ω are smaller than unity. When they are sufficiently small, the solution of the equations (1) of statistical equilibrium is

$$d = (I + \omega) r\rho \quad . \tag{7}$$

Hence, for the departure coefficient d_2 of the second level we obtain

$$d_2 = r_2 \rho_2 + \sum_{\ell \neq 2} \omega_{2\ell} r_\ell \rho_\ell \quad . \tag{8}$$

Now, the collisional coupling parameters $\omega_{2\ell}$ decay essentially as ℓ^{-3} times an exponential containing the temperature. The largest coefficient is ω_{23},

$$\omega_{23} \simeq \frac{\Omega_{23}}{R_{2k} + \Omega_{23}} \quad ; \tag{9}$$

ω_{24} is smaller by an order of magnitude. Hence, the departure coefficient d_2 depends mainly on the Balmer and Paschen "flux derivatives," ρ_2 and ρ_3. Since in B stars they are expected to have opposite signs, $\rho_2 < 0$, $\rho_3 > 0$, the coupling coefficient ω_{23} measures the approach to LTE of the hydrogen population in the second level.

A convenient measure for the approach to LTE of the population of the second hydrogen level is thus

$$\underline{P} = \frac{\Omega_{23}}{R_{2k}} \quad ; \tag{10}$$

$\underline{P}$ depends on the electron density n_e, the local kinetic temperature T, and the intensity of the radiation field in the Balmer continuum. Now, from the detailed model for an A0 star published by Lecar (1964), we find that the monochromatic intensity at the Balmer limit differs from the Planck function only by about 15%. We may therefore compute R_{2k} from the local kinetic temperature; $\underline{P}$ depends then only on n_e and T.

If the ratio $\underline{P}$ is to be a criterion for the effects of deviations from LTE on a stellar spectrum, it must be determined at a depth from which a significant fraction of the emergent flux is emitted. Since departures from LTE change the structure of a star only moderately, LTE models are adequate for estimating $\underline{P}$.

We have based the computations on the grid of LTE models published by Strom and Avrett (1965) and have determined $\underline{P}$ at the optical depth τ such that the flux $F(0, \tau, \lambda = 3647^-)$ emitted from layers above τ is the fraction ϕ of the emergent monochromatic flux $F_\lambda(0) = F(0, \infty, \lambda = 3647^-)$.

In Figure 1 the curves for log g = 4 and log g = 3 have been obtained from the grid of models. The curve for log g = 2 has been drawn by scaling; it should represent an upper limit on the ratio $\underline{P}$. The optical depths that correspond to the fractional fluxes $\phi = 2/3, \overline{1}/2, 1/3, 1/4$ are approximately $\tau = 0.9, 0.6, 0.3, 0.2$, respectively.

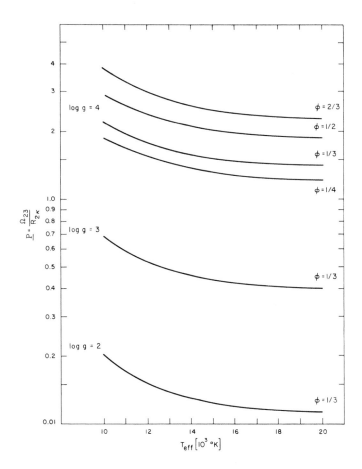

Figure 1. The ratio $\Omega_{23}/R_{2\kappa}$ of the collision fréquency (2-3) and the photo-ionization frequency $(2 - \kappa)$ as a function of surface gravity g and effective temperature T_{eff}; ϕ is the fraction of the emergent flux that is emitted above the layer for which the ratio is computed.

It is seen that the separation of curves for log g = 4 and different values of ϕ is small compared with the separation of curves for log g = 4 and log g = 3, with $\phi = 1/3$. Thus, the arbitrary choice of the depth for which $\underline{P}$ was computed has little effect on the results.

The collision rate has been computed with the aid of the Johnson cross section published by Mihalas (1967). We note that an error of a factor of 2 in the cross section would change the gravities defining the curves by an amount varying between Δ log g = 0.3 and 0.6.

When the ratio $\underline{P}$ is large compared with unity, the population of hydrogen atoms in the second level must be essentially in LTE at a depth at which the emergent flux is emitted. The Balmer discontinuity will then be the same as that determined under the assumption of LTE. Conversely, when the ratio $\underline{P}$ is small compared with unity at a depth at which a significant fraction of the emergent flux is formed, the population of the second level is determined mainly by photoionization, and it need not be in LTE. Then the Balmer discontinuity can be modified by deviations from LTE.

From the figure we see that stars with log g = 4 should have LTE populations in the continuum-forming layers, but stars with log g = 3 may show departures from LTE. For effective temperatures between 10,000°K and 20,000°K, the dividing line falls near log g = 3.5. We may conclude, therefore, that LTE models will be adequate for interpreting continuous spectra of main-sequence stars, but for giants and supergiants deviations from LTE should be taken into account.

REFERENCES

Lecar, M. 1964, NASA Tech. Note TN-D-2110, 107 pp.
Strom, S. E., and Avrett, E. H. 1965, Ap. J. Suppl., 12, 1.
Mihalas, D. 1967, Ap. J., 149, 169.

DISCUSSION

Strom: Would you wish to make a comparison between the results of Mihalas and your own results?

Kalkofen: There were two sets of investigations, the early results of Strom and myself and those of Mihalas. They were both for main-sequence stars, log g = 4. Mihalas found smaller departure coefficients than we did. He attributed this to the larger number of levels that he had taken into account, but from my analysis it appears that the number of levels has a negligible effect on the population of the second level. The reason for his small departure coefficient is not the channeling effect that he proposed, but the larger cross section that he used.

NON-LTE MODEL ATMOSPHERES

L. H. Auer[*] and D. Mihalas[†]

Joint Institute for Laboratory Astrophysics, Boulder, Colorado

Calculations are in progress to construct model atmospheres in hydro-static and radiative equilibrium in which atomic occupation numbers are determined by steady-state statistical equilibrium equations that are self-consistent with the radiation field. The atmospheres are assumed to be pure hydrogen, and the opacity consists of bound-free and free-free hydrogen absorption plus coherent isotropic scattering by electrons. The atomic models under consideration allow for departures from LTE in the first two or three levels; all higher levels are assumed to be in LTE. We are con-tinuing our calculations either assuming detailed radiative balance in the lines or allowing for line transfer. In the two-level models, only Lyman α is considered; in the three-level models, Lyman α, Lyman β, and Hα are accounted for.

Source functions that display explicitly the scattering and thermal source terms are obtained from the statistical equilibrium equations. The transfer equations are solved by the difference-equation method sub-ject to the constraint of radiative equilibrium, with use made of a lineari-zation technique (Auer and Mihalas, 1968). This temperature correction scheme is found to be extremely efficient, yielding fully converged (i. e., $|d\ell nF/d\tau| \lesssim 10^{-8}$ to 10^{-6}) non-LTE continuum models in only 9 iterations. The convergence is global and of a strongly damped oscillatory nature.

Detailed study of a non-LTE model with $T_{eff} = 15,000°K$ and $\log g = 4$ shows the boundary temperature to be determined by a balance between heat-ing in the Balmer continuum and cooling in the Paschen and higher (including free-free) continua. When only free-free absorption is considered at longer wavelengths than the Balmer edge (i. e., a strict two-level atom plus con-tinuum), the boundary temperature rises almost to T_{eff}. When an integral representation of the upper level opacities is introduced, the boundary tem-perature of the non-LTE model is about 1000°K higher than that of the LTE model, in qualitative agreement with the result of P. Feautrier (these pro-ceedings). When a spuriously high opacity is introduced by arbitrarily extrapolating all higher continua to zero frequency (the model presented at this meeting), a temperature drop of about 1000°K is found. These results serve to demonstrate the sensitivity of the boundary temperature to the details of the assumed atomic model and show the need for careful enforce-ment of radiative equilibrium in all continua.

Details of the method and a more comprehensive discussion of results will be published at a later time.

This research has been supported by the National Science Foundation, in part through grant no. GP-7761 and in part through grant no. GP-6595.

REFERENCE

Auer, L. H., and Mihalas, D. 1968, Ap. J., 151, 311.

---

[*]Now at Yale University Observatory.

[†]Now at the Department of Astronomy, University of Chicago.

DISCUSSION

Thomas: Explain what you just said about the definition of the b's.

Mihalas: Our b's were defined with respect to the local temperature, but our new local temperature is lower than that in the LTE model. Thus, if I compare the number densities in a non-LTE model with the number densities in the LTE model, I would get a much greater difference than the value of b that I find here. In other words, this b is not too illustrative a quantity for this particular case. I should really have plotted number densities.

Thomas: But it is the b's that enter the determination of the source function. I have difficulty seeing what the real trouble is here. If you prescribe the temperature distribution in the atmosphere, then the runs of the b's can be predicted very well. It seems to me that Cayrel's suggestion on the planetary nebula effect, namely, that it's the quality of the radiation field versus the quantity of the radiation field that counts, must be what controls the general nature of the result.

Mihalas: But the problem here is that the Lyman continuum dominates its own radiation field.

Thomas: But this is true of _any_ continuum. It can always make a self-consistent solution.
Now we know that the line-blanketing effect is small, so the crux of the matter, aside from computational difficulties, is how good is one's physical intuition, and the comforting thing about Miss Cuny's result is that it agrees with my intuition. We have two results and I can't claim that my intuition predicts both.

Mihalas: That's right. If this result is right, you're wrong; if you're right, we're wrong. I am saying that our results don't necessarily disagree with those of Feautrier until we examine the effects of the atomic model. [Note added after the meeting — The effect under debate here was caused by too crude a treatment of upper levels. Our reused results agree with Feautrier.]

Cayrel: I think that sometimes the b's give a very bad impression. After all, we are not so very much concerned if there is a very large change in the local temperature. We are only interested in the connection between the total flux and the actual population of the levels.

Thomas: Well, it's just a question of how your intuition has developed. If, by seeing the b's, you get an idea of what's going on, use them; if not, don't.

Mihalas: We personally feel that from now on it is just simpler to use number densities.

CONSTRUCTION OF NON-LTE EQUILIBRIUM ATMOSPHERES

P. Feautrier

Observatoire de Paris, Meudon, France

The following method was used to compute non-LTE radiative-equilibrium model atmospheres. We start from the transfer equation

$$\mu \frac{dI_\nu}{d\tau} = \kappa_\nu (I_\nu - S_\nu) \quad .$$

The source function is a function of the mean intensity J_ν since 1) we use the statistical-equilibrium equations to link J_ν and the departure coefficients, and 2) we impose the condition of radiative equilibrium to link J_ν to the temperature.

The transfer equation is to be solved subject to a two-point boundary condition; this can be done most easily if the equation is linear. In the present case, we express S_ν as a linear function of J_ν and neglect the dependence of κ_ν on J_ν.

Let us suppose for simplicity that only one transition (e.g., the Lyman continuum) is dominated by radiative processes. It is well known that for a two-level atom the source function is linear in the corresponding phototransition term (e.g., the photoionization term from the ground level).

In the case of a multilevel atom, this fact is still approximately valid, and we can express $S(\nu)$ as

$$S(\nu) = S_1(\nu) + S_2(\nu)\, P(J_\nu) \quad ,$$

where P is the phototransition rate coefficient. The easiest way to compute S_1 and S_2 is to solve the statistical-equilibrium equations for two values of P and to carry out a linear interpolation. This procedure was originally suggested by Y. Cuny.

Putting the resulting $S(\nu)$ in the condition for radiative equilibrium, we obtain

$$\int \kappa_\nu\, S_1(\nu)\, d\nu = \int \kappa_\nu\, J_\nu\, d\nu - P(J_\nu) \int S_2(\nu)\, \kappa_\nu\, d\nu \quad ,$$

from which we can write

$$S_1(\nu) = r(\nu) \left[\int \kappa_\nu\, J_\nu\, d\nu - P(J_\nu) \int S_2(\nu)\, \kappa_\nu\, d\nu \right] \quad ,$$

where

$$r(\nu) = \frac{S_1(\nu)}{\int \kappa_\nu\, S_1(\nu)\, d\nu} \quad .$$

This expression for $S_1(\nu)$ and the corresponding expression for $S(\nu)$ are linear in J_ν if the values of $r(\nu)$, κ_ν, and $S_2(\nu)$ are held fixed (i.e., are taken from the previous iteration). After we replace integrals over ν and μ by finite sums, we can solve the resulting set of coupled linear differential equations by the Fox-Goodwin method.

Three models were computed in this manner:

a. $T_{eff} = 25,000°K$, $\log g = 4$

b. $T_{eff} = 15,000°K$, $\log g = 4$

c. $T_{eff} = 5875°K$, $\log g = 4.44$.

Models a and b are hot main-sequence stars. The striking feature of the non-LTE models is a temperature rise near the surface. Preliminary investigations show that this phenomenon is restricted to the temperature range of 12,000°K, where there is not enough energy in the Lyman continuum to influence this temperature, to 30,000°K, where the first level of atomic hydrogen ceases to be overpopulated.

Model c is a solar model. It includes departures from LTE in the first 10 levels of atomic hydrogen and in the population of the negative hydrogen ion. In this case we find a temperature rise of about 500°K over the minimum temperature in the layers higher than $\tau = 10^{-7}$. The temperature and location of the temperature plateau are in good agreement with the BCA solar model.

DISCUSSION

Johnson: It seems to me Feautrier is getting a temperature rise just about where the Bilderberg continuum atmosphere rises, whereas yesterday Miss Cuny found the temperature rise much lower in the atmosphere.

Thomas: Cuny's results were by assumption on the distribution of T_e and n_e; Feautrier's are by calculation, assuming radiative equilibrium.

Johnson: That, I am sure, is your opinion, but I would like to have you comment on this, Miss Cuny.

Cuny: Inhomogeneities may be important.

Cayrel: This model by Feautrier is purely theoretical and in radiative equilibrium. On the other hand, Miss Cuny's was an attempt to interpret real observations in the ultraviolet solar spectrum. So the location of the plateau has an empirical origin.

Feautrier: It would be nice if we happened to get the same results.

Johnson: Yes, that's the point.

Thomas: No, excuse me. It seems to me that you have here three things:
a) What the sun actually is, and about this there is lots of argument.
b) What Feautrier has tried to produce on the basis of radiative equilibrium; the difference between what you have and what the actual sun does is presumably due to mechanical heating.   c) And then there are the attempts of Yvette Cuny and Heintze to construct a model, by assumption on the distribution of T_e and n_e, to compare with other observations. Yvette just adds the plateau to see whether it will explain the observations. So if you are right, Feautrier, then we evidently have something like 700° to make up with mechanical heating in the low chromosphere. So the really critical point still is, what is the actual temperature between $\tau_{5000} = 10^{-5}$ and $\tau_{5000} = 10^{-6}$.

Elste: If we want to compute an empirical model that comes close to the real situation, we should keep in mind that at layers high compared with the photosphere we should expect strong inhomogeneities probably increasing with height. So the question is the effect of these inhomogeneities on the observable quantities.

Thomas: The eclipse data on hydrogen in 1952 showed that for the continuum there was no evidence of inhomogeneities up to 1000 or 1200 km. Lew House's thesis, on the basis of the neutral metals, indicated inhomogeneities already at 700 km. And then there were the two-stream and three-stream models, in some of which the temperature fluctuation increased with height, while in others it decreased with height. So I personally don't know what the situation is with respect to inhomogeneities.

Elste: But you see enormous inhomogeneities in the hydrogen lines on the solar disk.

Thomas: The biggest problem is to know the height at which the disk features are actually formed. The second problem is to interpret what an observed inhomogeneity in radiation means in terms of inhomogeneity in T_e, n_e. Without the answers to both these questions, your observed fluctuations in hydrogen emission over the solar surface can't be translated into fluctuation in physical quantities at some level in the solar structure.

PART III

THE COMPARISON OF SYNTHETIC SPECTRA

WITH REAL SPECTRA

R. Cayrel, Chairman

COMPARISON OF SYNTHETIC SPECTRA WITH REAL SPECTRA

R. Cayrel

Observatoire de Paris, Meudon, France

1. INTRODUCTION

In the two preceding introductory talks we have heard a detailed account of theoretical work relevant to our topic and an enlightening discussion of the different types of quantitative measurements that have been made on the spectra of a large number of stars. It is our task now to bridge the two aspects of our problem and to arrive at a meaningful comparison between the synthetic spectra obtained from the computer and the real spectra obtained from observations.

This comparison is not absolutely straightforward, because the synthetic spectrum is affected by the simplifications introduced in the physics underlying the computations, whereas the observed spectrum is affected by lack of spectral resolution, uncertainties in absolute calibrations, etc.

Assuming that a proper comparison procedure has been found, we could either extract the required basic information (namely, effective temperature, gravity, chemical composition, etc.) from a successful fit or try to understand the physical cause of a misfit. It may be wise to mention here that a successful fit does not strictly guarantee that the proper parameters have been found. The simplifications of the physical processes in the computation of synthetic spectra might eventually produce the same synthetic spectrum as a more elaborate model for a different set of temperature, gravity, and chemical composition. This is actually occurring, in first approximation, when non-LTE or blanketing effects are taken into account. In other words, a good fit obtained with a poor theory must be treated with caution.

In the following we keep in mind that stellar classification is needed for distant and faint stars. Therefore, we shall stress the comparison of the continuous energy distribution and strong features of the spectrum rather than features that can be observed only at high resolution, i.e., only on bright objects.

2. OBSERVATIONAL DATA USEFUL FOR COMPARISON WITH SYNTHETIC SPECTRA

Yesterday Dr. Peat presented an elaborate reduction program in which the raw observations are transformed into effective temperature, gravity, chemical composition, projected rotational velocity v sin i, etc.

I wish to advocate[*] a more conservative position in which the observer supplies a clearly defined physical quantity that nearly always has the form

$$Q_i = \int_0^\infty F(\lambda) \cdot S_i(\lambda) \, d\lambda \quad , \tag{1}$$

---

[*]I realized after delivering my talk that I misunderstood Dr. Peat, who does not propose to give only the fully processed observations but also the physically observed quantities that have been processed.

where $F(\lambda)$ is the intrinsic spectrum of the star, and $S_i(\lambda)$ the overall sensitivity function of the channel: interstellar medium, earth atmosphere, optics of the telescope and photometer, filter, receiver.

It is worth noting that the radial velocity of the star shifts the wavelength scale of the intrinsic spectrum of the star with respect to the apparatus and that this phenomenon cannot be neglected in narrow-band photometry. In practice, the flux of the star is not actually observed alone but mixed with some background light:

$$Q'_i = \int_0^\infty (F_{star} + F_{background}) \, S_i(\lambda) \, d\lambda \quad . \tag{2}$$

The background light must be eliminated in the reduction of the data. The determination of $F_{background}$ may require that the angular distance of the moon to the optical axis of the telescope be known as well as the number of matches scratched during the exposure. Only the observer is fully aware of these details, and he must consider them when transforming the somewhat rough observation into a well-defined physical quantity of the type defined in equation (1) with an accurately specified sensitivity function $S_i(\lambda)$. Since the reflectivity of an aluminized surface deteriorates, the sensitivity function $S_i(\lambda)$ should be redetermined fairly often. I must repeat that when observed stellar spectra are being compared, the sensitivity function $S_i(\lambda)$ may be known up to a constant factor, but when an observed spectrum is being compared with a synthetic spectrum, this constant factor must be known with an accuracy as great as that required for the comparison.

3. FITTING THE CONTINUUM

Since we wish to discuss faint stars, we shall start with rather wide-band observations giving information on the energy distribution of the stars. We call what we are measuring a continuum, but it really is a continuum with lines, and possibly just as much resolution is needed to observe the actual continuum as to observe a single line because the spacing between the lines is not larger than the width of each line. This is a basic difficulty in the visible and the ultraviolet spectrum of late-type stars.

The widest band that can be imagined includes all photons from $\lambda = 0$ to $\lambda = \infty$. If we know also the distance of the star, we obtain the absolute bolometric magnitude of the star, or its luminosity:

$$L = \sigma \, T_{eff}^4 \cdot 4 \pi R^2 \quad , \tag{3}$$

where σ is the Stefan constant, T_{eff} the effective temperature of the star, and R its radius.

It would be a marvelous program to carry out from space. Unfortunately, the ultraviolet radiation of stars below the Lyman discontinuity is screened not only by the earth's atmosphere but also by the interstellar hydrogen, as Dr. Morton pointed out in an earlier session. It would then be necessary to fly a rocket up to the next H II region to get the integrated flux of a B star. As far as we know, the only star for which a direct measurement of the total luminosity has been made is the sun. Therefore, there is little to discuss except to state that for a star of known radius (spectroscopic binary star with a measurable angular diameter) L and R can be measured and T_{eff} derived from equation (3). This direct determination can be checked against other, more indirect methods (see Johnson, 1964, and Mihalas, 1966).

Multicolor photometry, scans, and spectrophotometry must now be considered. In early work using model atmospheres (Traving, 1955; Hunger, 1955; Cayrel, 1958), the Balmer jump and the blue and ultraviolet gradient measured by Chalonge and coworkers have been checked against predicted values from models whose parameters were otherwise determined, mainly by the study of ionization equilibria and the profile of the Balmer lines. Since these are much more sensitive to temperature and gravity than the blue gradient and the Balmer jump, the fact that a good self-consistency was obtained proved at least that the predicted continua were not affected by gross errors.

Melbourne (1960) computed a grid of gray models and compared the observed U, B, V colors on the Johnson and Morgan system with the same colors predicted from the models. He found a striking disagreement between the predicted and the observed colors (Figure 1, which is Melbourne's Fig. 11). He concluded that the discrepancy was due to erroneous sensitivity functions of the filters corresponding to corrections of $\Delta(B-V) = -0.17$ and $\Delta(U-B) = -0.23$.

Later, Matthews and Sandage (1963) showed that, assuming Code's energy distribution of Vega, the continua of stars scanned by Oke could be used to predict the U, B, V colors of these stars. They confirmed the interpretation of Melbourne and gave improved sensitivety functions for the U, B, V system. Much more sophisticated models computed by Mihalas (1966) in accurate nongray radiative equilibria and including the blanketing produced by the Balmer series lead to a nice fit of the observed and predicted (U-B, B-V) diagrams (see Figures 2, 3, and 4, which are Mihalas's Figures 6, 7, and 8) and for the continuum of Vega for an effective temperature not too far off the temperature determined from L and R (Equation 3) (R is obtained from interferometric measurements made by Hanbury Brown, Hazard, Davis, and Allen, 1964).

Strom, Gingerich, and Strom (1966) tried to reproduce the continua of Vega and Sirius. They also assumed the effective temperature derived from L and R direct measurements and concluded that it was necessary to increase the silicon content of Sirius by a factor of 10 to decrease the predicted Balmer jump to the observed value (see Figures 5 and 6).

Talbert and Edmonds (1966) found that they could not obtain a fit without some small residual discrepancies for Procyon.

Oke and Conti (1966, Figure 1) have successfully fitted unblanketed models with observed scans of the continua of Hyades stars (see Figure 7).

Recent nongray models computed by Feautrier (1967) have been checked against (D, ϕ_b, ϕ_{uv}) parameters obtained by Chalonge and coworkers.

The general impression is that the stellar continua are fairly well reproduced by classical models. But in view of the large uncertainties in the ultraviolet opacity and the ultraviolet blanketing, as well as non-LTE effects (probably small) and non-LTE blanketing correction, it is possible that the effective temperature derived from the fit is still uncertain by 5% or so. The gravity is determined with much less accuracy.

4. FITTING THE BALMER LINES

The continuous spectrum alone is not sufficient to supply the three independent parameters — temperature, gravity, and metal content — over the whole range of variation of these parameters. Strömgren's u b v y system takes the best of the information contained in the stellar continua in the A3 to G0 range.

However, for earlier types, the Balmer jump becomes gravity insensitive, and for later types the same Balmer jump becomes too small to be measurable. Furthermore, the continuous energy distribution is affected by interstellar absorption for stars at large distances or imbedded in local interstellar matter. Hence it is necessary to augment the information contained in the stellar continua. It is desirable that this addition is not too small a spectral feature so that the number of photons involved and the signal-to-noise ratio are as large as possible. The Balmer lines are of course the best

candidates in the spectral range B to F. It is possible either to measure the
equivalent width of a single line (for example Hβ, Hγ , or Hδ) by narrow-band
photometry or to measure the cumulative effect of several lines (Golay, 1964),
which allows a better limiting magnitude.

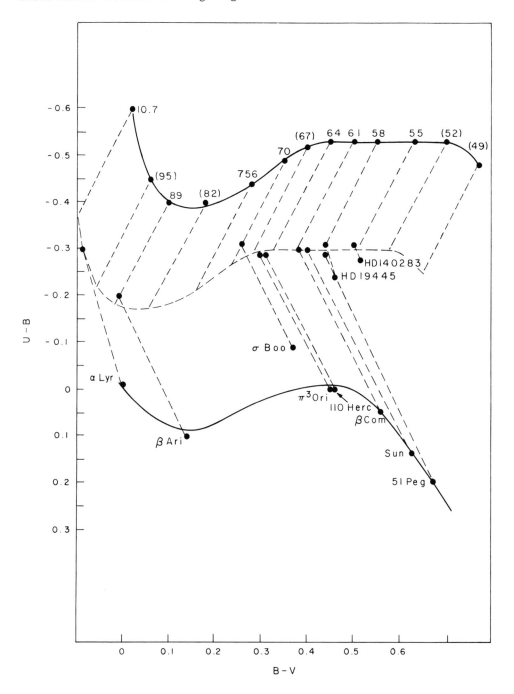

Figure 1. Observed and line-free U, B, V stellar colors; the observed
main-sequence relationship; the computed and translated
model colors. (From Melbourne, 1960.)

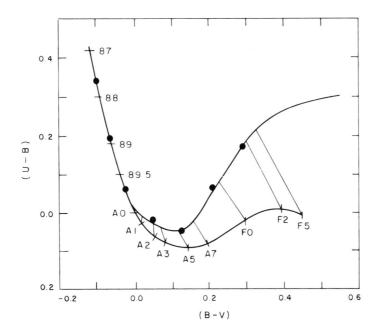

Figure 2. Heavy curve: observed locus of main sequence in two-color diagram
with spectral types indicated by short crossbars. Light curve: metal-
line unblanketed main sequence with blanketing vectors showing
unblanketed position of main-sequence spectral types. Dots:
computed colors of hydrogen-line-blanketed models with log g = 4.
(From Mihalas, 1966.)

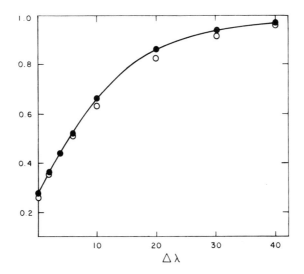

Figure 3. Observed profile of Hγ in α Lyr is shown as solid curve. Ordinate:
residual flux in units of continuum. Abscissa: distance from line
center in A . Solid dots: computed Hγ profile for model with
θ_e = 0.50, log g = 4. Open dots: computed Hγ profile for model
with θ_e = 0.55, log g = 4. (From Mihalas, 1966.)

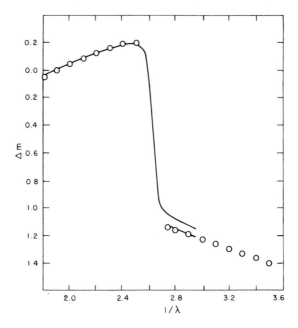

Figure 4. Solid curve is absolute-energy distribution of α Lyr as adopted by
Code and Oke. Lower straight line is flux distribution below Balmer
jump implied by Bahner's photoelectric measure. Open dots represent
flux distribution computed for model with θ_e = 0.525, log g = 4.
Ordinate: monochromatic flux on magnitude scale. Abscissa: $1/\lambda$,
where λ is in microns. (From Mihalas, 1966.)

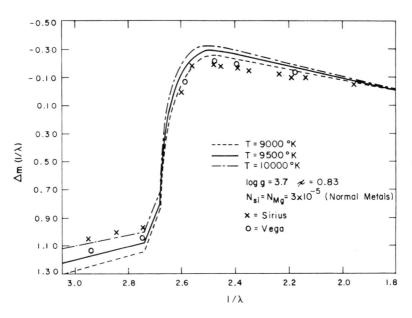

Figure 5. Comparison of spectrophotometric measurements of Vega and
Sirius with model stellar atmospheres having normal metal
abundances. (From Strom et al., 1966.)

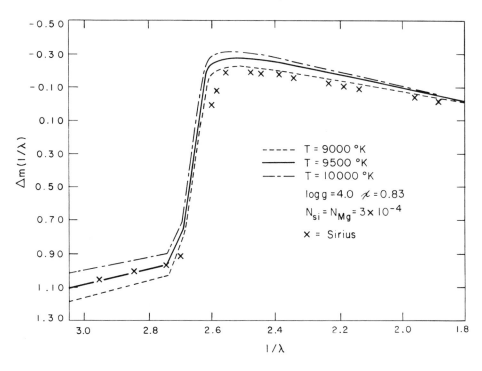

Figure 6. Comparison of spectrophotometric measurements of Sirius with model
stellar atmospheres having the abundances of Si and Mg increased by
a factor of 10. (From Strom et al., 1966.)

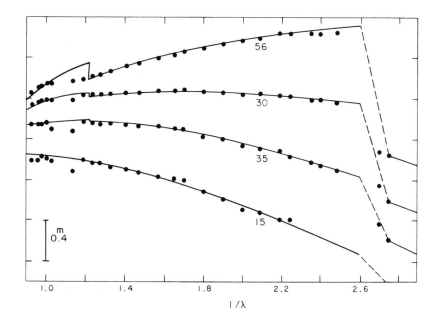

Figure 7. Absolute spectral energy distributions of the continuum for several
Hyades stars. The dots represent the observations, the solid curves
the fitted model atmospheres. (From Oke and Conti, 1966.)

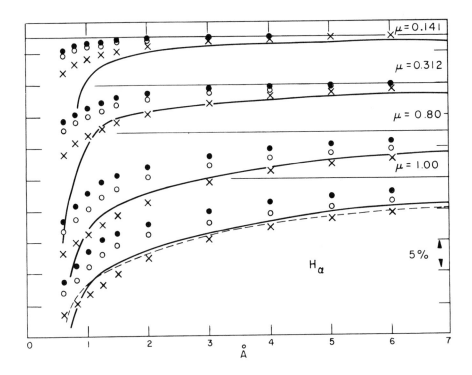

Figure 8. The wings of Hα in the solar spectrum. ⎯ : observed profiles. Filled dots: computed with quasi-static broadening by protons only. Open dots: computed including electron contribution with Griem theory. Crosses: computed with Griem theory plus self-resonance broadening. (From Cayrel and Traving, 1960).

In the early works already quoted (Traving, 1955; Hunger, 1955; Cayrel, 1958) the Balmer lines were used as criteria. Since that time, however, a great deal of work has been devoted to the theory of line broadening (Griem, 1964; Pfenning, Trefftz, and Vidal, 1966; Feautrier, Praderie and Van Regemorter, 1967), and more recent comparisons made with a better theory should now be considered. In 1959 Van Regemorter showed that a predicted Balmer jump − Hδ equivalent-width diagram was in better agreement with observed values when Hδ was computed according to Griem's theory. Searle and Oke (1962) have shown that the Hγ profile was a useful temperature criterion in F-type stars. The temperature can be derived either from the continuum or from the strength of Hγ. The authors found a good consistency in the temperatures derived from these two criteria for the star SU Draconis, but they found a small disagreement for RR Lyrae. Dr. Strom mentioned in the opening session that it may be difficult to disentangle the effect of temperature from the effect of gravity for F stars above the main sequence. On the main sequence, Hγ is virtually insensitive to gravity.

Most detailed analyses of stellar spectra have used the Balmer lines as a temperature — gravity criterion. At the approximation of LTE, which very likely is satisfactory for the wings of the Balmer lines, the correct theory of broadening is crucial.

The way in which the electrons contribute to the broadening has been some-what controversial. Still, when there is ambiguity regarding the prevalent type of broadening (quasi-static or impact), the various predictions are usually rather close, say within 20 to 30%. In view of the difficulty of locating the continuum with an accuracy better than 2% (which produces an error of 20% on a wing, 10% below the continuum), the remaining uncertainty does not impair the comparison between synthetic and real spectra to any great extent.

Mihalas (1966) found that the model giving the best fit for the observed continuum of Vega has an effective temperature of 9600°K, whereas the best fit for Hγ is obtained for log g = 4.0 and a temperature of 10,800°K. The difference could be removed by slightly lowering the gravity of Vega.

It has been found that in peculiar A stars and metallic-line stars (Mihalas and Henshaw, 1966; Praderie, 1968) the parameters that give a good fit for the continuum also lead to a satisfactory prediction of the Balmer lines. Similarly, Cayrel and Traving (1960, Figure 1) have shown that self-resonance broadening is important for Hα in the sun (see Figure 8). This is certainly even truer for cool subdwarfs such as Groombridge 1830, in which the effect is so strong that the equivalent widths of the Balmer lines increase from Hδ to Hα (Jugaku, 1960).

Departures from LTE in the hydrogen spectrum of the sun have been investigated by Cuny (1968). Her computation of Hα with five levels agrees with the profile observed by David (1961) within a few percent (Figure 9, which is Cuny's Figure 23).

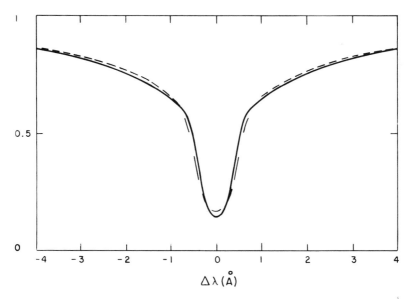

Figure 9. Profile of Hα. ——— de Jager model HA01 with a five-level atom.

 --- Profile observed by David (1961). (From Cuny, 1968.)

The residual intensity of Hα is considerably better accounted for by the non-LTE approach than by the LTE computation, which would predict a flat core according to the Bilderberg Model.

5. FITTING THE LINE SPECTRUM

Detailed analyses of stellar spectra have used, in addition to the continuous spectrum and the Balmer lines, the equivalent width of a large number of lines in order to improve the determination of the parameters — temperature, gravity, and metal or helium content — and to obtain the individual abundancies of the elements.

The ratio of lines of very different energy levels provides the most sensitive temperature criterion. When several criteria of this kind are used, they usually give slightly discrepant values for the temperature. The origin of these discrepancies is not known. They may arise from non-LTE effects, but possibly also from inaccurate values of the oscillator strengths. One may get a good insight into this aspect from Mihalas (1964) or Underhill (1966). The discrepancies we are talking about are not usually larger than 0.02 on the θ_{eff} scale.

It is also clear (Cayrel de Strobel, 1966, Figure 44) that strong resonance lines such as the sodium D lines or the magnesium D lines are deeper than predicted by LTE (see Figure 10). Chamaraux (1967) has obtained a successful fit for the solar D lines in the solar spectrum by non-LTE computations.

6. EFFECTS OF ROTATION

It has been shown in a recent series of papers (see, e.g., Roxburgh and Strittmatter, 1965; Collins, 1965; 1966; 1968) that the location of a star in the color-luminosity diagram is affected by rotation to a measurable extent. Photometric results with an accuracy of 1% must certainly include the rotation as a fourth independent parameter.

These recent papers have raised the hope that in the case of very rapid rotation the equatorial velocity and the angle of inclination of the axis of rotation over the line-of-sight could be determined separately, instead of only in their combined form v sin i.

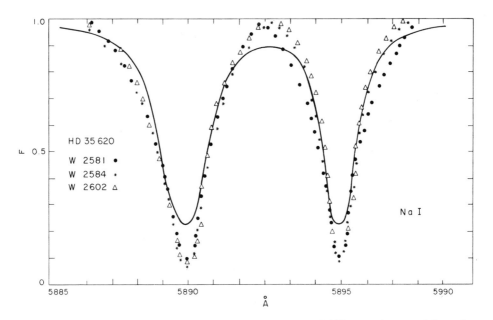

Figure 10. Profile of the sodium D lines for HD 35620. The model and abundances retained for this star are:

$$\theta_{eff} = 1.22; \ \log g = 1.0; \left[\frac{N_{Na}}{N_H}\right]_{\epsilon \ Vir} = -0.59. \quad \text{(From Cayrel de Strobel, 1966.)}$$

REFERENCES

Cayrel, R. 1958, Ann. d'Ap. Suppl., No. 6, 124 pp.

Cayrel, R., and Traving, G. 1960, Zs. f. Ap., 50, 239.

Cayrel de Strobel, G. 1966, Ann. d'Ap., 29, 413.

Chamaraux, P. 1967, Ann. d'Ap., 30, 67.

Collins, G. W, II. 1965, Ap. J., 142, 265.

_____. 1966, Ap. J., 146, 914.

_____. 1968, Ap. J., 151, 217.

Cuny, Y. 1968, Solar Phys., 3, 204.

Davis, K.-H. 1961, Zs. f. Ap., 53, 37.

Feautrier, P. 1967, Ann. d'Ap., 30, 125.

Feautrier, N., Praderie, F., and Van Regemorter, H. 1967, Ann. d'Ap., 30, 45.

Golay, M. 1964, Publ. Obs. Genève, Ser. A, No. 66, 19.

Griem, H. R. 1964, Plasma Spectroscopy (New York: McGraw-Hill Book Co.).

Hanbury Brown, R., Hazard, C., Davis, J., and Allen, L. R. 1964, Nature, 201, 1111.

Hunger, K. 1955, Zs. f. Ap., 36, 42.

Johnson, H. L. 1964, Bull. Ton. y Tac. Obs., 3, 305.

Jugaku, J. 1960, private communication.

Matthews, T. A., and Sandage, A. R. 1963, Ap. J., 138, 30.

Melbourne, W. G. 1960, Ap. J., 132, 101.

Mihalas, D. 1964, Ap. J., 140, 885.

_____. 1966, Ap. J. Suppl., 13, 1.

Mihalas, D., and Henshaw, J. L 1966, Ap. J., 144, 25.

Oke, J. B., and Conti, P. S. 1966, Ap. J., 143, 134.

Pfennig, H., Trefftz, E., and Vidal, C.-R. 1966, J. Q. S. R. T., 6, 557.

Praderie, F. 1968, Ann. d'Ap., 31, 15.

Roxburgh, I. W., and Strittmatter, P. A. 1965, Zs. f. Ap., 63, 15.

Searle, L., and Oke, J. B. 1962, Ap. J., 135, 790.

Strom, S. E., Gingerich, O., and Strom, K. M. 1966, Ap. J., 146, 880.

Talbert, F. D., and Edmonds, F. N., Jr. 1966, Ap. J., 146, 177.

Traving, G. 1955, Zs. f. Ap., 36, 1.

Underhill, A. B. 1966, in Vistas in Astronomy, ed. A. Beer and K. Aa. Strand (Oxford: Pergamon Press), 41.

Van Regemorter, H. 1959, Ann. d'Ap., 22, 681.

DISCUSSION

Popper: I should like to make one comment concerning the reference to wide-band photometry. Reference was made to the earlier difficulties in computing the colors and then to the improved sensitivity functions given in Matthews and Sandage [Ap. J., 138, 49, 1963]. I should like to say that all is not well in this respect. It is true that one can predict fairly well the B-V color from the energy distribution and the sensitivity functions. But what you don't get is V magnitude. That is, if you have the monochromatic magnitudes of stars of widely differing colors at different wavelengths, and you take the values at what is supposed to be the central wavelength of the V magnitudes, you do not get the observed magnitude differences of the wide-band V photometry. The deviation could be as great as 0.05 mag over a wide color range.

Underhill: Isn't that just the amount of blanketing, ignored in the models?

Popper: No, models are not involved. It is a matter on the one hand of observed wide-band magnitude differences converted to the V system in the usual way, and on the other hand of narrow-band (100 A, say) magnitude differences of the same stars. The question is, what central wavelength for the narrow-band observations gives agreement with the wide-band V magnitude differences? The answer is not at present known, but the suspicion is that it will not be the central wavelength of the Matthews-Sandage function. Because of the width and asymmetry of the V sensitivity function, there may not be a wide unique wavelength. You cannot, unfortunately, redetermine Johnson's sensitivity function. This problem was attacked by Willstrop without definitive results. Oke is now working on it, and the results will presumably come in. This will affect bolometric corrections. As any observer knows, there is always a systematic correction of the same sign to get from the observer to the Johnson V system. So this is a word of caution in relation to bolometric corrections computed with the usual sensitivity function.

Underhill: It was remarked at the beginning of the talk that theoreticians prefer the observers to give us their data not too well worked over. I think it is about time theoreticians begin to get smart enough to give us a little bit more detail. While the effective temperature is the basic parameter, it refers to the total luminosity. But, as you pointed out, we observe a small band, and I am firmly convinced that we would be better off if we picked theoretical parameters that are related only to the spectral region we observe. Now, this choice will depend on the spectral type. In the cooler stars you can observe most of the energy that is emitted. There the effective temperature is not too bad a theoretical parameter. But for a star of type A2 and earlier, effective temperature is too far from the actual interpretation we want. For the early-type stars effective temperature is not a useful parameter. We should try to choose something else — brightness temperature in a certain region, for example.

Cayrel: I fully agree with you. It would perhaps be better if you were working in some selected spectral region to replace the effective temperature by the brightness temperature at an accessible wavelength. There are various discrepancies that appear when you use an effective temperature and that cancel out when you use a brightness temperature.

Strom: I should like to comment briefly on the use of Hγ as a temperature indicator of late-type stars. For main-sequence stars there is good empirical correlation between the slope of the Paschen continuum and the hydrogen-line profile. Unfortunately, there are not good convective models yet available for quantitative calibration.

Gingerich: I should like to remark, apropos of Popper's comment, that the UBV system is really a theoretician's nightmare. It is a system that tends to be ignored by anyone who wants to do a very thorough analysis of one or just a couple of stars. The only reason people worry about it is that there is such a tremendous number of observations in that system. So here is a wealth of data that we feel somehow we ought to be able to do something with, but we really haven't. One of the reasons for our bringing together a conference of theoreticians and observers was to come to grips with problems of precisely this sort.

Jones: I would say that my own results are rather at variance with what Popper said. I have observed Strömgren's y magnitude and compared it with Johnson's V as observed by Cousins at the Cape. The root-mean-square difference is 0.032 mag, and most of that error is my own. If I did mine over, I am sure I would bring the scatter down, and the color equation is probably not significant. Cousins and I are cooperating in a program at the moment to try to wring a bit more data out of the UBV system.

Neff: I think perhaps this is the time to bring up a matter related to the calibration of intermediate bandpass systems. I have found that by folding my sensitivity functions into the scanner flux ratios I can find linear transformations with a scatter of 0.01 mag. I think this should also be true for most intermediate bandpass systems, but I don't know that it would be the case for narrow-band systems. It seems to me that this is probably the way to calibrate intermediate bandpass systems. I have compared a magnitude measured with a 200-A bandpass filter located at 5500 A with V magnitudes and I find a color term of 0.07 mag [Neff and Travis, Astron. J., 72, 48, 1967].

Buscombe: Since we're referring to observations that are sometimes used for galactic structure, where the UBV system is useful, may I make a beef about the photoelectric observers of eclipsing binaries. I find that, although the observations are very extensive over the light curve, this is the last group of objects that one can consider in galactic structure because local photometric systems are used in the discussion of most of the results. There is no way of finding out what the color excess or distance modulus is from the published observations.

Popper: Amen.

Thomas: I think Anne has slipped something over on you here. If I understand well, what you're talking about are normal spectra. You define a normal spectrum by radiative equilibrium, hydrostatic equilibrium. It seems to me that what she is saying is that it isn't enough to specify the gravity and the effective temperature after all. What you have to do is to start specifying color temperature in each region of the spectrum.
    You're giving up a lot. I hope you know exactly what you have given up, because this is much more violent than the standard non-LTE business. What do you mean, Anne? You had better explain yourself.

Underhill: What I mean is that I like narrow-band photometry, and I think the observers have tried to tell us that here is a star and the star says, "Look, I'm like this. You can't see the rest of me, it's all covered up. I have got a long nightgown on. All you can see is my head sticking out." What I'm interested in is a precise picture. I'm not too interested in using my imagination on what's under the nightgown.

Thomas: All you're really saying is that in astronomy I can only observe a limited number of things. But then, if there is such a thing as a normal star, and that is what is being debated here, presumably I shall be able to describe it in terms of gravity, effective temperature, etc. − in other words, some kind of measure of energy and of the density. If it turns out that we can't use these two parameters, then we had better stop talking about normal stellar spectra.

Underhill: I agree with you on this. And I think we are finding it more and more difficult to trace the spectral features of early-type stars back to these simple parameters.

Cayrel: Maybe I misinterpret Anne's thoughts − but the idea is that you simply cannot check the effective temperature against observations most of the time. If the star is observed in the ultraviolet, you can do it, but usually you cannot. You must find a substitute, and I suggest that the brightness temperature at a given wavelength would be sufficient.

Thomas: Roger, suppose I define a normal star by the gravity and the total flux. Now presumably I can make lots of predictions on the spectrum − not just one or two − and if it is indeed true that I can find some of these criteria that define the flux and the gravity, I really don't need to observe the <u>whole</u> range of the spectrum. Your check for that is whether I can find, say, six predictions, <u>all</u> of which determine the effective temperature. If they all agree, you have a normal star, and if they don't and if you have to specify the color temperature in every wavelength, then I assert that you have given up the search for a normal star.

<u>Morton</u>: I think Cayrel made the comment that we <u>do</u> get reasonable agreement between the observations and the models computed for a specific effective temperature, so I think the concept is still useful.

<u>Gray</u>: Is it really true that the discrepancy between the continuum and the hydrogen-line profiles is greater in the sun than in the stars?

<u>Böhm-Vitense</u>: I don't know the answer, but I think that was based on the old Griem theory, which is not quite correct. In addition, the temperature inhomogeneities were not taken into account, so we simply can't say.

<u>Cayrel</u>: The additional trouble with the sun is that you must know the structure of the convective zone, because the wings of the Balmer lines are formed rather deep, and it is very model-sensitive, and there was also some discussion of the modifications accordingly. Praderie can comment about that because it has been checked again with the hydrogen lines of the sun. Do you remember what you find for $H\gamma$ or $H\beta$?

<u>Praderie</u>: According to my computations there is no difference for a solar model on the $H\gamma$ profile because of the nature of the theory one uses (Griem or purely quasi-static for ions and electrons, with Mozer and Baranger distribution of the field), whereas there is a small discrepancy in the wings for a model with $T_{eff} = 10,000°K$, as is well known. Cayrel's slide shows a Holtsmark quasi-static profile, the broadening being due to the protons only.

I did not compute center-to-limb variations of the Balmer lines; but at the center of the disk, for $H\alpha$, the agreement is good at $\Delta\lambda \geq 4$ A between the profile observed by David and the profile computed with 1) broadening due to self-resonance plus quasi-static Stark effect and 2) the Utrecht 1964 model. So the structure of the deepest layers in that particular model allows one to reproduce the observed wings of $H\alpha$.

<u>Cayrel</u>: Could Spite make a comment?

<u>Spite</u>: The wings calculated by me with Mutschlecner's model and Praderie's method are in fair agreement with David's observations for both $H\beta$ and $H\gamma$. In fact, the measurements of de Jager shown on Cayrel's slide are also in fair agreement with David's one for $H\beta$, but disagree for $H\gamma$, so that the calculations on Cayrel's slide are also in fair agreement with David's observations for $H\beta$ and $H\gamma$.

The center limb variation was calculated for H only, with Mutschlecner's model, and is in fair agreement with David's one.

Krishna Swamy [Ap. J., <u>145</u>, 174, 1966; Ap. J., <u>151</u>, 1195, 1968] calculated the profiles of $H\alpha$, $H\beta$, $H\gamma$ at the center of the disk with Cayrel's solar model. These profiles are in good agreement with David's observations when $n_{He}/n_H = 0.095$.

<u>Weidemann</u>: You mentioned the dependence of the equivalent width of $H\gamma$ and the Balmer jump on luminosity. I should prefer to relate these to gravity.

<u>Böhm-Vitense</u>: If we give up talking about effective temperature, don't we have to give up talking to people who compute stellar evolution?

Cayrel: Yes, but it is not a question of _giving up_ the effective temperature. It is just more uncertain than the brightness in the spectral region where the opacity is well known.

Böhm-Vitense: I would like to defend Peat's approach. I think this is the only approach we can take. Sure, you _should_ observe more quantities than you need for interpreting the spectrum, but how else are you going to do it? After all, the observers do have the right to interpret.

Cayrel: It is very important to file the data without any loss of information.

Peat: Where is the information being lost?

Cayrel: In your computer program. When you process, you lose information.

Peat: I would rather lose information in a program than adopt a crude system that pretends there is no information to lose.

Gingerich: People would be reassured if they realized that you do publish the _original_ ratios as well as the results of processing.

Peat: Yes, we do.

[Editor's Note: After private discussion, Dr. Peat and Dr. Cayrel came to a full agreement on these matters.]

THEORY AND OBSERVATIONS OF ULTRAVIOLET SPECTRA
OF EARLY-TYPE STARS

D. C. Morton

Princeton University Observatory, Princeton, New Jersey

At Princeton we have continued the program of computing blanketed model atmospheres of hot stars begun in cooperation with Dimitri Mihalas (Mihalas and Morton, 1965; Adams and Morton, 1968). These models include the opacity of more than 100 of the strongest ultraviolet absorption lines evaluated at some 180 wavelength points shortward of 1700 A. The calculated emergent spectra of two models by Hickok and Morton (1968) are shown in Figures 1 and 2. They are believed to represent an O5 V star with $T_{eff} = 37,450°K$ and a B0V star with $T_{eff} = 28,640°K$. Both models have log g = 4.0 and He/H = 0.15 by number. Spectral types were estimated from the Balmer jumps, so that there is some uncertainty in the identification of the hotter model, which has a jump of only 0.049 mag.

If we compare blanketed and unblanketed models with the same Balmer jump, we find that the effective temperatures of the blanketed models are lower by 1000 or 2000°K. Consequently, line blanketing must be included in the hot models when they are compared with rocket and satellite observations of ultraviolet stellar fluxes, as shown recently by Bless, Code, and Houck (1968). For a model of a particular spectral type, the blanketing lowers the general level of the ultraviolet continuum; furthermore, when lines appear within the bandpass of the detector, there is an additional reduction in the predicted flux. Earlier comparisons (Chubb and Byram, 1963) neglected these effects as well as interstellar extinction and found significant deficiencies in the observed ultraviolet fluxes.

At Princeton, Gallagher and Morton have compared Smith's (1967) satellite observations in a 350-A band around 1376 A with our blanketed models, as shown in Figure 3. Smith has corrected the difference between ultraviolet and visual magnitudes $(m_{1376}-V)$ for interstellar extinction by comparing more and less reddened stars of the same spectral type to derive the ratio $E(m_{1376}-V)/E(B-V) = 7.6$. We see that the measured fluxes still lie a few tenths of a magnitude below the models, but this is nothing like the large discrepancies that had once been reported. The more thorough investigation by Bless et al. (1968) has shown agreement with the models from 1314 to 2800 A within 0.5 mag. The models tend to be a little bright in the ultraviolet, indicating that some opacity still may have been omitted, but small systematic errors are also possible in the detectors. Altogether, the agreement is rather encouraging and gives us considerable confidence in the blanketed models and the temperature scale derived from them by Morton and Adams (1968).

In Figure 3 it is of interest to note that of the three hottest stars with $(B-V)_0 = -0.32$, namely, χ Per (O7), S Mon (O7), and ζ Pup (O5f), it is the emission-line O star (f type) ζ Pup whose ultraviolet-to-visual flux ratio lies well below the extrapolation of the mean of the cooler stars.

Now we wish to compare the ultraviolet spectrum of a very hot star with the calculations. Unfortunately, a good spectrum of a normal O-type main-sequence star is not yet available, but Carruthers (1968), Stecher (1967), and Morton and Jenkins (1968) have obtained spectra of ζ Pup from recent rocket flights. The Princeton spectrum, shown in Figure 4, extends from 1100 to 1960 A with about 1-A resolution. As we might have expected for an Of star, the spectrum is not at all like that predicted in Figure 1 for an O5 V atmosphere.

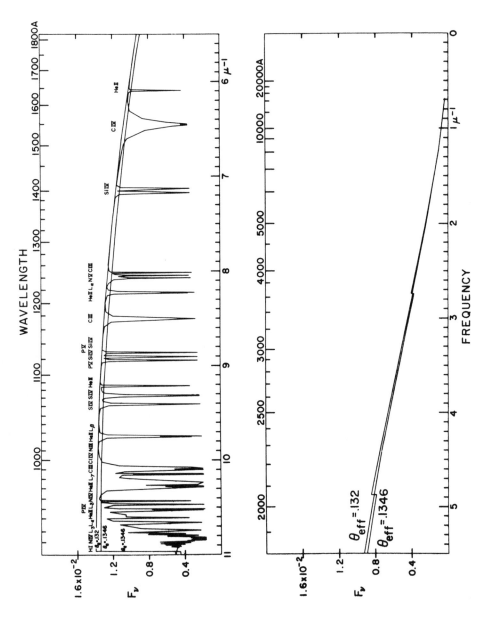

Figure 1.   Emergent flux from the line-blanketed O5 V model (θ_{eff} = 0.1346), with two unblanketed models for comparison.

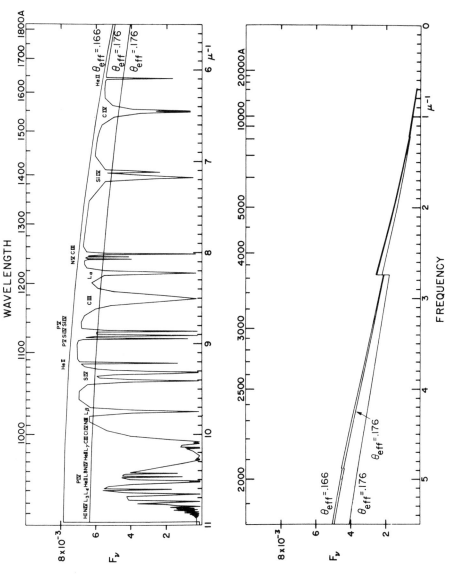

Figure 2.   Emergent flux from the line-blanketed B0 V model (θ_{eff} = 0.176), with two unblanketed models for comparison.

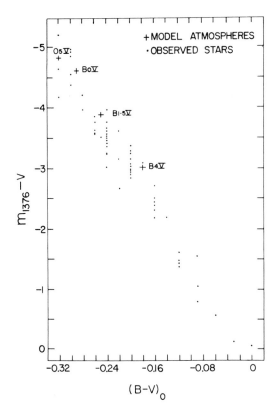

Figure 3. Comparison of observed ultraviolet-to-visual magnitude differences
with predictions from blanketed model atmospheres.

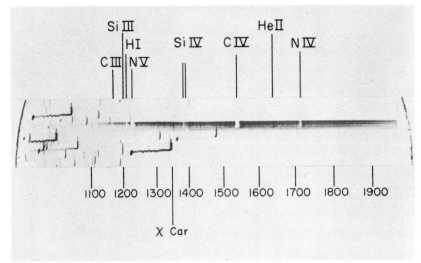

Figure 4. Ultraviolet spectrum of ζ Pup (O5f) from 1100 to 1960 A photo-
graphed with a NASA Aerobee rocket on November 1, 1967. The
zero-order images of other stars in the field of the objective
spectrograph fix the wavelength scale. The tails on some of the
images resulted from a failure of the pointing system toward the
end of the exposure.

Strong resonance absorption lines of C IV, N V, Si III, and Si IV and an excited line of C III (6. 5 ev) are present with velocity shifts averaging 1810 km sec^{-1} toward shorter wavelengths. These lines must be formed in an expanding shell escaping from the star, similar to the shells found in the hot supergiants in Orion. Excited lines of He II (40. 6 ev) and N IV (16. 1 ev) have lesser shifts of 380 and 980 km sec^{-1}, respectively, and therefore must be formed lower in the atmosphere in the acceleration region, where the density is higher and the excited states are more populated. Emission lines of He II, C IV, N IV, and N V are present close to their laboratory wavelengths.

Also present in the spectrum are at least 100 weaker lines that probably originate in the photosphere. Many of them have been identified as lines of C II, C IV, N III, N IV, O IV, Ar IV, and Fe V.

REFERENCES

Adams, T. F., and Morton, D. C. 1968, Ap. J., 152, 195.
Bless, R. C., Code, A. D., and Houck, T. E. 1968, Ap. J., 153, 561.
Carruthers, G. R. 1968, Ap. J., 151, 269.
Chubb, T. A., and Byram, E. T. 1963, Ap. J., 138, 617.
Hickok, F. R., and Morton, D. C. 1968, Ap. J., 152, 203.
Mihalas, D. M., and Morton, D. C. 1965, Ap. J., 142, 253.
Morton, D. C., and Adams, T. F. 1968, Ap. J., 151, 611.
Morton, D. C., and Jenkins, E. B. 1968, Astron. J., 73, S110 (abstract).
Smith, A. M. 1967, Ap. J., 147, 158.
Stecher, T. 1967, Astron. J., 72, 831 (abstract).

DISCUSSION

Underhill: This acceleration — do you expect it's chiefly due to radiation pressure in the strong lines?

Morton: Lucy and Solomon have made the suggestion that in the very strongest lines, such as in the C IV resonance lines, there can be momentum transfer from the photons to the particles. They made some calculations and obtained velocities comparable to those observed. The one possible objection to their point of view is the following. For this mechanism to operate — it's the old Milne mechanism for producing a solar wind from the calcium lines in the sun — the line must be formed at the very top of the photosphere so that as photons are absorbed and C IV is accelerated, for example, the velocity shift permits the ions to see the continuum photons of longer wavelengths that can be absorbed to keep pushing the ions out. The trouble is that if there is a strong C IV line in the photosphere, then there can be no flux coming out of that wavelength to be absorbed and no momentum can be transferred. That's why Milne's mechanism does not work in the sun; there is too much calcium in the photosphere.

Underhill: I think it works very well because from my calculations of O9 stars that line has an optical depth in the center of 10^6.

Morton: That's fine — 10^6 in the upper layers. The question is, what is the optical depth in the deeper layers where there is no motion?

LINE INTENSITIES FOR HYDROGEN AND HELIUM IN OB STARS

W. Buscombe[*]

Mount Stromlo Observatory, Canberra, Australia

ABSTRACT[†]

Coudé spectrograms of dispersion 150 μ/A have been exposed with the 74-inch reflector for 100 southern stars of types O5 to A3, including 25 supergiants. Equivalent widths for all distinct absorption features have been measured on direct-intensity tracings. All stars were observed in the interval 3800-4900 A, and some also from 5800-6700 A.

Although the absorption components may be weaker in the spectra of supergiants and emission-line stars, it is more usual for Hα to be shallower and broader than Hγ in absorption-line stars near the main sequence.

The central depths of the Balmer lines of H I for slow rotators increase monotonically with reciprocal temperature θ_{eff}, in close agreement with predictions based on atmospheric models computed by Mihalas.

A dispersion is observed in the relation of W(Hγ, δ) with M_V (adopted for the MK type); the lines tend to be stronger in the spectra of slow rotators.

Observed equivalent widths of lines of He I yield ratios for the diffuse triplet/singlet lines that are larger for supergiants than for dwarfs, but markedly smaller for the helium-rich stars.

For the stars from B3 to A0, the strength of the triplet lines He I 4026, 4471 follows a single relation with the intrinsic color-index Q. At class B1, however, the value of W(He I-3) ranges over a factor of 3, increasing from supergiants through stars near the main sequence to those with hydrogen lines in emission, and showing a positive correlation with $v_e \sin i$.

Indications from both triplet and singlet lines are of a He/H ratio being abnormally high in the atmospheres of ι Ori, γ Ara, ι Ara, and ε Cap, but low for δ Cen, κ Eri, β Tuc, and HD 1909.

[Editor's note: The discussion of Buscombe's paper is combined with that of Underhill's paper, which follows.]

---

[*]Present address: Astronomy Department, Northwestern University, Evanston, Illinois.

[†]Complete details have been submitted for publication in <u>Monthly Notices of the Royal Astronomical Society</u>.

259

A COMPARISON OF PREDICTED LINE PROFILES FROM MODEL ATMOSPHERES WITH OBSERVED LINE PROFILES IN MAIN-SEQUENCE O AND B STARS*

A. B. Underhill

Sonnenborgh Observatory, Utrecht, The Netherlands

ABSTRACT

Profiles of lines of H, He II, C II, O II, O III, Ne II, Mg II, Si II, Si III, Si IV, and Fe III in the part of the spectrum accessible from the surface of the earth have been computed for a series of model atmospheres representing main-sequence stars of types B6 to O9. The series of 28 unblanketed models have effective temperatures from 14,870°K (approximately type B6) to 33,965°K (approximately type O9). For most of them log g is 3.7 or 4.0. Not all the lines have been computed in all the models. Data are also available for two line-blanketed models. The structural changes in the model brought about by inclusion of line blanketing have rather little effect on the profiles of lines other than H and He. Thus the predictions made from unblanketed models serve for interpreting O- and B-type spectra.

The equivalent widths and central intensities of the lines vary along the sequence qualitatively in accord with observation, but when many lines predicted from one model are compared with the spectrum of a normal sharp-lined B or O star, the agreement in detail is unsatisfactory.

The broadening effects of microturbulence and of rotation have been investigated, and a comparison made with profiles in the spectrum of Zeta Draconis. This example demonstrates that factors other than abundance, effective temperature, and gravity may have a strong influence on the apparent strengths of lines strong enough to be used for spectral classification among the B stars.

DISCUSSION

Morton: Do I understand from Underhill's comments that she wants scattering to explain the lines?

Underhill: The essential problem is that I have entirely the wrong theory of line formation.

Morton: Is scattering the thing you want?

Underhill: What we want is a much more sophisticated theory of line formation. You must put more physics into the source function.

Morton: What interests me in these are not the strong resonance lines but the much weaker lines.

Underhill: Stop saying "relatively weak." The problem of B stars is that the lines may look weak, but if you compute with a model atmosphere what lines would be in the Doppler part of the curve of growth, you find that the equivalent width is 0.01 A, and we can't even observe that. We observe 0.025 to 0.050, so they are not weak.

---

*The full paper will be submitted to <u>Vistas in Astronomy</u>.

Thomas: May I ask Morton to make his point more specific. What is it that bothers you when you say it is <u>not</u> resonance lines?

Morton: I'm adopting the ancient point of view that the resonance lines are the ones most likely to have scattering problems and that the weaker lines are likely formed in LTE.

Thomas: But none of the developments of the last 10 years suggest that scattering effects are confined to the resonance lines.

Morton: Yes, well, I'm just supporting your point.

Peterson: Will Miss Underhill explain what line-broadening theory she used for the heavier elements?

Underhill: I used 10 times the classical damping. The profile seems to be much more strongly dependent on the Doppler motions than on damping, and it is these regions of the line profile that seem to be of greatest interest. But this is definitely a weakness of my work and is why I am summarizing it and cutting it off.

Praderie: I would like to quote a result on the Mg II 4481 doublet in α Pegasi, which is a β Cephei variable. Duchesne <u>et al</u>. [C. R. Acad. Sci. Paris, Ser. B., 265, 1213, 1967] observed this doublet with the Lallemand camera at a high spectral resolution and a small exposure time (10 min). He could separate the components of the doublet; this suggests that the effect attributed to microturbulence may be only a resolution defect in preceding observations, and an averaging of the profiles over too long a fraction of the period of the star.

Underhill: This is quite possible. We are probably getting an average of a lot of components. Furthermore, we found fleeting spectral differences in the spectrum of 10 Lacerta at Victoria. No periodicity was found.

Klinglesmith: In our calculation of the helium-line profiles we used the Griem broadening theory and found a shift of line center, which is a function of density. As you go down in the atmosphere, the density changes and there is an increased broadening due to the temperature structure of the atmosphere. This is an important kind of problem with the helium lines.

Cayrel: How large is this effect?

Klinglesmith: The shifts are very different for each line, a few tenths of an angstrom.

Hyland: I would like to remark that there has been some work by Hobbs and Cowley on the Mg II lines in B stars. They have recalculated the damping parameter for λ 4481 A, using Griem's value for the quadratic Stark effect and including quadrupole interactions, and when they do that it turns out that the predicted strength of the line, on the assumption of a normal solar abundance, comes much closer to what is actually observed.

Stibbs: Miss Underhill, you referred to a search for a non-LTE line-formation theory. Do you mean a theory in which lines are not formed in pure absorption?

Underhill: Yes. I do not wish to continue to use the Planck function for the source function of the line.

Stibbs: This is a vital point. For a very long time, many people have been seduced by the idea that pure absorption applies in this problem primarily because of the line depth. It is natural enough to think that this is in fact the physical process, because then the transfer equation can be solved very easily! It is then only a simple linear equation. I think we must go back to partly scattering and partly pure absorption and look into the details of the physics that is applicable.

Thomas: Welcome to the barricades.

Houziaux: Do you think that noncoherent scattering is likely to give a big enough effect?

Underhill: Yes, definitely.

Cayrel: How do you determine the effective temperature, Miss Underhill, from the continuum or from something else?

Underhill: I fit the $H\gamma$ profile with the observations of 10 Lacerta and I try to fit the Balmer jump. I haven't a really good fit, but it is a combination of both features. One cannot fit 10 Lacerta very well with a group of models at various effective temperatures; some models agree in some respects but not in others.

Cayrel: Can you remove part of the discrepancy by changing the effective temperature?

Underhill: No, not really. This is the best fit for a range of temperatures.

Heintze: I should like to ask Dr. Buscombe if the $H\gamma$ lines also show very sharp cores in the sharp-lined stars.

Buscombe: Yes, but it was the equivalent widths I compared with the calculated values, not the line depths.

Heintze: My experience from high-dispersion spectra is that if $H\alpha$ shows such a deep core, then its equivalent width is smaller than one expects from theory.

Peterson: Concerning Buscombe's paper, I would like to comment that in the northern hemisphere, which in this case is from Sirius on up, $H\alpha$ and $H\gamma$ agree with the computations, i.e., that $H\alpha$ is indeed smaller. The observations were made on the solar telescope and I'll discuss them in more detail in a few minutes.

THE TEMPERATURE SCALE OF B-TYPE STARS

J. R. W. Heintze

Sonnenborgh Observatory, Utrecht, The Netherlands

It is well known that it is impossible to derive the effective temperature and the gravity of a star from the Hγ profile only, even when the star is sharp-lined. Figure 1 shows two predicted Hγ profiles for two quite different models. These two profiles cannot be distinguished from each other when compared with observed profiles.

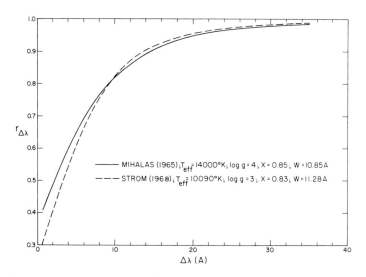

MIHALAS (1965); T_{eff} = 14000°K, log g = 4, X = 0.85, W = 10.85A
STROM (1968); T_{eff} = 10090°K, log g = 3, X = 0.83, W = 11.28A

Figure 1. A comparison of two theoretical Hγ profiles of nearly the same shape but calculated from models with quite different values for T_{eff} and log g.

Unfortunately, in the determination of the effective temperature from the comparison of observed and calculated Balmer jumps (BJ), some severe difficulties are encountered.

A. Observational Difficulties.

1. There are a number of uncertainties in the calibration of the BJ of the photometric standard α Lyr. Bahner (1963) has found that Oke's (1964) assumption of the BJ of Vega had to be increased by 0.07 mag; Whiteoak (1966) has found some observational evidence that an increase of 0.09 mag is necessary; Hayes (1967) has recently remeasured the energy distribution of α Lyr and found a correction of 0.15 ± 0.03 mag.

2. The uncertainties in the measurements amount to a few hundredths of a magnitude. To reduce the influence of these errors on the BJ as much as possible, the measurements of the monochromatic magnitudes for $1/\lambda = 2.2$ to $2.4\ \mu^{-1}$ and those for $1/\lambda > 2.7\ \mu^{-1}$ are represented by straight lines. These lines have been extended linearly to $1/\lambda = 2.7\ \mu^{-1}$, and the difference between the two magnitudes obtained in this way is defined as the BJ. The same is done with predicted energy distributions.

B.   Theoretical Difficulties.

1.   According to Mihalas (1967), deviations of LTE cause a decrease in the predicted BJ of about 0.02 mag at log g = 4 and 0.05 mag at log g = 3 in the temperature range 10,000 to 25,000°K. Kalkofen and Strom (1966) find even greater differences.

2.   Blanketing increases the predicted BJ in general.

3.   Rotation increases the predicted BJ (Collins, 1966; Hardorp and Strittmatter, 1968).

4.   Differences in metal content may also have an influence on the predicted BJ as shown by Gingerich (1964) and by Strom, Gingerich, and Strom (1966). The major effect arises from the photoionization of Si I. Since at an effective temperature of about 14,000°K nearly all silicon is ionized, this effect, if any, could be expected to be small for $T_{eff} > 14,000°K$.

5.   Fortunately, the predicted BJ does not depend too much on the gravity for $T_{eff} > 10,000°K$ and $3 \leq \log g \leq 4.5$. For example, a predicted BJ of a nonblanketed model equal to 0.78 mag corresponds to an effective temperature of 15,300 ± 250°K for log g = 4.0 ± 0.5.

However, the predicted shape of the Paschen continuum does not quite depend on gravity, non-LTE effects, and the abundance of the metals for effective temperatures around 15,000°K. Therefore, a nonrotating and nonreddened star with an effective temperature of about 15,000°K should be of great interest. Heintze (1968) has shown that it is very likely that the rotation and the reddening of the sharp-lined B5 IV star τ Her, if any, must be very small. The Hβ and Hγ profiles of this star have been determined very accurately. From these profiles and the last visible Balmer line, the effective temperature and the logarithm of the gravity could be estimated to be 14,500 ± 500°K and 3.5 ± 0.2, respectively.

The energy distribution of τ Her has also been determined very accurately from 3279 to 10,400 A. When calibrated with Oke's (1964) assumption of the energy distribution of Vega the effective temperature of τ Her should be around 11,000°K, whereas when calibrated with Hayes' measurements of the energy distribution of Vega, it should be a little higher than 15,000°K. There is evidence that Hayes' Paschen continuum of Vega in the interval $1.3 \leq 1/\lambda \leq 2.2$ is 0.02 mag too steep.

In Figure 2, U-B color indices are plotted against BJ's for both observations and models. The colors have been taken from Johnson, Mitchell, Iriarte, and Wiśniewski (1966). The observed BJ's have been obtained from spectral scans calibrated with Oke's (1964) standards. These BJ's have been increased by 0.07 mag. It is an interesting fact that a straight line can be drawn through the observations, which is parallel to straight lines that can be drawn through the calculated points. The 1965 and 1966 results of Mihalas have been taken for $T_{eff} \geq 12,000°K$ and $T_{eff} \leq 12,600°K$, respectively.

On the assumption that τ Her is a nonrotating and nonreddened star with T_{eff} = 14,500°K and log g = 3.5, the theoretical value of U-B differs by 0.025 mag from the observed one. The dashed line in Figure 2 represents the theoretical relation between the corrected U-B values and the BJ's for $T_{eff} \geq 12,600°K$. The difference between observed and calculated BJ's is now 0.14 mag for $T_{eff} \geq 12,600°K$. Part of this discrepancy (0.08 ± 0.03 mag) can be explained by assuming Hayes' energy distribution of Vega to be correct. The remaining difference of 0.06 mag can possibly be explained by B (1) and/or (2) above. For $T_{eff} \geq 12,600°K$, the BJ's calculated by Mihalas (1965) will be decreased by 0.06 when compared with the observed BJ's calibrated with Hayes' energy distribution of α Lyr. The linear (U-B) - BJ relation for the observations will be represented by

$$BJ = 1.20 (U-B) + 1.45 \text{ mag} \qquad (1)$$

for B-type stars.

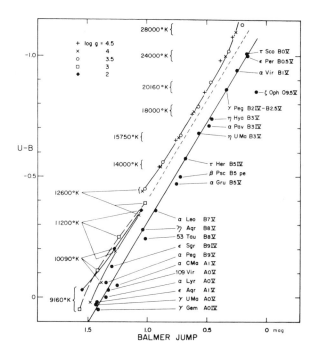

Figure 2. The observed (U-B)−BJ relation (closed circles) for B-type
stars compared with predictions from Mihalas (1965) (upper
part) and Mihalas (1966) (lower part).

With this formula BJ's have been determined from the observed U-B for
42 B stars. BJ's have also been determined for those stars for which spectral scans exist. These scans have been calibrated with Hayes' energy distribution of α Lyr. The results are given in Table 1 and are presented in
Figures 3 and 4. As shown by Heintze (1968), BJ's calculated by Mihalas
(1966) ($T_{eff} \leq 12,600°$K) need no correction. However, the predicted U-B
values according to these computations have to be corrected by 0.07_5 mag.

In Figure 3 the effective temperatures given in Table 1 are plotted
against spectral type, whereas in Figure 4 these temperatures are plotted
against the observed B-V. The effective temperatures found by Hanbury
Brown, Davis, Allen, and Rome (1967) are also plotted in these figures.
Figure 3 shows the results Smith (1967) obtained from UV fluxes measured
by a satellite experiment.

In general, the temperatures found by Hanbury Brown et al. agree nicely
with those derived in this investigation, except the effective temperature of
β Cru. Popper (1968) has found evidence that β Cru is a binary system.
Hence, the temperature measured by Hanbury Brown et al. must be too high,
because they received the light of two components instead of one. However,
β Cru does not lie on the line drawn in Figure 4 even when the lower temperature of β Cru found in this investigation is adopted. This fact could also be
explained by the binary structure of β Cru, since the binary systems α Vir
and ε Per lie to the right of the line drawn in this figure, and it is very unlikely
that α Vir is reddened. Of course, reddening will move the stars to the right
in this diagram. For example, B-V is 0.02 mag for the highly reddened star
ζ Oph (not plotted in Figure 4), whereas it should be about -0.26 mag. However, nothing can be said about the reddening of β Cru so long as the spectral
type of the other component is not known. The stars γ Ori, ε CMa, and
ε Ori are of class III, II, and Ia, respectively. The stars of these spectral
types could have quite different $T_{eff} - $ (B-V) relations from those of class
IV and V stars.

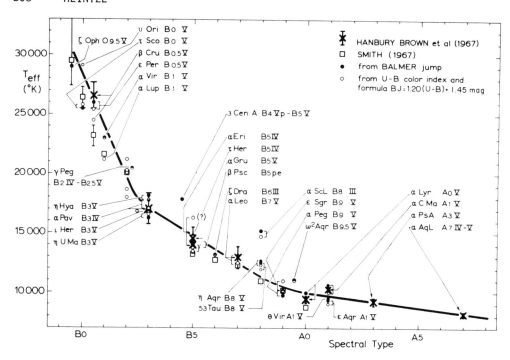

Figure 3. Effective temperatures of B-type stars as derived in this investigation plotted against spectral type. Results of Hanbury
Brown et al. (1967) and of Smith (1967) are also given.

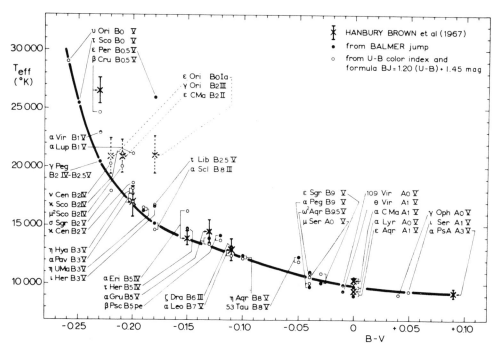

Figure 4. Effective temperatures of B-type stars as derived in this investigation and according to Hanbury Brown et al. plotted against
observed B-V color indices (Johnson et al., 1966).

Table 1.   Information concerning 42 B stars

1	2	3	4	5	6	7	8	9	10
					Balmer jump				
Name of star	Spectral type	U-B (mag)	B-V (mag)	sin i (km sec^{-1})	column 3 +eq(1)	Hayes (1967)	Ref for scan	T_{eff} (°K) from col. 6	T_{eff} (°K) from col. 7
ζ Oph	09.5 V	-0.85	+0.02	400	0.43	0.17	1,3	20399	29000
υ Ori	B0 V	-1.07	-0.26	37	0.17	--	3	29100	-
τ Sco	B0 V	-1.01	-0.25	13	0.24	0.25	3,4	25800	25500
ε Per	B0.5 V	-1.00	-0.18	150	0.25	0.24	2	25400	26000
β Cru	B0.5 V	-0.98	-0.23	-	0.27	--	-	24500	-
α Vir	B1 V	-0.94	-0.23	172	0.32	0.33	2,4	23000	23000
α Lup	B1 V	-0.88	-0.20	0	0.39	--	3	21200	-
κ Sco	B2 IV	-0.88	-0.21	99	0.39	--	3	21200	-
ν Cen	B2 IV	-0.85	-0.22	170	0.43	--	3	20300	-
μ^2 Sco	B2 IV	-0.84	-0.21	115	0.44	--	3	20100	-
κ Cen	B2 V	-0.78	-0.20	50	0.51	--	3	18700	-
σ Sgr	B2 V	-0.75	-0.22	225	0.55	--	3	18000	-
γ Peg	B2 IV -B2.5 V	-0.86	-0.23	5	0.42	0.42	2,3,8	20400	20500
τ Lib	B2.5 V	-0.69	-0.18	200	0.62	--	3	16800	-
η Hya	B3 V	-0.74	-0.20	130	0.56	0.54	1	17800	18200
α Pav	B3 IV	-0.71	-0.20	0	0.60	0.56$_5$	4	17100	17750
ι Her	B3 V	-0.69	-0.18	0	0.62	0.62$_5$	8	16800	16800
η U Ma	B3 V	-0.68	-0.19	215	0.63	0.65	2	16600	16300
3 Cen A	B4 Vp -B5V	-	-	0	--	0.56	5	-	17800
α Scl	B8 III	-0.57	-0.18	0	0.77	0.72$_5$	3,4	14700	15200
α Eri	B5 IV	-0.67*	-0.15	415	0.65	--	3	16300	-
τ Her	B5 IV	-0.57	-0.15	20	0.77	0.76	7	14700	14800
β Psc	B5 pe	-0.50	-0.12	140	0.85	0.80$_5$	3,8	13800	14250
α Gru	B5 V	-0.47	-0.13	285	0.89	0.83$_5$	4	13500	14000
ζ Dra	B6 III	-0.43	-0.11	23	0.93$_5$	0.94$_5$	8	13200	13100
α Leo	B7 V	-0.36	-0.11	370	1.02	1.01	1,2,4	12900	13000
η Aqr	B8 V	-0.28	-0.10	288	1.11	1.12	3,8	12300	12500
53 Tau	B8 V	-0.24†	-0.05†	0	1.16	1.09$_5$	6	12000	12400
ε Sgr	B9 V	-0.13	-0.03	-	1.29	1.38$_5$	3,4	11000	10200
α Peg	B9 V	-0.06	-0.04	157	1.38	1.43	2	10200	9800
ω^2 Aqr	B9.5 V	-0.13	-0.04	167	1.29	1.27	3,8	11000	11100
μ Ser	A0 V	-0.10	-0.04	83	1.33	--	3	10700	-
α CMa	A1 V	-0.05	0.00	0	1.39	1.33$_5$	4	10200	10600
58 Aql	A0 V	-	-	-	--	1.39	1,3	-	-
109 Vir	A0 V	-0.03	-0.01	335	1.41	1.41	1,3	10000	10000
α Lyr	A0 V	0.00	0.00	0	1.45	1.43	1	9600	9800
θ Vir	A1 V	+0.01	-0.01	10	1.46	1.46	1	9500	9500
ε Aqr	A1 V	+0.02	0.00	118	1.47	1.50$_5$	1,2,3	9400	9100(3,7)
ι Ser	A1 V	+0.02	0.05	106	1.47	-	3	9400	-
γ UMa	A0 V	+0.03	0.00	169	1.49	1.51$_5$	2	9150 (3,8$_5$)	9000(3,6)
γ Oph	A0 V	+0.04	+0.04	0	1.50	-	3,8	9150 (3,8$_5$)	-
γ Gem	A0 V	+0.05	-0.00	26	1.51	1.49$_5$	1	9150 (3,7)	9150(3,8)

*Hogg (1958)

†Osawa (1965)

1. Oke (1964)

2. Bahner (1963) calibrated with Oke's (1964) standards, except for the BJ

3. Willstrop (1965)

4. Aller, Faulkner, and Norton (1966)

5. Rogers, A. W., and Hyland, A. R., quoted by Hardorp (1966)

6. Jugaku and Sargent (1968)

7. Heintze (1968)

8. Heintze (1969)

The position of α Scl in Figure 4 is quite normal, but not so in Figure 3. It is evident that the classification of this star should be B5 rather than B8. The same could be true for 3 Cen A.

In another paper (Heintze, 1969), it will be shown that the effective temperatures of υ Ori and τ Her found in this investigation refer to effective temperatures derived from blanketed models. Therefore, it is believed that the effective temperatures found in this investigation refer to effective temperatures derived from blanketed models.

REFERENCES

Aller, L. H., Faulkner, D. J., and Norton, R. H. 1966, Ap. J., 144, 1073.

Bahner, K. 1963, Ap. J., 138, 1314.

Collins, G. W., II. 1966, Ap. J., 146, 914.

Gingerich, O. 1964, Proc. First Harvard-Smithsonian Conf. on Stellar Atmospheres, Smithsonian Ap. Obs. Spec. Rep. No. 167, 17.

Hanbury Brown, R., Davis, J., Allen, L. R., and Rome, J. M. 1967, Mon. Not. Roy. Astron. Soc., 137, 393.

Hardorp, J. 1966, Zs. f. Ap., 63, 137.

Hardorp, J., and Strittmatter, P. A. 1968, Ap. J., 151, 1057.

Hayes, D. S. 1967, Thesis, University of California at Los Angeles.

Heintze, J. R. W. 1968, Bull. Astron. Netherlands, 20, 1.

_____. 1969, Bull. Astron. Netherlands, in press.

Hogg, A. R. 1958, Mount Stromlo Obs. Mimeo., No. 2, 7 pp.

Johnson, H. L., Mitchell, R. I., Iriarte, B., and Wiśniewski, W. Z. 1966, Comm. Lunar Planet. Lab., 4, No. 63, 99.

Jugaku, J., and Sargent, W. L. W. 1968, Ap. J., 151, 259.

Kalkofen, W., and Strom, S. E. 1966, J. Q. S. R. T., 6, 653.

Mihalas, D. 1965, Ap. J. Suppl., 9, 321.

_____. 1966, Ap. J. Suppl., 13, 1.

_____. 1967, Ap. J., 149, 169.

Oke, J. B. 1964, Ap. J., 140, 689.

Osawa, K. 1965, Ann. Tokyo Astron. Obs., 9, Ser. 2, 123.

Popper, D. M. 1968, Ap. J., 151, L51.

Smith, A. M. 1967, Ap. J., 147, 158.

Strom, S. E. 1968, unpublished.

Strom, S. E., Gingerich, O., and Strom, K. M. 1966, Ap. J., 146, 880.

Whiteoak, J. B. 1966, Ap. J., 144, 305.

Willstrop, R. V. 1965, Mem. Roy. Astron. Soc., 69, 83.

[Editor's note: The discussion of Heintze's paper follows the next paper by Hyland.]

THE EFFECTIVE-TEMPERATURE SCALE AND BOLOMETRIC CORRECTIONS FOR B STARS

A. R. Hyland

Mount Wilson and Palomar Observatories, Carnegie Institute of
Washington, and California Institute of Technology, Pasadena, California

1. THE EFFECTIVE-TEMPERATURE SCALE

The fundamental importance of the effective-temperature scale for early-type stars is well known. Recently, Morton and Adams (1968) have used the best available line-blanketed model atmospheres (Hickok and Morton, 1968; Mihalas and Morton, 1965; Adams and Morton, 1968; Mihalas 1966) to derive a consistent scale of effective temperatures that agrees well with the interferometer results of Hanbury Brown, Davis, Allen, and Rome (1967). They note, however, that the least secure step in their procedure was in the identification of each model with a star of particular spectral type and color.

It is the purpose of this paper to overcome that problem by fitting the models directly to photoelectric scans of a large number of B stars and to present the results as an effective-temperature scale against the photometric quantity

$$Q = (U-B) - 0.72 (B-V) \quad ,$$

which is relatively insensitive to interstellar reddening.

Observations of 30 main-sequence B stars in the galactic clusters IC2391, IC2602, and M7 were made with a photoelectric spectrum scanner mounted at the cassegrain focus of the Mount Stromlo 50-inch telescope. Detailed observational results are being published elsewhere. Each star was observed with a resolution of 16 A from $\lambda\lambda$ 3300 A to 5840 A, and several were observed with a resolution of 32 A to λ 8200 A. Mean fluxes were measured in 50-A bands centered on the same wavelengths as Oke (1964) used and were corrected for the small effects of line blocking. The fluxes were calibrated with reference to the energy distribution of α Lyr proposed by Bessell (1967), but more recently were corrected to the new absolute calibration of α Lyr by Hayes (1967). Temperatures were derived by fitting the model predictions to the observed energy distributions, greatest weight being given to the region $\lambda\lambda$ 3300 A to 5000 A, as there is still some dispute regarding the slope of the Paschen continuum. Since the above calibrations differ by less than 0.02 mag below the Balmer discontinuity and are otherwise almost identical out to 5000 A, reciprocal effective temperatures θ_{eff} = 5040/T_{eff} derived in the latter case are larger than in the former by $\sim$ 0.005.

Effective temperatures have been derived for all the cluster stars observed, and also for those field B stars observed by Bahner (1963). There is a somewhat larger uncertainty in the latter case because the gravity of these stars is not known, whereas in the case of the cluster stars, log g and T_{eff} were determined simultaneously. For consistency, the field stars were fitted to models with log g = 4.0.

The results are shown in Figure 1, where the individual values of θ_{eff} are plotted against Q. The very tight relation obtained indicates that for early-type stars Q is indeed an excellent parameter for obtaining a first estimate of θ_{eff}. Table 1 gives the effective temperatures for the field stars. Individual results for the cluster stars are being published elsewhere. Reddening corrections derived from Q were applied to the energy distributions before the fitting procedure was attempted.

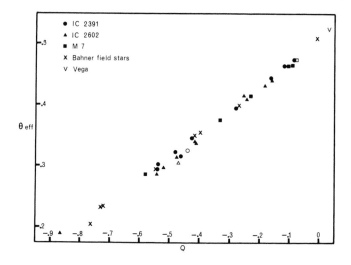

Figure 1.   The derived effective-temperature scale for bright B stars as a function of Q.   The ordinate is $\theta_{eff} = 5040/T_{eff}$.

Table 1.   Effective temperatures and colors of the field B stars observed by Bahner (1963)

Star	T_{eff} °K	θ_{eff}	B-V	U-B	E(B-V)
α Vir	24600	0.205	-0.23	-0.94	0.03
γ Ori	21450	0.235	-0.22	-0.88	0.03
γ Peg	20650	0.244	-0.23	-0.86	0.01
η U Ma	16750	0.301	-0.19	-0.68	0.00
δ Per	14400	0.350	-0.12	-0.51	0.03
β Tau	14050	0.359	-0.13	-0.49	0.01
α Leo	12650	0.398	-0.11	-0.35	0.00

Note: All the colors are taken from Johnson, Mitchell, Iriarte, and Wiśniewski (1966).

2. COMPARISON WITH OTHER RESULTS

Energy distributions are available in the literature for several of the stars observed by Hanbury Brown et al. (1967). It is important to compare the effective temperatures, obtained by fitting these with model atmospheres, with those derived from the monochromatic fluxes. The comparison is made in Table 2. It is evident that in every case but one the agreement is entirely satisfactory. The one discordant case, α Eri, is very puzzling since the broad-band colors indicate the same temperature as the photoelectric scans. This star is certainly worthy of more detailed study.

In Figure 2 the effective-temperature scale derived here is compared directly with the interferometer results. There is very good agreement for the complete temperature range covered in this investigation. Since the interferometer results depend on monochromatic fluxes while the scanner results depend only on the shape of the energy distribution, the agreement demonstrated here indicates that in regard to these parameters the models are perfectly consistent with the observations.

Table 2.   Comparison of effective temperatures derived from (a) mono-chromatic fluxes and (b) continuous energy distributions

Star	(a) T_{eff} °K	(b) T_{eff} °K	Reference
γ Ori	21000 ± 1500	21450	Bahner (1963)
α Pav	17100 ± 1300	18000	Aller, Faulkner, and Norton (1966)
α Eri	14000 ± 600	16525	Aller et al. (1966)
α Gru	14600 ± 950	14250	Aller et al. (1966)
α Leo	13000 ± 650	12650	Bahner (1963)

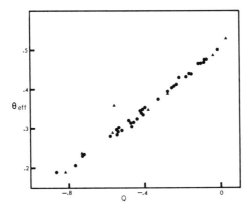

Figure 2.   The effective-temperature scale derived in this investigation compared with the new results (▲) of Hanbury Brown et al. (1967).

In Table 3 the temperature scale of Morton and Adams is compared with the present one, where now the models have been directly related to stellar colors through the observations. Between 25,000°K and 11,500°K the two scales are almost identical. Only between 11,500°K and 9600°K is the difference too large to be explained by experimental errors. Indeed, this region needs to be examined with more care since parameters other than the effective temperature become important in determining the shape of the stellar-energy distributions.

The present temperature scale represents the most complete attempt to date to combine the results of model-atmosphere computations with observed energy distributions and monochromatic fluxes for the range of B stars.

3.   BOLOMETRIC CORRECTIONS

Morton and Adams (1968) reported a puzzling disagreement between calculations of bolometric corrections, based on radiometric observations around F2V, and those based directly on solar quantities. For their final compilation they preferred the former scale. It is found, however, that unless the scale is based directly on the solar quantities, stellar radii determined from L and T_{eff} are inconsistent with those derived from absolute monochromatic fluxes in the manner described by Gray (1967). This is not really surprising, since one always relates L and T_{eff} to the corresponding solar values. It appears, therefore, to be more consistent to derive bolometric corrections directly related to the solar values, so that at least for the early-type stars the scale given by Morton and Adams should be corrected by -0.29 mag. This discrepancy certainly serves as a warning to those wishing to obtain stellar radii. Whenever possible it is far safer to use monochromatic fluxes rather than the more uncertain L, T_{eff}, and B.C.

Table 3. The present effective-temperature scale compared with that of Morton and Adams (1968)

U-V	B-V	Q	$T_{eff} °K$ Morton and Adams	$T_{eff} °K$ Present paper	θ_{eff}
-1.38	-0.30	-0.864	30900	28800	0.175
-1.29	-0.28	-0.808	26200	25600	0.197
-1.19	-0.26	-0.743	22600	22400	0.225
-1.10	-0.24	-0.687	20500	20500	0.246
-0.91	-0.20	-0.566	17900	17400	0.290
-0.72	-0.16	-0.445	15600	15300	0.330
-0.63	-0.14	-0.389	14600	14400	0.350
-0.54	-0.12	-0.334	13600	13450	0.375
-0.39	-0.09	-0.235	12000	12100	0.417
-0.25	-0.06	-0.148	10700	11100	0.455
-0.13	-0.03	-0.078	10000	10400	0.485
0.00	0.00	0.000	9600	9700	0.520

The major portion of this work was carried out at the Mount Stromlo Observatory while the author held an Australian National University Research Scholarship. Attendance at this conference was made possible in part by NSF Grant GP-7030.

REFERENCES

Adams, T. F., and Morton, D. C. 1968, Ap. J., 152, 195.
Aller, L. H., Faulkner, D. J., and Norton, R. H. 1966, Ap. J., 144, 1073.
Bahner, K. 1963, Ap. J., 138, 1314.
Bessell, M. S. 1967, Ap. J., 149, L67.
Gray, D. F. 1967, Ap. J., 149, 317.
Hanbury Brown, R., Davis, J., Allen, L. R., and Rome, J. M. 1967, Mon. Not. Roy. Astron. Soc., 137, 393.
Hayes, D. S. 1967, Thesis, University of California at Los Angeles.
Hickok, F. R., and Morton, D. C. 1968, Ap. J., 152, 203.
Johnson, H. L., Mitchell, R. I., Iriarte, B., and Wiśniewski, W. Z. 1966, Comm. Lunar Planet. Lab., 4, No. 63, 99.
Mihalas, D. 1966, Ap. J. Suppl., 13, 1.
Mihalas, D. M., and Morton, D. C. 1965, Ap. J., 142, 253.
Morton, D. C., and Adams, T. F. 1968, Ap. J., 151, 611.
Oke, J. B. 1964, Ap. J., 140, 689.

DISCUSSION

Popper: With respect to the position of β Cru in Heintze's Figure 4, there is still the question of a possible reddening of β Cru.

Heintze: Of course, reddening moves the position of a star in Figure 4 to the right. However, the binary systems α Vir and ε Per lie also to the right of the drawn line in Figure 4. For α Vir reddening is excluded. So the position of α Vir in Figure 4 has only to be explained by the binary structure of α Vir, and this could quite well be true for β Cru also.

Wallerstein: Heintze's result that Sirius is about a thousand degrees higher than Vega is interesting because Vega has λ 4471 of He I [Hunger, Zs. f. Ap., 36, 42, 1955] while Sirius does not. If they both have the same helium abundance and the lines are formed in similar layers, there is an inconsistency. However, if you're right that the temperature of Sirius is higher than that of Vega, then the weakness of the helium lines in Sirius may be similar to that in silicon stars, where the helium lines are too weak for the colors.

Heintze: Yes, the effective temperature of Sirius is definitely higher than that of Vega. The effective temperature of Vega derived in Paper I (10,000°K) has to be regarded as a maximum value. The maximum possible temperature of Sirius is about 10,800°K, whereas the lowest should be about 10,100°K (assuming a comparison C and 10 times more metals than normal). In Paper I a most probable value of 10,400°K is found.

Strom: I'd like to comment briefly on the differences between the analyses of my models and those of Mihalas, which appeared on Heintze's graph. He indicated that Mihalas' models are line-blanketed models and my models, while not blanketed, did include the effect of the silicon opacity. The effect of silicon opacity is to decrease the Balmer discontinuity, as I indicated yesterday, and this is in the direction of the difference between Mihalas' and my models. And the second point is that the inclusion of the Balmer lines tends to increase the temperature gradient in the atmosphere. The Balmer discontinuity measures not only the ratio of opacity but also the temperature gradient. This leads one to expect Mihalas' models without silicon opacity to show greater Balmer discontinuities than ours without silicon, and this is in fact the case.

Morton: In connection with the bolometric correction scale, I think that Hyland is probably right that for the hotter stars, A0 and earlier, we should apply this correction to our bolometric correction scale. We still don't understand why there is this discrepancy. I don't think that the correction should be applied to the A- and F-type stars. They just can't have bolometric corrections that large, and there must be something wrong with our corrections derived from the Mihalas models for F2 stars. Because of these problems with bolometric corrections, I think that whenever possible one should avoid using them. For example, in trying to determine the effective temperatures from Hanbury Brown's angular diameters, one way is through the bolometric correction and the other way, as Hanbury Brown actually did it, is to use the theoretical flux at a particular wavelength as a function of the effective temperature. In this way one can obtain the effective temperature entirely independently of the so-called bolometric correction. (This is the method advocated by Cayrel and Underhill.) On the other hand, we do want to compare with stellar interiors whenever possible, and then bolometric corrections are necessary. However, when you're using angular diameters, it is best to avoid bolometric corrections.

Gray: In regard to this comment, I should like to say that, if you do this, as they did, you can in fact avoid the bolometric correction but you still have to integrate that flux to get the T_{eff}.

Morton: But we think the models are okay so the integration would be all right. It is purely a question of the various means of deducing the bolometric correction from the theoretical model and relating it to the V magnitude, as you must.

Gray: With regard to Heintze's paper, we can also put in some astrometrical data, in the case of Sirius and some other stars, and in this way we can pin down the surface gravity quite accurately. This puts tremendous constraints on the situation, and I think that we should use these data. For Sirius the value I've calculated is log g = 4.1.

Heintze: I do not agree with your log g value of Sirius. Apparently you derived this value from your determination of the radius of Sirius. It could quite well be that if you used Hayes' energy distribution of Sirius, you would find a radius for Sirius very close to that found recently by Hanbury Brown et al. In that case, log g should be at least 4.3. The maximum value of log g I can get from the Hγ profile is 4.15, which yields a mass of 1.6 $M_\odot$ by assuming Hanbury Brown's radius. The observed mass is at least 2.1 $M_\odot$.

Gray: [addendum]: Heintze proves to be correct when the Sirius measurements of Hayes are used; log g = 4.28.

Underhill: The discussion of determining radii from visual absolute magnitude — if you look at the formulas, bolometric correction minus 10 log T_{eff} enters. To make yourself safe, all you should do is to take bolometric corrections computed from the same models. That combination — bolometric corrections versus log T_{eff} — is quite insensitive to the models.

Morton: It's the zero point in the bolometric correction that worries us.

Heintze: Can this correction for the bolometric correction be assumed constant from 6000° to 30,000°K?

Underhill: Yes.

Weidemann: It is just a question of the zero point. I proposed [Zs.f.Ap. 67, 415, 1967] using the zero point -0.1 and a reference blackbody in the case of the sun. If you do that, then you will be on a consistent scale and the models always give you the ratio of the flux in the wavelength region concerned to that of the blackbody. This is illustrated in the following equation:

$$B.C. = B.C._{bb} + 2.5 \log = \frac{\int F\lambda\, S\lambda\, d\lambda}{\int B\lambda\, S\lambda\, d\lambda} \quad .$$

The only remaining problem will be that of the filter function Sλ.

Popper: I should like to put this discussion into a different perspective, if I may. The models apparently aren't very much like the sun from this standpoint, if the numbers given in the models can be taken at their face value. If you compute the bolometric correction curve from the models, for example the Mihalas line-blanketed models, and take Oke and Conti's extrapolation of those [Ap. J., 143, 134, 1966], you get a minimum bolometric correction around F2. So you would expect, if the sun is anything like the models, that the bolometric correction for the sun would be numerically greater than that for an F2 star. But if you take the spectral-energy distribution of the sun, as published in Allen's Astrophysical Quantities [London, Athlone Press, 1963], you obtain a bolometric correction that is numerically smaller than for an F2 model. So even if the extrapolations of Oke and Conti represent the best in model atmospheres at these temperatures, they don't represent the sun. The difficulty referred to earlier by Morton [Morton and Adams, Ap. J., 151, 611, 1968] may have a similar origin.

Buscombe: I should like to come back to the effective-temperature scale for the very hot stars for a moment and point out that for the O stars there is a slacking of the slope of the effective-temperature scale with spectral class hotter than O8, and I'm interested in why this occurs, especially in view of the fact, as was pointed out last week, that if the scale is accepted one has to find a whole lot of O stars imbedded in the nebulosity, which are never seen at all, to account for this excitation of emission nebula.

Eggen: I should like to say I'm a bit disturbed, because you'll never get a better observation of mass than you have of Sirius. Why do you want to throw this away? There can't be another comparison. If you want to use observations at all, you must use Sirius.

Heintze: A wiggle in the orbit of Sirius has already been found by Volet and Zagar (1932) (see Paper I, section 5.2.6). The period should be about 6 years, whereas the period for the system Sirius A and B is about 50 years. Moreover, from the differences between the observed radial velocity and that calculated from the orbit of Sirius A and B, Campbell (1905) and Cambell and Moore (1928) concluded (see Paper I) that the radial velocity of Sirius A is variable. Volet has shown that the wiggle could be explained by a third component around A as well as around B. In the past several observers have claimed to have seen a third component. Because it was felt that a possible third component could not be seen close to the bright A component, it was supposed to belong to component B.

Bless: I should like to talk about the accuracy that can be achieved with absolute calibration. Although the effective-temperature scale is settling down in a nice consistent fashion, at least temporarily, I think it is worth reminding everyone that there is still the possibility of a large systematic error in the temperature scale. Hayes estimates that shortward of the Balmer jump there could be an error of 5% in his work. This is equivalent to 500° to 1000° for B stars. This accuracy may not be substantially improved by the work that is going on at Palomar in which a platinum blackbody is observed directly and compared against stars. There again one runs into fairly fundamental uncertainties as to the temperature of the platinum, and of the thermodynamic temperature scale itself, and these uncertainties, although as small as a few degrees, do introduce significant errors in the ultraviolet calibration on the order of 0.03 mag. So one is up against a fairly fundamental uncertainty in the effective-temperature scale. This is particularly serious in the rocket ultraviolet, where these uncertainties become much magnified and where it will be very difficult to do work better than to 10 or 20% accuracy. In this connection, at Wisconsin we are trying to redo an absolute ultraviolet calibration using synchrotron radiation, and we hope that we will be able to provide independent, and perhaps as accurate, or more accurate, calibrations as can be done now. Comparing ultraviolet (λ < 2800 A) filter photometry with blanketed model atmospheres and making appropriate corrections for reddening (which for the stars observed is relatively small), one gets agreement with the temperature scales as adjusted by Morton and Adams and by Hanbury Brown. You can say from these observations that the Harris scale is much too high. So the ultraviolet data, such as they are, agree reasonably nicely with these new temperature scales [Bless, Code, Houck, McNall, and Taylor, Ap. J., 153, 557, 1968; Bless, Code, and Houck, Ap. J., 153, 561, 1968].

LUMINOSITY EFFECTS IN THE BALMER LINES
OF EARLY-TYPE STARS

D. M. Peterson and S. E. Strom

Smithsonian Astrophysical Observatory and Harvard College Observatory,
Cambridge, Massachusetts

ABSTRACT

We discuss the application of the approximation of detailed balancing in the lines to the calculation of the Balmer-line profiles in early-type stars. The wings of the Balmer lines are shown to be affected by departures from LTE. In particular, the Hα equivalent width may be significantly decreased with increasing luminosity, compared with LTE calculations.

An observational test of these calculations in the late B stars confirms the predictions.

1. INTRODUCTION

The suggestion by Kalkofen (1964, 1966) that the bound-bound transitions in hydrogen could be assumed to be in detailed balance (the "saturation approximation") provided the key to the calculation of the departures from LTE in the photospheric layers of stellar atmospheres. The effects of the departures on the temperature structures and emergent continuum spectrum have been discussed elsewhere (Strom and Kalkofen, 1966, 1967; Kalkofen and Strom, 1966; Strom, 1967; Kalkofen, 1968; Mihalas, 1967a, b; Mihalas and Stone, 1968).

Following a suggestion by Kalkofen, we have undertaken an investigation to determine the effects of departures from LTE on the Balmer lines within the context of the saturation approximation. Initial calculations (Peterson, 1967) indicated that significant effects could result, particularly for Hα. This paper provides a discussion of the predictions for the Balmer lines of late B stars as well as the preliminary results of an observational test of those predictions.

2. THE SATURATION APPROXIMATION

At a first glance it may seem to be somewhat contradictory to investigate line formation within the context of an approximation that explicitly excludes lines. Indeed, we do not expect to understand how the lines are formed over the entire profile but only in the region beyond about 1 A from line center.

To see more precisely why we limit ourselves to the line wings, consider (following Kalkofen, 1966) a two-level atom representation of the Hα line. The basic concept in the approximation of saturation is that the bound-bound rates tend to cancel (i. e. , to approach detailed balance) much more strongly and at considerably smaller optical depths than the bound-free rates. Hence, below a certain depth, the chief nonequilibrium processes determining the populations of the hydrogen-bound states are the bound-free radiative transitions.

The depth at which saturation occurs is controlled by ϵ, the probability of thermalization of an Hα photon. Kalkofen (1966) has estimated that $\epsilon \simeq 0.03$ for late main-sequence B stars. This estimate is rather insensitive to density.

We know (Avrett and Hummer, 1965) for the case of the Doppler-emission profile that saturation "begins" at a line-center optical depth $\tau_{\ell_C} \sim 10 \, \epsilon^{-1}$. Because the net bound-bound rates fall off approximately exponentially with a scale of ϵ^{-1}, we expect that saturation will certainly be reached by $\tau_{\ell_C} = 10 \, \epsilon^{-1} = 300$. Moreover, since line-center optical depth is larger than the local continuum optical depth by a factor of 10^5, we see that a considerable portion of the atmosphere including the continuum formation region may be calculated with the saturation approximation.

3.   DEPTH OF WING FORMATION

Next, we consider the depth of formation of the wings of Hα. In particular, we wish to know at what distance from line center, $\Delta\lambda_S$, the emergent profile is formed at a line-center optical depth, $\tau_{\ell_C} = 300$. For $\Delta\lambda > \Delta\lambda_S$ the saturation approximation is valid in the computation of the effects of departures on the emergent profile.

This computation is straightforward, and we find that at 1 A from line center $\tau(\Delta\lambda \sim 1) = \tau_{\ell_C}/300$ for typical physical conditions. Furthermore, since the profile at line center is dominated by Doppler broadening, this estimate is rather insensitive to temperature and pressure.

The profiles of Hα for late B stars all show a rather marked change in their wavelength dependence at about 1 A, the residual intensity falling off much more rapidly toward line center than LTE calculations would predict. This is precisely the behavior of the line source function one would expect on the basis of two-level atom calculations and the above estimates.

Finally, we note that similar calculations for Pα and Bα (the Paschen and Brackett α lines) indicate that the assumption of detailed balance in the lines may be extended to higher quantum levels as well.

4.   THE CALCULATIONS

We have undertaken the computation of two grids of model atmospheres and Balmer-line profiles covering the region $10,000° \leq T_{eff} \leq 15,000°$, $4.0 \leq \log g \leq 2.5$. The first grid was computed under the assumption of LTE. The second grid was computed with the saturation approximation allowing departures from LTE in a five-level model of the hydrogen atom with the continuum and the sixth through fourteenth levels assumed to be in LTE. The "modified Bethe cross sections" suggested by Mihalas (1967a) have been used throughout for the collision rates.

Since our estimates indicate that we cannot calculate the line centers of the Balmer lines accurately, we will consider modified equivalent widths W, which are defined as

$$W = \int_{-16}^{-1} [1 - R(\Delta\lambda)] \, d(\Delta\lambda) + \int_{1}^{16} [1 - R(\Delta\lambda)] \, d(\Delta\lambda) \quad ,$$

where $R(\Delta\lambda)$ is the residual intensity. Furthermore, we do not feel that the broadening theory for the Balmer lines is well enough understood at present. As a result, we have calculated the widths, using both the theoretical profiles of Griem (1967) and the semiempirical profiles of Edmonds, Schlüter, and Wells (1967, hereafter referred to as ESW).

In Figure 1 we have plotted W(Hα) against W(Hγ) from the LTE and non-LTE grids. The ESW theory was used in computing the plotted points. We note that for W(Hα) $\lesssim$ 5 A the Griem theory predicts a relation virtually identical with the plotted ESW calculations. For larger values of W(Hα), the predictions of the broadening theories diverge somewhat. However, this divergence does not influence our testing for departures, since the largest departures from LTE occur for the smallest values of W(Hα).

From these considerations we see that if we restrict our attention to the higher luminosity or hotter late B stars, we can avoid for the most part the uncertainties of the broadening theory. Furthermore, in determining the effects of departures from LTE, we need not assign temperatures or gravities to the stars observed, since the LTE and non-LTE regions in Figure 1 are quite distinct.

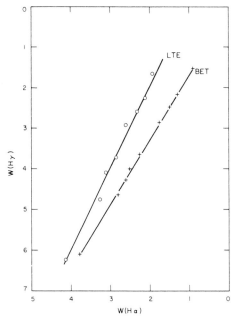

Figure 1.   The equivalent widths of Hγ plotted against the widths of Hα from a model atmosphere grid in the range: $10,000° \leq T_{eff} \leq 15,000°$, $4.0 \leq \log g \leq 2.5$.

5. THE OBSERVATIONS

The Hα and Hγ equivalent widths were measured from spectra of slowly rotating B stars. The spectra were obtained at the coudé focus of the Kitt Peak National Observatory 84-inch telescope during two nights in October 1967. Camera 3 was used in the second order to obtain IIa-O plates of Hγ (8.9 A/mm), and in the first order for 103a-E plates of Hα (17.8 A/mm). A list of the objects for which good plates of both Hα and Hγ were obtained is given in Table 1. These observations were reduced to yield W(Hα) and W(Hγ) directly comparable with the theoretical values. Where more than one profile was available, the scatter is indicated. These data were supplemented with photoelectric observations obtained at the McMath Solar Telescope at Kitt Peak. These observations are also presented in Table 1. The details of the observing and reduction procedures will be reported in subsequent papers.

6. RESULTS

In Figure 2 we have superposed the observational results on the W(Hα) - W(Hγ) relation discussed in section 4. We believe that this figure provides a confirmation of the calculated effects of departures from LTE on the wings of the Balmer lines.

Table 1.   Equivalent widths for B stars

Object	Spectral Type	Photographic		Photoelectric	
		Hγ	Hα	Hγ	Hα
67 Oph	B5 Ib	2.44 ± 0.11	1.68 ± 0.11	2.32 ± 0.10	1.43
η Leo	A0 Ib			2.96 ± 0.26	2.01 ± 0.14
π And	B5 V	4.48	3.25 ± 0.32		
β Tau	B7 III	4.54 ± 0.07		4.61	3.01
20 Tau	B7 III	4.81 ± 0.36	3.50		
ε Del	B6 III	4.97 ± 0.28	3.58 ± 0.43		

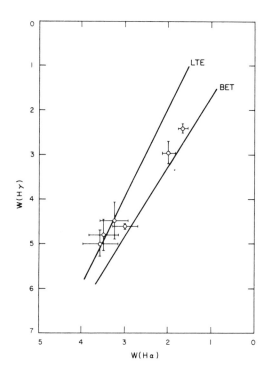

Figure 2.   A plot of the observations of Hα and Hγ. The approximate loci of the computed points and the internal errors in the observations are indicated.

Two final points should be made. First, we reiterate that this method of exhibiting the data is extremely convenient in that it eliminates the necessity of assigning temperatures and gravities to the stars. However, this plot tends to de-emphasize the really dramatic effects that occur in the wings of Hα. For example, at an Hγ equivalent width (W) of 4 A the equivalent width of Hα is reduced by a factor of 1.5 from the LTE theory. Second, for smaller surface gravities the effect is even more pronounced. In Figure 3 we have plotted the LTE and non-LTE Hα profiles for $T_{eff} = 13,000°$ and log g = 2.0. As can be seen, the wing of Hα is actually in emission near line center. The implications of such calculations in interpreting Hα emission in Ia supergiants are obvious and will be dealt with in a later paper.

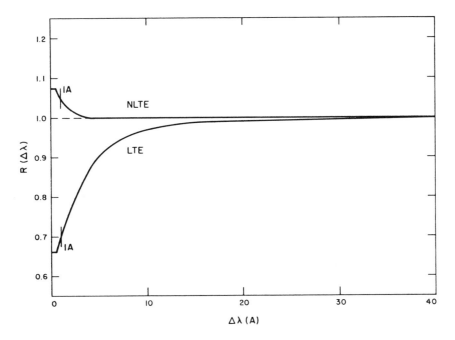

Figure 3.   Hα (Griem, 1967) for 13,000°, log g = 2.0, LTE and non-LTE
model atmospheres.

We take pleasure in acknowledging helpful discussions with Dr. W.
Kalkofen and Mr. R. L. Kurucz.   The model atmospheres were computed
with the program ATLAS written by Mr. Kurucz.   We were greatly assisted
in the photographic reductions by Mr. W. B. Persons.   The research was
supported in part by NASA grant NGR 22-024-001.   One of us (D. M. P.) would
like to acknowledge, with gratitude, fellowships from Harvard University
and the Smithsonian Research Foundation.

REFERENCES

Avrett, E. H., and Hummer, D. G.   1965, Mon. Not. Roy. Astron. Soc.,
    130, 295.
Edmonds, F. N., Jr., Schlüter, H., and Wells, D. C., III.   1967, Mem.
    Roy. Astron. Soc., 71, 271.
Griem, H. R.   1967, Ap. J., 147, 1092.
Kalkofen, W.   1964, Proc. First Harvard-Smithsonian Conf. on Stellar
    Atmospheres, Smithsonian Ap. Obs. Spec. Rep. No. 167, 175.
_____.   1966, J.Q.S.R.T., 6, 633.
_____.   1968, Ap. J., 151, 317.
Kalkofen, W., and Strom, S. E.   1966, J.Q.S.R.T., 6, 653.
Latham, D. W.   1968, Rev. Sci. Inst., in press.
Mihalas, D.   1967a, Ap. J., 149, 169.
_____.   1967b, Ap. J., 150, 909.
Mihalas, D., and Stone, M. E.   1968, Ap. J., 151, 293.
Olson, E. C.   1968, Ap. J., 153, 187.
Peterson, D. M.   1967, Astron. J., 72, 822 (abstract).
Strom, S. E.   1967, Ap. J., 150, 637.
Strom, S. E., and Kalkofen, W.   1966, Ap. J., 144, 76.
_____.   1967, Ap. J., 149, 191.
Strom, S. E., and Peterson, D. M.   1968, Ap. J., 152, 859.

DISCUSSION

Buscombe: You didn't get any stars on the main sequence, did you?

Peterson: No, we avoided the main sequence.

Buscombe: We're not having a meeting of minds, because my supergiants do what yours do but my main-sequence stars do something different from that.

Collins: I would like to make a small word of caution about the application of this mechanism to Be stars. If you accept the fact that the gravity is down, owing to rotation, you must also accept the fact that the temperature is down. Thus, the only way you might make it work is by use of a dilute radiation field originating at the poles and acting on a very low-gravity situation surrounding the star.

Peterson: I show these calculations only for 10, 000° to 15, 000°. They shouldn't be extrapolated beyond about 8000°.

Houziaux: This has been carried out for plane-parallel atmospheres, has it not? For Be stars you might have to take curvature into account.

Cayrel: I don't know why, but anytime you do a non-LTE calculation, the first time you find in emission what should be in absorption. I remember when Schatzman computed for the first time the effect of non-LTE in the Balmer jump. He found the Balmer jump to be in emission. I suppose he never published this.

SURFACE GRAVITIES AND HYDROGEN-LINE PROFILES IN ECLIPSING BINARY SYSTEMS

E. C. Olson

University of Illinois Observatory, Urbana, Illinois

A preliminary comparison has been made for eclipsing binary systems between surface gravities determined from light and radial velocity solutions and from Hγ and Hβ profiles observed in these systems. For the latter determinations, a grid of theoretical LTE profiles calculated by Strom and Peterson (1967) has been used. Calibrated 20 A/mm spectra of about two dozen bright B and early A binaries were obtained with the X spectrograph on the 60-inch Mt. Wilson reflector. These include single- and double-line systems, nine of which appear on lists (Wood, 1963, p. 377; Popper, 1967, p. 101) of binaries with well-determined dimensions or masses. In single-line cases, the surface gravity of the bright component is proportional to the assumed mass ratio, which can be estimated well enough to give log g to ±0.1. The line blending of the faint component is obviously easier to treat in these cases.

The Hγ profile for YZ Cas, shown in Figure 1, is fairly typical of the quality of the observations. The mean profile of the line in α Lyr obtained from three spectrograms agrees within 1% with Oke's photoelectric profile. The match between theoretical and observed profiles is made between points a few angstroms from the line center and at about the 95% level in residual flux. Thus, the influence of theoretical problems in the cores and the possible effects of satellite absorption features near the cores are reduced.

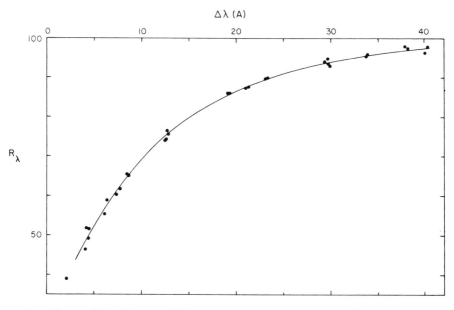

Figure 1. Hγ profile observed in YZ Cas from three 20 A/mm spectrograms.

Effective temperatures necessary for the gravity determinations were taken from the spectral-type—effective-temperature relation of Strom and Peterson (1968), which they estimate corresponds to an uncertainty of ± 0.1 in the log g for spectral types B3 to B9. In Figure 2, the gravities derived from profiles by means of the Griem (1967) and ESW (Edmonds, Schlüter, and Wells, 1967) broadening theories are compared with those from the binary solutions. With LTE profiles, the ESW theory is more nearly in accord with observations. No attempt has yet been made to treat interaction effects between the components. It should be noted, however, that no systematic difference is present between single- and double-line systems. The final results are summarized statistically in Table 1, where we further note the absence of a clear trend with r/a, the stellar radius in units of the semimajor axis of the relative orbit. In a few cases, single spectra were obtained at secondary minimum; changes in hydrogen-line profiles are not apparent at this phase.

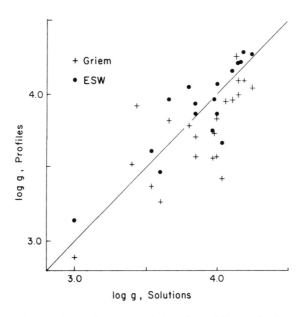

Figure 2. Comparison of surface gravities found from hydrogen profiles and from masses and radii obtained from photometric and radial-velocity solutions. The line is the 1:1 relation.

For stars between A0 and A2, the gravity determinations should be more reliable since the effective temperatures are better known. A-star results are collected separately in Figure 3, where only the comparison based on the ESW theory is shown. A and B spectral-type binary components yield consistent results. The most discordant star is β Aur, in which the eclipses are shallow and nearly equal. Part of the difficulty may be in the interpretation of the light curve.

Detailed results are presented in Olson (1968).

Table 1. Summary of gravity determinations

17 Systems with Griem, ESW Broadening:

$$\langle \log g \text{ (Griem)} - \log g \text{ (soln)} \rangle = -0.21 \pm 0.05 \text{ m. e.}$$

$$\langle \log g \text{ (ESW)} \quad - \log g \text{ (soln)} \rangle = +0.02 \pm 0.04 \text{ m. e.}$$

14 Systems, $r/a \le 0.30$:

$$\langle \log g \text{ (ESW)} - \log g \text{ (soln)} \rangle = +0.04 \pm 0.05 \text{ m. e.}$$

7 Systems, $r/a \le 0.22$:

$$\langle \log g \text{ (ESW)} - \log g \text{ (soln)} \rangle = -0.01 \pm 0.08 \text{ m. e.}$$

6 Systems, $r/a > 0.30$, omitting AO CAS:

$$\langle \log g \text{ (ESW)} - \log g \text{ (soln)} \rangle = +0.03 \pm 0.06 \text{ m. e.}$$

with AO Cas:

$$\langle \log g \text{ (ESW)} - \log g \text{ (soln)} \rangle = +0.10 \pm 0.09 \text{ m. e.}$$

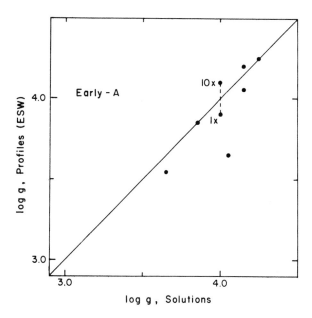

Figure 3. Comparison as in Figure 2 for early-A components. YZ Cas is a
borderline metallic star and the gravity determination is made with
normal Mg and Si abundances (1x) and ten times normal (10x).

This work was supported by the National Science Foundation.

REFERENCES

Edmonds, F. N. , Jr. , Schlüter, H. , and Wells, D. C. , III. 1967,
     Mem. Roy. Astron. Soc. , 71, 271.
Griem, H. R. 1967, Ap. J. , 147, 1092.
Olson, E. C. 1968, Ap. J. , 153, 187.
Popper, D. M. 1967, Ann. Rev. Astron. Ap. , 5, 85.
Strom, S. E. , and Peterson, D. M. 1967, private communication.
_____ . 1968, to be published.
Wood, F. B. 1963, in Basic Astronomical Data, ed. K. Aa. Strand (Chicago:
     University of Chicago Press), 377.

DISCUSSION

Eggen: How do you get the mass of a single-line binary?

Olson: One has to assume a mass ratio. Within the refinements that we are
talking about today, the surface gravity of the bright star is not terribly
sensitive to 0.1 or 0.2 in the log gravity. If we pick a mass ratio that gives
a reasonable mass for the bright star, we are probably in the right ball park.

Heintze: Did you correct for sin i?

Olson: Yes, it is preferentially smaller in the eclipsing systems.

Gray: Why does the mean error go down in the ESW case?

Olson: I don't know; in any case, the reduction is very small.

Buscombe: Is your investigation completed or would you care to have tracings
of Hγ for δ Orionis, η Orionis, ζ Phoenicis, δ Pictoris, and V Puppis, which
are all B-type eclipsers?

Olson: I would be happy to have any observations.

Neff: I am curious as to whether you're aware of the work Collins and
associates are doing on the reflection effects in the lines, because under
certain conditions you can get asymmetry of lines through reflection.

Olson: In the single-line systems, the line profiles are very symmetrical.
Most of these plates were obtained at maximum light with the stars in the
plane of the sky. For a few exposures near secondary minimum, I can't
detect any particular change in the profile.

Neff: The asymmetry would be greatest at maximum light.

Olson: Yes, apparently the single-line systems have very symmetrical lines.

Elste: What kind of model did you use?

Olson: These are theoretical calculations from Strom and Peterson.

Matsushima: How did you fit the profiles with the model profiles? Did you
interpolate for a particular star?

Olson: I interpolated in a grid of models.

Matsushima: Then you just chose the best-fitting interpolated value of T_{eff}
and g?

Olson: No, one has to adopt an effective temperature from spectral type. This is the only way to get the surface gravity with the present observations.

Vardya: What is the uncertainty in the log g that you obtained by the solution due to the mass distribution in the star?

Olson: This is one of the uncertainties — we don't know. I have tried to emphasize that there are apparently no systematic effects.

Underhill: I think it is remarkable that you have got these gravities at 45° differences — it is remarkable that the Hγ profile is significantly dependent on g. It is a fascinating thing to me. The g is defined by the geometric orbits.

Cayrel: It is a very remarkable and direct check, the spectroscopic determination of g. Can you also show that ESW is better than Griem?

Olson: Well, I think that in view of Deane Peterson's paper all bets are off.

Strom: What temperature did you use for β Aurigae?

Olson: About 9200°K.

Strom: If you went down another 400, it would probably be closer to where it really is.

Olson: I don't think this would really remove the discrepancy, which is about 0.4 in the log g.

STELLAR SURFACE GRAVITIES

D. F. Gray

The University of Western Ontario, London, Ontario, Canada

Some aspects of stellar-atmosphere theory can be tested by comparison of the observed and computed spectra of visual binaries. The advantage in using visual binaries is that we can make an accurate determination of surface gravity, thus fixing one of the main parameters of hydrostatic equilibrium models.

The method for determining g has been described by the author (Gray, 1967). First, we find the stellar radius by comparing the observed continuum with the continuum computed from a model. If we let F_ν erg sec^{-1} cm^{-2} per unit frequency be the stellar flux at the earth, and $\mathcal{F}_\nu$ erg sec^{-1} cm^{-2} per unit frequency be the computed flux at the star's surface, then the radius is given by

$$R = r(F_\nu / \mathcal{F}_\nu)^{1/2} \quad ,$$

where r is the distance to the star. Second, the mass of the star can be obtained from the astrometric measurements of the orbit by application of Kepler's third law. And third, the surface gravity can then be computed from

$$g = g_\odot \frac{M}{R_2}$$

where $g_\odot$ is the solar surface gravity and has a value of 2.74×10^4 cm sec^{-2}. This method has been applied to the stars listed in Table 1. The orbital data are from Worley (1963), and the parallaxes from Jenkins' catalog (1952, 1963). The values of surface gravity that result are shown in the last column. The probable errors are computed on the assumption that the errors in mass ratio, semimajor axis, and period are negligible compared with the errors in parallax, π, and the flux ratio, $F_\nu / \mathcal{F}_\nu$. The formal parallax errors in by Jenkins. The error in F_ν is associated with uncertainty in the visual magnitude, while the error in $\mathcal{F}_\nu$ comes from uncertainty in choosing the proper model spectrum. These flux errors have been taken to be 3% and 13%, respectively. If we assume gaussian error distributions, the random error in g is given by

$$\frac{dg}{g} = \left[\left(\frac{d\pi}{\pi}\right)^2 + \left(\frac{d\mathcal{F}_\nu}{\mathcal{F}_\nu}\right)^2 + \left(\frac{dF_\nu}{F_\nu}\right)^2 \right]^{1/2} \quad .$$

We note that for the vast majority of stars the parallax error will dominate and the percentage error in g will be approximately equal to the percentage error in parallax.

One important application is the following. Various spectral lines are sensitive to the pressure structure of the atmosphere. If these features are computed theoretically and compared with the observations, we can test the precision of the theory if the helium abundance is known in some other way, or we can determine the helium abundance if we are sufficiently confident of the theory.

Table 1. Stellar surface gravities and related data

Star	B-V	Sp	π	Mass	$g \pm$ p. e. unit = 10^4 cm sec^{-2}	References[†]
Sirius	+0. 01	A1V	0. 374 ± 4	2. 15	1. 90 ± 13%	Hayes, 1968
γ Vir A	+0. 35	F0V	0. 092 ± 7	1. 08	1. 06 ± 15%	Gray, 1966
Procyon	+0. 40	F5IV-V	0. 283 ± 4	1. 83	1. 05 ± 14%	Bessell, 1967
ξ Sco AB	+0. 45	F5IV	0. 036 ± 4	1. 92	0. 88[*]± 17%	Willstrop, 1965
η Cas A	+0. 57	G0V	0. 169 ± 4	0. 96	2. 13 ± 14%	Gray, 1967
HD 158614	+0. 73	G8V-IV	0. 051 ± 4	1. 94	2. 7[*] ± 20%	Whiteoak, 1967
ξ Boo A	+0. 75	G8V	0. 149 ± 4	0. 86	2. 55 ± 14%	Gray, 1967
70 Oph A	+0. 86	K0V	0. 193 ± 4	0. 99	2. 05 ± 14%	Gray, 1967

[*]Average of the AB components that are photometrically difficult to separate but appear to be identical.
[†]Observed energy distributions are from the sources cited.

REFERENCES

Bessell, M. S. 1967, Ap. J. , 149, L67.
Gray, D. F. 1966, Unpublished data.
_____. 1967, Ap. J. , 149, 317.
Hayes, D. 1968, private communication.
Jenkins, L. F. 1952, General Catalogue of Trigonometric Stellar Parallaxes
    (New Haven, Conn.: Yale University Press).
_____. 1963, Supplement to the above π catalogue.
Whiteoak, J. B. 1967, Ap. J. , 150, 521.
Willstrop, R. V. 1965, Mem. Roy. Astron. Soc. , 69, 83.
Worley, C. E. 1963, Publ. U. S. Naval Obs. , 18, Part 3, 74 pp.

A COMPARISON OF PHOTOMETRIC DETERMINATIONS OF RADII AND SURFACE GRAVITIES FOR A, F, AND G STARS WITH VALUES DETERMINED FROM ECLIPSING BINARY STARS

J. S. Neff

Department of Physics and Astronomy
University of Iowa, Iowa City, Iowa

1. INTRODUCTION

Theoretical photometric radii and surface gravities are obtained by comparison of measurements made with an intermediate bandpass photometric system with fluxes computed from a grid of model atmospheres. These theoretical results are compared with radii and surface gravities obtained from eclipsing binary stars with well-determined orbital elements. The intermediate bandpass system employed is the Morgan-Neff or chi system, which was described by Neff and Travis (1967), with additional observations by Neff (1968). The model atmospheres are those of Mihalas (1966) and Strom (1967). The author is grateful for the opportunity to compute additional models, using Strom's program.

2. THE PROPERTIES OF THE PHOTOMETRIC SYSTEM

The properties of the photometric system are summarized in Table 1. In the lower part of Figure 1, the sensitivity functions of the UBV system (dotted lines) are compared with those of the chi system (solid lines). The U filter and the 0.37-μ filter are influenced by the blending of the Balmer lines near the Balmer limit. The B filter is influenced by Balmer lines to a lesser degree. The V filter and the 0.55-μ, 0.47-μ, and 0.34-μ filters are not influenced by Balmer lines. A flux distribution predicted by a Balmer-line-blanketed model atmosphere is shown in the center of Figure 1. The open circles are predicted flux ratios for the chi system obtained by calculation of the mean flux in each filter bandpass from the predicted flux distribution. The sensitivity functions were tabulated by Neff and Travis (1967). The schematic flux distributions for ι U Maj and α Lep are based on the intermediate bandpass measurements of these stars and the flux-ratio calibration given by Neff and Travis (1967). An improved calibration has been given by Neff (1968).

3. METHODS FOR DETERMINING T_{eff}, g, ΔB, AND R

Several methods for determining T_{eff}, g, and a blocking index ΔB have been employed. The method for determining R is very similar to that used by Gray (1967). The method for finding T_{eff}, g, and ΔB used in this study was to construct a set of diagrams where the coordinates of a star on each plane were given by a pair of flux ratios or spectral parameters obtained from the schematic flux distribution. Defined in Figure 1, $\Delta (0.37\ \mu)$ and $\Delta (0.33\ \mu)$ are examples of spectral parameters that are functions of temperature, surface gravity, and line-blocking effects in the filter bandpasses. In addition, $\Delta (0.37\ \mu)$ is influenced by the broadening mechanism for the Balmer lines. A set of companion planes was constructed for the model atmosphere with $T_{eff} > 4500°K$, where the coordinates of the model were the predicted flux ratios or spectral parameters. The predicted mean flux for the 0.37-μ bandpass requires a Balmer-line-blanketed model atmosphere, so that the flux ratios and spectral parameters based on this bandpass could not be predicted for all models.

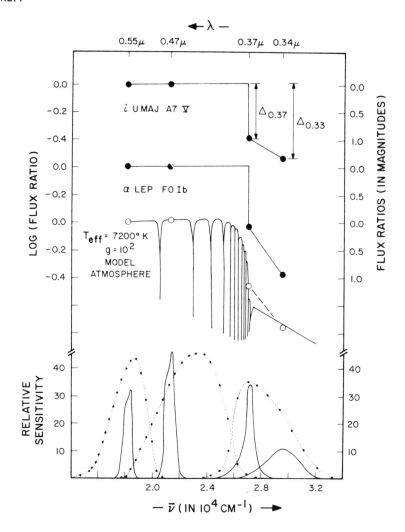

Figure 1. The upper part of the figure shows log $F\nu/F\nu_0$ versus wave number for two stars and a Balmer-line-blanketed model atmosphere. The relative sensitivities of the UBV system (dotted lines) and the chi system (solid lines) are shown below. The significance of the various symbols and parameters is explained in text.

If there were no line-blocking effects and no systematic errors in the flux-ratio calibration or in the model atmospheres, the observed and theoretical planes could be made to coincide and T_{eff} and log g could then be read off for each star. In practice, the line-blocking effects can act in the same manner as changes in the T_{eff} or log g, or a combination of changes in the two parameters. For example, on the log $F(0.33\ \mu)/F(0.55\ \mu)$ versus $\Delta(0.33\ \mu)$ plane, the line-blocking effects act to give an apparent decrease in surface gravity for G and F stars. The effect of blocking on the temperature determination is fairly small in this region of the plane. In contrast, the relations on the $\Delta(0.33\ \mu)$ versus $\Delta(0.37\ \mu)$ plane are straight lines with the same slope. The intercept depends only on surface gravity; the $\Delta(0.33\ \mu)$ and $\Delta(0.37\ \mu)$ are very strongly correlated with respect to temperature and line-blocking effects by non-Balmer lines, but the surface-gravity effects are different owing to the influence of the Balmer lines on the $\Delta(0.37\ \mu)$ parameter.

Table 1. Characteristics of the detectors

λ (μ)	0.330	0.370	0.470	0.550	λ of peak transmission
$\Delta\lambda$ (in A)	343	145	193	217	Equivalent rectangular bandpass
σ_0 (in 10^4 cm^{-1})	2.949	2.702	2.129	1.818	Constant-energy wave number
$\overline{E}_{h\nu}$ (in ev)	3.656	3.350	2.639	2.254	Mean photon energy
$\Delta E_{h\nu}$ (in ev)	0.291	0.155	0.082	0.068	rms width of detectors
*$F^0_{h\nu}$ (photons cm^{-2} sec^{-1})	1.60×10^5	1.27×10^5	1.95×10^5	1.69×10^5	Photon flux for a mean A0V star with $m_{0.55} = 0.000$
$F_{h\nu}$ (photons cm^{-2} sec^{-1})	5.20×10^{15}	4.03×10^{15}	7.46×10^{15}	8.99×10^{15}	Solar photon flux at 1 AU

*$F_{h\nu} = F^0_{h\nu}(\sigma) \, 10^{-0.4 m_\sigma}$

Hence, we have to find a set of values for T_{eff}, log g, and ΔB that will give a self-consistent position for the star on all of the relevant relations between flux ratios or spectral parameters. This was done by an iterative procedure. Most of the stars treated had parallaxes greater than 0.030 arcsec so that interstellar reddening could be neglected. For the supergiants, which are presumably reddened by interstellar matter, several solutions for T_{eff}, log g, and ΔB were obtained for a number of assumed values for the color excess.

The method for determining R, which was based on Gray's method, is illustrated in Figure 2. The ordinate is the absolute monochromatic magnitude at 0.55 μ for stars with $\pi > 0\overset{''}{.}03$ and measured colors on the chi system. The solid line is the relation for a theoretical star the size of the sun normalized to the observed absolute monochromatic magnitude of the sun and the observed effective temperature of the sun. The relation is rather insensitive to variations in surface gravity. For a fixed T_{eff}, the difference between the plotted position of the star and the theoretical curve is proportional to $2 \log (R_*/R_\odot)$.

4. COMPARISON OF PHOTOMETRIC RADII AND SURFACE GRAVITIES WITH THOSE OBTAINED FROM ECLIPSING BINARY STARS

In the lower part of Figure 3, photometric radii obtained in this study (open and filled symbols with error bars) and those obtained by Gray (1967) (open and filled symbols without error bars) are compared with radii tabulated by Popper (1967) for eclipsing binary stars with well-determined spectroscopic and photometric elements (open rectangles). The temperatures of the eclipsing binary stars were estimated from their spectral types by means of the Morton and Adams (1968) calibration. The filled symbols denote main-sequence stars and the open denote subgiants. The large error bars for some stars are due to large errors in their trigonometric parallaxes. With the exception of AR Aur, the two relations appear to coincide. Five stars had photometric radii determined by both Gray and the writer, and in all cases were found to agree within 5%. This is significant because the observations, the model atmospheres, the method of fitting the models to the observations, and the absolute flux calibration differed.

In the upper part of Figure 3, photometric surface gravities are compared with surface gravities obtained from eclipsing binary stars (large open rectangles) and visual binaries (small filled squares) by Gray (1967). The error bars indicate differences in two photometric methods for determining log g and log T_{eff}. The four F supergiants in the upper part of the diagram have large uncertainties owing to a lack of accurate knowledge of their color excesses and the fact that 2 was the smallest value for log g in the grid of model atmospheres. It was therefore necessary to extrapolate to find log g and log T_{eff}. Photometric surface gravities are poorly determined by this method for $T_{eff} > 9500°K$ and $T_{eff} < 5500°K$. It appears that the photometric surface gravities are systematically larger than those obtained from eclipsing binary observation by about 0.3 in the logarithm.

5. SUMMARY

Observations made with an intermediate bandpass photometric system are used in conjunction with a grid of model atmospheres to derive T_{eff}, log g, and an index of line blocking in the various filter bandpasses. Photometric radii were determined from the measured monochromatic magnitudes and the derived value of T_{eff}. The radii were compared with those determined by Gray by a similar method, and with those obtained from eclipsing binary stars. The agreement between the various methods was within the uncertainty of the measurements, which suggests that the systematic errors are small relative to the measurement errors. It should be noted that this method is relatively insensitive to errors in log g.

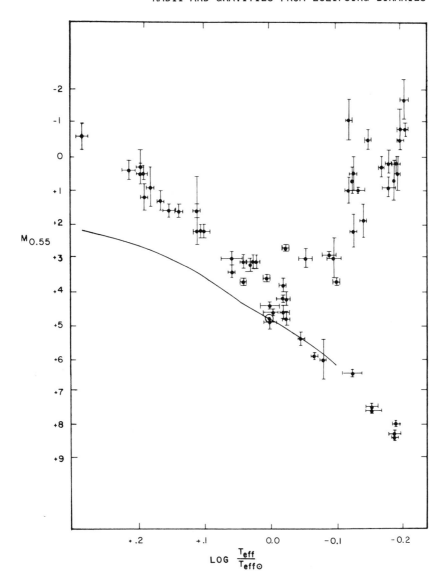

Figure 2.   The relation between the absolute monochromatic
magnitude at $\lambda = 0.55~\mu$ and $\log (T_{eff}/Te_{\odot})$ for
stars with $\pi > 0.030$ arcsec.

The photometric surface gravities were compared with surface gravities
obtained from eclipsing and visual binary stars.   Because of the necessity of
determining the line-blocking corrections, the errors of the determination
were larger than expected.   The photometric values of $\log g$ appear to be
systematically larger by 0.3 than the values obtained from binary stars.

These comparisons are being repeated with a new grid of model atmospheres and a new flux calibration to see if better agreement can be obtained.

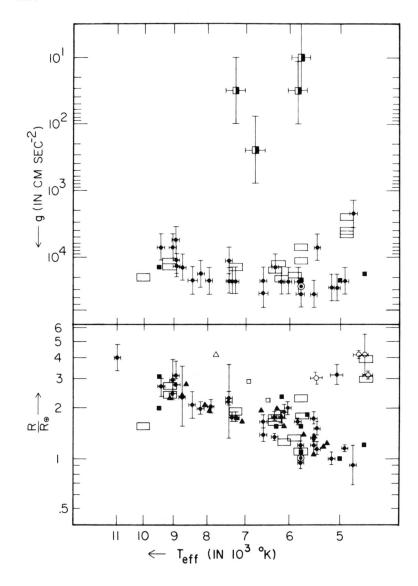

Figure 3.   The upper relation is between the surface gravity (log scale)
and T_{eff} (log scale) for A, F, and G stars.   Half-filled rectangles denote little reddened supergiants; filled circles lower
luminosity stars near the sun; filled squares visual binary
stars; and large open rectangles eclipsing binary stars.
The lower relation is between $R/R_{\odot}$ (log scale) and T_{eff}
(log scale).   The notation is the same as for the upper relation except subgiants are represented by small open symbols.
Open and filled triangles denote Hyades members.   Symbols
with and without error bars are explained in the text.   The
eclipsing binary below and to the right of the main sequence
is AR Aur.

REFERENCES

Gray, D. F. 1967, Ap. J., 149, 317.
Mihalas, D. 1966, Ap. J. Suppl., 13, 1.
Morton, D. C., and Adams, T. F. 1968, Ap. J., 151, 611.
Neff, J. S. 1968, Astron. J., 73, 75.
Neff, J. S., and Travis, L. D. 1967, Astron. J., 72, 48.
Popper, D. M. 1967, Ann. Rev. Astron. Ap., 5, 85.
Strom, S. E. 1967, private communication.

DISCUSSION

Wallerstein: When you refer to stars within 50 parsecs of the sun, do you really mean stars within 50 parsecs or do you mean stars with parallaxes measured to be greater than 0!'02? They are not the same.

Neff: Yes.

Buscombe: There was the same kind of color term in the transformation in the photometric work of the late Arthur Hogg for 244 bright southern stars. He has a pretty good transformation for the main sequence, but he was in great difficulty for the supergiants. This was overlooked in some of the other early work. Unfortunately, "UBV photometry" got a rather bad start in Australia because both Hogg and Westerlund used filters for supergiants that were not quite the usual ones. They were off in opposite directions.

Neff: From the point of view of empirical classification it is a very sensitive method for recognizing late A and F supergiants.

RELATIONS BETWEEN COLORS ON THE UBV SYSTEM AND BALMER-LINE-FREE COLORS OF AN INTERMEDIATE BANDPASS PHOTOMETRIC SYSTEM

J. S. Neff and K. R. Honey[*]

Department of Physics and Astronomy
University of Iowa, Iowa City, Iowa

1. INTRODUCTION

Departures from a linear transformation between a related intermediate bandpass color index $(m_{0.47\,\mu} - m_{0.55\,\mu})$ and B-V and the corresponding departures from a linear transformation between a second color index $(m_{0.33\,\mu} - m_{0.47\,\mu})$ and U-B are examined. The departures r (B-V) and r (U-B) are found to be sensitive luminosity indicators for stars in the A-G spectral region. The intermediate bandpass filters employed are the three filters of the Morgan-Neff or chi system that are not influenced by Balmer lines. This comparison can be considered as an indicator of differences between Balmer-line-free UBV colors and measured UBV colors. This is an oversimplification because the differential shift in effective wavelengths of the broad and intermediate bandpass filters and the blocking effect of non-Balmer lines also play roles. For the B5 - F8 spectral region, the influence of the Balmer lines is the dominant effect. Thus, this comparison is relevant to the discussion of the validity of other Balmer-line-free UBV colors, in particular the Matthews and Sandage (1963) calibration of the UBV system.

2. PROCEDURE

The departures from a linear relation between B-V and U-B are defined below:

$$r(B-V) = (B-V) - C_1 (m_{0.47\,\mu} - m_{0.55\,\mu}) \quad ,$$

$$r(U-B) = (U-B) - C_2 (m_{0.34\,\mu} - m_{0.47\,\mu}) \quad ,$$

where C_1 and C_2 are scale factors,

$$C_1 = \frac{\Delta\sigma\,(B-V)}{\Delta\sigma(m_{0.47\,\mu} - m_{0.55\,\mu})} \quad ,$$

$$C_2 = \frac{\Delta\sigma\,(U-B)}{\Delta\sigma\,(m_{0.34\,\mu} - m_{0.55\,\mu})} \quad ,$$

and $\Delta\sigma$ indicates the difference in wavenumber for the various combinations of filters. All the color indices are normalized to a mean A0V star so that r (B-V) = r (U-B) = 0 for a mean A0V star.

Little-reddened B0V and O5 stars have very smooth energy distribution and hence one would expect r (B-V) = r (U-B) = 0 for these stars. But, because of the normalization procedure, the departures are r (B-V) = + 0.10 mag and r (U-B) = - 0.10 mag for them. We can conclude, therefore, that the Balmer lines influence the colors of an A0V star by 0.10 mag in both B-V and U-B.

---

[*] Now at the University of Missouri.

3. SUMMARY AND CONCLUSIONS

The principal results of the comparison of the Balmer-line-free colors
and the measured B-V and U-B colors are shown in Figure 1. The notation
for the various types of stars is given in the figure. The term "scaled"
means that the intermediate bandpass colors have been multiplied by the
scale factors C_1 and C_2. The observed and scaled colors for each star are
connected in the figure by a line. The length and slope of the line are functions of r (B-V) and r (U-B).

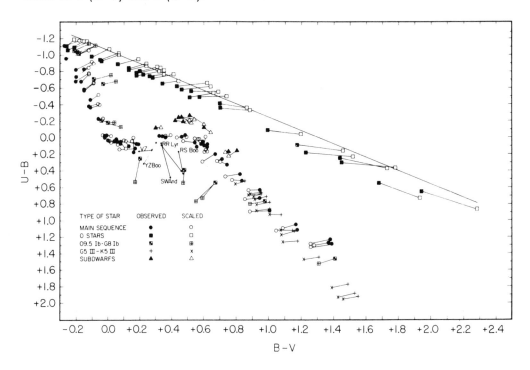

Figure 1.   The relation between the observed B-V and U-B colors
            and the scaled B-V and U-B colors.

The interstellar reddening line is delineated by observation of thirty-one
O5-O8 stars having a wide range of color excess. The solid line is the reddening line for monochromatic measurements made at the constant energy
wavelengths of the U, B, and V filters (Code, 1960). The slope of this line
was obtained from Whitford's interstellar reddening law (1958). The scaled
colors for the O stars scatter about this line because the 0.47 μ intermediate
bandpass filter is at a larger wavelength than the wavelength (λ = 4200 A)
where the reddening law changes slope. The observed colors on the UBV
system show the well-known curvature of the reddening line due to the
increase in the effective wavelengths of the filters with increasing interstellar reddening and the curvature of the law of interstellar reddening.

For main-sequence stars the influence of the Balmer lines on the B-V
and U-B colors increases from B0 to A0, but because of the normalization
convention the opposite behavior is shown in Figure 1. For stars later than
A0, r (B-V) is larger in magnitude than r (U-B). For main-sequence stars,
r (B-V) changes sign between F5 and G5.

The most striking differences between observed and scaled colors is for
the RR Lyrae stars and the A-G supergiants. Inspection of Figure 1 in the
paper by Neff (these Proceedings) shows that this is due to the influence
of the Balmer lines on the U and B filters. This effect is a very sensitive

luminosity indicator of practical importance for the study of the space distribution of high-luminosity stars.

Mr. Keith R. Honey has found combinations of colors based on all four filters of the CHI system that predict B-V and U-B with standard deviations of ±0.02 mag. He has obtained relations for the following types of stars: O5-O8, B0V-A0V, A0V-G0V, G0V-K7V, Gs III-M5 III, and F-G subdwarfs. He has been unable to find a single relation that would predict either B-V or U-B with sufficient accuracy over the entire observed region of the two-color plane. His results are given in Table 1.

This work is relevant to the discussion of the Matthews and Sandage calibration of the UBV system. Their calibration is based on five stars in the spectral interval O9V-G2V and two F subdwarfs. Their calculated U-B and B-V colors are based on Balmer-line-free flux distributions. The work reported here implies that their calibration will not hold for stars or other objects that have flux distributions differing significantly from those of their calibration stars. Thus, it appears unlikely that their calibration can be applied to the study of white dwarfs, RR Lyrae stars, or A-G supergiants.

REFERENCES

Code, A. D. 1960, in Stars and Stellar Systems, Vol. VI, Stellar Atmospheres, ed. J. L. Greenstein (Chicago: University of Chicago Press), 50.

Matthews, T. A., and Sandage, A. R. 1963, Ap. J., 138, 30.

Whitford, A. E. 1958, Astron. J., 63, 201.

Table 1. Transformation coefficients from the intermediate bandpass system to B-V and U-B colors

$$B-V = A1\,(m_{0.47\,\mu} - m_{0.55\,\mu}) + B1\,(m_{0.37\,\mu} - m_{0.47\,\mu}) + C1\,(m_{0.34\,\mu} - m_{0.37\,\mu}) + D1$$

$$U-B = A2\,(m_{0.47\,\mu} - m_{0.55\,\mu}) + B2\,(m_{0.37\,\mu} - m_{0.47\,\mu}) + C2\,(m_{0.34\,\mu} - m_{0.37\,\mu}) + D2$$

Type of star	Number	σ(B-V)	σ(U-B)	A1	A2	B1	B2	C1	C2	D1	D2
O5 - O8	31	$0^{m}.02$		1.004		0.274		-0.059		0.179	
			0.02		0.121		0.595		0.510		-0.161
B0 V - B9 V	19	0.02		1.212		0.246		-0.283		0.023	
			0.02		-0.111		0.771		0.442		0.010
A0 V - G0 V	30	0.03		1.329		0.330		-0.302		0.016	
			0.02		-0.106		0.811		0.153		0.011
G1 V - K7 V	19	0.02		0.632		0.295		0.332		0.295	
			0.02		-0.134		0.821		0.449		-0.006
Subdwarfs	9	0.01		1.289		0.262		0.057		0.056	
			0.01		-0.407		1.018		-0.948		0.066
G5 III - M5 III	23	0.01		0.950		0.206		-0.255		0.282	
			0.03		-0.028		0.641		1.232		-0.026

COMPUTED AND OBSERVED OI LINE PROFILES IN THE SPECTRA OF B3-A0 STARS

L. Houziaux[*]

Université de Liège, Liège, Belgium

ABSTRACT

In the spectral type range B3 to A0, oxygen can be detected only through the OI lines. Although there are numerous OI transitions in the range λ 950 to λ 10,000, only a few of them are suitable for study of the oxygen spectrum. Transitions between ground and excited levels fall in the region λ 990 to λ 1300 and are blended with strong stellar lines or with strong interstellar lines of OI itself.

The appearance of lines whose upper levels lie above the first ionization limit has never been reported in astronomical spectra. This fact is not quite explained. Most of the lines have rather high excitation potentials and some of the levels may be subject to auto-ionization. Hence, only the transitions between excited levels whose limit is $S_{1/2}$ are seen, and among those the strongest transitions fall at λ 13,254, λ 11,299, λ 11,287, λ 9260, λ 8446, and λ 7772. Other lines belonging to the same series occur at 7254, 7002, 6456, 6158, 4368, and 3847 A, but they are considerably weaker and in many cases blended with stronger features.

Because of numerous and important atmospheric absorptions and experimental difficulties in observing in the infrared, only the lines at λ 8446 and at λ 7772 can at present be clearly recorded. Between these two multiplets, λ 7772 is to be preferred since, owing to the large width of the Paschen lines of hydrogen, λ 8446 falls in the wing of P_{18}, a line for which the absorption coefficient is difficult to compute with accuracy.

In this paper we study the sensitivity of computed profiles for the triplet at λ 7772 to various atmospheric parameters, and the comparison of observed profiles (in bright stars) with computed ones.

1. THEORETICAL RESULTS

Four models in radiative equilibrium are used. For all of them, log g = 4. The first three models are due to Underhill (1962), and the fourth was proposed by Strom (1964). Effective temperatures are 12,689°, 15,326°, 19,186° (with X, hydrogen abundance by weight, = 0.68), and 10,000° (with X = 0.56). They are intended to simulate the atmospheres of B8, B5, B3, and A0 main-sequence stars, respectively.

Continuous absorption takes into account the contributions of H, H⁻, He, He⁺, and electron scattering.

The line absorption coefficient includes the effect of overlapping the multiplet components. The contributions of hydrogen atoms, helium atoms, ions, and electrons are considered in the computation of the collisional damping constants. Line profiles are computed at 46 wavelengths with the Kirchhoff-Planck function as source function.

---

[*]Present address: Départment d'Astrophysique, Faculté des Sciences, Mons, Belgium.

Results are shown in Figure 1 for six values of the oxygen abundance A, ranging from A = 2.5×10^{-4} to 8×10^{-3} (by a factor of two). Except for model 4, no overlapping of the components appears and line centers are almost completely saturated. Lines are formed between the highest layer in the atmosphere and a Rosseland depth of 0.222 (model 1), 0.308 (model 2), 0.480 (model 3), and 0.5 (model 4), where $\tau_\nu = 1$ in the local continuum. Central intensities are rather large since the continuous opacity in this spectral range is larger than in the visible region.

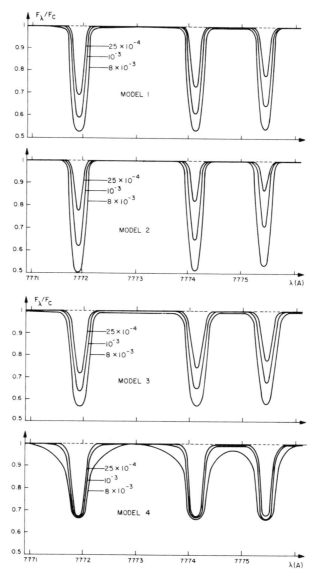

Figure 1.   Predicted residual intensities for the infrared OI triplet. The parameters specifying the models 1 to 4 are described in the text.

The related curves of growth indicate that, for the abundances considered, the lines lie in the intermediate portion of the curve. Equivalent widths range between 0.1 and 0.5 A.

As most of the B3-A0 stars are rotating, it is of interest to see how the appearance of the multiplet is changed when rotation occurs. Results show that a rotational velocity of 10 km sec^{-1} is sufficient to reduce to one half the depth of the line, and at 150 km sec^{-1} the multiplet is practically washed out.

Microturbulence has been introduced as a supplementary parameter in the atmospheres. As shown in Figure 2, it has a considerable influence on intensity, width, and shape of the lines, which is understandable since lines are located on the intermediate portion of the curve of growth.

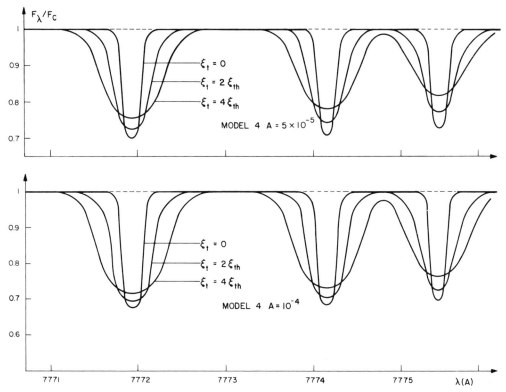

Figure 2.   The effects of microturbulence on the predicted OI line profiles.

2.   COMPARISON WITH OBSERVATIONS

Good profiles are not easy to obtain, because of the lack of bright, non-rotating main-sequence B stars, at least in the northern hemisphere. So far we have obtained spectra for only Sirius, Vega, α Cygni, and β Orionis. These coudé spectrograms were obtained with the 193-cm telescope of the Haute-Provence Observatory. A glance at Figure 3 shows that profiles computed directly from the models (see Figures 1 and 2) bear little relation to the observed line shape (average of two tracings) for Sirius. The model atmosphere adopted (T_{eff} = 9500°K, log g = 4; metal abundance = 10 × solar abundance) was taken from Strom, Gingerich, and Strom (1966). The main constraints to meet in order to fit the observed profiles are:

A.   The central depths should not be too large and should reproduce the observed ratios between the three components of the triplet.

B.   The equivalent widths should match the observed values.

C.   The line shape should be well reproduced, especially the blend of the two weaker lines at λ 7774.14 and λ 7775.35.

Of course, one must keep in mind that A and C are fairly sensitive to effects due to the finite resolving power of the spectrograph. Our projected slit width is 29 μ, corresponding to 0.185 A, i.e., about 1/10th of the total width of a component of the multiplet.

The observed equivalent width in the case of Sirius is 0.445 A. In order to reconcile this intensity with the observed central depths, it is necessary to keep the abundance as low as 3×10^{-5} with respect to hydrogen and to introduce a microturbulent velocity amounting to three times the thermal velocity. A great improvement in fitting the line shape is obtained by adding a small axial rotation (v sin i = 10 km sec^{-1}).

For Vega, with use of a (9000°, 4; solar abundance) model, the best fit is obtained with the following parameters: $A = 8 \times 10^{-5}$, $\xi_t = 3\xi_{th}$, v sin i = 15 km sec^{-1}.

The aspect of the line at λ 8446 confirms that some additional broadening mechanism is required to explain the observed line shape. Both in Vega and in Sirius, this line appears as single, while it is in fact a doublet. Computations without microturbulence show a doublet structure, and instrumental broadening is not sufficient to blend completely the two components.

In the case of supergiants like α Cygni or β Orionis, equivalent widths are much larger and suggest that the lower level of the transition ($5S^0$), which is metastable, is overpopulated in these atmospheres. In the spectrum of β Orionis, the two components at λ 7774.14 and λ 7775.35 are completely blended (Figure 3) and the general aspect of the multiplet is quite different from what it is in main-sequence objects of the same temperature.

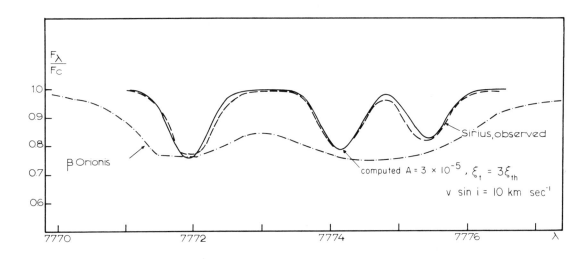

Figure 3.   Comparison between the observed OI triplet for Sirius and the predicted profile having microturbulent velocity three times the thermal velocity. A rotational velocity of 10 km sec^{-1} is found to improve the fit.

REFERENCES

Strom, S. E. 1964, Thesis, Harvard University.
Strom, S. E., Gingerich, O., and Strom, K. M. 1966, Ap. J., 146, 880.
Underhill, A. B. 1962, Publ. Dominion Ap. Obs., Victoria, B. C., 11, 433.

DISCUSSION

Strom: What is the abundance that you get for Sirius?

Houziaux: Well, for the 7772 line it is 3×10^{-5}, but I should say that the line is extremely insensitive. You can raise the oxygen and nothing happens because it lies on this flat part of the curve of growth. This is unfortunate, because it's the only line that is unblended.

Strom: What does 10^{-4} represent? Is that an upper limit to the abundance?

Underhill: Is that by number?

Houziaux: An upper limit by number.

Strom: So it is lower than GMA?

Houziaux: Yes, by at least an order of magnitude. I don't see any way of reducing these central depths without touching the abundance. If you keep a higher abundance, you are bound to reach saturation.

Strom: In non-LTE it would probably still be an overestimate of the abundance.

Cayrel: Do your calculations help us understand the oxygen abundance anomaly in Ap stars found by Sargent and his coauthors?

Houziaux: No. They refer to a lower temperature range. I go from about 10,000 to 19,000°. There are lots of things that may happen to these lines, especially when you go to the abnormal objects.

Underhill: You said oxygen had some doubly excited levels. I know those lines, but you were talking about some other lines and you said that you didn't see them. You need to have less oxygen than the standard amount in order to get your lines. Now suppose that oxygen had got more ionized than you expected by the Saha-Boltzmann relation that you have used. Of course, then the oxygen would all be sitting up in the ionized state and you couldn't see it. What I suspect is wrong, where the problem is with these supergiants, is actually with the ionization. How much oxygen is up in the unobservable Of stage? That ratio can be very touchy, quite different in supergiants than in main-sequence stars.

Strom: If more is ionized, you will still overestimate the abundance; 10^{-4} is still an upper bound.

Underhill: Of how much is neutral, but still no estimate of the total abundance, because you don't know how to divide into the two stages.

Wallerstein: I think it is very difficult to fool around with the state of ionization and to change the strength of the lines that way, because when you change the state of ionization of oxygen you'll find yourself changing the state of ionization of hydrogen since those lines have exactly the same ionization potential. Therefore, you would change the continuous opacity by the same amount that you would change the line opacity.

Houziaux: The structure of the hydrogen atom is not the same as the structure of the oxygen. You may have peculiar effects that affect the ionization of neutral oxygen without having the same effects for hydrogen.

Elste: As we tried to understand the solar spectrum in a little bit more detail when we obtained higher dispersion spectra, it was obvious that we needed to take inhomogeneities into account. Now we are looking at spectra of giants, and it may be necessary to take into account the inhomogeneities in giants too. I think of cool and high temperature regions in these atmospheres, which could produce a strengthening of certain lines.

Strom: What is the abundance you got for Vega?

Houziaux: It is 8×10^{-5}. It's lower than the GMA abundance.

Strom: But about twice what you got for Sirius.

THE ATMOSPHERE OF BD + 10° 2179

K. Hunger and D. A. Klinglesmith

Goddard Space Flight Center, National Aeronautics and Space
Administration, Greenbelt, Maryland

ABSTRACT

Preliminary analysis of the 10-mag star BD + 10° 2179, sometimes referred to as Klemola's Star, yields a carbon abundance of
30% by mass. Hence, carbon provides the major source of opacity.
In order to fit a model for this star, we have the two usual parameters log g and T_{eff}, plus the hydrogen x, the mass fraction of
carbon z_c, and the turbulence. Most of the observed lines of carbon
fall on the flat portion of the curve of growth and are more or less
proportional to the turbulence. At the beginning it looked to us completely hopeless to disentangle these five parameters, but we think
we have obtained a unique solution. If x is smaller than 10^{-3} and
z_c is larger than 10%, then hydrogen can be ignored.

Since from previous analyses we knew that the temperatures
would run between 16,000° and 18,000° and that log g was in the
range 2.5 to 3.0, we chose four grid points. We computed at each
grid point with various abundances of carbon to make 12 models.
Then we computed the carbon lines in these models; we integrated
over the profiles to get the equivalent widths. We have done this
not only for every model at each grid point, but also for microturbulent velocities of 0, 5, and 10 km sec^{-1}. There are a number
of good lines for carbon in the spectrum. We calculated the standard deviations and looked for the minimum. This is analogous to
the classical shifting of the curve of growth to get the best fit.
Fortunately, there is a clear-cut minimum. If we agree to this
method, then we infer from the data a zero microturbulent velocity,
and, for example, at 18,000° and log g = 2.5 the abundance of carbon
is 66%. From then on the analysis proceeds in the usual way.

We next computed more models in this vicinity and established
the temperature from the ratio of the silicon II to silicon III lines.
Then we looked at the helium-line profiles in the hope of determining
log g from the shapes of the lines. Since the hydrogen lines are
weak—0.2 to 0.3 in equivalent width—and broadened by rotation, we
used the helium lines. What a hydrogen line is to an ordinary star,
a helium line is to a helium star.

In going from log g = 2.5 to 3.0, the z_c was also changed in order
to fit the equivalent widths of the lines. Thus we found that the
helium lines are not a log g indicator. We checked on an earlier
study of the UBV colors. They turned out to be rather insensitive
to changes in the abundances and hence did not give a further clue.
The models predicted several discontinuities, but it was hard to
compare these with observations because of the line blanketing and
the problem of finding the true continuum.

The basic parameters for this star turn out to be: T_{eff} = 16,000°,
log g = 3 ± 0.5, z_c = 0.48, and x = 2 × 10^{-4}, and the microturbulence
is zero.

DISCUSSION

Buscombe: I am very pleased that the sharpness of line profiles has shown up in this spectrum. This agrees with what little experience we had with HD 96446, a hydrogen-deficient star for which I hope some day Aller will bring out some further results. What has appeared so far is not enough. I think it was disappointing that we did not have a model atmosphere. I'm wondering if Dr. Stewart could enlighten us about the hydrogen-deficient star that he has discussed in his thesis, which has not yet seen the light of day.

Stewart: Well, there is not much relationship between the stars that I discussed and this one. The star that I analyzed was much less helium rich. It still had more hydrogen than helium in it and, as I recall, there was nothing peculiar about the carbon.

Buscombe: Were the line profiles quite sharp?

Stewart: Yes, very sharp.

Peterson: I was wondering at what instrument and what resolution?

Hunger: These plates were taken by Herbig with the 120-inch telescope at Lick at 16 A/mm.

Nariai: The difficulty in your analysis is almost the same as that I met with in HD 30353. The difficulty was how to find a first guess, for if I made a mistake in the first guess, the way would be very long because there are too many parameters to choose. How did you get your first guess?

Hunger: The only first guess that we make is to choose this array of grid points.

Klinglesmith: I might say as far as the first guess is concerned on this star, Klemola and Hill have both made a coarse analysis and they derived $T_{eff} = 18,000°$, $\log g = 3$. We started out with their values and ended up where we are.

Fischel: I should only like to comment that it was really just fortuitous that you hit a grid point. It could have been more laborious in the center of your grid.

Weidemann: Where did you get your opacities for carbon?

Hunger: Carbon opacities come mostly from C II; we have used the unpublished quantum-defect computations of Miss Peach for C II and C III.

Stibbs: Did you take into account the apparatus function of the spectrograph in dealing with the line profiles?

Hunger: No, we don't have to do it when we deal with the broad helium lines. It has no effect.

Peat: How do your results compare with those for the other helium-rich stars?

Hunger: The only data we can compare with so far are from Klemola and Hill, and those are based on a coarse analysis. Coarse analysis might be all right for a standard star, but it is no good for a helium star. There are large discrepancies between our results and their results.

Matsushima: Is there any evidence of anomaly of the abundance of elements other than helium and carbon in this star?

Hunger: Yes, we have completed the analysis and all of the other elements are down by a factor of 5 at least.

Matsushima: No, I mean did you compute the profiles for the other lines?

Hunger: Yes.

Matsushima: Do you have other general comments as to the effect of other opacities on the continuum?

Hunger: The other elements are way down, so there are no other sources of opacity. In our analysis compared with the previous analysis, helium is of course up and hydrogen is way down. By number we have 75% helium and 25% carbon. By mass it is 50-50. Nitrogen is still up, but we have no quantum-defect calculations for N II. We looked at the absorption edges and feel that they have no influence. Oxygen is down — nobody really knows why. Magnesium is normal. Aluminum seems to be up. Silicon is normal. Phosphorous seems to be up. Argon is normal. Calcium is normal, and iron, based on one line, seems to be up.

Stibbs: Would you care to comment on the fact that the computed profiles are all deeper than the observations, and also on the effect you find outside the Doppler core?

Hunger: We have some rotation in these stars. We estimate v sin i = 40 km/sec. That takes care of it besides the non-LTE effect.

Stibbs: But we have the outstanding feature in the comparison between theory and observation that the line cores theoretically are deeper than the observed cores.

Hunger: Yes, because the observed cores are washed out by rotation; it flattens them out.

Stibbs: So why not put rotation into your theoretical curves?

Hunger: We could have done so.

Stibbs: Then of course you will have changed the wings.

Hunger: Not appreciably, not on the helium lines. You would only change the very faint lines.

HYDROGEN ABUNDANCE IN WHITE DWARFS

K. Nariai[*] and D. A. Klinglesmith

Goddard Space Flight Center, National Aeronautics and Space
Administration, Greenbelt, Maryland

Kodaira (1967) recently obtained a spectrogram of Sirius B, one of
the two white dwarfs with known mass, at the coudé focus of the 188-cm
reflector at Okayama. The other star of known mass is 40 Eri B. Masses
and the bolometric magnitudes for these two white dwarfs are given in
Table 1.

Table 1. Mass and bolometric magnitude for Sirius B and 40 Eri B

	Mass	M_{vis}	B.C.	M_{bol}
Sirius B	1.05 ± 0.04	+11.1	1.5	+9.6
40 Eri B	0.43 ± 0.04	+11.0	1.5	+9.5

Terashita and Matsushima (1966) calculated colors and Balmer-line
profiles of DA-type white dwarfs for various values of temperature and
gravity with the use of model atmospheres with the normal abundances. The
temperatures and gravities for Sirius B and 40 Eri B according to Terashita-
Matsushima's scale are shown in Table 2 (Kodaira, 1967).

Table 2. Log g and T_{eff} obtained with the model atmospheres

		Weidemann	Terashita-Matsushima
Sirius B	log g	8.0	7.2 ± 0.2
	T_{eff}	14800	17000 ± 1000
40 Eri B	log g	7.5	7.0 ± 0.1
	T_{eff}	12900	16000 ± 500

The radius of a star can be calculated in two ways:
with the effective temperature T_{eff} and the bolometric magnitude M_{bol} as

$$\log (R/R_{\odot}) = -0.2 (M_{bol} - M_{bol\odot}) - 2 \log (T_{eff}/T_{eff\odot}) \quad ; \qquad (1)$$

or with the mass M and the gravity g as

$$\log (R/R_{\odot}) = 0.5 \log (M/M_{\odot}) - 0.5 \log (g/g_{\odot}) \quad . \qquad (2)$$

Table 3 gives the radii for Sirius B and 40 Eri B calculated by Kodaira (1967).
Kodaira (1967) has made clear that the Terashita-Matsushima models
give inconsistent values of the radii for Sirius B and 40 Eri B; the radii calculated with the mass and the gravity are larger than those calculated with the
effective temperature and the bolometric magnitude. Terashita and Matsushima
recently revised their models, taking the effect of Lyman and Balmer lines
into account. But the inconsistency has not been removed completely.

---

[*]Holder of an NRC-NASA Resident Research Associateship.

Table 3.   Radius obtained with the Terashita-Matsushima models

	log $(R/R_{\odot})$	
	from T_{eff}	from g
Sirius B	-1.96 ± 0.4	-1.38 ± 0.75
40 Eri B	-1.89 ± 0.2	-1.45 ± 0.2

This inconsistency, however, can be removed if we assume that the abundances of DA-type white dwarfs are not normal. We calculated the change of the equivalent width of $H\gamma$ with hydrogen abundance for T_{eff} = 16,000°K and log g = 8.0 (Figure 1).

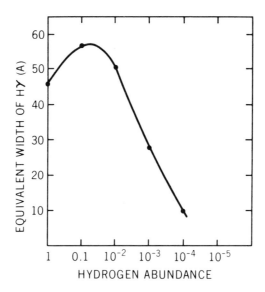

Figure 1.   Equivalent width of $H\gamma$ in white-dwarf spectra (log g = 8, T_{eff} = 16,000°K).

As the hydrogen abundance decreases, the equivalent width of $H\gamma$ becomes larger until the continuous absorption due to hydrogen becomes smaller than that due to helium, and then the equivalent width decreases. Therefore, when the deficiency in hydrogen is not large, it shows a "quasi-gravity effect." This effect was first studied analytically by Strömgren (1944). Osawa (1959) and Böhm-Vitense (1967) calculated helium-rich models for main-sequence stars and confirmed this effect. If the temperature is not high enough, the helium lines are not observable, and hence it is hard to separate the effect of hydrogen deficiency from the effect of gravity. The results are shown in Figure 1, and the models are illustrated in Figure 2. The results for very small hydrogen abundances may vary if we take into account the opacity sources, other than H, H^-, He, He^+, and σ, such as C, N, O, and metals. The low hydrogen-abundance models become convectively unstable, but we made our calculation in radiative equilibrium.

In the calculation of the $H\gamma$ profile, the Stark broadening function, which includes the correlation of shielding (Edmonds, Schlüter, and Wells, 1967), was used. Greenstein (1958) has shown that the equivalent width of $H\gamma$ for DA-type white dwarfs is a function of U-V color. The dispersion of the equivalent width, which amounts to 50%, has hitherto been interpreted to be the dispersion in

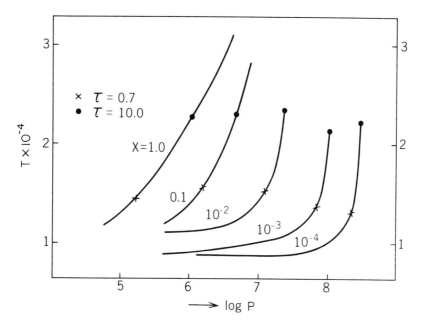

Figure 2. P-T diagram for white-dwarf models (X = hydrogen abundance;
log g = 8, T_{eff} = 16,000°K).

the gravity. But we can also interpret the dispersion to be due to the difference in the hydrogen abundance. Therefore, if the available data are UBV
colors and the equivalent width of Hγ, we cannot determine whether DA-type
white dwarfs have the same composition and different values of gravity, or
whether they have different compositions and the same value of gravity. But
Weidemann (1968) showed that the former case might be appropriate because
the deviations from the standard line on the UBV diagram show a correlation
with the radii (and thus gravity) derived from white dwarfs with known
parallaxes, which indicates that white dwarfs have different radii and, consequently, different gravities.

The authors wish to express their thanks to Dr. V. Weidemann for his
discussions. One of the authors (K. N.) wishes to thank Dr. K. Kodaira for
discussions and his kindness in showing his results to the author before
publication.

REFERENCES

Böhm-Vitense, E. 1967, Ap. J., 150, 483.
Edmonds, F. N., Jr., Schlüter, H., and Wells, D. C., III. 1967, Mem.
    Roy. Astron. Soc., 71, 271.
Greenstein, J. L. 1958, Handbuch der Phys., 50, 161.
Kodaira, K. 1967, Publ. Astron. Soc. Japan, 19, 172.
Osawa, K. 1959, Publ. Astron. Soc. Japan, 11, 253.
Strömgren, B. 1944, Publ. Obs. Univ. Copenhagen, No. 138, 85 pp.
Terashita, Y., and Matsushima, S. 1966, Ap. J. Suppl., 13, 461.
Weidemann, V. 1968, Ann. Rev. Astron. Ap., 6, 351.

DISCUSSION

Weidemann: The point is you find a correlation between gravity derived from the color diagram and the radius derived from the distance. And if you believe the general theory that radii and gravity are coupled, this indicates that the scatter is really a gravity-scatter graph and not caused by abundance differences.

Another remark I would like to make is that if you would include in these models calculations of helium lines, then they would show up very early. Yours is essentially a pure case that was calculated without opacities other than hydrogen or helium. In fact, for the white dwarfs that show only helium lines these lines are much too weak. You cannot assume a pure hydrogen-helium mixture — you have to include carbon, oxygen, and nitrogen. This of course brings down not only the strength of the helium lines as observed in these stars but also the strength of the hydrogen lines, so that in true white dwarfs corresponding to the one that was presented, the curve would look quite a bit different.

Klinglesmith: This is just the beginning. Give us time and we will put the other opacities in.

Matsushima: May I comment on the work of John Graham from Kitt Peak? He is making color plots based on the Strömgren four-color system. At least those doing work on white dwarfs should look into these observations. The point is that some types are very well separated, whereas for Eggen and Greenstein's they were well mixed. I wonder if Dr. Eggen has any comment about this.

Eggen: In the UBV system the DA's, DB's, and DC's all fall along the black-body curve. If this is correct, the DA's separate out, that's all.

Matsushima: If we take an iso-g curve, then those stars must have tremendous g values.

Weidemann: I explained that in the blanketing conference two years ago. If you go to line-free white-dwarf atmospheres, then you get the hotter DA white dwarfs and main sequence coinciding. If you add the line blanketing, then you go back to the blackbody curve. So that explains why the DA white dwarfs are close to the blackbody curve. There are two effects compensating each other; this is shown very nicely by Graham's observations. In the UBV two-color diagram you have a line-blanketed case. So there is no physical reason that Graham's DA stars show deviations from the blackbody curve.

REMARKS ON THE DETERMINATION OF EFFECTIVE TEMPERATURE AND SURFACE GRAVITY FROM HYDROGEN-LINE PROFILES

S. Matsushima

Department of Astronomy, Pennsylvania State University,
University Park, Pennsylvania

Since the Stark profile depends on both temperature and electron density, it is generally not possible to determine the effective temperature and the surface gravity simultaneously from hydrogen-line profiles alone. The situation is well demonstrated in Figure 1, where the observed profiles of the $H\gamma$ (upper) and $H\delta$ (lower) lines of 40 Eri B are compared with a grid of theoretical profiles. It is apparent that the theoretical profile increases if either T_{eff} or log g increases. As a result, it appears that the models falling approximately along a diagonal line from the upper left to the lower right in the figure predict equally good agreement. Hence, it is necessary to know either T_{eff} or log g in order to determine the other quantity. However, if we have reliable data on both luminosity and mass (or radius) of a star, we can use the observed mass to make simultaneous determinations of T_{eff} and log g from line profiles or equivalent widths only. The following is an example of such a method being applied by the writer and Yoichi Terashita (1969) in the analysis of white-dwarf spectra.

The method should be straightforward, especially for stars of T_{eff} greater than 10,000°K, because the bolometric correction ΔM_b is practically independent of log g and there is a linear relation between ΔM_b and log T_{eff}. Bolometric corrections computed for a number of line-blanketed model atmospheres of white dwarfs show little differences between the models with log g = 7 and 8, and, in effect, they are quite similar to the values for main-sequence stars. Moreover, if we plot ΔM_b against log T_{eff}, we find that for $T_{eff} > 12,000°K$, the following expression gives ΔM_b within an accuracy of 2%:

$$\Delta M_b = 24.060 - 6.078 \log T_{eff} \quad . \tag{1}$$

A similar relationship seems to exist down to $T_{eff} = 10,000°K$ for main-sequence stars. Substituting the above equation into the usual relation between mass, radius, and surface gravity, we have

$$\log \frac{M}{M_\odot} = -0.4 M_v - 1.57 \log T_{eff} + \log g + 2.87 \quad , \tag{2}$$

where M_v denotes the absolute visual magnitude. For 40 Eri B, $\log M/M_\odot = -0.366$ and $M_v = 11.0$ mag, so that

$$\log g = 1.57 \log T_{eff} + 1.16 \quad . \tag{3}$$

Another T_{eff} − log g relationship can be found for models that predict the best-fitting profiles to observations. For this purpose, we determine a grid of theoretical profiles with small divisions of log g and T_{eff} through interpolations from those computed for a certain number of basic models; that is, we take an interval of 0.2 in log g and 500°K in T_{eff}. The grid of profiles is obtained through linear interpolations from nine basic models given by combinations of T_{eff} = 14,000, 15,000, and 16,000°K and log g = 7.0, 7.5, and 8.0. By use of a Cal-Comp plotter, the resulting profiles are plotted

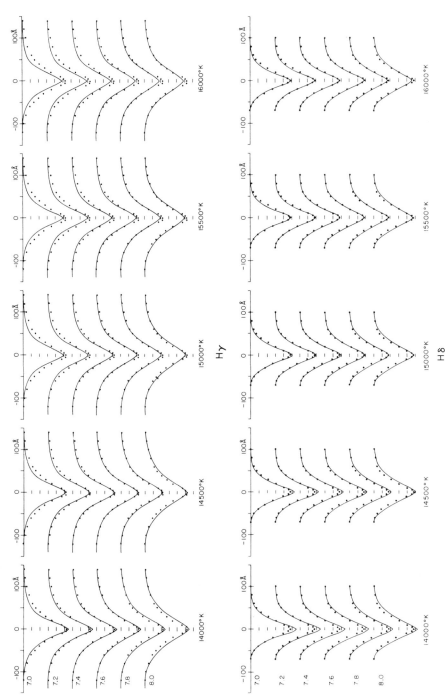

Figure 1.   Comparison between the observed profiles of Hγ (upper) and Hδ (lower) lines and the theoretical predictions computed from a grid of pure-hydrogen model atmospheres. The numbers at the left end give the values of log g for the five profiles along each row. The dots represent the photoelectrically scanned profiles by Oke (1963).

along with the observed profiles in Figure 1. The best-fitting value of log g is then determined for each temperature. Due to the considerable asymmetry of both the observed and the theoretical profiles, the determination of log g for a given T_{eff} is made for the short- and long-wavelength sides of the line center independently. The asymmetry occurs mainly owing to the difference in overlapping of the wing portion with neighbor lines, which defines a 'continuum' in the case of white-dwarf spectra. The mean of the two log g values thus obtained for each T_{eff} is plotted in Figure 2 for both Hγ and Hδ lines, with the lower and upper values given by vertical bars on each point. We then find that the mean values can be well represented by a straight line in Figure 2. In the same figure, equation (3) is represented by the solid line; thus, the intersections of this line with the two broken lines give the best values of T_{eff} and log g as determined from Hγ and Hδ profiles. Or we can express the broken line for Hγ by

$$\log g = 7.434 \log T_{eff} - 23.250 \quad , \tag{4}$$

and for Hδ,

$$\log g = 11.834 \log T_{eff} - 42.028 \quad . \tag{5}$$

By solving equations (3) and (4) simultaneously, we find for Hγ, $T_{eff} = 14,550°K$ and log g = 7.696, and similarly for Hδ, $T_{eff} = 16,100°K$ and log g = 7.767. The large difference in T_{eff} found between the two determinations is a direct result of the nearly parallel nature of the two broken lines, whereas the small variation of log g is due to the slow increase of the solid line with respect to log T_{eff}. Ideally, the three lines should intersect at one point, defining one pair of values for T_{eff} and log g. The discrepancy may be due partly to the inaccuracy in theoretical values, especially for Hδ profiles for which it is often necessary to extrapolate available data of the Stark-broadening function for higher ion density. It should be noted that the hydrogen-line profiles depend considerably on the hydrogen content assumed for the basic models. The models used in the present investigation are for 100% hydrogen atmospheres.

The preceding method may also be applied to comparisons of the equivalent widths, yielding similar values of T_{eff} and log g, as expected. In fact, in this case the procedure is much simpler since for $T_{eff} > 12,000°K$, W(Hγ) or W(Hδ) is nearly linear with respect to log g for a given T_{eff}. Further details of this investigation will be published elsewhere.

This work was supported in part by the National Science Foundation under Grant NSF-GP8058.

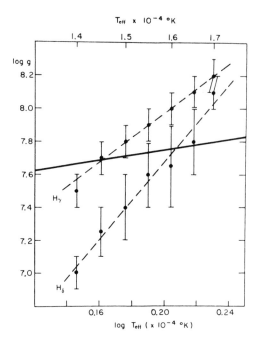

Figure 2.   Values of T_{eff} and log g that give the best fit to the
observed line profiles of Hγ (upper broken line) and Hδ
(lower broken line) for 40 Eri B.   The heavy solid line
represents the combination of T_{eff} and log g that
gives the observed mass of the same star.

INFRARED STELLAR SPECTROSCOPY

P. Connes

Centre National de la Recherche Scientifique (CNRS), Meudon-Bellevue, France

Presented by L. D. Kaplan

Up to now there has been a considerable reduction in the resolving power from available spectra of astronomical sources at wavelengths greater than 1.1 μ. The basic reason for this has not been any decrease in energy but a lack of efficient recording techniques. Below 1.1 μ, photographic plates and image tubes are practical; they make full use of the available light over the entire spectral range of interest. Above 1.1 μ, the only way of recording a spectrum has been spectral scanning with a monochromator, the output of which is fed to a single-channel receiver.

Multiplex spectroscopy, a technique in which the output of a two-beam interferometer is recorded and then transformed into the spectrum by Fourier inversion, solves the problem completely. Early difficulties that prevented either attaining high resolving powers or working on astronomical sources through atmospheric turbulence are now solved.

In the laboratory, emission and absorption spectra have been recorded with resolving powers and accuracies of wavenumber calibration exceeding those of the largest grating spectrometers (5×10^{-3} cm^{-1} resolution and 10^{-4} cm^{-1} accuracy). Most of the astronomical efforts have been devoted so far to producing near-infrared (1 to 2.5 μ) planetary spectra with the Haute Provence Observatory telescopes (193 to 150 cm). The wavenumber resolution has been 0.08 cm^{-1} on Mars and Venus and 0.3 cm^{-1} on Jupiter. The gain in resolving power is 100 compared with that of scanning spectrometers. Most of the planetary atmospheres lines can be easily separated from telluric and solar absorptions. The main results, so far, have been the detection of CO, HCl, and HF on Venus, and CO on Mars. A great deal of information can be extracted from the CO_2 data on Venus, which compare favorably with the best laboratory spectra of CO_2 and are currently used to determine molecular constants.

Because of limited telescope time only a small effort has been made to produce stellar spectra; even so, about a dozen relatively bright red stars (from α Orionis to Y Canum Venaticorum and O Ceti) have been recorded with resolutions between 0.1 and 0.5 cm^{-1}. Figure 1 shows a small fraction of a spectrum (including a CO band) from α Orionis. The resolution is about twice as good as that of the Michigan Solar Atlas. Figure 2 shows the resolving power that can be expected with a given S/N ratio, and the minimum detectable equivalent line width for stars of various classes and visual magnitudes with a 150-cm telescope. Some improvement can still be expected from better detectors; so far, however, high resolving power is clearly limited to a moderate number of infrared bright objects. Clearly, telescope time and size are now the main obstacles to acquiring data.

A new situation arises since an interferometer can tolerate much greater image spread than a slit spectrometer and telescope accuracy requirements are much relaxed. This opens the possibility of building very large instruments for a moderate cost. Spectroscopic resolving power will be proportional to the square of telescope diameter, and figures of the order of 15 to 25 m appear attainable. As a preliminary demonstration, the construction of a 4.2-m light collector has been undertaken at Bellevue (Figure 3) with extremely modest means. It is built basically like a radio telescope, on the surface of which are added small optical mirrors individually controlled.

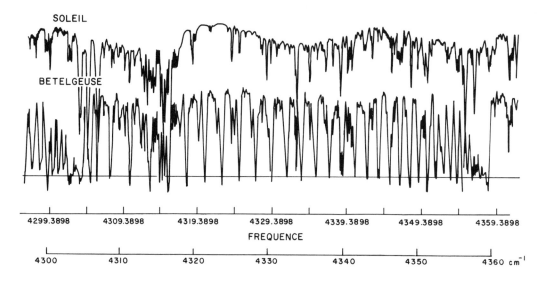

Figure 1.   The spectra of the sun and Betelgeuse in the region of 2.3 μ.

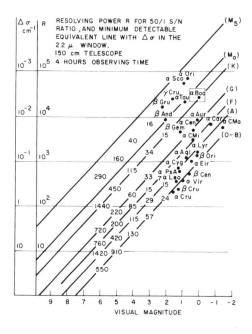

Figure 2.   Resolving power for 50/1 S/N ratio and the minimum detectable
equivalent-width line with a 60-inch telescope.

Extrapolation to a much larger instrument will be justified only if there
is enough interest in high-resolution infrared spectra. Groups already
engaged in analyzing the existing stellar data are those of R. Bouigue
(University of Toulouse), H. Spinrad (Berkeley University), and F. Edmond
(University of Texas).

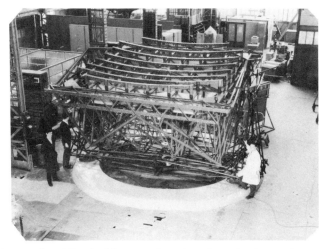

Figure 3.   The half-completed 4.2-m light collector.

DISCUSSION

Buscombe: For years Swings and others have been telling us about the possible control that carbon monoxide would have over the separation of M stars from carbon stars. It whets my curiosity to know what you found in Y Canum Venaticorum.

Kaplan: The one I showed is the only stellar spectrum that I have looked at. I have not looked at the others.

Buscombe: The chances are that you would not see these lines resolved in Y Canum Venaticorum, for example.

Elste: Have you got an estimation of the carbon-isotope ratio?

Kaplan: Yes, I tried to get a number. It was around 4 or 5.

Auman: I am not very familiar with the units you are using. Have you by chance calculated the number of CO molecules above the photosphere?

Kaplan: No. Let me tell you what I have done. First of all, I have assumed there is no self-emission, but I have calculated the number of molecules as if they were purely Doppler broadened because this assumption of no emission gives a lower limit and the Doppler broadening gives too low a value. The assumption of Doppler rather than collisional broadening gives an error in the other direction. I don't know how far away we are from the center of the star, so I don't know what the pressures are. One thing I wanted to show you is that if you have a good background, it looks as if you're getting close to the continuum. Thus, it looks as if the pressure is not very high. They do look like Doppler-broadened lines with some turbulent spread. We have actually looked at the line shapes.

Auman: I have calculated the theoretical spectrum of the vibration-rotation bands of the H_2O molecule at the temperatures that are found in the atmosphere of late-type stars. I find that for $T < 3000°K$ and for solar abundances there are so many faint H_2O lines that for $\lambda > 1\,\mu$ they completely overlap, even between the bands, so that there are no frequencies where you have only the continuous opacity. Therefore, I would expect that the continuum that you see may in fact be composed of the radiation of a great many faint H_2O lines.

Kaplan: These are high J lines from the 1.9-μ band of water vapor?

Auman: The 1.9- and 2.7-μ bands will overlap in this region. One other point I should like to make is that on the basis of calculations of M-type giant and supergiant atmospheres (at least as far as water is concerned, but I would expect the same thing would be true for CO), transitions between vibrational states will be determined by radiative rather than by collisional processes. This would mean that you do not have LTE in the giants and supergiants and that it may not be safe to use a "vibrational temperature" to find the relative strengths of separate vibrational bands. On the other hand, transitions between rotational states that are in the same vibrational state will probably be dominated by collisions, so that "rotational temperatures" will have some meaning.

Kaplan: The maximum J is somewhere around 20, which again suggests a low rotational temperature. The intensities change very little and it is hard to pick it out. I'm used to planetary atmospheres, where you don't have to worry about this. In the first slide there is very little sign of water vapor. There is nothing that would seem to mess this up a great deal in this region. In the CO regions in particular there are practically no features that I can't identify. So I think there is only very little water-vapor interference.

Auman: Well, this spectrum can certainly be used to determine an abundance for water. You will know where some of the stronger water-vapor lines are, and you can look for them.

Kaplan: I haven't looked for them, but Spinrad is looking for them in this very spectrum.

Auman: In making your analysis I think you should at least allow for the possibility that the star has temperature fluctuations across the surface due to convection, so that you may have to use distribution of effective temperatures rather than a single effective temperature. You would expect temperature inhomogeneity due to convection.

Kaplan: One thing I should say is that I am not really competent to do this analysis since I am not a stellar astrophysicist. This sort of spectrum is something that should be used by a stellar astrophysicist who does know all the processes involved, when they have LTE or when they don't have LTE.

Anonymous: They don't know [general laughter].

Auman: You are claiming more knowledge for me than I am claiming for myself.

Gingerich: Let me show a slide related to a model atmosphere of roughly that temperature. It is computed for a dwarf star, so it isn't completely relevant for Betelgeuse. We can see the strength with which the water-vapor bands enter the opacity. We can see that in that region of roughly 2.3 μ there is an immense amount of water-vapor absorption. The next slide will show us that on the assumption that we have GMA abundances we get these very large chunks taken out of the predicted continuum because of the water-vapor bands. The thinnest line here shows what happens if the metals — and by the metals we mean carbon, nitrogen, and oxygen — are reduced to 1/100. The bands come out much smaller. You would have to make the carbon abundance proportionately much higher than the oxygen in order to get such a preponderance of carbon monoxide over the water vapor. Otherwise, with the GMA abundance, one finds that the carbon monoxide and water vapor run quite close to each other as far as the abundance is concerned. These are calculations made by myself, Duane Carbon, and David Latham.

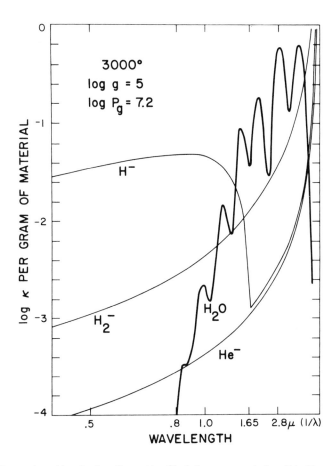

Opacities at optical-depth unity (1μ) for a model with T_{eff} = 2500°.

Kaplan: I know that if there were water vapor in amounts comparable to the carbon monoxide, they would show up as very common features in this part of the spectrum even on the basis of the low J water-vapor lines that I happen to have noticed.

Gingerich: That's what makes the results you have reported so extremely interesting.

Auman: Vardya and Spinrad, when they assumed 3000° effective temperature, got an O/C ratio of 1.05, which, because CO always forms until you run out of oxygen or carbon, means that there was very little oxygen left over to form water. If the effective temperature of Betelguese is really 3000° and if oxygen is low compared with the solar values, this could explain why there is no water. If the actual effective temperature of Betelguese were as high as 3600° or in that range, then you would get rid of the water because of the higher temperatures, while still retaining the CO because of its higher dissociation energy.

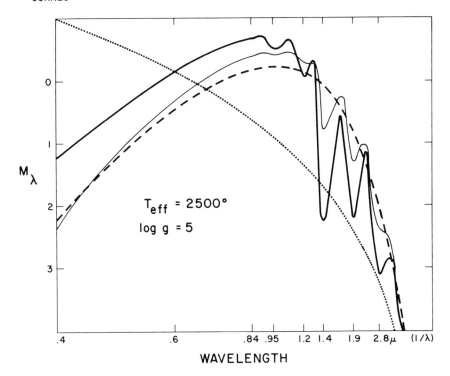

Emergent fluxes (ergs/cm^3 sec ster) in magnitudes for T_{eff} = 2500°, log g = 5: dashed line — model without H_2O; heavy solid line — model H_2O and normal abundances; light solid line — model with H_2O and 0.01 normal metal abundances; dotted line — constant energy curve proportional to 2.5 log λ^2.

Cayrel: Auman, what was the result of the Stratoscope experiment?

Auman: They saw water.

Cayrel: And now that we have high-resolution spectra, we don't see any water?

Kaplan: At high temperatures there should be strong water-vapor lines here. I agree on that.

Vardya: I would just say here that our result of the C/O ratio was 1.05. It is really very tentative because I think one should also have the other critical ratios. Another thing with respect to the slide that Gingerich just showed is that the water vapor may be so abundant there because of the very high pressures that they used.

Popper: I should like to ask a question about the first paper. I hear stories or rumors about the almost magical abilities of the Connes to operate this equipment, and I wonder if anyone can tell me whether the equipment is now to be considered more routine so that anyone can use it?

Fehrenbach: Oh, yes, now it is completely routine, but of course it works better for Connes [laughter].

Kaplan: There is a duplicate instrument being built at JPL by Dr. Reinhard Beer. He expected it to be working a few months ago, although it is not yet completely working to get this sort of resolution. Some people say that

Connes must be at the controls in order to do it, but I think that is not quite true. Connes himself is very optimistic about it being usable. But I think the person who can best answer this sort of question is in the audience — Larry Mertz. He knows what the state of the art is.

Mertz: I have never tried to use his instrument. Although I can't say how difficult it would be to use, I can't see that it would be overwhelmingly difficult. It seems one could go ahead, struggle away, and get some results.

Fehrenbach: One of the problems is the connection between the instrument itself and the Observatory. It takes a rather large computer. That is one of the difficulties — if we had a computer at Haute Provence, it would be easier.

Kaplan: But Dr. Popper is asking about the instrument itself.

Cayrel: May I answer on this point to Dr. Popper. In fact, the mechanical quality of the instrument has been extremely high, and I know that Dr. Connes controls the position of the mirror of the interferometer with an accuracy of 1 A. It is a stepped-progress mirror.

GENERAL DISCUSSION ON PART III

<u>Cayrel</u>: Are there any other questions relating to Dr. Kaplan's paper? If not, we still have time for a general discussion, and I call for questions from the audience.

<u>Buscombe</u>: We've heard from most of the narrow- and intermediate-band systems today, but I thought that it might interest the group to report that Walraven predicted the identification of a number of early-type southern supergiants from his five-band system in South Africa, and I was able to confirm a fair number of these objects as supergiants from slit spectrograms.
      Now may I ask a question on a completely different subject? What happens to the Teller-Inglis formula after the revision to Stark broadening by Griem?

<u>Cayrel</u>: Mihalas computed the number of the last Balmer lines with the Griem theory on a number of models.

<u>Strom</u>: But at that time there was considerable uncertainty in the broadening theory for the higher quantum levels. I think it acts to decrease the opacity and therefore to decrease the last one.

<u>Cayrel</u>: Yes, but it's interesting to see that various papers — the number one by Elsa van Dien [<u>Ap. J.</u>, <u>109</u>, 452, 1949]—give a very high discrepancy between the computed and predicted last Balmer lines, but I don't think a difference in computation changes it by more than one line [showing slide of Figure 9, <u>Ap. J. Suppl.</u>, <u>13</u>, 1, 1966]. On the ordinate you have the number of the last visible Balmer line. The curves are what you get from the observations. The dots are what you get from the computed models. The solid dots are for log g = 4, and the open circles are for log g = 3. For the main sequence we don't have a strong agreement.

<u>Nariai</u>: I wish to make a comment for those who are studying non-LTE problems. As I understand, those who are studying non-LTE are studying the non-LTE problem in A-type or B-type stars. But even in normal stars there are indications that a temperature reversal occurs owing to acoustic energy dissipation. Therefore, rather than study non-LTE with radiative equilibrium, we should study non-radiative equilibrium with LTE for the stars later than F [laughter].

<u>Underhill</u>: How?

<u>Matsushima</u>: I'm afraid nobody understood your last point.

<u>Nariai</u>: Well, the non-LTE with radiative equilibrium becomes important at very thin optical depths. But the dissipation of acoustic energy changes the temperature distribution at the depth if it ever exists. It seems to me that many stellar astronomers do not attack this problem for the reason that only a few things are known about the mechanism in the chromosphere. But, thanks to solar astronomers, much is known about the dissipation mechanism, much more than a stellar astronomer guesses.

<u>Cayrel</u>: If I understand you very well, you say that the gain in accuracy that we have by a non-LTE treatment may be entirely spurious because we don't know the temperature distribution beyond 10^{-2}.

<u>Nariai</u>: Yes, and this is important for stars later than F.

Thomas: I think you are confusing two effects. It was very popular 10 or 15 years ago to say that we only need to study non-LTE effects in those stars with chromospheres; the point missed in such a comment is that there are two kinds of non-LTE effects. First, one comes simply because the star has a boundary; the latest thing I hear is that all stars have boundaries [laughter]; so one has non-LTE effects in all stars. This first kind of effect simply arises because of the scattering term in the source function. Second, there is the effect that comes from adding a chromosphere, and a lot of complications that various people have been studying in order to try to tell the difference in a line that comes from a star with or without a chromosphere. One of the earliest predictions made by the first kind of non-LTE effects was that if I have an absorption line in a star without a chromosphere, I should have a deeper line than predicted by LTE theory. The second kind of non-LTE effect depends upon an outward rise in T_e. The anomaly in the temperature distribution or the uncertainty in the temperature distribution is precisely why we try to do non-LTE calculations, so that we can distinguish between three kinds of lines; those predicted by: 1) LTE (which doesn't exist), 2) non-LTE without a chromosphere, 3) non-LTE with a chromosphere. If there is a difference between the last two, we can use it to say if there is a chromosphere.

Cayrel: Lines that are dominated by photoionizations are so little sensitive to this temperature rise that, practically, it does not matter if this rise does or does not exist. Of course, this is not true for lines that are collisionally dominated. We have made some experiments. Chamaraux, at Meudon, computed the D lines with a temperature rise and without a temperature rise in a solar model atmosphere. He found that with a temperature rise he had a central intensity that was 4.8%, and without the temperature rise, 2.4%.
  Are there other questions?

Strom: I should like to add very quickly a fourth non-LTE problem, which is to find out the gross effects on continuum and line strengths used in T_{eff} and g determinations. In other words, is there enough change in the temperature structure where the continuum or the wings of strong lines are formed to affect the temperature and gravity determinations? I think this problem has already been considered for a pretty wide range of spectral types from solar up through the B stars.

Cayrel: We have to distinguish between the different things. I should perhaps state that with respect to the continuum the work heard so far is such that, on the main sequence, non-LTE credits LTE. I can probably say that, when the continuous absorption is due to H^-, for luminosity class IV-V and G-K stars. From what we have heard here it seems that we cannot apply this statement to A and B stars and to high luminosities.

Stewart: Does one not also have to say that it does not apply for the Lyman continuum?

Cayrel: Of course, yes. I was considering ground-based observations and the continuum. I am quite sure that for the wings of some lines most of the time non-LTE effects are not very large, but it is absolutely clear that at the center some lines display important effects. So that's what we can summarize about LTE and non-LTE right now.

Peterson: Although there haven't been a lot of computations in the hydrogen spectrum in cooler stars, the departure coefficients are not held down, so to speak, by the constraint of flux constancy. In hotter stars the hydrogen continua are carrying the flux, and this tends to diminish the non-LTE effects. The initial calculations at cooler temperatures indicate that you can really get surprisingly large departures, say for the second and third levels of hydrogen, when you go to reasonably low densities. Although the Hα wings aren't very large, the populations still can deviate considerably from LTE, and one would expect to see quite significant effects there too. I think that Ted Simon is looking into this.

Strom: The same reasoning that applies to the hydrogen lines also applies to the Balmer discontinuity. As Peterson says, you can get pretty large departures for the second level, and these affect the deduced values of surface gravity, as I pointed out earlier. I would estimate that you have to consider departures from LTE even for main-sequence stars of type late A and F if you were using the Balmer discontinuity as a gravity criterion.

Cayrel: This seems to be somewhat in conflict with the Mihalas computations.

Strom: He has only worked with stars hotter than A0. The point is that the flux is increasing so rapidly as you go in that you get a tremendous depopulation of the second level. I should also point out that when Mihalas and I use the same collision cross sections, we come up with identical results for the main sequence and a few low-gravity objects hotter than A0. Therefore, I would suggest that the computational integrity of both our programs is good.

Peterson: You don't see the effects in the cooler stars so much because H⁻ is the opacity.

Underhill: Yes, it is the difference in having the H⁻ there and the real fact that early-type atmospheres down to the A's have almost a factor of 100 lower density atmospheres than you think of for the sun. This is a considerable factor.

Cayrel: It is important to know when LTE is justified because tomorrow we can express a preference for computing a grid on a large scale with LTE models in view of the quantity of computing time required for non-LTE.

Strom: No, it doesn't actually take that much more computing time.

Matsushima: From those who are computing the non-LTE we would like to know how the departure coefficients go over the entire spectral range.

Strom: For the F and G stars, b_2 is less than unity, b_3 is greater than unity. Quantitatively, this results in a change of a factor of 2 in the estimated surface gravity. The sense is that you would choose (on the basis of non-LTE models) lower gravities by a factor of 2 when judged on the Balmer discontinuity. The effect on the slope of the Paschen continuum is negligible, so the temperatures that you deduce are not affected. For the B stars, again the Balmer discontinuity is decreased. The effect on the choice of T_{eff} for main-sequence stars is a few hundred degrees at the most. If you go to lower log g values (or higher luminosities) the effects can become as large as one or two thousand degrees, and the effects become noticeable on the Balmer lines. We have done calculations covering the T_{eff} range from roughly 20,000° to 5000°K for a variety of luminosities. The basic uncertainty lies in the assumed collision cross-section values.

Cayrel: There is another effect that has not been much discussed — that is, the blanketing effect. And of course the state of the theory is such that it is now possible to compute models and radiative equilibrium taking into account the blanketing from the lines. Strom has a very elaborate program to do that. Since LTE is a rather poor approximation in lines, it is a crucial matter to include the non-LTE effects in the line calculations to find out what is the real temperature distribution.

Strom: We discussed that problem at Heidelberg. What you would like to do as a limiting case is to use the temperature distributions that you calculate for a LTE non-blanketed model $\tau = 0.1$. For large τ values you just take into account the blocking. Is that correct?

Cayrel: What I said was that the backwarming effect is about the same for LTE or non-LTE, but that the temperature drop that you obtain at the surface when you compute in LTE is probably spurious.

Strom: There are two cases, one where you have pure LTE and another where you have the LTE unblanketed. Would you agree that those are the limiting cases?

Cayrel: Yes.

Auman: I have made a few calculations of the relative frequencies of radiative and collisional transitions of the H_2O molecules in red dwarfs. It would seem on the basis of these calculations that a H_2O molecule in a given vibration-rotation state will be much more likely to leave that state by a collisional than by a radiative transition. Therefore, since the H_2O opacity is the dominant opacity, it should be possible to assume LTE when the structure in these atmospheres is being determined. If the M dwarfs were only brighter so that we could get observations, we would have something.

PART IV

ASTRONOMICAL PROBLEMS INFLUENCING
THE SELECTION OF PARAMETERS FOR
A PROPOSED GRID OF
MODEL STELLAR ATMOSPHERES

B. Strömgren, Chairman

ASTRONOMICAL PROBLEMS INFLUENCING THE SELECTION OF PARAMETERS FOR MODEL ATMOSPHERES

B. Strömgren

The Observatory, Copenhagen, Denmark

In the previous session, the problem of deriving effective temperature and gravity from the observations of spectra by means of model atmospheres was discussed, so we can count on being able to derive T_{eff} and g for a number of stars. In this morning's session we shall discuss the applications to problems that are important in various areas of astrophysics and stellar astronomy. In particular, we shall discuss problems of evolution of our Galaxy.

In a general way, of course, it has been realized for quite some time that if we know the basic parameters of a star, we can then, in principle, follow this star through its series of equilibrium situations, so that as a function of the age for a specified star we derive radius and luminosity, and from these, T_{eff} and g. If we had the atmospheric abundance parameters and the rotational velocity, and the parameter that we keep as an extra parameter for the time being, the microturbulence, then we might, in principle, compute the intensity of the continuous spectrum and the line profiles. From these quantities, through a folding process, we might compute any of the observational indices (provided, as has been emphasized, the observer specifies precisely what he has observed).

There are areas in the Hertzsprung-Russell diagram where we can carry out calculations of this kind to a reasonable precision. For example, on the main sequence, from stars of spectral type, say B2, to the A stars, the evolutionary sequences through the hydrogen-burning stages appear to be well established. It is in this range that the calculation of the continuous spectrum and the strengths of the hydrogen lines appear to be in good shape. Hence, we might say that for this range of the Hertzsprung-Russell diagram it is possible, given the mass, the age, and the abundance parameters, to go all the way and compute indices such as the Hβ index and the c_1, or some other form of the Balmer discontinuity index. This means that in the observational diagrams we could draw the calibration curves and read off mass and age if we could specify the abundance parameters.

For this discussion we assume that we have already derived the T_{eff} and g. The question now is how do we use these quantities to derive the basic parameters of the star.

Since it is not possible in a short talk to cover all of the regions of the Hertzsprung-Russell diagram, I should like to discuss three problems that I think are important and are also good examples of the requirements from the theory of model atmospheres.

1. POPULATION I MAIN-SEQUENCE STARS, SPECTRAL TYPES B, A, AND F

The evolutionary sequences for this range have been computed by a number of people; there is good agreement, and as a consequence we do have good calibration of these parameters.

Just to remind ourselves of the lines of constant mass and the isochrones in such a diagram, Figure 1 illustrates the diagram derived by Kelsall and myself (1966), on the basis of evolutionary sequences in this range. This covers the temperature range from about 20,000°K to 8,000°K. The ordinate is ΔM_{bol}, which is the difference between the bolometric magnitude of the star at a given time and the bolometric magnitude on the zero-age main sequence at the same effective temperature. Lines of constant mass are also shown. At the top of the main sequence we note that the mass for the same effective temperature generally is slightly different from that of the zero-age main sequence, by about 0.1 in the logarithm of the mass. It is, of course, important to take this into account when we translate this diagram into the T_{eff}-g diagram.

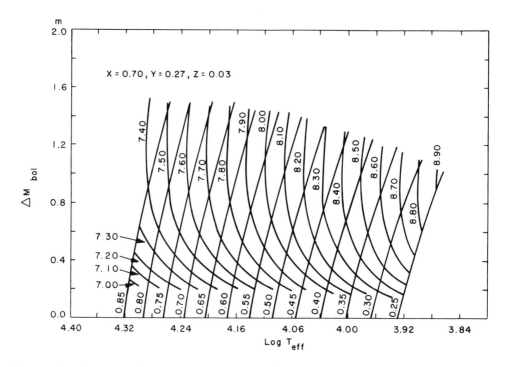

Figure 1. Curves of equal mass and equal age in the log T_{eff}-M_{bol} plane for X = 0.70, Y = 0.27, Z = 0.03. The upper limit of the calibrated region corresponds to X_c = 0.1.

An important property of this diagram is that the isochrones in the upper half (i.e., above the 0.7 mag) are very nearly vertical. Thus any age determination based on the observed magnitude and effective temperature is practically a function only of the effective temperature, if we have determined that the star actually lies in the upper half of the diagram.

Such diagrams have been constructed for a number of combinations of the original composition X, Y, and Z, and in this range it happens that the age is practically independent of X, and for these stars the scatter in the age is estimated to be less than 10%.

Now, if we look at the problem as one of determining age when T_{eff} and g are given, then we can go one step further and determine gravity from our L and R. The result is, as is well known, almost constant gravity along the zero-age main sequence (ZAMS), and on the upper edge of the main sequence the gravity is smaller by about a factor of 3.

This then gives us the background for estimating the kind of accuracy we can hope to get from the accuracy attributed to the T_{eff} and g. From the slopes of the curves we find the following ratio in the upper half of the main-sequence band:

$$\frac{\Delta \log \text{Age}}{\Delta \log \text{g}} = 3.5 \quad .$$

Of course, we can compare this way of looking at the problem with Strom's discussion in his opening talk, where he considered the derivation of T_{eff} from, say, a color index measuring the Balmer discontinuity. He also considered these derivatives and arrived at an estimate of the accuracy of T_{eff}. Hence, we may conclude that, if we can be sure the star is in the upper half of the main-sequence band, we can derive ages with an accuracy of 3.5% if the T_{eff} is accurate to 1%.

Let us recall what we have been hearing these past few days about the accidental and systematic errors of the temperature scale. The calibration curves at present available agree quite well, so I gather that we are now fairly confident that the temperatures can be accurate by a couple of percent in this range of spectral type. This means that the corresponding age errors are less than 10%. Then, if we look at the accidental errors, with the aid of the derivatives discussed by Strom, we get a change in the age with color index. As an example, among the B-type stars, if we measure better than a hundredth of a magnitude in the color index, the corresponding accidental error is no more than 2% in the age. In spite of the good agreement among various temperature scales in this range, it is, of course, well to keep in mind that there may be systematic errors, and here is where we can see if the state of the art in the determination of effective temperatures is good enough to determine T_{eff}, g, and age.

To summarize what we have been hearing: First, in this range with log g between 4.0 and 3.5, it has been established through the work of Kalkofen, Strom, Mihalas, and others that the deviations from LTE, as far as we are interested in them for the calculation of the continuous spectrum, are quite small. At the top of the main sequence we are getting into the region where they become important, but for the main sequence they are not yet important. According to Morton's investigations, we must pay attention to the blanketing effect in the far ultraviolet, and in this connection it is of course necessary to consider the accuracy with which we can compute these blanketing corrections.

Next, we are interested in the run of the blanketing correction, ΔF, for the optical depths that control the continuous spectrum. The peculiar difficulties that are encountered when we compare synthetic spectra with observed spectra in this spectral-type range, as discussed by Anne Underhill and others, should not concern us too much, because we want the blanketing corrections at great depths. However, in this case we have to assume abundances, and the question is can we indeed trust these derived abundances to be accurate enough to give the blanketing effect. The damping and the microturbulence that we have to use in these calculations also come into question, but since the corrections are not very large, I gather that the accuracy of the blanketing corrections is quite good.

Another particular problem is the influence of silicon. We have heard that, for a star like Sirius, if we increase the silicon content by a factor of 10, obtaining a type of silicon star, then the transport of energy is affected to the extent that the Balmer discontinuity is changed. The effect is quite noticeable, but, given the silicon abundance, it can be corrected for. In the case of Sirius, where this effect occurs in a mild form, I think we can say that if we neglected this effect the errors would be less than 10%.

I think these are the main considerations regarding the requirements for the determination of T_{eff} in this range of spectral type. Let us now look at g. On the basis of g, can we indeed decide that the star is in the upper half of the main-sequence band where the isochrones are essentially vertical? There are at least three questions that we must discuss.

First, if we assume different initial conditions, we will arrive at a different ZAMS and, therefore, for a given log g we derive different locations relative to the ZAMS. How big is this effect? On the basis of evolutionary sequences, when we calculate the location of the ZAMS in the T_{eff}-g diagram we find that there is a small effect — very small — due to a change in the hydrogen-helium ratio. Second, the effect of a change of the Z is somewhat larger; however, with the expected range of the variation of the Z for Population I stars, this will not throw us off. That is, the effect is small compared with the width of the main sequence: 0.5 in the log g.

Finally, there is the very important problem that has been referred to a number of times regarding B stars — the influence of rotational velocity on the spectrum. From calculations that have been carried out by Collins and others, we have an expectation regarding the magnitude of the effect. We can also look at this in an empirical way, and there are now available material on observations at Kitt Peak by Abt and Osmer on the subject and a recent unpublished investigation by D. L. Crawford. What we look at is the scatter of the points and we seek a systematic correlation with rotational velocity in the plot of Balmer discontinuity versus Hβ. This, of course, corresponds to a T_{eff}-g diagram. If we do this, and exclude rapid rotations with v sin i > 200 km sec^{-1}, then the scatter in Hβ is small, i.e., 0.012, which is about 1/6th of the width of the main-sequence band, and hence the scatter in log g is about 0.1. Thus we can be fairly confident of the placement of a star in the upper half of the main-sequence band. But, of course, the effect of rotation has to be kept in mind because, if it were, in fact, a couple of times larger, then it would mean that we could not distinguish between evolutionary and rotational effects without knowing something about v sin i. But if we exclude the rapid rotators and stay with v sin i ≤ 200, it seems that we have satisfactory accuracy.

At this point I should like to make a brief comment that goes a little beyond the subject of this morning's discussion. What do we mean when we say that it is fine to determine an age with an accuracy of 10 or 15%? What will we want to use these ages for? In the study of galactic evolution, in particular the studies of star formation, there is, as has become clear in the last several years, an important field of application of ages in this range. We can combine the space velocities and galactic orbits with these ages and find the positions of formation of these stars. And for these applications, the errors of ages must be small compared with the epicycle period of the Galaxy—about 200 million years. If we are off by one quadrant, then it would be quite serious, so we would like to have ages with an accuracy of 40 million years, and in the range from zero to 200 million years this means an accuracy of about 15%. Hence, for this particular application, I believe that this indicates the requirement for satisfactory determinations of both T_{eff} and g.

I should like now to discuss the question of the atmospheric helium content. I mentioned that the ZAMS is practically the same if we change the initial helium-hydrogen ratio, going from X = 0.6 to 0.7; this makes hardly any difference in the T_{eff}-g diagram. There is, however, an additional effect that can be estimated by the aid of a convenient equation derived by Auer and Mihalas for the range of the late B-type stars. If we add helium to an atmosphere, we add weight and pressure, but very little opacity in this range. Hence, for the late B-type stars we have the result that the addition of helium changes the effective gravity. This change is a good deal larger than the change of the g because of the change of the interior hydrogen-helium ratio. We should therefore emphasize that the relative helium content in the atmosphere is an important parameter. If we can be confident that it is not very different in the atmosphere from the original hydrogen-helium ratio, then we can pursue this program; otherwise it becomes important to increase the accuracy of the helium abundance determination for the B stars.

2. POPULATION I MAIN-SEQUENCE LATE F STARS

As we go to the middle F stars, two things change that are important in this context. The rotational velocities become much smaller, and we have an outer convective zone. We compare this situation with that in the earlier type stars. We find that if we want to determine ages from the location in the T_{eff}-g diagram, using T_{eff}'s and g's derived from the atmospheres, then the situation is less favorable. First, the evolutionary sequences, i.e., L and R as functions of time, are less accurate because of the well-known difficulty in connection with the outer convection zone. Models have been computed with the mixing-length theory by a number of groups, and the uncertainty can be judged by a comparison of models that have been computed by using mixing lengths of one or two times the scale height. This is the range we expect to be of importance, and it would give the order of magnitude of the uncertainty. When we look at the results, we find that up to about F3, or log T_{eff} of 3.82 or 3.83, the situation is not bad. In fact, as an example, if we take a star in the upper half of the main-sequence band, where again the isochrones are nearly vertical (according to computations by Peterson and by Reiz and his group in Copenhagen), we find that it makes a difference whether we use models computed with $\ell/H = 1$ or $\ell/H = 2$; the uncertainty in the age, however, is still no more than 15%. For stars later than F3 it becomes more difficult because the width of the main sequence is no more than 0.35 in log g for the later F's. A further difficulty is that $\Delta \log \text{Age}/\Delta \log T_{eff}$ is now larger for the late F's and becomes something like 7 (compared with the 3.5 for the B stars), and as we approach the age of 10^{10} years, this factor reaches 10. This shows that quite accurate log T_{eff}'s are required.

We might discuss the influence of changes of X and Z on the results. The changes of the hydrogen-helium ratio, X, do not affect the location of the ZAMS, but there is again the atmospheric effect, which is slightly different from the case discussed by Auer and Mihalas, and the correction is somewhat smaller. However, there is a fairly large dependence on Z. The ZAMS curve in the T_{eff}-g plane changes with the assumed Z and, in view of the narrowness of the main sequence, this is an important effect. The magnitude is about 0.1 in log T_{eff} if we go from Z = 0.04 to Z = 0.02. In this connection, I should like to refer to the results shown by Crawford (in the Discussions following paper by Landi Dessy) regarding the distribution of stars in the c_1 - (b-y) diagram. The ordinate corresponds to log g_e measured by c_1. Crawford and I found a couple of years ago that there is a difference in c_1 (or in the u-b, which also measured the Balmer discontinuity) between the Hyades and Coma. We have two Population II clusters of about the same age in which the unevolved sequences differ by values as much as 0.06 mag in u-b. In Crawford's diagram this was shown to be quite general—the lower envelope for the field stars differs from the distribution for the Hyades. It is more like that for Coma. This is for B stars. Hence, if we interpret this in terms of separation in Z, we get, within Population I stars younger than 1 billion years, a reduction in Z of a factor of a little less than 2.

If we had in this range a reliable index of the metal content, we could correct for its effect. But I think it is fairly obvious that while the available indices (for example, the m_1 index) clearly single out the intermediate Population II and, of course, the extreme Population II where the Z is down by a factor of 3 or more, it is difficult to determine Z in the range from the Hyades and down by a factor of 2. The question is whether this analysis can lead to relative work with an accuracy that really allows the T_{eff} to be determined in this difficult range. The situation is illustrated by the comparison of the Hyades and Coma by Crawford and myself. The difference in the ΔM is 0.01 mag as against 0.06 in the u-b effect.

Now, if we compare this with what we find in going from the Hyades to the intermediate Population II, we have a different situation. In the Hyades the effects, say, for a factor of 5 reduction in Z are 0.18 in u-b and 0.06 in m_1. It is a very clear effect and the ratio of the two values is 3. I do not think it is surprising that the relation is not a linear one. If we look at what happens when we reduce the metal content from the Hyades value by a factor of 2, we see there are two effects. First, there is the effect that the ZAMS lines change; when we decrease the Z by a factor of two, the log g increases by 0.1. Second, there is the blanketing effect. Also, when we come to interpret these small quantities, it is quite clear that we must worry about the microturbulence.

Now the answer may be given through an increase in the accuracy of the relative determinations of abundance. Another possibility is that we might be able to choose more favorable observational indices, and I would like to refer just briefly to some recent work of Nielsen of the Ole Rømer Observatory in Aarhus. He has carefully reviewed the Utrecht atlas and has chosen one region, a few angstroms wide, that contains eight lines all on the linear, steep part of the curve of growth. In another region, he has nine lines that are all on the flat part of the curve of growth. The hope is that when he determines these indices carefully, he will be able to separate observationally the microturbulence from the abundance. If that were possible, we could then correct for this change in the ZAMS, and it would assist us in determining whether a star is in the upper half of the main-sequence band.

Until we have such an index, and I think Crawford and I agree that the m_1 is not a suitable index for this particular class, we must leave this as an additional scatter. In c_1 the scatter has a total range of something like 0.03 mag or 0.04 mag, and this would correspond to a standard error that is somewhat over 0.01 mag. Compared with widths of 0.10 mag, this is still not too serious when deciding whether a star is in the upper half of the main-sequence band.

The difficulties of age determination as we go later than F3 are the effects I have just mentioned; in addition, there is the calibration difficulty. So, if we increase the accuracy of the effective temperature determination in this range — and I think that what Strom had to say on the use of red and visual intermediate-band measures is quite encouraging in that respect — and if we can correct for the "Z effects," and if we trust the helium content to be fairly constant, which is by no means certain, and if we have a good mixing-length theory, then I think that these determinations may become possible with reasonable accuracy up to the 10 billion years that we are interested in. For the time being, of course, the age determination for clusters follows another line. There, as it is shown by the work of Demarque and his collaborators and that of Iben, we can use the luminosity of the turn-off point in clusters. It has been shown that this luminosity is not sensitive to the effect of the convection zone. When we move to Population II, then again this work shows that the age for the turn-off point is sensitive to X and Z.

3. POPULATION II; THE HYDROGEN-HELIUM PROBLEM

I shall not discuss in detail the problems of Population II stars right now. However, the hydrogen-helium ratio is quite a crucial problem in cosmology. In this connection we can ask whether we can use T_{eff} and g to contribute to the solution of that problem.

The hydrogen-helium ratios have been determined for Population II stars in the following ways: First, there is the method of locating the ZAMS, and this is, of course, sensitive to the X and Y, but the sensitivity is not great. Because of the uncertainties in the parallaxes of the extreme Population II stars, we are not in a very good position to fix the ZAMS. In a globular cluster we can fix it, if we trust the RR Lyr stars, but perhaps not with very great accuracy. And if we add to this the difficulty that the ZAMS depends on the mixing length, even among the earliest stars of the extreme Population II, then we must admit that the results are not very conclusive, although they do indicate that the helium content is probably not far from that of the sun. That is, we probably do not have pure hydrogen stars.

Next, we have already heard from Strom on the revision of the effective temperature scale. This puts the extreme Population II stars further down and indicates a little more clearly, therefore, the high helium content corresponding to the solar value. We also have Dr. Cayrel's results, and I think that a little further improvement might lead to a definite result. In view of the enormous importance of the conclusions that are drawn from this, it is clear that it is important to look for other ways to determine X and Y. If we had just one very good mass and luminosity, then, as is well known, we could determine the X and Y. Since it is found quite generally that an L for a given M is quite insensitive to the choice of the mixing length, this is the perfect method. However, we are not in this position, although a case like μ Cas may yield something, but we cannot today derive X and Y by directly observing mass and luminosity. So the question arises whether the state of the art in the determination of T_{eff} and g has improved to the point that we can get mass, etc. that are useful in this connection. I am not able to judge this, but my impression is that the requirements of the accuracy of the mass for this application are so high that we cannot quite reach it yet, though it might still be possible. In this connection, it is important to realize that when we locate a star in the upper half of the main-sequence band what we are doing is to determine relative values of log g, but here it is the absolute value that is required. Furthermore, when we look at it a little more closely we see that we determine T_{eff} and g from a combination of L and M, so that the uncertainty in L, from the parallaxes, is very important. I wanted to mention this because it might be quite an important method for the future.

The method of determining mass from T_{eff} and g, for example, for the horizontal branch stars is becoming more and more important, and this means of course that good model atmosphere grids that cover these regions are of great importance.

Finally, I should like to mention briefly one point concerning the role of stellar atmosphere calculations for the theory of stellar interiors for the cool stars. I refer to stars in which we have a convective core that extends almost out to the atmosphere. It is clear that, for the calculation of stellar interiors, the starting point is the atmosphere, so good model atmospheres are important in this context.

REFERENCE

Kelsall, T., and Strömgren, B. 1966, in Vistas in Astronomy, Vol. VIII, ed. A. Beer and K. Aa. Strand (Oxford: Pergamon Press), p. 167.

DISCUSSION

Thomas: Let me rephrase Strömgren's comments on how one gets the parameters used in discussing evolutionary problems, in order to make more clearly explicit the assumptions implicit in this procedure. Then let me comment on these assumptions from the standpoint of what some of us have been trying to do over a number of years in the way of making more coherent the spectroscopic diagnostics of stellar atmospheres.

Strömgren's aim is to infer from the observed stellar spectrum those parameters characterizing the structure and evolution of the star. Such inference has four stages: (i) a decision on what these structural-evolutionary parameters are; (ii) a decision on what parameters fix the structure of the atmosphere; (iii) a decision on what parameters determine the spectrum produced by a specified atmosphere; and (iv) a specification of the procedures for going back and forth between these three sets of parameters.

To focus attention on the problems as they exist in even the simplest case, consider a nonrotating star, which has no "intrinsic" magnetic field and whose origin goes back to whatever gas cloud collapsed to form the star. Then Strömgren takes (i) the structural-evolutionary parameters to be mass, age, and composition or, equivalently at any epoch, mass, radius, luminosity, and composition. As the parameters necessary and sufficient (ii) to fix the atmospheric structure and (iii) to predict the spectrum, he takes T_{eff}, g, composition, and microturbulence. (ii) Under the assumptions of LTE, hydrostatic equilibrium, and radiative or convective equilibrium, T_{eff}, g, and composition give atmospheric structure through values of T_e, n_e, ρ at each point. (iii) Under the assumption of LTE for the method of formation of the spectrum, T_e, n_e, and microturbulence at each point suffice to specify source-function, occupation numbers, and line-broadening parameters at each point. Consequently, (iv), the inversion problem of passing from observed spectrum to spectroscopic parameters to atmospheric parameters to structural-evolutionary parameters is, conceptually, <u>almost</u> unambiguous. The only spectroscopic parameter not predictable from the structural-evolutionary parameters is the microturbulence, which is, indeed, introduced empirically simply as a "fix" factor of unknown origin. And historically, one can regard this "turbulence" parameter as an "uneasiness" parameter that introduces doubt as to whether all is really so straightforward in the diagnostic scheme just summarized.

In the actual stellar atmosphere:

Regarding (ii), it is not always true that in the region of line-formation, radiative or convective equilibrium can be applied to predict T_e. We are beginning to recognize that many, rather than exceptional, classes of stars have atmospheric regions where T_e is conditioned by mechanical energy dissipation associated with velocity fields.

Regarding (iii), it is generally not true that a knowledge of T_e, n_e at a point suffices to predict source-function and occupation numbers at that point. Generally, these two quantities depend upon the radiation field at a point; and the radiation field at that point depends upon the distribution through the atmosphere of what are called source-sink terms for the line. The interpretation of a line profile depends first upon the identification of what are the source-sink terms, then a determination of their distribution through the atmosphere. Neglect of this dependence of source-function and occupation numbers upon radiation field can introduce a spurious microturbulence into the analysis of the line, an underestimate of the height in a specified atmosphere where the line is formed, and an underestimate of an empirical value of T_e.

The problem is not simply that an erroneous T_e-model or a misapplied LTE analysis leads to erroneous <u>values</u> of the spectroscopic parameters, thence to erroneous values for atmospheric parameters and structural-evolutionary parameters. The problem is that an <u>a priori</u> decision to force our diagnostics to lie within a framework based on LTE, hydrostatic equilibrium, and radiative or convective equilibrium may prescribe a completely inadequate set of atmospheric parameters and an erroneous set of spectroscopic parameters. The history of the analysis of stellar atmospheres exhibits many such examples: e.g., the state of the solar chromosphere as being described by a recombination spectrum in a low-temperature, supersonic-turbulent atmosphere; the profiles of the Ca^+ H and K lines as being formed in an atmosphere in which T decreased outward through the photosphere, then increased outward, then again decreased outward so that the atmospheric model was a series of hot and cold clouds; the presence of emission lines requiring either the Schuster mechanism, or fluorescence effects, or an extended atmosphere in which it suffices to have T_e decreasing monotonically outward.

Probably what we must do is look at the problem of inferring structural-evolutionary parameters from the atmospheric spectrum in something like the following way. Above the energy-generating region of the star, and below those atmospheric regions where the radiation field becomes inhomogeneous because of the presence of the boundary of the star, we can describe the stellar structure reasonably well by a macroscopic set of LTE parameters. Ideally, the atmospheric spectrum would give us boundary values of these macroscopic LTE parameters; unfortunately, because of the presence of the boundary of the star, the atmosphere acts as a filter — the parameters describing its transmission are not those describing the structure of the interior. Possibly, in time, our sophistication will be great enough to go back and forth unambiguously between these two kinds of parameters; at present, it is not. When we approach the surface, in the atmosphere, two kinds of effects occur. On the one hand, any instability can produce velocity fields whose size, dissipative effects, and possibly momentum transfer can amplify relative to the radiation field because of the small mass in the atmospheric regions. The parameters (ii) describing the state of the atmosphere must describe these effects. At our present stage of sophistication, we can in general not predict these effects from T_{eff}, g, and composition; so until we can, we must enlarge the set of atmospheric parameters. On the other hand, we must prescribe the source-sink parameters for any line for which a transfer equation must be solved to predict its emergent intensity, so the parameters (iii) must include this set. Presumably, in time we will be able to predict these source-sink terms from a given distribution of T_e, n_e, composition — by treating the prediction of the whole line spectrum simultaneously, or at least the whole spectrum of a given atom (because of the radiation field from one line sometimes occurring in the source-sink term of another line). But in the meantime, we must be aware that a direct inversion of the line profile to obtain T_e, n_e, composition is not always obvious, even in those cases where we can invert the line profile to obtain the depth-distribution of the source function.

Until these problems have been solved, I think we must be prepared to admit the presence of a considerable ambiguity of interpretation of spectrum to obtain structural-evolutionary parameters.

Strömgren: I'd like at this point to make a comment on microturbulence as a parameter. I think we would expect from general considerations that this is not a free parameter, but I think we have to qualify this. And if we add additional parameters, such as instability, I think we should also add a magnetic parameter. But we don't know whether this really is an independent parameter.

I would like to ask one question in regard to Thomas' remarks. Would you say that, by pursuing these studies of what goes on in the outer layers of the extended atmosphere, we might also get further insight and direct information on the velocity fields in the deeper layers?

Thomas: Yes, very definitely. It seems to me that if we are given all the initial parameters of the star, including its angular momentum, we must expect to be able to predict all these phenomena in the extreme outer layers.

Strömgren: Yes. The star has, after all, only one course.

Thomas: Right. But for the moment we must do it empirically and try to determine what are the velocity fields.

Strom: I would like to make one point in regard to the helium content. The work of R. F. Christy has shown that the location of the blue end of the instability strip is a sensitive indicator of the helium abundance. The higher the effective temperature, the higher is the helium content. The recent revision of temperature scale on the basis of my atmospheric calculations suggests an increase of the effective temperatures by about 200°. This implies an increase in the helium abundance by about 0.06 by mass fraction. So this line of evidence also indicates a higher helium abundance for the Population II stars than we thought before.

Underhill: So far, the discussion has concentrated on single stars, isolated in space. We do know, however, that the majority of stars are double. This is, indeed, a complication that should be taken into account, although I know perfectly well why we have preferred to ignore it. We have consciously tried to avoid some of the difficulties in order to make some progress and obtain an understanding of the phenomena. But I would like to point out that the question of age determination is very closely related to the problem of close double stars. The point is that with reasonably close double stars one can have a significant mass exchange before the more luminous star has evolved from the main-sequence band. Clearly, if this happens, a star can, in a sense, be rejuvenated after it starts its evolution. If one really requires age determinations of 10 to 15%, for cosmological designs, I wonder whether this may not be a serious factor, considering the very large fraction of stars that are known to be doubles.

Strömgren: In that connection I think that the important question to ask is whether we can recognize these stars spectroscopically.

Underhill: That's the problem; I doubt if we can. There is the problem of stars like Sirius and its companion and also the blue stragglers in clusters; I just don't know how to detect very close binaries spectroscopically when one component is faint.

Böhm-Vitense: It seems to me that if it's only very close binaries that we have to worry about, then the fraction is certainly not 50%.

Underhill: Well, that depends on the mass, but I think that periods up to about 10 days could lead you into trouble here.

Strömgren: I tend to agree with what Mrs. Böhm-Vitense implies. The number of cases where we have this type of mass exchange, and where we don't recognize it, is probably a fairly small fraction of the stars. Otherwise, I quite agree that we are in a difficult position, and I am reminded of the story that I have quoted in another context. An old lady told a friend that she was always afraid when she heard noises, because she thought it meant there were burglars around. And then the friend said, "But burglars don't make noises." Then the lady was always afraid when she didn't hear a noise.

Demarque: I would like to comment on the three main points of Strömgren's talk. First, with regard to the Population I stars — talking from the point of view of stellar interiors, and therefore regarding the atmospheres as a boundary condition rather than something of intrinsic interest. It is well known that for early-type stars we don't need to have very accurate boundary conditions; the zero-surface boundary conditions give us about the right radii. But if one compares the models constructed with zero-surface boundary conditions and models constructed with fairly accurate boundary conditions using model atmospheres, the structure of the subphotospheric region is different even though the radius is the same. In the case of the ionization zone, for example, there might be differences that could be important for the study of variable stars. If the pulsation mechanism is related to the ionization zone, then this question will be of importance.

I should mention also that John Percy at the University of Toronto has extended the Kelsall-Strömgren computations to higher masses, going all the way up to 20 solar masses, with the same composition using the opacities computed by A. N. Cox. He has been interested in the β Cephei stars, and we have been attempting to determine something about the helium content of the young stars, because after all this can tell us something, as can the helium content of the old stars. We thought that we could find reasonable agreement with the value of Y = 0.30. Our agreement depended quite strongly on the slope of the log T_{eff} — (B-V) relation, and this is something that perhaps stellar-atmosphere calculations can provide. This would be a very valuable thing to do.

Regarding the late-type main-sequence stars, I should mention that we have made new evolutionary computations for late-type Population I stars, this time using Cox's opacities rather than the Keller-Meyerott opacities used in earlier computations. With these new opacities, we find that we can reproduce the gap in the color-magnitude diagram of NGC 188 a little better, although one has to use quite a high value of Z. The reason for this is that, when you use the Los Alamos tables, the convective core, which does not exist on the main sequence, suddenly appears as the star moves off the main sequence, thus giving rise to a hydrogen-exhaustion phase.

Another effect is to change the mass-luminosity law slightly and to reduce rather substantially the age estimates for clusters. It now looks as though we should adopt a value of about 6×10^9 years for NGC 188.

Concerning the Population II stars, we find that the whole structure of the envelope, and therefore the radius of the star, depend very critically on the atmosphere and the hydrogen convection zone. (This problem also occurs to a smaller extent among the Population I stars.) You cannot construct an interior model unless you have a good atmosphere and envelope. Clearly this is an implicit problem, because you can't construct a good atmosphere and envelope unless you know the effective temperature and the gravity. Thus, it looks as though one would have to construct a very large grid of atmospheres in order to satisfy all the requirements of the interiors. Therefore, I suggest that the best approach would be for atmosphere workers interested in this problem to get in touch with the stellar-interiors investigators; otherwise, I think things will get out of hand.

The other point is that, if convection is important in Population II stars, it seems to me that the same sort of mixing-length theory should be used in both the interior and the atmosphere calculations.

Underhill: Did you say you have tried to determine the helium content of the Canis Majoris stars?

Demarque: Well, one can do it this way. It seems that the position of the hydrogen-exhaustion phase depends critically on the hydrogen content of the star. The star turns back toward the left when the hydrogen content of the interior has reached a few percent, and clearly this will happen sooner when the hydrogen content is originally low. So the position of the instability strip should be a function of the original hydrogen content. This argument, of course, is based on the assumption that δ Cephei instability occurs during the hydrogen-exhaustion phase.

Strom: How great is the change in the effective temperature?

Demarque: Unfortunately, it is quite small, amounting to about 0.03 log T_{eff}, but one can compare the colors of the main-sequence stars with those at the time of the turn-back and perhaps determine the change of the effective temperature as a function of the hydrogen content. As long as we know the slope of the log T_{eff} − (B-V) relationship, we don't need to know the absolute values.

Cayrel: I would like to support Nissen's use of photoelectric photometry, because to achieve comparable accuracy by high-dispersion photographic spectroscopy would require considerably more time at the telescope and data-reduction time at home. In fact, the big advantage of photometry is that you integrate over an interval of say 10 A and you obtain exactly what you want. To convince my students that integration gives you a greater accuracy than the quantity you are integrating, I give them a simple exercise of evaluating the logarithm of 2 by integrating 1/X using Simpson's rule and 100 ordinates. I ask them to round the numbers to the 5th decimal place and they obtain the result to 6 exact decimal places. And I imagine that any system that involves integration has this advantage.

When you use empirical photometric systems, you always meet the trouble that the different indices you can deal with are not quite independent. They may, in fact, be mathematically independent, but in the presence of errors of observation, and noise in the system, they are no longer physically independent. I think it is quite important to devise photometric systems that try to measure something definite; for example, weak-line absorption for abundances, to measure lines on the flat part of the curve of growth for turbulent velocities. It seems to me also that there should be a type of band that could measure strong iron lines that show damping wings and that are therefore sensitive to the density because they are collisionally important. I think that such a system is what we need for comparing synthetic spectra with real spectra.

Strömgren: I, of course, quite agree with what Cayrel has just expressed so clearly. The question of high-dispersion photographic photometry versus photoelectric photometry in a well-defined narrow band is, of course, just a question of observational technique. Although the progress in the accuracy of photographic photometry has been quite impressive, still, you cannot beat the accuracy of 0.002 mag that you get right away from the comparison of a 3-A band containing, for example, well-defined but weak lines, with a broader wavelength region. In this connection, we might think also of the type of echelle spectrograph that Walraven uses in his radial-velocity work. He has been able to use a template that singles out a large number of well-defined spectroscopic features, and in this way one can add dozens of very weak lines and still obtain the well-defined index.

Buscombe: I think there is bad news for Demarque. Isn't it true that there are some rapidly rotating β Canis Majoris stars?
But seriously, I am very perturbed by the idea, expressed in Strömgren's review just now, that one has to reject the early-type B stars that are rotating more rapidly than 200 km sec^{-1} to make the diagram look neat. That would remove more than half of the stars.

Strömgren: Well, this is the type of question where we can set the limit where we want. I would be against using only sharp-lined stars, but if you are safe when you set the limit at 200 km sec^{-1} and if you get one-half of the stars or even one-third of them, this is all right. For our purposes, that is.

Buscombe: I would be interested to know whether there is some filter system that serves as a trap for fast rotators without going through the tedious procedure of determining line profiles.

Strömgren: I believe that the answer is no. However, I understand that Sinnerstad and Furenlid, in Stockholm, are continuing their earlier work combining equivalent widths and center intensities of hydrogen lines and are now adding a temperature parameter so that the central intensity can measure the rotational velocity.

Sinnerstad: We don't yet know whether the method will work, because we have so very few stars.

Strömgren: I have no doubt that it will work in view of the very large range of central intensities shown in Buscombe's work.

Crawford: Perhaps I could show this diagram concerning the influence of rotational velocity. I have used the published Hβ and UBV photometry in the upper Scorpio-Centaurus region, and Slettebak's recent rotational figures for these stars. The individual points in the diagram are for the stars in this association from B0 to about B9. The root-mean-square scatter in the data around the mean line is about 0.012 mag, as compared with 0.010 mag for the strictly observational errors. The range in v sin i is from about 30 km sec^{-1} to about 350 km sec^{-1}. Collins has looked at this diagram and agrees that the scatter evident here is about what one would expect from the theory. So maybe the only statement we can make now is that there are very few stars rotating rapidly enough so that the effect shows up in a diagram such as the Hβ versus intrinsic color diagram. Therefore, it appears that the percentage of stars that will give trouble in the determination of ages is quite small.

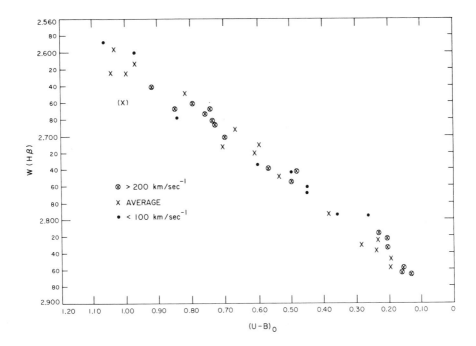

Underhill: We have just heard the advantage of integration in determining spectral indices, but I would strongly caution against the implied assumption that these indices can be linearly independent from each other. We are, after all, still looking for spectroscopic features that are directly related in a one-to-one way with the stellar parameters, and we haven't yet been too successful in that. From this point of view, I would express some reservation about the value of integrated quantities; clearly they are practical, but the question is: What is their physical significance?

Wallerstein: Of course, we have to remember that we are stuck with integrated quantities in any case, and here it is simply a question of making the painful decision as to what the band width should be. If we wish to apply these techniques to very faint stars, for example the stars in the Magellanic clouds, we clearly have to go to very wide bands and admit that we will lose some information.

Underhill: Yes, I know the observations have to be integrated, but I object to integrating our interpretations.

INTERMEDIATE-BAND PHOTOMETRY OF RR LYRAE VARIABLES

D. H. P. Jones

Radcliffe Observatory, Pretoria, South Africa

I am pursuing a program of intermediate-band photometry of the RR Lyr variables and analogous Population I stars. The design of this program springs from three recent investigations.

A. Strömgren and his associates have demonstrated the value of an intermediate-band u v b y system of photometry. The function of the b-y color is to measure color temperature, the v (violet) band to measure line blanketing, and the u band to measure the Balmer jump. Measures of bright stars indicate a significant dispersion in line blanketing at type G0, but only a small dispersion at type F0. This happens because the lines that provide the blanketing are intrinsically weaker at the earlier type.

B. Preston has conducted a survey of the RR Lyr variables in which he was able to classify their metal abundance from the difference in spectral type of their Ca II and H lines. This program could be extended to fainter stars with larger telescopes, but it is doubtful if the precision attained by Preston can be improved on.

C. Oke has observed light curves of Cepheids and RR Lyr variables with a spectral purity of 10 to 50 A, using a spectrum scanner. High-dispersion spectra were used to correct the scans for line blanketing. They were then compared with model atmospheres to determine the run of temperature and pressure through the cycle.

The filters were selected with the u, b, and y bands identical with those of Strömgren. To replace the v band, a k band was substituted centered on the K line, thus increasing the amount of line blanketing and measuring a parameter similar to that of Preston. The filters are:

	wavelength	whole half-width
u	3500 A	240 A
k	3933	60
b	4674	166
y	5490	248

Note that k, unlike v, has no Balmer line within its half-power points.

A ring of 16 standard stars has been set up with the use of these filters, and about 50 constant stars compared with it. The observations correlate with UBV and u v b y in the manner expected. About 50 variables, mostly RR Lyr stars but including dwarf Cepheids and δ Scuti stars, have also been measured. Most have been observed at two phases, but complete light curves exist for a few.

The majority of stars define a close relationship in b-y, k-b, but metal-deficient stars, e.g., HD 140283, HD 122563, show a k-b excess of up to half a magnitude (Figure 1). Mild subdwarfs, e.g., τ Ceti, HD 134439/40, show no excess. The reddening vector runs nearly parallel to the intrinsic relationship; all stars are corrected following Sharov, but the amounts are mostly small. As RR Lyr variables follow their light cycle they describe extremely narrow loops in this plot, again nearly parallel to the intrinsic relationship. The k-b excess of the RR Lyr correlates with Δs; the correlation varies with the phase of the light curve used and the way in which stars of different temperatures are combined. Stars in the globular cluster ω Cen show a significant k-b excess. It is planned to observe more stars in globular clusters in the near future.

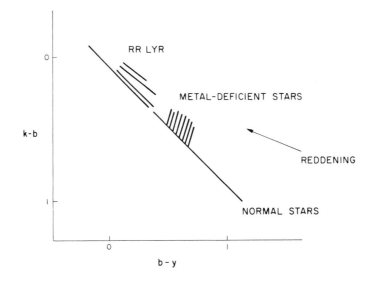

Figure 1.   The relationship b-y versus k-b for various stars.

The acquisition of data is proceeding in a straightforward manner. The difficulties in interpretation are as follows:

A.   What are the blanketing coefficients for this system? Line-blocking coefficients have been published for a number of stars by several authors, but the results are surprisingly discordant, especially for u and k. Approximate values for a star with B-V = 0.50 are 0.30, 0.35, 0.08, and 0.05 for u, b, k, y, respectively. In principle, it is possible to estimate the blanketing empirically for stars of known T_{eff}. For HD 140283 one derives 0.07 mag and 0.70 mag for b-y and k-b, respectively. Assuming reliable line-blocking and backwarming coefficients to be available, we should be able to define two functions,

$$u-b - A(k-b) ,$$
$$b-y - B(k-b) ,$$

which are independent of metal abundance, at least to the first order.

B.   How does one map these two parameters on to the P_e, T_{eff} plot? Do model atmospheres of sufficient accuracy exist for the domain of the RR Lyr stars? Is it really possible to fit a series of static models to the moving atmosphere of an RR Lyr variable?

This short note does not warrant a complete bibliography. Strömgren's work has been reviewed by himself (1963), as has that of Oke (1965) and Preston (1964). Sharov's (1964) work has appeared in an English translation.

REFERENCES

Oke, J. B. 1965, Ann. Rev. Astron. Ap., 3, 23.
Preston, G. W. 1964, Ann. Rev. Astron. Ap., 2, 23.
Sharov, A. S. 1964, Sov. Astron. -AJ, 7, 689.
Strömgren, B. 1963, in Basic Astronomical Data, ed. K. Aa. Strand
    (Chicago: University of Chicago Press), p. 131.

ZERO-ORDER ABUNDANCES AS INPUT PARAMETERS FOR THE STUDY OF HYDROGEN-POOR STARS

G. Wallerstein

University of Washington, Seattle, Washington

In the past few years great progress has been made toward the solution of problems of stellar atmospheres and the construction of models. The question arises where should we go from here.

There are a number of problems that have barely been investigated by the modern methods now available. For example, the magnetic stars have not been discussed from the standpoint of model atmospheres. The magnetic pressure, which we can compute from the observed magnetic fields, is comparable with, or greater than, the gas pressure, and the magnetic field can control the motions of ionized material. Hence, we would no longer have isotropic turbulence. Supergiants have been investigated only very little, and in these stars the line profiles suggest that the turbulent pressure may be comparable to the gas pressure. Variable stars have hardly been investigated at all, and in these there will be velocity gradients causing a host of new phenomena, such as asymmetric line profiles. Work on the cool stars has begun, although the complications involve a large number of molecular bands that have to be included as part of the continuous opacity. In fact, on these stars one sees no real continuum at all.

Furthermore, there are stars of peculiar chemical composition, and these are the ones I would like to mention here. I would now like to show three photographs of stellar spectra. The first slide was published some 10 years ago (Burbidge, Burbidge, Fowler, and Hoyle, 1957). It presents a normal A-type supergiant α Cyg, and below it υ Sag. The latter shows very weak hydrogen lines, and evidently it has a very low hydrogen abundance. The low opacity, due to the low hydrogen abundance, leads to a great strengthening of all the other lines in the spectrum. The next slide (Bidelman, 1950) shows υ Sag, the normal F5 supergiant α Per, and HD 30353 in the middle. The latter has been analyzed by Nariai and by three of us at the University of Washington. It is rather similar in some respects to υ Sag, although it is somewhat cooler; below this is the star HD 25878, which is a R Cor Bor star. This star shows enhanced carbon I lines while υ Sag does not. We evidently have several types, all of which show extremely weak hydrogen lines. The next slide (Klemola, 1961) presents Klemola's star BD + 10° 2179, which was analyzed in detail by Hunger and Klinglesmith (these proceedings). This star shows a vast number of carbon lines and, indeed, they found carbon to be extremely overabundant. The Hδ and Hγ Balmer lines are hardly visible.

There are substantial differences among these hydrogen-poor stars, and although they are almost always helium-rich, there appears to be one subgroup that is carbon-rich, and there certainly is a group that is nitrogen-rich. Thus, the abundances of hydrogen, helium, carbon, nitrogen, and oxygen cannot be specified without some preliminary analysis. Unfortunately, a model cannot even be constructed until one knows the zero-order abundances, because once hydrogen is deficient by a factor of 10^3, these other atoms become the primary sources of opacity. Thus, we need the absorption coefficients and the relative abundances of these atoms to within at least a factor of 2 or 3. Interesting departures from LTE are bound to arise in these models because helium, in particular, has a well-known metastable state that can become overpopulated, and similarly there are metastable states in various ions of carbon, nitrogen, and oxygen.

Now, if we write the number of parameters that are necessary for an analysis of hydrogen-poor stars, we find that, besides T_{eff} and g, we have the abundances of hydrogen, helium, carbon, nitrogen, oxygen, an average for the metals, and a turbulent velocity. This looks like 9 parameters, but, since we do know that in stellar atmospheres the sum of all the abundances must be one, we really have 8 parameters. If we wish to compute a net of models covering all the ranges of these parameters, we would take, for example, 20 values of the temperature, 5 values of the gravity, and 5 values for each of the abundance parameters, and we will require 1.5×10^6 models. If the computation requires about 10 min per model, we will need 30 years of continuous operation without allowing any time to fix the computer; the cost, at \$3 per minute, would be about 50 million dollars!

For these reasons, it is necessary to do some sort of very rough preliminary analysis on some particular stars. Hunger has already described how it is necessary to compute 16 models just to analyze one star, even after two preliminary analyses. Table 1 summarizes some recent results of abundance analyses in these hydrogen-poor stars. Klemola's star serves as a particularly good example of the use of successive approximations in the analysis of atmospheres with peculiar compositions. Hill (1965) improved upon the analysis by Klemola, and Hunger and Klinglesmith have used model atmospheres to make a further improvement upon the analysis of Hill. It is clear from the composition obtained by Hunger and Klinglesmith that the opacity of carbon must be taken specifically into account in constructing the model. It is also interesting to note that the abundances of magnesium, silicon, and sulphur now seem to be normal, while Klemola had indicated that magnesium, at least, and other elements not shown in the table are substantially deficient. In the case of HD 160641, we see that helium is high but that carbon, nitrogen, and oxygen are very nearly equal. For R Cor Bor, on the other hand, we find a similarity in some respects to Klemola's star, while HD 30353 is very different. It shows a very high abundance of nitrogen, a very low abundance of carbon, and quite a low abundance of oxygen, so that the primary source of opacity is He^-, photoionization of neutral nitrogen, and about 10% by electron scattering. Similar abundances might apply to υ Sag, except that it is hotter so that the neutral helium rather than the negative helium ion might be important. Finally, the indications are that many of these stars have abundances of elements beyond neon that are quite normal, so at least that parameter can be left at the normal value.

Table 1. Abundance determinations

log N	Normal	R Cor Bor	HD 30353	HD 160641	Klemola (1961)	Hill (1965)	Hunger and Klinglesmith (these proceedings)
H	12	8.5	7.6	-	9.1	8.5	8.4
He	11.2	11.6	11.6	11.6	11.6	11.6	11.2
C	8.7	9.6	6.2	8.7	8.0	9.5	10.7
N	8.0	7.6	9.2	8.8	8.2	8.7	8.9
O	9.0	8.0	7.5	8.9	<6.8	<8.2	<8.2
Ne	8.5	-	8.5	-	-	-	-
Mg	7.6	7.2	-	-	6.3	7.2	8.0
Si	7.6	7.2	7.6	-	-	7.4	7.5
S	7.3	-	6.8	-	-	-	7.4

The table columns Klemola (1961), Hill (1965), and Hunger and Klinglesmith (these proceedings) fall under the heading: BD + 10° 2179

It is clear that zero-order abundances from curves of growth are indeed zero-order and no better, but they are needed to construct first-order model atmospheres. When the derived first-order abundances show sufficient differences from the zero-order abundances in the opacity-producing elements, second-order model atmospheres must be computed, and new abundances must be derived using the new atmospheres.

I present these comments not as new results but rather as an indication of the type of work we will be involved in during the coming years.

REFERENCES

Bidelman, W. P. 1950, Ap. J., 111, 333.

Burbidge, E. M., Burbidge, G. R., Fowler, W. A., and Hoyle, F. 1957, Rev. Mod. Phys., 29, 547.

Hill, P. W. 1965, Mon. Not. Roy. Astron. Soc., 129, 137.

Klemola, A. R. 1961, Ap. J., 134, 130.

DISCUSSION

Buscombe: I am sorry but I don't agree that the element abundances for the heavy elements are normal in the helium-rich stars. There are very strange superabundances of certain of the heavy elements in the helium-rich stars, krypton for example.

Wallerstein: Remember, though, the continuous opacity is very low.

Underhill: But how do you know what is overabundant? Remember, Ga is one of those elements with a very strong autoionization between the first and second spectra. Indeed, this seems to be true of practically every element that at some time or another has been stated to have an abnormal abundance.

Wallerstein: I agree with Anne Underhill. The fact that lines look strong, in a hydrogen-poor star, cannot be taken as evidence of high abundance. The opacity is down by a factor of a thousand, so many lines you would not ordinarily see will appear.

DETERMINATION OF ATMOSPHERIC PARAMETERS
BY INTERMEDIATE-BAND PHOTOMETRY

J. A. Williams

Department of Astronomy
University of Michigan, Ann Arbor, Michigan

Presented by G. H. E. Elste

Conti and Deutsch (1967) have shown that Δm_1, the line-strength indicator in the Strömgren four-color system, is about twice as sensitive to the microturbulent velocity as to the metal/hydrogen ratio for stars of solar type. However, Barry (1967) and McNamara (1967) have presented observational evidence that Δm_1 indicates metal/hydrogen, not microturbulence. This is understandable since Wallerstein (1962) found in a study of approximately 30 G dwarfs that, except for subdwarfs, Fe/H deviated from the solar value by factors as large as 3, while the microturbulent velocity deviated from the solar value by not more than 20%. Thus, if Wallerstein's results are typical of solar-type stars, then even though Δm_1 is more sensitive to microturbulence than to composition, most of the observed variation would be due to composition.

Wallerstein also found some correlation between Fe/H and microturbulent velocity. It may be that this correlation is due to both quantities having some dependence on the age of the star, Fe/H through the gradual enrichment of the interstellar medium and microturbulent velocity through the decay of turbulence, the latter related perhaps to the decay of chromospheric activity. To check this hypothesis and to investigate the relative importance and distribution of Fe/H and microturbulent velocity in a larger sample of stars, observations are planned of the 80 F8 to G2 dwarfs in the Strömgren-Perry Catalog. The photometric system is indicated in Figure 1.

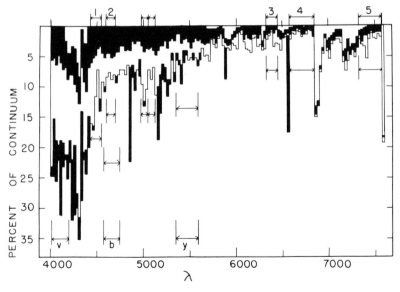

Figure 1. Line absorption in 25-A intervals of the solar spectrum against wavelength. From the top, the first light area is molecular absorption or absorption in the earth's atmosphere. The first dark area is weak-line absorption, the light area is medium-strong-line absorption, and the final dark area is strong-line absorption.

This figure is a plot of the line absorption in 25-A intervals of the solar spectrum against wavelength. The wavelength regions marked 1 and 2 have equal weak-line absorption, about 5%, but the medium-strong-line absorption differs by a factor of 2. One should be able to determine relative metal/hydrogen ratios and microturbulent velocities from measurements of these regions together with a continuum point, region 4, and two regions, 3 (near λ 5800 A, rather than as indicated in the figure) and 5, with well-matched absorption and a large wavelength difference to give a continuum temperature. Very accurate photometry is required since the expected variation in the abundance index is only about ± 0.03 mag, and in the turbulence index is about ± 0.01 mag. A two-channel pulse-counting monochromator will be used to help achieve the accuracy and freedom from systematic errors required for this project. The monochromator and electronics are being constructed at the University of Michigan, and observations should begin this summer.

REFERENCES

Barry, D. C. 1967, *Ap. J.*, *148*, L87.
Conti, P. S., and Deutsch, A. J. 1967, *Ap. J.*, *147*, 368.
McNamara, D. H. 1967, *Ap. J.*, *149*, L133.
Wallerstein, G. 1962, *Ap. J. Suppl.*, *6*, 407.

DISCUSSION

Cayrel: Under what assumption is the photometric variation only 0.01 mag with a variation of microturbulence? Is this on the basis of calculations like those of Wallerstein and of Deutsch and Conti?

Elste: According to J. A. Williams, this estimate is based on the range of microturbulent velocity found by Wallerstein and the expected sensitivity of the photometric indices to this quantity. The expected sensitivity is based on a calculation similar to that used by Conti and Deutsch to discuss the sensitivity of the Strömgren metal index to microturbulent velocity.

Strom: I think that if you believe that blanketing has an effect on the temperature structure, in the sense of depressing the boundary temperature, it is possible to argue [Strom, *Publ. Astron. Soc. Pacific*, *80*, 269, 1968] that the variation of hydrogen/iron abundance ratio should influence the temperature distribution in such a way as to mimic the "observed" variation of microturbulence with abundance. In fact, a blanketing analysis carried out by Judith Cohen and myself on the extreme subdwarfs HD 19445 and HD 140283 resulted in normal values for the turbulence as opposed to the very low value obtained previously.

Also, I might indicate that Fred Chaffee at the University of Arizona has been carrying out a study of the turbulence parameter for stars between A7 and G0. He finds that microturbulence does not seem to be correlated with c_1 or m_1. He finds a maximum value of the turbulence parameter for stars near F5.

Demarque: Actually, the following comment is from Dr. Iben, who is unable to attend this session.

He is interested in interpreting the color-magnitude diagram of globular star clusters, and he has been computing interior models for stars near the turnoff point and on the horizontal branch. His interpretation is based on the fact that the magnitude difference between the turnoff point and the horizontal branch is always approximately 3.5 mag with (B-V) in the range 0.38 to 0.40 mag. He would like to stress the urgency of having models for these two regions, so that it would be possible to transform from the bolometric magnitudes and effective temperatures to apparent magnitudes and the colors. These calculations would be for metal-poor stars such as those found in the old globular cluster M3. I certainly agree wholeheartedly with him on this point.

I also wish to make one comment concerning the helium content of the subdwarfs. There is now some evidence due to Cayrel and the Stroms that the subdwarfs are situated below the Hyades main sequence in the theoretical H-R diagram, in other words, that the term subdwarf is not a misnomer after all. This result is interpreted as evidence for a high helium content ($Y \simeq 0.25$) in subdwarf interiors. I just want to remark that even if the subdwarfs were actually observed superimposed on the Hyades sequence, as was previously held, the Population II ZAMS would have to be placed 0.3 mag below the Population I ZAMS because of evolutionary corrections in the Population II stars, which are very old. And this would still mean a "high" helium content in subdwarf stars.

Matsushima: Is there any indication that the continuous opacity in R Coronae Borealis is less than normal by a factor of 10 or so?

Myerscough: I have been working on R Coronae Borealis, and I think that the source of opacity is probably C^- if the high abundance of carbon is correct.

Vardya: I would like to add that in some of these giants and supergiants, Rayleigh scattering due to helium may also be important, in addition to the band opacities. The contribution of the metal opacities is sensitive to temperature, and at very low temperatures they will not be important.

Myerscough: I did not find Rayleigh scattering to be important except at fairly short wavelengths in my carbon-helium supergiant models. The continuous opacity is lower than that in stars like the sun by a factor of a hundred, but of course it is not so low compared to normal supergiants. The temperatures of my models were around 6000°K.

THE HELIUM ABUNDANCE IN NORMAL B STARS AND SILICON STARS

A. R. Hyland

Mount Wilson and Palomar Observatories,
Carnegie Institute of Washington, and
California Institute of Technology, Pasadena, California

1. INTRODUCTION

Since helium is the second most abundant element, an accurate knowledge of the value of the helium/hydrogen ratio in a large variety of objects is essential for many reasons, which will not be discussed here. Relevant to this session, however, is the choice of a helium/hydrogen ratio to be incorporated into the proposed grid of model atmospheres. This ratio can be determined indirectly from the comparison of observations with the predictions of stellar-interior theory, or directly from the spectra of emission nebulae and early-type stars.

The best recent attempts at a direct evaluation of N_{He}/N_H from stellar spectra were made by Mihalas (1964) for the OB stars, for which he found $N_{He}/N_H = 0.15 \pm 0.05$, and by Scholz (1967), who found $N_{He}/N_H = 0.13$ for τ Sco. In determinations from the spectra of such early-type stars, the temperature distribution in the outer layers of the atmosphere is important and leads to a fairly large uncertainty in the results.

A notable feature of the previous direct determinations of stellar helium abundances has been the neglect of the later B-type stars. These are just the stars where the weaker neutral-helium lines are formed fairly deep in the atmospheres, and are thus more amenable to a simple interpretation. This apparent neglect has stemmed from basic uncertainties in 1) the theories of the Stark broadening of these lines, and 2) the effective-temperature scale. The first of these has been alleviated by the detailed Stark-broadening calculations of Griem, Baranger, Kolb, and Oertel (1962). The second objection follows from the high (21-ev) excitation potential of the lower levels of the helium transitions, and the consequent extreme sensitivity of the helium-line strengths to temperature. The effective-temperature scale presented earlier in this conference (Hyland, these proceedings) appears reliable enough to ensure that errors from this source are within those due to inaccuracies in the equivalent widths.

2. OBSERVATIONS

It is the purpose of this short paper to present some results of a study of late B and silicon-rich stars in two southern galactic clusters IC 2391 and IC 2602. The observations consisted of spectra in the blue wavelength region at 6.7 A/mm, obtained with the coudé spectrograph of the Mount Stromlo 74-inch telescope. Eleven B stars in the clusters proved to have neutral-helium lines whose equivalent widths could be accurately measured, i. e. , to within ± 10%. Three of the weaker He I lines, λ 4713 A, λ 4121 A, and λ 4438 A, were chosen for this study. In general, these met the requirements that they should be measurable over a large range in spectral type (B2-B8), yet not be so strong as to be formed in the outermost layers of the atmospheres. Hardorp (1966) has shown that these lines are relatively insensitive to the temperature distribution near the surface at B3, and the same result almost certainly holds for later-type stars.

The projected rotational velocities of the stars in the two clusters differ widely (Hyland, 1967a). Hence, before I discuss the derived abundances, it is necessary to consider the importance of stellar rotation, since Deeming and Walker (1967a, b) have suggested that rotation plays a major role in determining the neutral-helium line strengths in B stars. Hyland (1967a), however, showed empirically that for stars later than B2, the helium-line strengths depend almost wholly on the effective temperature and little if at all on the rotational velocity. The results of Hardorp and Strittmatter (1968) have since shown that this is just what is expected theoretically. It should therefore be possible to obtain meaningful helium abundances, regardless of the rotational velocities of the stars.

3. RESULTS FOR THE NORMAL STARS

Analysis of the equivalent widths of the measured lines was carried out with the standard assumptions of pure absorption, LTE, and zero microturbulence, with atmospheric parameters chosen from the absolute energy distributions and H_γ profiles of the stars. It is acknowledged that such a technique can only be approximate, but it is hoped that, by the choice of lines and the use of late B stars, errors introduced by the assumptions are minimized. The homogeneity of the results over the temperature range 17,500°K to 14,500°K is thus very satisfying.

In Table 1 the mean values of N_{He}/N_H are given for the two clusters. The numbers quoted are those derived from the equivalent widths of the λ 4713 A line only, since this was the most reliable line to study. A few remarks on the other two lines are in order. The λ 4121 A line lies in the wing of Hδ and, hence, was difficult to measure accurately, especially in stars of high rotation such as most of those in IC2602. In the stars where it could be measured with an accuracy comparable to that of the λ 4713 A line, values of N_{He}/N_H were derived by our neglecting the problem of blending, and were found to be slightly lower than the λ 4713 A values. The λ 4483 A line is considerably weaker than either of the others and could only be measured with sufficient accuracy in four stars. The derived values of N_{He}/N_H were found to be slightly higher than those for the λ 4713 A line. The mean value obtained from all three lines was in each case close to that of the λ 4713 A line alone, and for consistency this latter result was adopted as the final value.

Table 1. Mean values of N_{He}/N_H for two clusters

Cluster	N_{He}/N_H	No. of stars
IC2391	0.098	5
IC2602	0.107	6

One remarkable result is the consistency of the helium/hydrogen ratio not only within each cluster but between the two clusters. Only one of the 11 derived values deviates by more than 0.02 from the mean. The adopted value, $N_{He}/N_H = 0.105 \pm 0.03$, is somewhat lower than the values obtained by Mihalas and Scholz for the early OB stars. In Table 2 the direct stellar determinations are compared with the values obtained from gaseous nebulae and those indirectly obtained from the mass-luminosity relation, the β Cep stars, and the sun. From the table it can be seen that there is a large amount of evidence for a helium/hydrogen ratio of 0.10 for young objects within the Galaxy, although the most recent solar value obtained by Lambert (1967) is 0.06.

Table 2.   Helium-abundance determinations

	N_{He}/N_H	Reference
τ Sco	0.13	Scholz (1967)
OB stars	0.15 ± 0.05	Mihalas (1964)
Cluster B stars	0.105 ± 0.03	Hyland, this paper
Emission nebulae	0.11	Mathis (1962)
Carina Nebulae	0.12	Faulkner and Aller (1965)
M-L relation	0.09 0.11	Harris (1963) see Percy and Cester (1965) Demarque
β Cep stars	0.10	Percy and Demarque (1967)
Sun	0.096	Gaustad (1964)

4.   THE SILICON STARS

Three Si λ 4200 stars lie in these clusters and possess very weak helium lines for their effective temperatures — a feature that is common to all silicon-rich stars.   The question arises as to whether the weak lines are due to a low atmospheric abundance of helium or to some abnormality of the atmospheric structure.   I shall here briefly summarize an argument that favors the former interpretation.

From the line strengths alone one can deduce that $N_{He}/H_H < 0.01$. Now the distribution of electron pressure throughout an atmosphere can be found by fitting computed Hγ profiles to the observed ones.   This distribution is related to the gravity of the star through the equation of hydrostatic equilibrium, and the gravity can be determined if the helium/hydrogen ratio is known.   In particular, the value of log g derived from a given electron pressure is 0.1 larger if $N_{He}/N_H = 0.01$ rather than 0.1.   With larger values of log g, the silicon stars almost obey the same mass-luminosity relation as the normal stars; i.e., their interior abundances are the same as those of the normal stars.   If, on the other hand, we believe that the helium/hydrogen ratio is really normal in the atmospheres of these stars, then they lie well off the mass-luminosity relation defined by the normal cluster stars.   Their position can then be explained only if we assume that their interiors possess a much larger abundance of helium than $N_{He}/N_H = 0.10$ implies.   In this case, the interior should have a high helium abundance while the surface is normal. Observations of silicon stars in clusters of different ages show that they evolve in a manner identical with that of the normal stars (Hyland, 1967b). This is strong evidence for supposing that the interiors are normal and that therefore the surface is depleted as indicated by the line strengths.

5.   ACKNOWLEDGMENTS

This research was done at the Mount Stromlo Observatory while the author held an Australian National University Research Scholarship.   Attendance at this conference was supported in part by NSF Grant GP-7030.

REFERENCES

Deeming, T. J., and Walker, G. A. H. 1967a, Nature, 213, 479.
————. 1967b, Zs. f. Ap., 66, 457.
Faulkner, D. J., and Aller, L. H. 1965, Mon. Not. Roy. Astron. Soc., 130, 393.
Gaustad, J. E. 1964, Ap. J., 139, 406.
Griem, H. R., Baranger, M., Kolb, A. C., and Oertel, G. 1962, Phys. Rev., 125, 177.
Hardorp, J. 1966, Zs. f. Ap., 63, 137.
Hardorp, J., and Strittmatter, P. A. 1968, Ap. J., 153, 465.

Hyland, A. R. 1967a, Nature, 214, 899.
_____. 1967b, in The Magnetic and Related Stars, ed. R. C. Cameron
    (Baltimore: Mono Book Co.), p. 311.
Lambert, D. L. 1967, Nature, 215, 43.
Mathis, J. S. 1962, Ap. J., 136, 374.
Mihalas, D. 1964, Ap. J., 140, 885.
Percy, J. R., and Demarque, P. 1967, Ap. J., 147, 1200.
Scholz, M. 1967, Zs. f. Ap., 65, 1.

DISCUSSION

Underhill: I certainly would like to congratulate Dr. Hyland for his fine work. But I think you should not assume, because you get the same hydrogen-helium ratio in these strong-line stars as in the nebulae, that you are safe. It is a very tricky business to compare the nebular results with the stellar results, because the conditions are so different and there may be hidden errors in the analysis. In other words, I attach no great significance to consistency. Another point is that the silicon stars give me the impression of being cool stars, say around A0, with hot chromospheres. Of course, it's not quite clear how this would generate the very blue continuum.

Hyland: I should like to take issue with regard to your comment on the silicon stars, because the continuous spectrum all the way from 3000 A to 8000 A is identical to that for stars that are of this photometric color and that are apparently normal. It seems to me that you need to heat up the entire atmosphere and not just the chromosphere.

Elste: How did your Balmer lines compare in the silicon stars and in the ordinary B stars?

Hyland: They agreed almost identically, except that in the silicon stars they had slightly deeper cores.

GENERAL DISCUSSION ON PART IV

Strömgren: I should now like to ask for general questions on the subject of this session.

Elste: There has been some confusion about the term "line blanketing," and the meaning of line-blanketing corrections. We must distinguish between the line blanketing as employed in the construction of models and the line-blanketing corrections applied to the measures made with intermediate and wide passbands.

Neff: I think this suggestion has been made before, and I have tried carefully to use the word "blocking" in discussing photometric corrections.

Whitney: In closing this morning's session, I should like to comment that, during the afternoon session, we should discuss whether a grid of models should be computed and, specifically, what sort of models should be included. I think we should also follow Demarque's suggestion, a point that was in fact one of the motives for organizing this conference, and encourage the internists and the externists to talk to each other a bit about their common needs and their capabilities.

It may be that we will decide not to compute any models at all for a grid. It is rather pointless to compute a small percentage of a type of grid that George Wallerstein proposed earlier, because without the whole grid there is probably little value in the portions of the grid. And besides, I'm not sure that if we computed the whole grid George would pay any attention to it anyway. Clearly some models are important and needed, and the point will be to decide which.

PART V

THE GRID OF MODEL STELLAR ATMOSPHERES

DISCUSSION ON THE TABULATION OF MODELS

A. B. Underhill, Chairman

Underhill: The title of this session is "The Choice of Parameters for the Output and Tabulation." The subject is: Shall we make a definitive grid of models — definitive for what? — and what shall we have to define the models by? I have xeroxed a few notes I made to start our discussion, which should be concentrated on the following five points:

I.   Purpose of the grid
II.   Introduction to the tables
III.   Data to be tabulated
IV.   Choice of parameters
V.   Scope of proposed tabulation.

The purpose of the grid: Is it to be a small grid of three or four check models so that anyone who is setting up a new program could always have a check to compute in order to see if his own program is working right? Also as a check to see if a new idea is going to work or not? If you need a model, here is one. Or is our purpose going to be to provide some special-purpose models for a particular part of research that has not been covered by a particular grid of models yet made?

Under introduction to the tables, I have made obvious remarks that the editors of the tables must take into account: List the equations that are actually used for getting the models in the spectrum, continuous and line. Don't forget that a model is a model, and in spite of my remarks about non-LTE it is perfectly all right to make an LTE model. It is a model, but we haven't said what it is a model of — that's a different question. In any event, since I think it is important to know where to find the details, list explicit references or equations for absorption coefficients and list compositions as mass fractions and as numbers. I find it extremely useful to have both, and I don't think it takes much more paper.

We may want more discussion on what should be tabulated. Under the emergent continuous spectrum, I think that's obvious; I prefer F_ν. If you want F_λ, you can just list c/λ^2. Special-purpose data such as the tabulation of the emergent spectrum may require a lot of space. I think books of tables are just about as difficult to read as a telephone directory and about as interesting, so I suggest you do not list such things as opacity (mean or monochromatic) as a function of depth, degrees of ionization, etc. If you really need them, you can compute them from the model.

The choice of parameters is of course closely related to the purpose of the grid. The scope of the proposed tabulation depends, of course, on the economic situation here at the Smithsonian. I have mentioned a few points — economic limitations and scientific needs — and particularly the second point we may wish to discuss. I propose that in this discussion people should mention under which point their remark comes home. In order that we may have something to discuss, I shall ask Owen Gingerich to explain the proposal that he has prepared for us to consider.

Gingerich: Let me first give some of my philosophy as to why this particular grid has been proposed, and then we can discuss it on more general terms. If you decide we need these models, we can discuss more specifically how they ought to be set up. First of all, I propose that we have three standard models, differing somewhat from the standard models proposed for the conference. A comparison of those standard models took place the other day among some of the model-computing people, and it indicated that those standard models computed in advance of the conference were very closely matched by astrophysicists in various places computing with the same specifications. So it is very clear that there is some agreement on how to compute.

It is also fairly clear that there are lots of quite reliable computing programs around, so that most people who want to take a detailed look at a particular star will, in fact, have access to some facility in order to make up their own detailed grid of models.

I believe there is, however, some merit in printing at least three models in perhaps excessive detail, the purpose being that if you want to start your own model atmosphere program, it is desirable to have something to check against. If you disagree with the one that has been published, it is extremely frustrating to try to find out why unless there is a fair richness of detail to enable you to check what is different about what you are doing. Therefore, I have suggested that among other things the actual opacity routines might be published in their FORTRAN versions, which will make it very easy for someone to get quickly a standard opacity to put into his own program to check. If he disagrees about the way to compute the opacity, he can then change it to suit himself, but at least he will know that the differences he is now getting in the model are due to his disagreement about the opacity rather than to some numerical constant or the integration of hydrostatic equilibrium, for example.

Now it seems to me that these details are one thing that might be needed, and that is completely apart from the remainder of the models I have proposed. Because of the intended usage, there are some peculiar things about the standard models. The standard models are not computed, for example, with hydrogen-line blanketing, because in general that is a rather messy thing to include if you simply want to match with somebody else's computation. For the same reason, I have left the H_2^+ opacity out, and I have suggested that in the coolest model the opacity should be computed as absorption rather than as scattering.

I have also proposed that at those three standard grid points we might want to get some additional information. Since we have the facility here in Strom's program to compute a non-LTE hydrogen model with the first five levels in non-LTE and the sixth and higher levels in LTE, we could compute such models so that people could see quickly how the LTE and non-LTE ones differ. We would want to publish details about the non-LTE model that would not be relevant for other models. Similarly, we can have a variation in gravity and in metal abundances; for this latter reason, the silicon and magnesium opacities have been put in the standard models even though not everyone may agree with the way we put them in. At least this way we can get a derivative as to what happens when the metal abundance is changed.

I also have a philosophy about the third section. It is a skeletal grid with many models, and these models can be easily supplied by us — we have them already, or nearly, so that the computing time would not be too long. If we would publish something like 25 depths, I estimate it would require approximately one-half page per model. This means that in terms of actual publication space we get an awful lot of models onto comparatively few pages. Even if these models are not the most useful, they don't take up much space. We must keep this in mind as part of the economy. I know that Anne Underhill believes that it is a little redundant to do the hotter ones similar to those already published, but I would say that there might be some justification for redoing them because the space involved is so small and it puts all the models into one very handy source.

These are what I would call the "quick-look" models; they are available to someone who doesn't have an atmosphere program and who just wants to know how a model with given parameters might look. It is not supposed to be used for subtle analyses of stars; obviously, the grid will be no good at all for helium and carbon stars, low-metal abundance stars, and things like that. But it would give an idea of how things go in the main sequence. These are not going to be the kinds of model that will do stellar-atmosphere computing people much good, because there are not enough of the details that they would get when they do it themselves. They could be used to furnish a starting iteration to save time in converging the model. For this reason we might publish the specifications of card formats that give the temperature and pressure, which you could get from us to feed directly into your own computer. And these would give many more optical depths than we would actually publish.

Because some people would use these models in the way some astrophysicists use Allen's Astrophysical Quantities, that is, without using much discrimination about the models they're looking at, it seems to me best to try to construct them with as much sophistication as possible within the general agreement of the conference. This conference is more or less the swan song of normal-type models, and this will be the last publication of a big grid of normal models. Afterward, people will be concentrating on much more advanced problems and peculiar stars. Hence, we shouldn't print the complete quick-look set by standards that could already have been achieved at the First Stellar Atmospheres Conference, but we ought to compute them with the state of the art right now, which means that for the hotter models there should be hydrogen- and Lyman-line blanketing, and for the cooler stars it means putting in some blocking; that is, these should be blanketed models.

The blocking can be accounted for reasonably easily by a statistical procedure similar to the one I proposed at Heidelberg. That is, you examine some representative lines, between 50 and 100, in the stars of these cooler temperatures, and from this you decide how line opacity behaves as a function of depth. Then you simply multiply this by some arbitrary N until you produce in the observed spectrum a chunk removed from the observed spectrum that matches some carefully specified blocking. In other words, I would suggest that we might want to specify an η as a blocking parameter from 4500 A to 3600 A and then another η from 3600 A to 2000 A. In the case of the sun, we can say that the first η is 10% and the second η is perhaps as high as 50%. Some of the observers might help us to specify these blocking parameters. They won't have to resemble real stars too precisely, but it would be a first approximation to make the temperature-pressure distribution in the models more reasonable. Since in this way we can make a reasonable approximation to the solar photosphere, we have at least one point to know that we are doing it rather well. I would say that models from 9000° down to 4000° should be done this way.

Furthermore, in the range around 7000°, where convection is important, we should put convection in. We have a mixing-length theory, and we can compute a model that has the proper flux constancy. It is reasonable to compute with a mixing length like two, which is perhaps extreme, and then compare this with no convection at all; thus, we would have bracketed the actual situation. Then when nonstellar-atmosphere people come to use the grid, they would get an idea of how bad the approximations are from the variations.

Underhill: Thank you. We will want to discuss this proposal in more detail. Before going any further, this η was roughly the way I did my line-blanketed models, and I think it is very important to have the η a thing that varies with depth according to the Saha-Boltzman equation.

Gingerich: My η is an observed blocking parameter, which is reproduced in the observed spectrum.

Underhill: Right, in the output spectrum, but if you go down deeper in the star at 10,000° it changes its character.

Gingerich: I already said that. The η is not the N with which you multiply the opacity. Since we have already set up the opacities so that they vary with depth the way the representative lines do, you just try to reproduce the observed spectrum.

Underhill: There are so many details to discuss that we may want to form a committee when we have decided what to do. I think we should now go alphabetically to discuss this.

<u>Cayrel</u>: For me, one of the quite useful purposes is that these models should allow us to compute a great many things according to the choice of the worker. That is, I am not so much interested in the quantities that will be tabulated with the models, but rather in the possibility of computing other things. For example, to compute some strong ion line with the model implies something about the presentation of the model. In particular, I should like to have the density because if we are going to compute a line, we will always need it.

<u>Lecar</u>: I would like to pass on a request from Iben, who says that he finds it virtually impossible to go from B-V to effective temperature. He thinks that some attempt should be made to arrange for a theoretical calibration.

<u>Strom</u>: That's not in hand yet — another six months.

<u>Underhill</u>: That is difficult, because we can't decide how to compute B-V colors directly. That is a subsidiary thing, but if we had a grid, anyone could compute for himself from the continuous spectrum flux.

<u>Lecar</u>: Model stellar atmospheres can now be calculated in less than one minute on a 7094-class computer. So, even to derive a single quantity, it is usually more efficient to recompute an atmosphere than to interpolate from a grid.

I should like to suggest that we collaborate with researchers in stellar interiors to produce a set of models from which one could deduce mass, radius, and age from observations of luminosity, color, and perhaps a few line profiles.

<u>Strom</u>: I should take issue with Mike Lecar's point of view that you can compute a model in half a minute. In many cases, people have machines that are somewhat slower than the machines available to us, and it is a little bit more difficult for them.

<u>Peat</u>: I have just one or two comments to make. First, I can't say that I really agree with the comments Dr. Lecar has made. From our point of view, we intend to compute many hundreds of models in any case, but we're not going to expect that many models be tabulated. What we should particularly like would be more variation models to be published. We should like to see a very large range in metal abundances; we should like to see several times the solar abundance and one that is two orders of magnitude less than the solar abundance. Our own particular purposes may not coincide with those of other people.

<u>Underhill</u>: Let us hear more about purpose.

<u>Matsushima</u>: I'm not so certain whether one needs extensive tabulation of a bunch of models. You carefully specify the boundary conditions; then, whoever wants to check on it can use exactly the same boundary conditions and published flux and get the same models.

Also, I think that since the method of opacity calculations is continually changing, if you publish a very extensive grid of models this summer, how long would it last? If the Harvard people want to make such a table, it is nice to have it, but I'm not certain.

<u>Underhill</u>: I also feel this way. I have published about 8 or 10 models in considerable detail (and I have a paper coming out that publishes the references to them). No one ever refers to them because they are too hard to find — everyone refers to Mihalas' or Strom's list, both of which are actually more comprehensive, but we all cover the same region and where we have the same log g we're almost identical. So I'm sure that down to type A we don't need any more models.

Matsushima: The greatest part of model-atmosphere calculations is an integration to find a final T-τ or T-p relation; other than that, just computing a model for a given temperature distribution is so very simple that this morning George Wallerstein proposed a grid of 10^7 models.

Wallerstein: I did not propose it!

Matsushima: It shows the number of models that can be produced from the same T-τ relation, then scale it and compute the flux. So I think one has to be very careful to set up the boundary conditions for the kind of tables you want to reproduce.

Underhill: As far as scaled models are required, there are a considerable number calculated by de Jager and Neven. They have two sets and have actually used a scaled T-τ relation, and they compute everything else, which is very simple. In fact, it is not very difficult to compute a model or its flux if you assume that the source function is B_ν and you don't have to iterate. If you iterate to correct the temperature, it is a little more complicated but not too difficult, as Owen and Strom and all these others know. They have got the routines that work fast.

Feautrier: Why don't you just distribute the program?

Matsushima: That doesn't help either — you can't change it from one machine to another.

Strom: Several of us, including Owen and Bob Kurucz, have programs that are rather straightforward to use. A description of these computing engines, as Owen calls his routine, and how to use them would be helpful to people interested in such programs. As far as I am concerned, I should be willing to distribute copies of the program and the description, which would be useful for people who want to start.

Underhill: I agree with what you say.

Thomas: I have been going to sleep back here...

Underhill: I expect you to.

Thomas: Can you just tell me as a simple man what are the assumptions on which this discussion is based? Can you just write them down so I will know?

Underhill: You mean the assumptions for making a model by hydrostatic equilibrium, radiative equilibrium? These are standard practical models, so therefore it's LTE.

Thomas: What are you doing about line blanketing?

Whitney: I think Owen has covered that. You have slept through the description, Dick.

Underhill: These are details of procedure that are in the hands of the people who will compute them. I said at the beginning that we realize these are models, not stars.

Strom: I think that scaling from the models in this really coarse grid will be quite useful for some people who are not interested in doing detailed calculations centered about a given effective temperature. It's foolish to publish a very extensive grid covering the entire HR diagram for the simple reason that every time you find you want to analyze a particular problem, you have to make up a grid appropriate to that particular problem. By scaling from published temperature and pressure, people with slower machines can start and make some progress in a reasonable amount of time.

Wallerstein: As a user, to confirm what a lot of other people are saying, I wish to say that these things can be very useful for those of us who need scaled models, who want to look at a few lines or look how the ionization of some element will behave. This type of grid is just the type of thing we should like.

Cayrel: I agree with Strom's comment that many people will just borrow the T-τ relation from your grid of models and recompute a new model with a slightly different T_{eff} or a different g. Now in connection with Dick Thomas' remark, I would say that I should appreciate very much if blanketing effect is taken into account in two ways: first in LTE, and second assuming scattering in all lines with a frequency independent function and $\epsilon = 10^{-2}$.

Underhill: We will have to set aside a special investigation for that unless the Harvard people are further along in their development than I think they are. That is an important point but an extremely large problem.

Thomas: Tell me, maybe I am naive here — suppose I have lots of grids of models and suppose I haven't got what Cayrel just suggested; then I make a few calculations along the lines he suggests and find it makes a big difference. What do you do then — junk all the calculations made ignoring the blanketing effect?

Underhill: Sure.

Thomas: Well, then, why not make the test calculations first?

Underhill: We're not going to make the second group, but we are going to make the first group.

Thomas: Are you sure? A lot of people seem to be ready for the second group.

Underhill: Who here can make a model atmosphere and compute noncoherent scattering for all the lines? Is there anyone here with such a program?

Thomas: For a selected number of lines.

Underhill: Do you see any hands lifted?

Thomas: Mihalas tells me this is not difficult to do.

Underhill: All right, Smithsonian is paying for this group. How about having JILA support the next group and we'll compare them and have the next step forward?

Thomas: I'm not arguing; I'm just asking which way the sheep are drifting today.

Underhill: Well, it's what we're aiming for, but not what we're ready to do, unfortunately. I wish we were.
    May I summarize as far as purpose is concerned. We want a few models to be used on an experimental basis to give an idea how to set up your own program as a control, and against which you can test the effects of various factors. This means that at this moment we are considering section 1 of

Gingerich's suggestion. The change I should like to make is that I would suggest no models hotter than 10,000°, because there are already models between 10,000° and 30,000° from Mihalas and myself and what Strom has published. That information is in the literature. So let's save our paper and our money for the region where we need more, 10,000° down. I would go to 10,000°, 8,000°, 6,000°. I should like more log g's — perhaps three, four, five; if I did that I would have nine models. This would be for a minimum grid. I would suggest that the Smithsonian people consider choosing their grid within that range.

I think it would be particularly helpful also if we would give more attention to the applications of the stellar-structure people. They appear to be most interested in being able to relate colors to effective temperatures in this region. This grid would give you a region of interpolating — to get an idea of where to do your special-purpose grid. You can already do it for the hotter ones. The cooler ones? That is a region I don't know. Gingerich, what is your reason for not going cooler?

Gingerich: I went down to 4,000°. The reason I suggested stopping at 4,000° and not 2,000° or 3,000° is that at 3,000° the water vapor begins to show sufficiently in the emergent flux so that you don't get the structure unless you have the molecular stuff in; at 4,000° you do get into the formation of the H_2, but it isn't yet important and the convection is not close enough to the surface to throw the wrench into it, so it's really a good stopping point. I discussed this with Auman, and he also thinks that is the end of the safe region and getting into the frontier area.

Whitney: They are only models in any case; the frontier is at every temperature. The purpose of such a grid might be to represent some stones to be dug up by archeologists — representing the state of the art right now for classical model atmospheres. I'm not in charge of this discussion, but I don't want it to drag on past the point of effectiveness. Since these things can really be decided by a committee, all we can do is take a sense of the meeting. I should like to bring this to a head by suggesting that Owen's list and his earlier description be taken as a focus.

Underhill: Let's discuss this a little more before we make a committee.

Heintze: I would suggest that we also put 15,000° on the list, to see how magnesium and silicon affect the Balmer jump and also to see how Balmer-line-blanketed models behave.

Gray: It seems to me that as long as the circulation of these models is relatively restricted to the publication of the proceedings of this conference we ought to accept virtually anything that people are willing to publish. I'm going to put it on my book shelf, and if I don't want to use it, I don't have to. Other than that I don't see why we should be restricted.

Underhill: He who pays the piper calls the tune.

Jones: I have two observational points. We ought to have the fluxes where they have been observed from the ultraviolet out to 12 μ. Some of the published stuff is much too restricted. Second, could we have the fluxes to 3- or 4-figure accuracy? Models such as those of de Jager and Neven can't be differentiated accurately.

Underhill: I'm glad you mentioned that point. I think the fluxes must be listed at all the integration points, probably from 912 A to 8000 A or longer at every 1000 A. We must have the whole spectrum from one end to the other, but I don't know how many integration points you have.

Gingerich: About 30 or 40.

Underhill: We must be able to interpolate the continuum fairly accurately over the parts that we would observe from 912 A to 12 μ. I think it is worth considering putting in more than the integration points.

Matsushima: If Harvard has enough only to publish all of this grid...

Whitney: We don't. [General laughter.]

Matsushima: Then we should decide what data we want rather than what data we don't want.

Underhill: I think everybody agrees we should include what I have listed as the basic data here. But what data do we want in terms of spectrum output? First, we want the continuum points. The next item that comes up is which lines are useful for us, computed in a very simple manner as model lines, not real lines. They show what a star would look like if it only knew how to make lines the way textbooks and astrophysicists say it should. At least it gives a defined basis of reference. Since the blackbody is not a detailed enough basis for reference for a stellar spectrum, I suggest we want Lyman α. I have never computed Lyman α in LTE, but somehow I have a feeling that it would be a shocking object to see. It might even give a real feeling as to what a star would do if it were in LTE. When I say lines, I always mean LTE lines because I think that's the only rapid way to compute them, and I feel that it would be unjustified in this sort of covering problem to suggest computing them correctly — those are real research jobs that are worthy of detailed and extensive study.

Matsushima: I should wish to have some other metal lines because their computation is based on a different method than the hydrogen lines.

Cayrel de Strobel: I should like to suggest sodium, magnesium, and some very strong iron lines for the models below 5000°K. I should like lines in the red so they won't be blended. Also, perhaps calcium. All these lines can be calculated in LTE, because even from LTE models you can extract a lot of information. Of course, it would be interesting to calculate these lines with non-LTE models and then show the difference.

Fischel: If we're going to do models, we should do them with hydrogen-line blanketing anyhow, so that if we want to have those lines, we would have them in at least a crude manner from what we already have in computing the continuum. I should like to see in this scheme something to look at — to see if we are doing it right. Some lines should be in detail.

Gingerich: I tend to be rather skeptical about doing a big flock of lines because I don't think that these models should be used that closely for comparison. The actual stellar spectra are only rough ball-park models, and therefore I feel that if somebody wants to get these lines out, he can compute them himself because the temperature-pressure relations are all given.

Underhill: I disagree with you from the point of view that not everyone has a computer in his backyard. Actually it is very handy to have these crude representations of lines on the crude set of models, but I think maybe I had better investigate this more thoroughly.

Whitney: There is too much agreement.

Underhill: I now retire from the chair.

THE GRID OF MODEL STELLAR ATMOSPHERES FROM 50,000° TO 11,000°

R. Kurucz

Smithsonian Astrophysical Observatory and
Harvard College Observatory, Cambridge, Massachusetts

The models with effective temperatures greater than 10,000°K were calculated with the model-atmosphere computer program ATLAS, which will be described in the forthcoming Smithsonian Astrophysical Observatory Special Report 309.

The opacities included were all those available in ATLAS, namely, H I, He I, He II, H^-, He^-, H_2^+, Si I, Mg I, Al I, C I-V, N II-V, O II-V, Ne II-V, Rayleigh scattering from H and He, electron scattering, and hydrogen lines. The carbon, nitrogen, oxygen, and neon opacities were derived from Hidalgo (1968), Henry and Williams (1968), and Flower (1968); the others are referenced in the next paper. (The opacities Si, Mg, Al, H_2^+, and He^- could have been omitted because they are irrelevant at higher temperatures.) Hydrogen-line blanketing was approximated by using a frequency spacing chosen to give triangular profiles for Lyman and Balmer lines as described by Carbon and Gingerich in the next paper. The hydrogen-line routine was supplied by Deane Peterson.

Only hydrogen and helium were considered in the equation of state. Partition functions depending on temperature and electron number were adopted from Drawin and Felenbok (1965). A helium abundance of 0.10 by number relative to hydrogen has been adopted throughout the grid. The other abundances used were the Lambert-Warner solar values except that neon was set equal to carbon. The abundances with their references are given in the following table:

Element	Abundance	Reference
H	0.909	
He	0.091	
	Log abundance	
C	-3.49	Lambert (1968)
N	-4.11	Lambert (1968)
O	-3.27	Lambert (1968)
Ne	-3.49	= C
Mg	-4.56	Lambert and Warner (1968a)
Al	-5.64	Lambert and Warner (1968b)
Si	-4.49	Lambert and Warner (1968c)

The models were calculated against τ_{5000}, with 40 τ values spaced six per decade. The final depth was chosen large enough so that the last τ_ν at the most transparent frequency was greater than 10. The tabulated values were obtained by interpolation.

The source function, including scattering, was computed by a matrix solution of the integral equation (Kurucz, 1969). The temperature-correction scheme, substantially modified from Avrett and Krook, will be described in the write-up of program ATLAS.

The flux accuracy of the models is of the order of 0.1%. The flux derivative error, the ratio of the flux derivative to the total flux, is everywhere less than 1% and usually about 0.1%.

REFERENCES

Drawin, H. R., and Felenbok, P. 1965, <u>Data for Plasmas in Local Thermo-</u>
<u>dynamic Equilibrium</u> (Paris: Gauthier-Villars).

Flower, D. R. 1968, in <u>IAU Symp. No. 34</u>, ed. D. E. Osterbrock and C. R.
O'Dell (Dordrecht-Holland: D. Reidel) 77.

Henry, R. J. W., and Williams, R. E. 1968, <u>Publ. Astron. Soc. Pacific</u>,
<u>80</u>, 669.

Hidalgo, M. B. 1968, <u>Ap. J.</u>, <u>153</u>, 981.

Kurucz, R. 1969, <u>Ap. J.</u>, <u>156</u>, 235.

Lambert, D. L. 1968, <u>Mon. Not. Roy. Astron. Soc.</u>, <u>138</u>, 143.

Lambert, D. L., and Warner, B. 1968a, <u>Mon. Not. Roy. Astron. Soc.</u>,
<u>140</u>, 197.

_____. 1968b, <u>Mon. Not. Roy. Astron. Soc.</u>, <u>138</u>, 181.

_____. 1968c, <u>Mon. Not. Roy. Astron. Soc.</u>, <u>138</u>, 213.

THE GRID OF MODEL STELLAR ATMOSPHERES FROM 4000° TO 10,000°

D. F. Carbon and O. Gingerich

Smithsonian Astrophysical Observatory and
Harvard College Observatory, Cambridge, Massachusetts

ABSTRACT

Fifty constant-flux model stellar atmospheres with
$10,000° \geq T_{eff} \geq 4000°$ have been computed with the SOURCE program. For $8500° \geq T_{eff} \geq 4000°$, a mixing-length theory of convection has been incorporated into the program, as well as a statistical line-blanketing theory. Predictions from these theoretical models are compared with empirical data from the sun and Procyon. Temperature distributions in the grid appear to validate a scaling procedure for obtaining models cooler than $T_{eff} = 6500°$.

The 50 models for the lower effective temperatures, from 10,000°K to 4000°K, have been computed with a CDC 6400 program, SOURCE, at the Smithsonian Astrophysical Observatory. An earlier version of this program has been described by Gingerich (1963), and we plan to publish complete details of the particular configuration used for these models in the forthcoming Smithsonian Special Report 300.

1. REMARKS ON THE COMPUTATIONAL METHODS

1.1 Equation of Hydrostatic Equilibrium

In SOURCE the equation of hydrostatic equilibrium is integrated (by Hamming's method) with τ_{5000} as the independent variable. All the opacities described below are used when generating the opacity at 5000A. The electron contribution from each of the eight elements listed in Table 1 has been considered in solving the equation of state: the Saha equation has been used with the temperature-independent partition functions listed in the table. A helium abundance of 0.10 by number relative to hydrogen has been adopted throughout the grid. The metal abundances are taken from Lambert and Warner, with the references indicated.

Table 1. Elements entering equation of state

Element	Relative abundances	Ionization potentials	Log $2U_{II}/U_I$	References
H	1.0000	13.595	0.000	
He	0.1000	24.581	0.602	
C	3.548E-04	11.256	0.110	(Lambert, 1968)
Na	1.514E-06	5.138	-0.010	(Lambert and Warner, 1968a)
Mg	3.020E-05	7.644	0.600	(Lambert and Warner, 1968c)
Al	2.512E-06	5.984	-0.470	(Lambert and Warner, 1968a)
Si	3.548E-05	8.149	0.080	(Lambert and Warner, 1968b)
Fe	3.236E-06	7.870	0.490	(Warner, 1968)

Radiation pressure, though unimportant for log g = 4 in this temperature range, is nonetheless significant at lower gravities. It has been included by using an effective gravity when integrating the equation of hydrostatic equilibrium, where

$$g_{eff} = g - \frac{\sigma}{c} T^4_{eff} \bar{\kappa} \quad .$$

The flux mean opacity, $\bar{\kappa}$, comes from the radiation analysis of the preceding iteration.

All the models were calculated with 20 steps per decade in optical depth. In order to avoid any difficulties associated with the starting steps, an additional decade of shallower layers has been calculated but is not shown in the tabulations. Occasionally, the models were extended to greater depths than shown in the tabulations so as to treat properly unusually transparent wavelength regions.

1.2 Convection

For the models with $T_{eff} \leq 8500°$, we have included the effects of convective energy transport. We have adopted, with some modifications, the particular formulation of the mixing-length theory described by Mihalas (1967). In the expressions that follow we use the notation:

κ = a representative opacity (e.g., κ_{5000}),
C_p = the specific heat at constant pressure,
a = an arbitrary efficiency parameter equal to the mixing-length/scale-height ratio,
g = the local gravitational acceleration,
μ = the mean molecular weight,
ρ = the mass density,
∇ = the ambient temperature gradient,
∇_e = a gradient characterizing the temperature changes experienced by the convecting elements, and
∇_s = the ambient adiabatic gradient.

All the gradients are locally determined quantities of the form $\partial T / \partial \tau$.

Letting P be the total thermodynamic pressure (i.e., the sum of the gas pressure and the radiation pressure), we define the pressure scale height, H, as

$$H^{-1} \equiv \frac{\partial \ln P}{\partial x} = \frac{\rho g}{P} \quad .$$

The flux carried by the convecting elements, assumed to be moving with a mean velocity v, is

$$\pi F_{con} = \frac{1}{2} \frac{\kappa \rho v}{g} a \, P \, C_p (\nabla - \nabla_e) \quad , \tag{1}$$

where

$$v^2 = \frac{P \kappa}{T} \frac{a^2 QH}{8} (\nabla - \nabla_e) \tag{2}$$

and

$$Q \equiv 1 - \frac{\partial \ln \mu}{\partial \ln T}\bigg|_P \quad .$$

Since equations (1) and (2) are indeterminate, it is necessary to provide a third relation involving the unknowns. Consideration of the efficiency with which the convective elements transport energy in the presence of radiative losses yields this additional relation:

$$\frac{\nabla_e - \nabla_s}{(\nabla - \nabla_e)^{1/2}} = \frac{16\sqrt{2}\ \sigma\ T^3}{a\,(HQ)^{1/2}\ C_p\ \rho} \left(\frac{T}{P\kappa}\right)^{1/2} \theta \quad , \tag{3}$$

where

$$\theta = \frac{a\,H\kappa\rho}{1 + \dfrac{3}{4\pi^2}\,(a\,H\kappa\rho)^2} \quad .$$

Equation (3), adapted from the work of Henyey, Vardya, and Bodenheimer (1965), differs from the Mihalas form in that it attempts to account for the variations in optical thickness of the convecting elements during their motion. Equations (1) to (3) were used to evaluate the convective fluxes at each depth for the subsequent radiation analysis. The mixing-length/scale-height ratio was set equal to 1.5 unless otherwise indicated. This choice was a compromise among the values obtained by various investigators doing stellar interior calculations (e.g., Demarque and Percy, 1964; Demarque and Roeder, 1967; Torres-Peimbert, Simpson, and Ulrich, 1969).

1.3 Radiative Fluxes and Intensities

All the models we present include the effects of blocking from hydrogen Balmer lines. Since a detailed representation of the hydrogen-line profiles would have been prohibitively expensive, we endeavored to approximate the lines by using a small set of carefully selected wavelengths. These were chosen so that they removed the same amount of energy from the continuum as would a more detailed representation. Figure 1 illustrates how the wavelengths were chosen for grid models with $T_{eff} \geq 7000°$K. The dotted lines are the detailed flux distributions in the Paschen continuum for (10,000,4) and (10,000,2) models (Strom, 1969, private communication). These may be compared with the solid lines, which represent computations with the abbreviated wavelength set actually used for the grid. The area (integrated flux) beneath the solid curve is equal to that under the dotted curve to within 1.5% for both gravities, despite the change in shape of the hydrogen lines. For stars cooler than 7000°K it was possible to choose an even smaller set of wavelengths to represent the Balmer-line blocking. The complete wavelength sets used to calculate the models are tabulated in Table 2. The asterisks mark the wavelengths of the division points for the blanketing weights discussed in Section 3.

Calculation of the monochromatic fluxes, mean intensities, and specific intensities has been achieved to an accuracy of a few hundredths of a percent; details of the procedure and its tests will be included in the aforementioned Smithsonian Special Report 300. Scattering has been considered explicitly in the source function, both for deriving the models and in calculating the limb-darkening tables.

The integrals over frequency are accurate to a fraction of a percent; in the final tables of monochromatic fluxes, all the wavelengths used are shown. The integration constants, based on low-order Newton-Cotes rules, are also shown in Table 2, scaled downward by 10^{14}. Further details on the subdivision of the frequencies and weights will be found in Section 3 on the statistical line-blanketing procedures.

Table 2.   Information for frequency integrations.

(7000) TO (4000)

WAVELENGTH	FREQUENCY	WEIGHT
911.763 +	3.288024E+15	2.81318
1099.990 -	2.725388E+15	2.81318
1100.010 +	2.725339E+15	1.52853
1238.990 -	2.419632E+15	1.52853
1239.010 +	2.419593E+15	1.71735
1443.990 -	2.076122E+15	1.71735
1444.010 +	2.076094E+15	.56546
1527.202 -	1.963001E+15	.56546
1527.204 +	1.962999E+15	.57999
1623.118 -	1.847001E+15	.57999
1623.119 +	1.846999E+15	.32999
1683.267 -	1.781001E+15	.32999
1683.269 +	1.780999E+15	1.36499
1987.996 -	1.508001E+15	1.36499
1987.999 +	1.507999E+15	.39625
2137.795	1.402333E+15	1.18874
2312.005	1.296667E+15	1.18874
2517.126 -	1.191001E+15	.39625
2517.131 +	1.190999E+15	.46124
2807.019	1.068001E+15	1.38372
3172.368	9.450039E+14	1.38372
3647.053 -	8.220062E+14	.53767
3716.159 B	8.067201E+14	.12346
3759.999 B	7.973140E+14	.07420
3785.801 B	7.918800E+14	.17245
3930.002 B	7.628240E+14	.22993
* 4019.204 B	7.458940E+14	.08465
* 4019.204 +	7.458939E+14	.36104
* 4449.999 -	6.736855E+14	.36104
* 4450.001 +	6.736852E+14	.19753
4879.188	6.144260E+14	.79012
* 5399.999 -	5.551668E+14	.19753
* 5400.001 +	5.551666E+14	.23729
6094.659	4.918897E+14	.71186
6994.423	4.286129E+14	.71186
* 8205.869 -	3.653361E+14	.23729
* 8205.873 +	3.653359E+14	.26639
10503.515	2.854187E+14	1.06556
14588.208 -	2.055016E+14	.26639
14588.215 +	2.055015E+14	.12330
17790.504	1.685112E+14	.49320
22794.079 -	1.315210E+14	.12330
22794.096 +	1.315209E+14	.06698
26904.500	1.114275E+14	.26791
32823.479 -	9.133401E+13	.06698
32823.483 +	9.133400E+13	.22833
65646.765	4.566714E+13	.45667

(10000) TO (7500)

WAVELENGTH	FREQUENCY	WEIGHT
911.763 +	3.288024E+15	2.81318
1099.990 -	2.725388E+15	2.81318
1100.010 +	2.725339E+15	1.52853
1238.990 -	2.419632E+15	1.52853
1239.010 +	2.419593E+15	1.71735
1443.990 -	2.076122E+15	1.71735
1444.010 +	2.076094E+15	.56546
1527.202 -	1.963001E+15	.56546
1527.204 +	1.962999E+15	.57999
1623.118 -	1.847001E+15	.57999
1623.119 +	1.846999E+15	.32999
1683.267 -	1.781001E+15	.32999
1683.269 +	1.780999E+15	1.36499
1987.996 -	1.508001E+15	1.36499
1987.999 +	1.507999E+15	.39625
2137.795	1.402333E+15	1.18874
2312.005	1.296667E+15	1.18874
2517.126 -	1.191001E+15	.39625
2517.131 +	1.190999E+15	.46124
2807.019	1.068001E+15	1.38372
3172.368	9.450038E+14	1.38372
3647.053 -	8.220062E+14	.69038
3862.387 B	7.761780E+14	.26366
3897.051 B	7.692740E+14	.06684
3930.074 B	7.628100E+14	.07834
3978.073 B	7.536060E+14	.08458
* 4019.198 B	7.458950E+14	.03856
* 4019.204 B	7.458940E+14	.03906
4061.738 B	7.380830E+14	.08216
4109.736 B	7.294630E+14	.07128
4141.736 B	7.238270E+14	.14548
4280.470 B	7.003670E+14	.17206
4348.469 B	6.894150E+14	.09549
4400.471 B	6.812680E+14	.18940
* 4601.295 -	6.515340E+14	.14867
* 4601.302 +	6.515330E+14	.14867
4821.333 B	6.217990E+14	.17932
4869.329 B	6.156700E+14	.05074
4901.332 B	6.116500E+14	.32799
* 5450.019 -	5.500715E+14	.30789
* 5450.020 +	5.500714E+14	.23092
6137.038	4.884929E+14	.69276
7022.249	4.269145E+14	.69276
* 8205.869 -	3.653361E+14	.23092
* 8205.873 +	3.653359E+14	.26639
10503.515	2.854187E+14	1.06556
14588.208 -	2.055016E+14	.26639
14588.215 +	2.055015E+14	.12330
17790.504	1.685112E+14	.49320
22794.079 -	1.315210E+14	.12330
22794.096 +	1.315209E+14	.06698
26904.500	1.114275E+14	.26791
32823.479 -	9.133401E+13	.06698
32823.483 +	9.133400E+13	.22833
65646.765	4.566714E+13	.45667

KEY TO THE DISCONTINUITIES:

```
1100- C I
1239- C I
1444- C I
1527- SI I
1623- MG I
1683- SI I
1987- SI I
2517- MG I
3647- H I
8206- H I
14588- H I
22794- H I
32823- H I
```

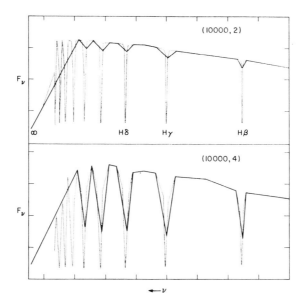

Figure 1.   Flux distribution in the Paschen continuum for (10,000,4) and
(10,000,2) models: dotted lines represent flux distributions using
Strom's detailed wavelength set; solid lines represent flux distributions for abbreviated set used in grid.

1.4 Temperature Correction

The temperature distribution in each of the models has been established
by iterative use of the Avrett-Krook correction procedure (Avrett, 1964).
The Avrett-Krook equations were modified for convection by including the
most important terms derived by Mihalas (1967). For the atmospheric layers
where energy transport was purely radiative, the flux errors were driven to
less than 1%. In the convectively unstable layers, this convergence criterion
is often unrealistically strict. The convective fluxes are quite sensitive to
local values of the temperature gradient; errors of only a few degrees in
temperature may cause large flux errors. This behavior is particularly
acute in the cooler dwarfs. The convective models we present have been
driven to better than 1% flux error in the atmospheric layers responsible
for the emergent continuum. In deeper layers, larger flux errors were
tolerated only when it was apparent that the associated temperature changes
were unimportant. The ripples occasionally appearing in the tabulation of
the convective fluxes are local features; they do not represent significant
irregularities in the temperature distributions. (We are able to supply more
extensive tables of F_{con} vs. τ_{5000} for those who might require for smooth-
ing purposes depths not tabulated here.)

2. OPACITIES

Copies of the FORTRAN routines used to calculate all the opacities
(with the exception of the Balmer lines and electron scattering) are explicitly
reproduced at the end of this paper. They are shown below.

Opacity Source	Routines	Authority[*]
H	AVHYD	Menzel and Pekeris
H	GAUNT	Karsas and Latter
H$^-$	AVHM	John, Geltman
H$_2{}^+$	AVH2P	Bates et al.
H lines		Griem (Peterson, 1969)
C	AVCAR	Peach
Mg, Si	AVMET	Gingerich and Rich (1968)

Scattering

Electron		
H Rayleigh	AVRAY	Dalgarno
H$_2$ Rayleigh	AV2RY	Dalgarno and Williams

Each routine yields the absorption per neutral hydrogen atom (scaled upward by 10^{26}) when multiplied by the appropriate quantities specified in the comments at the beginning of the various routines. The routines all include the effect of stimulated emission.

In Table 3, we give the monochromatic opacities for the (10,000,4) model as generated from these subroutines. They have been included primarily as an aid to other workers in checking out their own programs. The opacity is given per neutral hydrogen atom, and the final column (e. g. , C EDGE) gives the opacity ratios on the two sides of the absorption edge. The individual opacities have been normalized to show their proportional contributions; the part not shown arises from electron scattering.

3.   BLANKETING

In recent years, various investigators (e. g. , Swihart, 1966; Strom and Kurucz, 1966; Gingerich, 1966; Parsons, 1969) have attacked the difficult computational problems presented by the metal-line blanketing in stars of intermediate and late spectral types. The most ambitious of these efforts was that of Strom and Kurucz. To calculate a blanketed model atmosphere for Procyon, they used a statistical scheme to represent the effects of over 30, 000 atomic lines. This particular approach, while instructive, is far too costly at present to permit its use for calculating extensive grids of blanketed atmospheres. Economic considerations have led us to search for less expensive, but equally accurate, techniques for handling metal-line blanketing. In this investigation, we based our representation for the line blanketing on a triple-picket approximation. This is an extension of the method used to construct a theoretical blanketed model for the sun (Carbon, 1969). We shall first review the construction of the blanketed solar model and then describe the procedure for choosing the appropriate blanketing parameters for the models in the grid.

3. 1   Depth Dependence of the Picket Opacity

For the solar model, the three steps of the picket were chosen to represent opacities characteristic of the continuum, line cores, and line wings. The relation between line-core and line-wing opacities was established on the assumption that the metallic lines responsible for blanketing have doppler profiles and by dividing the profile between line core and line wing. We arbitrarily chose as the dividing point the wavelength at which the line opacity

---

[*]Citations to the original references are given in the routines and have generally been given by Gingerich (1964).

Table 3. Monochromatic opacities for (10, 000, 4).

911.8+

TAU 5000	TAU NU	OPACITY	H	H-	H2+	SI	MG	C	RAY S
.001	1.44E-01	2.682E-22	.0010	.0000	.0000	.0006	.0000	.9823	.0149
.010	2.28E+00	4.781E-22	.0009	.0001	.0000	.0007	.0000	.9896	.0084
.100	1.18E+01	4.729F-22	.0035	.0003	.0001	.0011	.0000	.9861	.0085
1.000	1.78E+01	3.344E-22	.1010	.0009	.0002	.0023	.0000	.8707	.0120
10.000	2.00E+01	1.228E-21	.5021	.0004	.0001	.0028	.0000	.4007	.0033

1100.0-

TAU 5000	TAU NU	OPACITY	H	H-	H2+	SI	MG	C	RAY S	C EDGE
.001	1.49E-01	2.794E-22	.0017	.0000	.0000	.0013	.0000	.9869	.0089	10.86
.010	2.38E+00	4.996E-22	.0015	.0001	.0000	.0016	.0000	.9915	.0050	10.90
.100	1.24E+01	4.953E-22	.0058	.0003	.0001	.0025	.0000	.9858	.0050	8.99
1.000	1.87E+01	3.725E-22	.1572	.0010	.0003	.0048	.0000	.8183	.0067	3.39
10.000	2.14E+01	1.707E-21	.6266	.0004	.0001	.0045	.0000	.3017	.0015	1.32

1239.0-

TAU 5000	TAU NU	OPACITY	H	H-	H2+	SI	MG	C	RAY S	C EDGE
.001	1.36E-02	2.485F-23	.0271	.0006	.0003	.0263	.0001	.8879	.0459	7.51
.010	2.13E-01	4.542E-23	.0226	.0011	.0006	.0314	.0001	.9160	.0251	9.34
.100	1.21E+00	5.592E-23	.0731	.0033	.0015	.0382	.0001	.8594	.0204	6.00
1.000	2.23E+00	1.345E-22	.6167	.0032	.0011	.0233	.0001	.3149	.0085	1.43
10.000	4.19E+00	1.741E-21	.8702	.0004	.0001	.0076	.0001	.0571	.0007	1.06

1444.0-

TAU 5000	TAU NU	OPACITY	H	H-	H2+	SI	MG	C	RAY S	C EDGE
.001	2.10E-03	3.345E-24	.3145	.0051	.0027	.3010	.0007	.1566	.1307	1.19
.010	2.63E-02	5.524E-24	.2906	.0110	.0055	.3976	.0009	.1893	.0792	1.23
.100	1.82E-01	1.213E-23	.5268	.0186	.0082	.2714	.0007	.1199	.0361	1.14
1.000	7.26E-01	1.418E-22	.9137	.0037	.0013	.0340	.0001	.0135	.0031	1.01
10.000	3.36E+00	2.506E-21	.9441	.0003	.0001	.0081	.0001	.0026	.0002	1.00

1527.2-

TAU 5000	TAU NU	OPACITY	H	H-	H2+	SI	MG	C	RAY S	SI EDGE
.001	1.87E-03	2.904E-24	.4263	.0064	.0031	.3536	.0010	.0000	.1077	1.45
.010	2.24E-02	4.688E-24	.4029	.0140	.0066	.4780	.0013	.0000	.0667	1.71
.100	1.63E-01	1.177E-23	.6386	.0207	.0087	.2858	.0009	.0000	.0266	1.32
1.000	7.61E-01	1.629E-22	.9363	.0035	.0012	.0305	.0001	.0000	.0019	1.03
10.000	3.82E+00	2.918E-21	.9540	.0003	.0001	.0072	.0001	.0000	.0001	1.01

1683.3-

TAU 5000	TAU NU	OPACITY	H	H-	H2+	SI	MG	C	RAY S	SI EDGE
.001	1.59E-03	2.319E-24	.7076	.0092	.0040	.0747	.0001	.0000	.0767	1.08
.010	1.64E-02	3.322E-24	.7537	.0226	.0096	.1173	.0001	.0000	.0535	1.13
.100	1.33E-01	1.140E-23	.8744	.0245	.0094	.0566	.0001	.0000	.0156	1.06
1.000	8.54E-01	2.087E-22	.9686	.0031	.0010	.0057	.0000	.0000	.0009	1.01
10.000	4.83E+00	3.806E-21	.9686	.0003	.0001	.0017	.0001	.0000	.0000	1.00

2000.0

TAU 5000	TAU NU	OPACITY	H	H-	H2+	SI	MG	C	RAY S
.001	2.07E-03	3.101E-24	.8699	.0087	.0030	.0003	.0003	.0000	.0224
.010	2.19E-02	4.461E-24	.9228	.0213	.0072	.0005	.0005	.0000	.0155
.100	1.90E-01	1.716E-23	.9552	.0206	.0065	.0004	.0003	.0000	.0040
1.000	1.34E+00	3.377E-22	.9836	.0024	.0007	.0002	.0001	.0000	.0002
10.000	7.77E+00	6.149E-21	.9813	.0002	.0001	.0002	.0001	.0000	.0000

2517.1-

TAU 5000	TAU NU	OPACITY	H	H-	H2+	SI	MG	C	RAY S
.001	3.59E-03	5.592E-24	.9314	.0066	.0016	.0003	.0032	.0000	.0039
.010	4.03E-02	8.324E-24	.9548	.0157	.0037	.0005	.0054	.0000	.0026
.100	3.57E-01	3.258E-23	.9706	.0150	.0034	.0004	.0031	.0000	.0007
1.000	2.55E+00	6.454E-22	.9904	.0017	.0004	.0002	.0006	.0000	.0000
10.000	1.48E+01	1.162E-20	.9897	.0001	.0000	.0002	.0003	.0000	.0000

3647.1-

TAU 5000	TAU NU	OPACITY	H	H-	H2+	SI	MG	C	RAY S	H EDGE
.001	9.41E-03	1.517E-23	.9753	.0040	.0005	.0003	.0001	.0000	.0003	19.73
.010	1.10E-01	2.296E-23	.9823	.0093	.0012	.0006	.0001	.0000	.0002	21.62
.100	9.89E-01	9.063E-23	.9870	.0088	.0012	.0005	.0001	.0000	.0000	19.49
1.000	7.10E+00	1.791E-21	.9962	.0010	.0002	.0002	.0000	.0000	.0000	12.92
10.000	4.04E+01	3.129E-20	.9961	.0001	.0000	.0002	.0001	.0000	.0000	6.93

Table 3 (Cont.)

```
4000.0
 TAU 5000 TAU NU   OPACITY      H     H-    H2+    SI    MG   RAY S
    .001 7.19E-04  8.927E-25  .5753 .0765 .0088 .0036 .0010 .0030
    .010 6.30E-03  1.280E-24  .6674 .1878 .0218 .0068 .0020 .0021
    .100 5.96E-02  5.778E-24  .7817 .1545 .0187 .0048 .0016 .0005
   1.000 5.59E-01  1.776E-22  .9609 .0115 .0016 .0012 .0005 .0000
  10.000 5.57E+00  5.754E-21  .9789 .0005 .0001 .0008 .0004 .0000

5000.0
 TAU 5000 TAU NU   OPACITY      H     H-    H2+    SI    MG   RAY S
    .001 1.00E-03  1.365E-24  .7101 .0642 .0054 .0017 .0008 .0008
    .010 1.00E-02  2.099E-24  .7671 .1470 .0126 .0030 .0014 .0005
    .100 1.00E-01  9.973E-24  .8493 .1147 .0105 .0022 .0010 .0001
   1.000 1.00E+00  3.225E-22  .9767 .0081 .0009 .0006 .0003 .0000
  10.000 1.00E+01  1.026E-20  .9880 .0004 .0001 .0005 .0002 .0000

6500.0
 TAU 5000 TAU NU   OPACITY      H     H-    H2+    SI    MG   RAY S
    .001 1.64E-03  2.435E-24  .8290 .0451 .0029 .0008 .0004 .0001
    .010 1.83E-02  3.913E-24  .8559 .0987 .0064 .0015 .0007 .0001
    .100 1.90E-01  1.932E-23  .9074 .0741 .0053 .0011 .0005 .0000
   1.000 2.00E+00  6.524E-22  .9874 .0050 .0004 .0003 .0002 .0000
  10.000 2.01E+01  2.067E-20  .9939 .0003 .0000 .0003 .0001 .0000

8205.9-
 TAU 5000 TAU NU   OPACITY      H     H-    H2+    SI    MG   RAY S  H EDGE
    .001 2.70E-03  4.193E-24  .8980 .0290 .0016 .0005 .0002 .0000   2.84
    .010 3.17E-02  6.839E-24  .9114 .0627 .0035 .0009 .0004 .0000   2.85
    .100 3.34E-01  3.433E-23  .9430 .0467 .0029 .0007 .0003 .0000   2.76
   1.000 3.59E+00  1.174E-21  .9925 .0032 .0002 .0002 .0001 .0000   2.36
  10.000 3.58E+01  3.639E-20  .9964 .0002 .0000 .0002 .0001 .0000   1.85

14588.2-
 TAU 5000 TAU NU   OPACITY      H     H-    H2+    SI    MG   RAY S  H EDGE
    .001 3.17E-03  4.987E-24  .9222 .0162 .0012 .0007 .0003 .0000   1.65
    .010 3.82E-02  8.320E-24  .9425 .0358 .0026 .0013 .0006 .0000   1.64
    .100 4.30E-01  4.655E-23  .9641 .0279 .0019 .0010 .0003 .0000   1.59
   1.000 5.58E+00  2.004E-21  .9951 .0022 .0001 .0003 .0001 .0000   1.44
  10.000 6.90E+01  7.620E-20  .9980 .0001 .0000 .0003 .0001 .0000   1.29

16400.0
 TAU 5000 TAU NU   OPACITY      H     H-    H2+    SI    MG   RAY S
    .001 2.52E-03  3.913E-24  .9020 .0193 .0015 .0011 .0005 .0000
    .010 3.00E-02  6.580E-24  .9291 .0432 .0032 .0019 .0008 .0000
    .100 3.48E-01  3.851E-23  .9559 .0342 .0023 .0014 .0006 .0000
   1.000 4.91E+00  1.837E-21  .9942 .0027 .0001 .0004 .0002 .0000
  10.000 6.75E+01  7.742E-20  .9979 .0002 .0000 .0003 .0001 .0000

22794.1-
 TAU 5000 TAU NU   OPACITY      H     H-    H2+    SI    MG   RAY S  H EDGE
    .001 5.11E-03  8.247E-24  .9448 .0174 .0006 .0008 .0004 .0000   1.30
    .010 6.42E-02  1.419E-23  .9484 .0381 .0013 .0014 .0006 .0000   1.29
    .100 7.52E-01  8.333E-23  .9648 .0301 .0010 .0010 .0005 .0000   1.26
   1.000 1.06E+01  3.933E-21  .9960 .0024 .0001 .0003 .0001 .0000   1.20
  10.000 1.43E+02  1.632E-19  .9988 .0002 .0000 .0003 .0001 .0000   1.13

32823.5-
 TAU 5000 TAU NU   OPACITY      H     H-    H2+    SI    MG   RAY S  H EDGE
    .001 8.80E-03  1.447E-23  .9576 .0204 .0003 .0009 .0004 .0000   1.16
    .010 1.14E-01  2.542E-23  .9478 .0438 .0007 .0015 .0006 .0000   1.16
    .100 1.36E+00  1.526E-22  .9628 .0338 .0005 .0010 .0005 .0000   1.14
   1.000 1.99E+01  7.522E-21  .9963 .0026 .0000 .0003 .0001 .0000   1.11
  10.000 2.82E+02  3.262E-19  .9991 .0002 .0000 .0003 .0001 .0000   1.07

65646.8
 TAU 5000 TAU NU   OPACITY      H     H-    H2+    SI    MG   RAY S
    .001 2.99E-02  5.008E-23  .9694 .0234 .0001 .0008 .0004 .0000
    .010 3.99E-01  8.976E-23  .9471 .0491 .0001 .0014 .0006 .0000
    .100 4.88E+00  5.518E-22  .9609 .0371 .0001 .0010 .0004 .0000
   1.000 7.39E+01  2.848E-20  .9967 .0028 .0000 .0003 .0001 .0000
  10.000 1.10E+03  1.293E-18  .9994 .0002 .0000 .0002 .0001 .0000
```

drops to one-half its peak value. The line-wing opacity was extended to the point where the opacity had dropped to 1% of its peak value. For this division of a doppler profile, the peak line opacity, the mean core opacity, and the mean wing opacity are in the ratios of 1.000:0.807:0.159, and the widths of the core and wing are in the ratio of 1:1.6. This two-step representation is shown schematically in Figure 2.

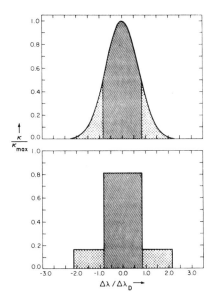

Figure 2.   Above: The doppler profile of the line opacity showing division between core and wing. Below: The two-level representation of the line opacity used in the picket.

We established the absolute levels of the core-like and wing-like steps semiempirically by selecting over 150 representative metal lines from the Utrecht and Brückner solar atlases. These lines were chosen so that they had the same strengths as the lines responsible for the major fraction of blanketing in each spectral region. The line-core opacity was calculated as a function of depth for each sample line. For this purpose we used in a (5781, 4.44) model the gf values and excitation potentials for the sample lines from the extensive tabulation of Strom and Kurucz. We also calculated line profiles and compared them with the observed profiles in the solar atlases in an effort to eliminate lines whose gf values were blatantly in error. The majority of the lines in the final set had similar depth dependences for the line-core opacities, although there was a systematic difference between neutral and ionized species. Figure 3 shows the mean depth dependences for the neutral lines in different wavelength regions (normalized to the surface value); also shown is the mean curve for all the ionized lines in the sample. The flattening of the curves at large depths is due to the influence of convection. The simple behavior of the line opacity, κ^ℓ, with depth and wavelength $\Delta\lambda_i$ suggested that it could be approximated by

$$\kappa^\ell(\tau, \ \Delta\lambda_i) = a(\tau) \cdot \zeta(\Delta\lambda_i) \quad .$$

The underlying assumptions of this method are 1) that the depth dependence of the lines is identical throughout the entire spectrum, and 2) that the line widths are independent of depth. The curve for the lines with $\lambda < 4019$ A was chosen to represent the function $a(\tau)$. The specification of $\zeta(\lambda)$ will be discussed below.

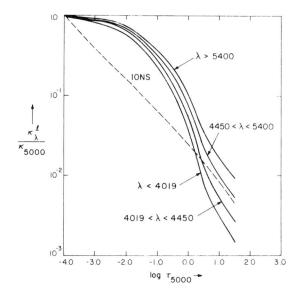

Figure 3.   Ratio (normalized to the surface value) of line center opacity to
κ_{5000} plotted versus continuum optical depth. Solid curves are
for neutral species in different wavelength regions; dashed curve
is the mean for all the ionized lines in the sample.

3.2 Frequency Weights of the Picket Opacity

To determine the relative frequency weights of the core-like and wing-
like opacities, consider the amount of blanketing in different spectral regions.
A useful observational quantity for this purpose is the integrated blanketing
or, occasionally "blocking" coefficient, $\eta(\lambda_1, \lambda_2)$, defined by the relation

$$\eta(\lambda_1, \lambda_2) = 1 - \frac{\int_{\lambda_1}^{\lambda_2} F_\lambda \, d\lambda}{\int_{\lambda_1}^{\lambda_2} F_\lambda^c \, d\lambda} \quad ,$$

where F_λ is the actual emergent flux, and F_λ^c is the continuum flux expected
at wavelength λ in the absence of blanketing. Tabulations of η (e.g., Labs
and Neckel, 1968; Iwanowska, these Proceedings) indicate that the energy
removed by metal lines is a decreasing function of wavelength. It is convenient to represent the average behavior of the blanketing coefficients by dividing the spectrum into five wavelength intervals and assigning the following
mean blanketing coefficients:

$$
\begin{aligned}
\lambda &\leq 4019^- & \eta &= 0.50 \\
4019^+ \leq \lambda &\leq 4450^- & \eta &= 0.25 \\
4450^+ \leq \lambda &\leq 5400^- & \eta &= 0.10 \\
5400^+ \leq \lambda &\leq 8206^- & \eta &= 0.03 \\
\lambda &\geq 8206^+ & \eta &\leq 0.01.
\end{aligned}
$$

The ultraviolet blanketing coefficients involve some uncertainty since no
reliable values are available for the short-wavelength regions. In lieu of
better information, we have assumed that the value of η for the interval from
911 to 4019 A was equal to the value observed in the near ultraviolet.

In order to convert the mean blanketing coefficients into weights for the picket levels, it is necessary to adopt some model for the distribution of spectral lines in both strength and frequency. Examination of the solar spectrum shows that in each of the five wavelength regions the lines responsible for most of the blanketing have roughly the same strength and do not appreciably overlap one another. Consequently, we have assumed that the lines blanketing a particular wavelength interval can be represented by a series of nonoverlapping doppler profiles of constant depth. We believe that this approximation, though incorrect, goes the greater part of the way toward producing realistically blanketed models. We will show later that the photospheric temperature distributions of the final models are not particularly sensitive to these assumptions.

Given this model, then, the blanketing coefficient is related to the number of lines, η, in the interval (λ_1, λ_2) of residual flux $\bar{r}$ and width $\Delta\lambda_D$ by

$$\eta(\lambda_1, \lambda_2) = \pi^{1/2}(1 - \bar{r}) \left[\frac{n\Delta\lambda_D}{\lambda_2 - \lambda_1} \right] .$$

The quantity in square brackets is a measure of the fraction of the spectral interval occupied by line opacity. If w_1, w_2, w_3 are the fractions of the wavelength interval occupied by the continuum, core-like and wing-like opacities, respectively, then the two-step representation for the line absorption coefficient requires the frequency weights

$w_2 = 0.935 [\eta /(1 - \bar{r})]$

$w_3 = 1.59 w_2$

$w_1 = 1.0 - (w_2 + w_3) .$

For the sun we find the frequency weights shown schematically in Figure 4.

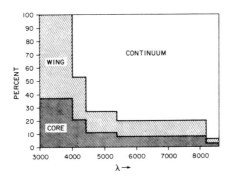

Figure 4.   Relative frequency weights of the three picket levels versus wavelength.

3.3 The Strength of the Picket Opacity $\zeta(\lambda)$

Once the blanketing weights and the depth dependence of the line opacities have been established, the only remaining unknowns are the values of $\zeta(\lambda)$. These line strengths were used as free parameters in the construction of the blanketed models. The value of $\zeta(\lambda)$ was varied in each spectral region until the calculated η for that region corresponded to the observed value. In the solar case this proved to be a self-consistent procedure. That is, the level of the picket for each wavelength interval in the converged model agreed reasonably well with the $\bar{r}$ values adopted in determining the w's; at the same time, reasonable agreement was obtained between the final picket opacities and those of the representative lines.

Figure 5 shows that a blanketed solar model constructed in this fashion compares favorably in the photosphere with the most recent empirical determinations of the solar temperature distribution (Gingerich and de Jager, 1968; Gingerich, 1969).   An unblanketed model of the solar effective temperature and gravity is plotted for reference.   The discrepancies between the observed and the theoretical temperature distributions in the shallow layers will be discussed at the end of Section 3. 5.

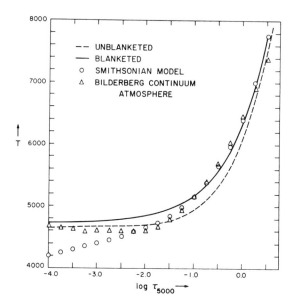

Figure 5.   A comparison of theoretical and empirical solar models.

3. 4   Extension to the Grid Models

The approach just outlined for the solar case was extended in an attempt to simulate the effects of line blanketing for all the models in the grid with $T_{eff} \leq 8500°K$.   This effort was hampered by the paucity of detailed blanketing information in the literature for stars other than the sun.   Nevertheless, in view of the large differences between models with and without blanketing, we felt that even a rough approximation for the blanketing was better than none at all.   We endeavored to select the best set of blanketing parameters that could be gleaned from the available observational material; since others might have chosen different parameters, we will suggest later how the models might be appropriately scaled.   It should be noted that more detailed information on the blanketing over a wide range of effective temperatures and gravities is urgently needed.   Any model-atmosphere technique attempting to reproduce the effects of blanketing must yield agreement with observed values of $\eta(\lambda_1, \lambda_2)$.

The blanketing coefficients that we used for the grid models were taken from Milford (1950), Wildey, Burbidge, Sandage, and Burbidge (1962), and Parsons (1969).   Since these studies did not cover all the temperatures and gravities represented in the grid, considerable interpolation and extrapolation were required.   The resulting η's for the various wavelength intervals are shown in Table 4.   Because of uncertainties both in the measured η's and in the assignment of appropriate temperatures and gravities for the stars, these results will undoubtedly change as more and better observational material becomes available.   The frequency weights were derived from these η's on the assumption of solar values of $\bar{\tau}$ in the different wavelength intervals.   In a few instances, w_2 was found to increase toward longer wavelengths rather than decrease. When this occurred, the value of w_2 for the preceding interval was adopted. Since practically no data are available for the η's in the interval $\lambda > 8206$, the frequency weights for that region were scaled in the same proportion as those of the preceding wavelength interval. The depth-dependent opacity function, $a(\tau)$, was established separately for three ranges in T_{eff}. Sample lines, selected from high-dispersion spectra of Procyon and Arcturus, were used, as in the solar case, to establish a characteristic $a(\tau)$ relation for both stars (see Figure 6). The Procyon $a(\tau)$ relation was used for all blanketed models with $T_{eff} \geq 6500°$, the solar relation for $6000° \geq T_{eff} \geq 5000°$, and the Arcturus relation for $T_{eff} \geq 4500°$. As before, the absolute strengths of the picket opacities were adjusted until the calculated η's (see Table 5) for each wavelength interval agreed reasonably with those in Table 4. Our confidence in the procedure was buoyed by the fact that only minor adjustment of the opacity level in the pickets was required to obtain the desired blanketing. The variations in $\zeta(\lambda)$ were often only a few percent from one effective temperature to another; the variation of $\zeta(\lambda)$ with gravity was usually negligible.

Table 4. Observed blanketing coefficients, $\eta(\lambda_1, \lambda_2)$

$(T_{eff}, \log g)$	η_1 $\lambda \leq 4019$	η_2 $4019 \leq \lambda \leq 4601$	η_3 $4601 \leq \lambda \leq 5450$	η_4 $5450 \leq \lambda \leq 8206$
(8500, 4)	0.05	0.05	0.035	0.015
(8500, 3)	0.05	0.05	0.035	0.015
(8000, 4.5)	0.07	0.06	0.05	0.02
(8000, 4)	0.07	0.06	0.05	0.02
(8000, 3.5)	0.07	0.06	0.05	0.02
(8000, 3)	0.07	0.06	0.05	0.02
(8000, 2)	0.07	0.06	0.05	0.02
(7500, 4)	0.09	0.06	0.05	0.02
(7500, 3)	0.11	0.10	0.06	0.02
(7000, 4)	0.17	0.10	0.06	0.02
(7000, 3)	0.21	0.16	0.08	0.03
(7000, 2)	0.24	0.21	0.10	0.04

$(T_{eff}, \log g)$	$\lambda \leq 4019$	$4019 \leq \lambda \leq 4450$	$4450 \leq \lambda \leq 5400$	$5400 \leq \lambda \leq 8206$
(6500, 4)	0.29	0.16	0.08	0.02
(6500, 3)	0.32	0.23	0.10	0.04
(6000, 4.5)	0.42	0.24	0.10	0.03
(6000, 4)	0.42	0.24	0.10	0.03
(6000, 3.5)	0.47	0.32	0.13	0.05
(6000, 3)	0.47	0.32	0.13	0.05
(6000, 2)	0.53	0.40	0.17	0.07
(5500, 4)	0.58	0.32	0.13	0.05
(5500, 3)	0.63	0.40	0.17	0.06
(5000, 4)	0.72	0.39	0.17	0.05
(5000, 3)	0.78	0.49	0.22	0.09
(5000, 2)	0.84	0.60	0.28	0.12
(4500, 4)	0.84	0.46	0.20	0.07
(4500, 3)	0.90	0.56	0.27	0.11
(4000, 4)	0.95	0.53	0.24	0.08
(4000, 3)	0.99	0.63	0.33	0.14
(4000, 2)	0.99	0.75	0.43	0.18

3.5 Discussion

Since the procedure outlined above is a considerable simplification of the blanketing problem, it is important to determine the sensitivity of the final models to some of our assumptions. The form of $a(\tau)$ has some influence on the final temperature structure. Figure 7 shows the effect of different choices of $a(\tau)$ on a $(6000, 4)$ atmosphere. Each of these models has the same set of frequency weights and η's. In one variation, $a(\tau)$ has no depth dependence; in the other, $a(\tau)$ follows a $1/\sqrt{\tau}$ dependence that is more characteristic of ionized than neutral lines. Good agreement is obtained in the region where the visible continuum is formed, despite the large differences in representation of $a(\tau)$. Appreciable divergence of the temperature distributions occurs only in those layers that are near the surface and optically thin in the continuum.

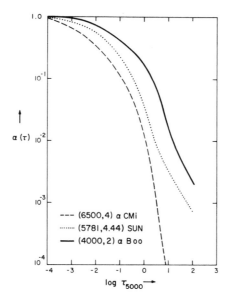

Figure 6.   The $a(\tau)$ relations selected for Procyon, the sun, and Arcturus.

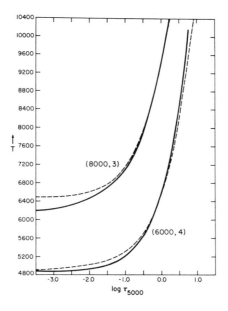

Figure 7.   For the $(6000, 4)$ models: solid curve, $a(\tau)$ = constant; dashed curve, $a(\tau) \propto 1/\sqrt{\tau}$. For the $(8000, 3)$ models: solid line, case I, dashed line, case II.

Table 5. Calculated blanketing coefficients, $\eta(\lambda_1, \lambda_2)$

Model	η_{total}	η_1	η_2	η_3	η_4	η_5
(8500, 4)	0.033	0.055	0.045	0.034	0.018	0.000
(8500, 3)	0.033	0.054	0.047	0.036	0.018	0.000
(8000, 4.5)	0.042	0.075	0.057	0.044	0.024	0.000
(8000, 4, L/H = 0)	0.043	0.076	0.061	0.047	0.025	0.000
(8000, 4)	0.043	0.075	0.060	0.047	0.025	0.000
(8000, 3.5)	0.041	0.072	0.061	0.047	0.024	0.000
(8000, 3)	0.041	0.071	0.063	0.049	0.025	0.000
(8000, 2)	0.042	0.067	0.066	0.054	0.027	0.000
(7500, 4)	0.046	0.107	0.059	0.046	0.015	0.000
(7500, 4, L/H = 2.5)	0.047	0.108	0.058	0.046	0.015	0.000
(7500, 4, L/H = 0)	0.048	0.109	0.061	0.046	0.015	0.000
(7500, 3)	0.057	0.115	0.098	0.065	0.024	0.000
(7000, 4)	0.075	0.207	0.093	0.062	0.015	0.000
(7000, 4, L/H = 0)	0.076	0.208	0.093	0.062	0.015	0.000
(7000, 3)	0.088	0.227	0.147	0.077	0.036	0.000
(7000, 2)	0.098	0.229	0.201	0.093	0.046	0.001
(6500, 4)	0.107	0.337	0.148	0.080	0.024	0.000
(6500, 3)	0.120	0.359	0.221	0.101	0.052	0.001
(6000, 4.5)	0.133	0.469	0.240	0.091	0.024	0.001
(6000, 4)	0.127	0.460	0.242	0.092	0.024	0.001
(6000, 4, L/H = 0)	0.127	0.460	0.244	0.092	0.024	0.001
(6000, 3.5)	0.146	0.503	0.329	0.117	0.040	0.001
(6000, 3)	0.138	0.488	0.333	0.118	0.040	0.001
(6000, 2)	0.155	0.506	0.429	0.164	0.059	0.002
(5500, 4)	0.144	0.612	0.337	0.120	0.035	0.001
(5500, 3)	0.155	0.612	0.430	0.167	0.052	0.002
(5000, 4)	0.155	0.808	0.440	0.176	0.045	0.002
(5000, 3)	0.163	0.795	0.523	0.224	0.083	0.003
(5000, 2)	0.161	0.771	0.554	0.292	0.111	0.004
(4500, 4)	0.149	0.929	0.514	0.216	0.076	0.006
(4500, 3)	0.159	0.922	0.544	0.297	0.118	0.010
(4000, 4)	0.102	0.947	0.567	0.272	0.088	0.009
(4000, 3)	0.122	0.942	0.593	0.375	0.155	0.016
(4000, 2)	0.123	0.926	0.624	0.431	0.185	0.021

Also in Figure 7 we demonstrate the effect of changing the blanketing weights. In these (8000, 3) models the weights are assigned as follows:

$$\lambda \leq 8206 \begin{cases} \text{Case I} & w_1 = 0.87 \quad w_2 = 0.05 \quad w_3 = 0.08 \\ \text{Case II} & w_1 = 0.05 \quad w_2 = 0.08 \quad w_3 = 0.87 \end{cases}$$

$\lambda \geq 8206$ Both $w_1 = 0.97$ $w_2 = 0.012$ $w_3 = 0.018$

The depth dependence, $a(\tau)$, is the same for both models, and both are iterated to yield the same values for the η's. Once again the final temperature distributions agree well in the lower photosphere, diverging only in the shallower layers. These examples illustrate that the atmospheric layers responsible for the visible continuum are not strongly affected by the choice of blanketing parameters provided that the iterated models have the appropriate η values. The reason for this behavior will be discussed in Section 5.

It is apparent that the surface temperatures of these models are not so well determined as those at greater depths. As the line frequencies become optically thin, they begin to contribute to the net outward flux. This additional flux contribution causes the temperature to drop by an amount necessary to maintain a zero flux derivative. The extent of the temperature drop is dependent upon the strength of the blanketing opacity and its frequency distribution. Experimental computations have shown that including the opacities representative of very strong lines in LTE will noticeably influence the surface temperatures even though the lines occupy such small intervals in frequency space that the blanketing η's are barely affected. Finally, although we have assumed throughout that the line-formation process is in LTE, this has not been proved to be the case. It should be noted that several investigators (e. g., Thomas, 1965; Frisch, 1966; Cayrel, 1966) have argued that when non-LTE processes are taken into account, the surface temperature drop is less than in LTE.

4. SOME COMPARISONS WITH OTHER STUDIES

The elaborate Procyon model calculated by Strom and Kurucz (1966) may be directly compared with our (6500, 4) model in Figure 8. The Strom-Kurucz model has systematically higher temperatures in the photosphere and a lower surface temperature than does ours. This would seem to indicate that their model is the more heavily blanketed. Although it is not possible to compare calculated blanketing coefficients, some information may be gained by comparing the blanketed flux distributions for the two models with the spectral-energy distribution reported by Bahner (1963) (see Figure 9). Bahner's results are uncorrected for line blocking and represent the average flux through a 54-A bandpass; they have been adjusted to agree with Hayes' (1967) calibration of Vega. Our grid model is more successful in matching the observed fluxes in the interval $\lambda\lambda 4000$ to 6000 A, thereby reducing the discrepancy between the earlier theoretical and observed Paschen continuum pointed out by Underhill (1966).

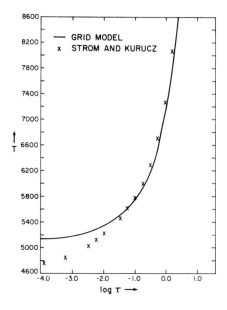

Figure 8.   Blanketed temperature distributions for Procyon. The grid model is plotted against a τ_{5000} scale while the Strom-Kurucz model is plotted against a τ_{4939} scale (Strom, 1969, private communication).

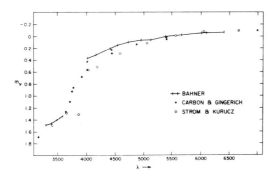

Figure 9.   Predicted and observed flux distributions for Procyon in magnitudes per unit frequency interval.   The curves have been normalized to agree at 5560 A.

In Figure 10, we compare two recent empirical determinations of the solar temperature distribution (Gingerich and de Jager, 1968; Gingerich, 1969) with two models from the grid, (6000,4) and (5500,4).   Plotted for reference is an unblanketed (5781,4.44) model without convection.   The differences between the two empirical models are representative of the uncertainties that still remain in the solar temperature distribution.   It is apparent, however, that the grid models are considerably more successful than the unblanketed in reproducing the photospheric T(t) relation.

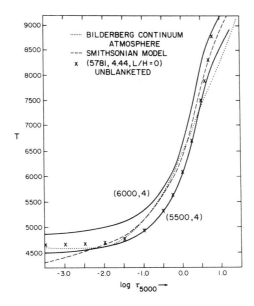

Figure 10.   Temperature distributions from two blanketed grid models compared with empirical solar models and an unblanketed solar model.

Finally, in Figure 11, a (6000,2.4) model constructed by Parsons (1969) is shown along with two models from the grid.   Since Parsons' blanketing technique is similar to our own in that he iterates to an observed η, his model shows a temperature structure very similar to that shown by ours.

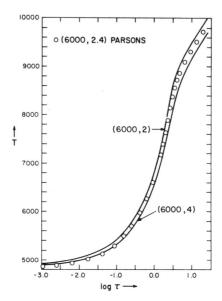

Figure 11.   Blanketed temperature distributions from the grid compared
with a model by Parsons.   The grid models are plotted against
a τ_{5000} scale while Parsons' model is plotted against a τ_{4950}
scale (Parsons, 1969, private communication).

5.   SCALING THE TEMPERATURE DISTRIBUTIONS

The total blocking coefficient for the entire spectrum can be written as

$$\eta_{total} = 1.0 - \frac{\int_0^\infty F_\lambda \, d\lambda}{\int_0^\infty F_\lambda^c \, d\lambda} \quad .$$

The integral of F_λ^c can be used to define a temperature T_c such that

$$\frac{\sigma}{\pi} T_c^4 = \int_0^\infty F_\lambda^c \, d\lambda \quad .$$

So specified, T_c is a "characteristic temperature" representative of the
important continuum-forming depths of the blanketed model.   The relation

$$\frac{T_c}{T_{eff}} = \left(\frac{1}{1 - \eta_{total}} \right)^{1/4}$$

shows that the characteristic temperature is related to the effective temperature by a simple scaling factor in η_{total}.   In view of this relation, it is not
surprising that the pairs of models in the examples of Section 3.5 have similar $T(\tau)$ distributions at the continuum-forming depths.

The foregoing equation provides an obvious approximation for deriving
models with blanketing η's different from those we have chosen.   This scaling
procedure can also be used with some success to scale from an unblanketed

photospheric distribution to a blanketed, or vice versa. It should be noted that the sense of the scaling will be correct only for the deeper continuum-forming layers.

A common technique (e. g. , Cayrel and Cayrel, 1963; Cayrel and Jugaku, 1963; Krishna Swamy, 1966) for producing the temperature distributions of intermediate spectral-type stars consists of scaling an empirical solar $T(\tau)$ relation by the ratio $T_{eff}^{star}/T_{eff}^{\odot}$. Some estimate of the validity of this approach can be obtained by use of the models in the grid. In Table 6 we compare the temperature distributions determined by directly calculating blanketed models and by scaling a (6000, 4) model in this manner. The tabulated quantities are the differences between the directly calculated and the scaled temperatures in the sense T (calculated) - T (scaled). The scaling relation fails in going to higher values of T_{eff} because of the shift from H^- to H as the dominant continuous opacity and the increased importance of the hydrogen-line blocking. Scaling to cooler temperatures involves appreciably less error since H^- remains the principal opacity source. These results suggest that a solar model scaled to 4000° effective temperature will differ comparatively little from the explicitly calculated grid models. Hence, the scaling technique, which has commonly been used without thorough examination, has been validated.

Table 6. Temperature differences between calculated and s called models

$\log \tau_{5000}$	(8000,4)	(7500,4)	(7000,4)	(6500,4)	(5500,4)	(5000,4)	(4500,4)	(4000,4)
-4	-233	-204	-176	-115	39	- 5	- 36	- 8
-3	-244	-217	-182	-113	39	- 5	-42	- 25
-2	-174	-159	-123	- 67	35	11	-41	- 44
-1.5	-136	-136	- 97	- 45	31	16	-52	- 66
-1.0	- 75	-108	- 74	- 24	26	16	-62	- 90
-0.5	88	- 37	- 43	- 15	20	11	-67	-120
0.0	543	258	86	5	5	3	-44	-149
0.5	248	80	121	42	-63	-68	29	-187

We should first like to thank Miss Barbara Welther for her patient help at every stage of this investigation; she often fell heir to unpleasant tasks of tabulation and cross-checking and did them well. Professor C. A. Whitney's continued interest and support over the many years of development of this stellar-atmosphere program has given us inestimable encouragement. We thank Dr. S. E. Strom and R. K. Kurucz for informative discussions on the Procyon problem. Dr. Frank Edmonds very kindly provided us with unpublished data on the infrared blanketing in Arcturus. Finally, we thank those members of the SAO Computing Center staff who did whatever they could to make a seemingly interminable computational task possible; Richard Kozlowski deserves special mention for his care in running the reproduction output used for these tables.

REFERENCES

Avrett, E. H. 1964, Smithsonian Ap. Obs. , Spec. Rep. No. 167, 83.
Bahner, K. 1963, Ap. J. , 138, 1314.
Carbon, D. F. 1969, Bull. Am. Astron. Soc., in press.
Cayrel, G. , and Cayrel, R. 1963, Ap. J. , 137, 431.
Cayrel, R. 1966, J. Q. S. R. T. , 6, 621.
Cayrel, R. , and Jugaku, J. 1963, Ann. d'Ap. , 26, 495.

Demarque, P. R. , and Percy, J. R. 1964, Ap. J. , 140, 541.

Demarque, P. , and Roeder, R. C. 1967, Ap. J. , 147, 1188.

Frisch, H. 1966, J. Q. S. R. T. , 6, 629.

Gingerich, O. 1963, Ap. J. , 138, 576.

_____. 1964, Proc. First Harvard-Smithsonian Conference on Stellar Atmospheres, Smithsonian Ap. Obs. , Spec. Rep. No. 167, 17.

_____. 1966, J. Q. S. R. T. , 6, 609.

_____. 1969, to be published.

Gingerich, O. , and de Jager, C. 1968, Solar Phys. , 3, 5.

Gingerich, O. , and Rich, J. C. 1968, Solar Phys. , 3, 82.

Hayes, D. S. 1967, unpublished thesis: An Absolute Calibration of the Energy Distribution of Twelve Spectrophotometric Standard Stars (University of California, Los Angeles).

Henyey, L. , Vardya, M. S. , and Bodenheimer, P. 1965, Ap. J. , 142, 841.

Krishna Swamy, K. S. 1966, Ap. J. , 145, 174.

Labs, D. , and Neckel, H. 1968, Zs. f. Ap. , 69, 1.

Lambert, D. L. 1968, Mon. Not. Roy. Astron. Soc. , 138, 143.

Lambert, D. L., and Warner, B. 1968a, Mon. Not. Roy. Astron. Soc. , 138, 181.

_____. 1968b, Mon. Not. Roy. Astron. Soc. , 138, 213.

_____. 1968c, Mon. Not. Roy. Astron. Soc. , 140, 197.

Mihalas, D. 1967, in Methods in Computational Physics, vol. 7: Astrophysics, ed. B. Alden, S. Fernbach, and M. Rotenberg (New York: Academic Press).

Milford, N. 1950, Ann. d'Ap. , 13, 243.

Parsons, S. B. 1969, Ap. J. Suppl. No. 159, 18, 127.

Peterson, D. M. 1969, Smithsonian Ap. Obs. , Spec. Rep. No. 293, 199 pp.

Strom, S. E., and Kurucz, R. L. 1966, J. Q. S. R. T. , 6, 591.

Swihart, T. L. 1966, Ap. J. , 143, 358.

Thomas, R. N. 1965, Ap. J. , 141, 333.

Torres-Peimbert, S. , Simpson, E. , and Ulrich, R. K. 1969, Ap. J. , 155, 957.

Underhill, A. B. 1966, J. Q. S. R. T. , 6, 607.

Warner, B. 1968, Mon. Not. Roy. Astron. Soc. , 138, 219.

Wildey, R. L. , Burbidge, E. M. , Sandage, A. R. , and Burbidge, G. R. 1962, Ap. J. , 135, 94.

```
      FUNCTION AVMET(THETA,FREQ,ASI,AMG)                        AMET  10
C-----SILICON AND MAGNESIUM OPACITY PER NEUTRAL H ATOM, SCALED BY E26.  AMET  20
C-----ARGUMENTS ARE RECIPROCAL TEMPERATURE, FREQUENCY,          AMET  30
C-----ABUNDANCE OF NEUTRAL SILICON PER NEUTRAL HYDROGEN ATOM,   AMET  40
C-----ABUNDANCE OF NEUTRAL MAGNESIUM PER NEUTRAL HYDROGEN ATOM. AMET  50
C-----GINGERICH AND RICH, SOLAR PHYSICS 3,82,1968.             AMET  60
      DIMENSION FLIM(12)                                        AMET  70
      DATA FLIM/19.63E14,18.47E14,17.81E14,15.08E14,14.43E14,11.91E14,  AMET  80
     *7.99E14,7.73E14,6.00E14,4.29E14,3.5E14/                  AMET  90
      TK=.43429/THETA                                           AMET 100
      IF(FREQ-FR) 30,50,30                                      AMET 110
   30 FR=FREQ                                                   AMET 120
      HV=4.1352E-15*FR                                          AMET 130
      A=0.                                                      AMET 140
      B=0.                                                      AMET 150
      C=0.                                                      AMET 160
      E=0.                                                      AMET 170
      G=0.                                                      AMET 180
      V=0.                                                      AMET 190
      W=0.                                                      AMET 200
      X=0.                                                      AMET 210
      Z=0.                                                      AMET 220
C-----ALLOW FOR NJ**-14 DEPENDENCE OF MG TRIPLET P, 2515 A      AMET 230
      WFR=(11.91E14/FR)**11                                     AMET 240
C-----ALLOW FOR NJ**-1.5 DEPENDENCE OF SILICON SINGLET S, 1986 A  AMET 250
      CFR=SQRTF((FR/15.08E14)**3)                               AMET 260
      DO 40 I=1,10                                              AMET 270
      IF(FLIM(I)-FR) 50,50,40                                   AMET 280
   40 CONTINUE                                                  AMET 290
      I=11                                                      AMET 300
      GO TO 11                                                  AMET 310
   50 I=I                                                       AMET 320
      GO TO (1,2,3,4,44,5,6,7,8,10,11),I                        AMET 330
C-----GROUND STATE 3P2 TRIPLET P FROM SI. (1526A).             AMET 340
    1 IF(FR-22.20E14) 101,101,102                               AMET 350
C-----RICH EXPERIMENTAL VALUE, 40 MGBARNS AT LIMIT,CONSTANT WITH FREQ.  AMET 360
  101 A=(FR/19.63E14)**3                                        AMET 370
      GO TO 2                                                   AMET 380
  102 A=(22.20E14/FR)**2*1.446                                  AMET 390
C-----RICH GUESS OF NU**-5 FALL OFF BELOW 1350A                 AMET 400
C-----GROUND STATE 3S2 SINGLET S FROM MG. (1620A).             AMET 410
C-----DITCHBURN AND MARR EXPERIMENTAL VALUE, 1.18 MGBARNS AT LIMIT.  AMET 420
    2 V=0.0282                                                  AMET 430
      I=3                                                       AMET 440
C-----3P2 SINGLET D FROM SI.    (1682A).                       AMET 450
C-----RICH EXPERIMENTAL VALUE, 35 MGBARNS AT LIMIT, CALCULATED NU**-3  AMET 460
    3 B=0.444/EXPF(.78/TK)                                      AMET 470
C-----3P2 SINGLET S FROM SI.    (1986A).                       AMET 480
C-----RICH QUANTUM DEFECT CALCULATION, 46.5 MGBNS AT LIMIT, NU**-1.5 DEPAMET 490
    4 C=CFR*0.0628/EXPF(1.91/TK)                                AMET 500
C-----GROUND STATE OF AL.        (2076A).                      AMET 510
C-----PARKINSON AND REEVES EXPERIMENTAL VALUE, 22 MGBARNS.      AMET 520
   44 G=0.230                                                   AMET 530
C-----3P TRIPLET P ODD FROM MG. (2515A).                       AMET 540
C-----BOTTICHER EXPERIMENTAL VALUE, 45 MGBARNS, NU**-14 DEPENDENCE.  AMET 550
    5 W=WFR*2.32/EXPF(2.71/TK)                                  AMET 560
C-----3P SINGLET P ODD FROM MG. (3750A).                       AMET 570
    6 X=0.065/EXPF(4.33/TK)                                     AMET 580
C-----CONTINUUM FROM 4S FOR SI. (3880A).                       AMET 590
    7 D=.974*TK/EXPF(5.0/TK)                                    AMET 600
C-----CONTINUUM FROM 4S FOR MG. (5000A).                       AMET 610
    8 Y=1.90*TK/EXPF(5.24/TK)                                   AMET 620
C-----3D TERMS FROM SI.                                         AMET 630
      E=.069/EXPF(5.7/TK)                                       AMET 640
C-----3D TERMS FROM MG.          (7000A).                      AMET 650
   10 Z=.090/EXPF(5.9/TK)                                       AMET 660
      GO TO (13,13,13,13,13,13,13,13,12,11,11) , I              AMET 670
C-----HIGHER TERMS FOR MG.                                      AMET 680
   11 Y=.962*TK/EXPF((7.64-HV)/TK)                              AMET 690
C-----HIGHER TERMS FOR SI.                                      AMET 700
   12 D=.493*TK/EXPF((8.14-HV)/TK)                              AMET 710
C-----SCALE TO GET 10000-DEGREE PARTITION FUNCTIONS.           AMET 720
   13 SUM=ASI*(A+B+C+D+E)*.825 +AMG*(V+W+X+Y+Z)*.725            AMET 730
C-----CONSTANT FROM ALLEN ASTROPHYSICAL QUANTITIES, P. 89.     AMET 740
      AVMET=(1.E26/FR)*SUM*(1.-EXPF(-HV/TK))*(2.815E29/FR/FR)  AMET 750
C-----FINAL EQUATION INCLUDES STIMULATED EMISSION.             AMET 760
  100 RETURN                                                    AMET 770
      END                                                       AMET 780
```

```
      FUNCTION AVH2P (TH,V)                                          AH2P   10
C-----H2+ OPACITY PER NEUTRAL HYDROGEN ATOM AND PER H+, SCALED BY E26. AH2P 20
C-----BASED ON BATES, J.CHEM.PHYS. 19,1122,1951--M.N. 112,40,1952--   AH2P   30
C-----PHIL.TRANS.R.SOC.LONDON A 246,215,1953.                        AH2P   40
C-----SEE GINGERICH SAO SPECIAL REPORT 167,21,1964.                  AH2P   50
C-----12 FEBRUARY 1962 - 29 JANUARY 1964.                            AH2P   60
      DIMENSION FR(46),US(46),UP(46),E(46)                           AH2P   70
      DIMENSION N(1),ELIM(46)                                        AH2P   80
      DATA ELIM/1.E36,2.716,2.437,2.156,1.898,1.660,1.445,1.254,1.086, AH2P 90
     D .9354,.8109,.7004,.6051,.5229,.4518,.3904,.3374,.2915,.2517,  AH2P  100
     D .2171,.1871,.1611,.1386,.1190,.1021,.08758,.07498,.06390,.05434, AH2P 110
     D .04629,.03950,.03357,.02840,.02397,.02029,.01722,.01456,.01225, AH2P 120
     D .01028,.008657,.007312,.006153,.005149,.004300,.0 /           AH2P  130
C----PRESET N FOR FIRST ENTRY ONLY.                                  AH2P  140
      DATA N/20/                                                     AH2P  150
      DATA E/3.0,2.852,2.58,2.294,2.023,1.774,1.547,1.344,1.165,1.007, AH2P 160
     D.8702,.7516,.6493,.5610,.4848,.4189,.3620,.3128,.2702,.2332,   AH2P  170
     D.2011,.1732,.1491,.1281,.1100,.09426,.0809,.06906,.05874,.04994, AH2P 180
     D.04265,.03635,.0308,.026,.02195,.01864,.01581,.01332,.01118,.00938AH2P 190
     D,.00793,.00669,.00561,.00469,.00392,.0033/                     AH2P  200
      DATAUP/85,,9.99465,4.97842,3.28472                             AH2P  210
     D       ,2.41452    , 1.87038    ,  1.48945   ,  1.20442   ,.98279AH2P 220
     D       ,.80665     ,  .66493    ,  .54997    ,  .45618    ,.37932AH2P 230
     D       ,.31606     ,  .26382    ,  .22057    ,  .18466    ,.15473AH2P 240
     D       ,.12977     , 1.08890E-01, 9.14000E-02, 7.67600E-02,6.44  AH2P 250
     D500E-02, 5.41200E-02, 4.54000E-02, 3.81000E-02, 3.19500E-02,2.67 AH2P 260
     D600E-02, 2.23700E-02, 1.86900E-02, 1.56100E-02, 1.30200E-02,1.08 AH2P 270
     D300E-02, 8.99000E-03, 7.45000E-03, 6.15000E-03, 5.08000E-03,4.16 AH2P 280
     D000E-03, 3.42000E-03, 2.77000E-03, 2.21000E-03, 1.78000E-03,1.45 AH2P 290
     D000E-03,1.24000E-03,1.14000E-03/                               AH2P  300
      DATA US/-85,,           -7.1426,       -2.3984,   -.99032,-.39105 AH2P 310
     D     , -.09644    ,    .05794   ,    .13996   ,  .18186   ,.20052AH2P 320
     D     ,  .20525    ,    .20167   ,    .19309   ,  .18167   ,.16871AH2P 330
     D     ,  .15511    ,    .14147   ,    .12815 ,    .11542   ,.10340AH2P 340
     D     ,  .09216    , 8.18000E-02, 7.22900E-02, 6.36700E-02,5.58   AH2P 350
     D400E-02, 4.88400E-02, 4.25700E-02, 3.69900E-02, 3.20700E-02,2.77 AH2P 360
     D500E-02, 2.39400E-02, 2.06100E-02, 1.77200E-02, 1.52200E-02,1.30 AH2P 370
     D500E-02, 1.11900E-02, 9.58000E-03, 8.21000E-03, 7.01000E-03,6.00 AH2P 380
     D000E-03, 5.11000E-03, 4.35000E-03, 3.72000E-03, 3.22000E-03,2.86 AH2P 390
     D000E-03,2.63000E-03/                                           AH2P  400
      DATAFR/0.,4.30272E-18,1.51111E-17, 4.02893E-17, 8.89643E-17, 1.70 AH2P 410
     D250E-16, 2.94529E-16, 4.77443E-16, 7.25449E-16, 1.06238E-15, 1.50 AH2P 420
     D501E-15, 2.08046E-15, 2.82259E-15, 3.76256E-15, 4.93692E-15, 6.38 AH2P 430
     D227E-15, 8.17038E-15, 1.02794E-14, 1.28018E-14, 1.57371E-14, 1.91 AH2P 440
     D217E-14, 2.30875E-14, 2.75329E-14, 3.27526E-14, 3.85481E-14, 4.52 AH2P 450
     D968E-14, 5.18592E-14, 5.99825E-14, 6.92092E-14, 7.94023E-14, 9.01 AH2P 460
     D000E-14, 1.01710E-13, 1.14868E-13, 1.29969E-13, 1.46437E-13, 1.63 AH2P 470
     D042E-13, 1.81440E-13, 2.02169E-13, 2.25126E-13, 2.49637E-13, 2.73 AH2P 480
     D970E-13, 3.00895E-13, 3.30827E-13, 3.64140E-13, 3.99503E-13, 4.34 AH2P 490
     D206E-13/                                                       AH2P  500
C                                                                    AH2P  510
      IF(V-VEE) 38,52,38                                             AH2P  520
   38 VEE=V                                                          AH2P  530
C----GIVEN V AND T, GET K(H2P).                                      AH2P  540
      EV=3.03979E-16*V                                               AH2P  550
      N=N                                                            AH2P  560
      IF(EV-ELIM(N  )) 39,44,42                                      AH2P  570
   39 N=N+1                                                          AH2P  580
   40 IF(EV-ELIM(N)) 39,44,44                                        AH2P  590
   42 N=N-1                                                          AH2P  600
      IF(EV-ELIM(N  )) 39,44,42                                      AH2P  610
   44 FV=PARA(E(N),1,FR(N),1,EV)                                     AH2P  620
      VP=PARA(E(N),1,UP(N),1,EV)                                     AH2P  630
      VS=PARA(E(N),1,US(N),1,EV)                                     AH2P  640
C----V CONSTANT, GIVEN T, GET K(H2P).                                AH2P  650
   52 TK=0.0319273/TH                                                AH2P  660
      AVH2P =ABSF(FV*(EXPF(VS/TK)-EXPF(-VP/TK)))                     AH2P  670
      RETURN                                                         AH2P  680
      END                                                            AH2P  690


      FUNCTION AVRAY(FR)                                             ARAY   10
C-----RAYLEIGH SCATTERING PER NEUTRAL HYDROGEN ATOM, SCALED BY E26.   ARAY   20
C-----DALGARNO (SEE GINGERICH, SAO SPECIAL REPORT 167,20,1964)       ARAY   30
      FY=AMIN1(FR,29.22E14)                                          ARAY   40
      W2=(2.9979E18/FY)**2                                           ARAY   50
      AVRAY=(5.799E13*(W2**2)+1.422E20*W2+2.78E26)/(W2**2)**2        ARAY   60
      RETURN                                                         ARAY   70
      END                                                            ARAY   80
```

```
      FUNCTION AVHYD(THETA,FREQ)                                    AHYD  10
C-----HYDROGEN OPACITY PER NEUTRAL ATOM, SCALED BY E26.             AHYD  20
C-----BASED ON MENZEL AND PEKERIS, M.N. 96,77,1935.                 AHYD  30
C-----GINGERICH - 21 JULY 1960 - 21 MARCH 1967.                     AHYD  40
      DIMENSION G(8)                                                AHYD  50
      T=5040.4/THETA                                                AHYD  60
      U=157779./T                                                   AHYD  70
      EHVKT=EXP(-4.79895E-11/T*FREQ)                                AHYD  80
      D=0.                                                          AHYD  90
C-----IF FREQUENCY REPEATS, DO NOT RECOMPUTE GAUNT FACTORS.         AHYD 100
   10 IF(FR-FREQ) 11,15,11                                          AHYD 110
   15 ASSIGN 16 TO MISS                                             AHYD 120
      GO TO 111                                                     AHYD 130
   11 ASSIGN 14 TO MISS                                             AHYD 140
      FR=FREQ                                                       AHYD 150
      WAVE=2.9979E18/FR                                             AHYD 160
      IF(FR-5.5E14)101,103,103                                      AHYD 170
  101 ASSIGN 112 TO NG3                                             AHYD 180
      GO TO 104                                                     AHYD 190
C-----WAVELENGTH BELOW 5450A, SET FREE-FREE GAUNT FACTOR = 1.       AHYD 200
  103 ASSIGN 113 TO NG3                                             AHYD 210
      G3=1.                                                         AHYD 220
  104 N=0                                                           AHYD 230
      IF(FR-0.5137538E14)17,12,12                                   AHYD 240
C-----TEST SEQUENCE DETERMINES N, LOWEST ATOMIC LEVEL CONSIDERED.   AHYD 250
   12 ITEM=IFIX(3.2880242E15/FR)+1                                  AHYD 260
    9 N=N+1                                                         AHYD 270
      N2 = N*N                                                      AHYD 280
      IF(N2-ITEM) 9,13,13                                           AHYD 290
   13 NPRIME= MINO(8,N+2)                                           AHYD 300
C-----CONTROL COMES HERE IF FREQUENCY IS REPEATED.                  AHYD 310
  111 GO TO NG3, (112,113)                                          AHYD 320
C-----WAVELENGTHS ABOVE 58,400 A.                                   AHYD 330
   17 N=9                                                           AHYD 340
C-----K STROM≠S FIT TO KARSAS AND LATTER FREE-FREE GAUNT FACTORS.   AHYD 350
  112 AA=1.0828+3.865E-6*T                                          AHYD 360
      BB=7.564E-7+(4.920E-10-2.482E-15*T)*T                         AHYD 370
      CC=5.326E-12+(-3.904E-15+1.8790E-20*T)*T                      AHYD 380
      G3=AA+(BB+CC*WAVE)*WAVE                                       AHYD 390
  113 IF(NPRIME-N) 18,20,20                                         AHYD 400
C-----KALKOFEN≠S INTEGRAL FOR LONG WAVELENGTHS.                     AHYD 410
   18 D=(1./EHVKT-1.+G3)/2./U                                       AHYD 420
      GO TO 22                                                      AHYD 430
C-----SUMMATION OF CONTRIBUTION FROM N TO NPRIME RELEVANT LEVELS.   AHYD 440
   20 DO 16 K=N,NPRIME                                              AHYD 450
      XK=K                                                          AHYD 460
      SQXN=XK*XK                                                    AHYD 470
      GO TO MISS, (16,14)                                           AHYD 480
   14 G(K)=GAUNT(FR,XK)                                             AHYD 490
   16 D = G(K)/(SQXN)*EXPF(U/SQXN)/XK + D                           AHYD 500
C-----ASYMPTOTIC RELATION FOR LEVEL NPRIME + 1 AND HIGHER.          AHYD 510
      PRIME=(NPRIME+1)**2                                           AHYD 520
   19 D=D+(EXPF(U/PRIME)-1.+G3)/2./U                                AHYD 530
C-----NUMERICAL CONSTANT FROM ALLEN, ASTROPHYSICAL QUANTITIES, P. 89 AHYD 540
   22 AVHYD=(1.E26/FR)*( D   /EXPF(U))*((1.-EHVKT)/FR*2.815E29/FR)  AHYD 550
      RETURN                                                        AHYD 560
      END                                                           AHYD 570


      FUNCTION GAUNT(FREQ,X)                                        GAUNT 10
C-----GAUNT FACTORS FOR BOUND-FREE HYDROGEN,GINGERICH,MARCH 1964    GAUNT 20
C-----K IS LEVEL FOR GAUNT FACTOR                                   GAUNT 30
      K=X                                                           GAUNT 40
      W=2.9979E15/FREQ                                              GAUNT 50
      IF(K-6)90,6,7                                                 GAUNT 60
   90 GO TO (1,2,3,4,5),K                                          GAUNT 70
    1 GAUNT=0.9916+(9.068E-3-0.2524*W)*W                            GAUNT 80
      GO TO 100                                                     GAUNT 90
    2 GAUNT=1.105+(-7.922E-2+4.536E-3 *W)*W                         GAUNT100
      GO TO 100                                                     GAUNT110
    3 GAUNT=1.101+(-3.290E-2+1.152E-3 *W)*W                         GAUNT120
      GO TO 100                                                     GAUNT130
    4 GAUNT=1.101+(-1.923E-2+5.110E-4*W)*W                          GAUNT140
      GO TO 100                                                     GAUNT150
    5 GAUNT=1.102+(-0.01304+2.638E-4*W)*W                           GAUNT160
      GO TO 100                                                     GAUNT170
    6 GAUNT=1.0986+(-0.00902+1.367E-4*W)*W                          GAUNT180
      GO TO 100                                                     GAUNT190
    7 GAUNT=1.                                                      GAUNT200
  100 RETURN                                                        GAUNT210
      END                                                           GAUNT220
```

```
      FUNCTION AVHM(THETA,FREQ)                                       AHM    10
C-----H- OPACITY PER NEUTRAL HYDROGEN ATOM AND UNIT ELECTRON PRESSURE, AHM    20
C-----SCALED BY E26.  ALL OPACITY ROUTINES INCLUDE STIMULATED EMISSION. AHM   30
C-----FREE-FREE BASED ON JOHN, M.N. 128,93,1964.                       AHM    40
C-----BOUND-FREE BASED ON GELTMAN, AP.J. 136,935,1962.                 AHM    50
C-----SEE GINGERICH, SAO SPECIAL REPORT 167,19-20,1964.                AHM    60
      TH=THETA                                                        AHM    70
      IF(FREQ-FR) 30,50,30                                            AHM    80
   30 FR=FREQ                                                         AHM    90
      HVK=9.5210E-15*FR                                               AHM   100
C-----W IN MILLIANGSTROMS.                                            AHM   110
      W=2.9979E15/FR                                                  AHM   120
      XL=16.419-W                                                     AHM   130
      IF(XL) 34,36,36                                                 AHM   140
C-----FREE-FREE ONLY.                                                 AHM   150
   34 ASSIGN 70 TO NFF                                                AHM   160
      GO TO 50                                                        AHM   170
C-----BOUND-FREE PART ALSO EXITS VIA STATEMENT 60.                    AHM   180
   36 ASSIGN 60 TO NFF                                                AHM   190
      IF(W-14.2) 44,44,40                                             AHM   200
   40 XK=(0.269818+(0.220190+(-0.0411288+0.00273236*XL)*XL)*XL)*XL    AHM   210
      GO TO 50                                                        AHM   220
   44 XK=0.00680133+(0.178708+(0.16479+(-0.0204842+5.95244E-4*W)*W)*W)*WAHM  230
C-----FREE-FREE.                                                      AHM   240
   50 A=0.0053666+(-0.011493+.027039*TH)*TH                           AHM   250
      B=-3.2062+(11.924-5.939*TH)*TH                                  AHM   260
      C=-0.40192+(7.0355-0.34592*TH)*TH                               AHM   270
      FF=A+(B+C*W)*W/1000.                                            AHM   280
      GO TO NFF,(60,70)                                               AHM   290
   60 BF = XK*0.4158*TH**2*SQRT(TH)*EXP(1.737*TH)*(1.-EXP(-TH*HVK))    AHM   300
      AVHM=BF+FF                                                      AHM   310
      RETURN                                                         AHM   320
   70 AVHM=FF                                                         AHM   330
      RETURN                                                         AHM   340
      END                                                            AHM   350


      FUNCTION AVCAR(TH,FR)                                           ACAR   10
C-----BOUND-FREE CARBON I OPACITY, PER NEUTRAL CARBON, SCALED BY E26. ACAR   20
C-----FIT TO PEACH (1967) MEM.R.A.S.,71,1.                            ACAR   30
C-----THE HIGHER LEVELS ARE FREQUENCY INDEPENDENT FITS TO THE DATA.   ACAR   40
C-----THE COEFFICIENTS IN THE EXPONENTIALS HAVE BEEN ADJUSTED TO GIVE ACAR   50
C-----APPROXIMATELY CORRECT TEMPERATURE DEPENDENCE.                   ACAR   60
C-----GINGERICH, 18 NOVEMBER 1968.                                    ACAR   70
      IF (FR.LT.2.9979E18/1444.) GO TO 13                            ACAR   80
      HV=0.43429/        TH                                          ACAR   90
      IF (FR.GE.2.9979E18/1100.) GO TO 11                            ACAR  100
      IF (FR.GE.2.9979E18/1239.) GO TO 7                             ACAR  110
C-----WAVELENGTH BETWEEN 1240 AND 1444.                               ACAR  120
      AVCAR=0.898E8/EXP(2.56/HV)                                     ACAR  130
      RETURN                                                         ACAR  140
C-----WAVELENGTH BETWEEN 1101 AND 1239.                               ACAR  150
    7 AVCAR=4.86E8/EXP(1.18/HV)                                      ACAR  160
      RETURN                                                         ACAR  170
C-----WAVELENGTH 1100 OR LESS.                                        ACAR  180
   11 AVCAR=0.776E9+0.75E24/FR                                       ACAR  190
      RETURN                                                         ACAR  200
   13 AVCAR=0.                                                       ACAR  210
      RETURN                                                         ACAR  220
      END                                                            ACAR  230


      FUNCTION PARA(X,JX,Y,JY,ARGX)                                  PARA   10
C     PARABOLIC INTERPOLATION - 6 JULY 1962 - GINGERICH              PARA   20
C-----GIVES PARA, THE INTERPOLATED VALUE IN THE X ARRAY FOR ARGUMENT PARA   30
C-----  ARGX.                                                        PARA   40
C-----FOR ZERO JX, THE A, B, C CONSTANTS WILL BE UNCHANGED.          PARA   50
C-----MODIFIED FOR 6400 WHICH ACCEPTS NEGATIVE AND ZERO SUBSCRIPTS   PARA   60
      DIMENSION X(1),Y(1)                                           PARA   70
      IF(JX) 1,2,1                                                   PARA   80
    1 XLOW=X(-JX+1)                                                  PARA   90
      X12=XLOW-X                                                     PARA  100
      X23=X-X(JX+1)                                                  PARA  110
      X31=X(JX+1)-XLOW                                               PARA  120
      Y12=Y(-JY+1)-Y                                                 PARA  130
      Y23=Y-Y(JY+1)                                                  PARA  140
      Y31=Y(JY+1)-Y(-JY+1)                                           PARA  150
      C=(X12*Y23-X23*Y12)/(X12*X23*X31)                             PARA  160
      B=Y12/X12-C*(XLOW+X)                                          PARA  170
      A=Y(-JY+1)-(B+C*XLOW)*XLOW                                    PARA  180
    2 PARA=A+(B+C*ARGX)*ARGX                                        PARA  190
      RETURN                                                        PARA  200
      END                                                           PARA  210
```

Sixty-eight nongray constant-flux model atmospheres are presented in the following tables. The 18 models from 50,000° to 11,000° have been calculated by Robert Kurucz, and the 50 from 10,000° to 4000° by Duane Carbon and Owen Gingerich. The 10,000° models were in fact calculated with both programs as a control on computational techniques.

We first tabulate the models themselves, that is, the τ-T-P-etc. relations. These tables are followed by the monochromatic fluxes (F_ν, not πF_ν), the monochromatic limb darkening, and the Hγ profiles using both the Griem and the Edmonds, Schlüter, and Wells theories. The models are always listed in order of descending effective temperature, and for a given effective temperature in order of descending gravity. The notation (10000,4) refers to T_{eff} = 10,000°, log g = 4.

For certain models, variations have been provided. At 10,000° and 6000°, models with 10 times the normal metal abundances have been included to show the effect both of decreasing electron contributors and of lowered silicon and magnesium opacities. Note that the helium and carbon abundances remain normal for these variation models.

Beginning with the (8500,4) model, convection with a mixing length/scale-height ratio of 1.5 has been included, with the following exceptions: convection is not important for (8500,3) or (8000,2) and hence has been omitted with no special designation; at (8000,4), (7500,4), (7000,4), and (6000,4), designated models have been included without convection; and at (7500,4) one model has been included with ℓ/H = 2.5. Finally, we note that models (6000,4,10X) and (6000,4,1/10X) have neither convection nor blanketing.

In Figure 1 we have graphed the temperature distributions of five models in order to show the effects of convection.

Beginning at (8500,4), the models include the statistical theory of line blanketing described in the preceding paper by Carbon and Gingerich, with the exception of the (6000,4,10X) and (6000,4,1/10X) models noted above and two others specially designated models at (6000,4). The key to the special designations is given twice within the tables.

Hence, 11 of the 68 atmospheres may be properly called variation models, although in a certain sense the six log g = 3.5 and 4.5 models might also be considered variations. In the original conception of the grid, it was proposed also to include some models with the hydrogen handled by a non-LTE solution. At log g = 4, where the variations are imposed, the non-LTE calculations made no difference in the results, and consequently they are not included here.

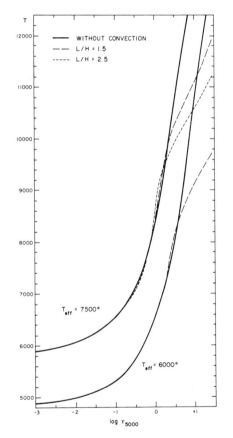

Figure 1. Convective and non-convective temperature distributions compared.

Unfortunately, since the non-LTE programs are still highly experimental, the results remain uncertain at the lower gravities, where significant non-LTE effects can occur. For this reason, we have chosen to omit non-LTE models from the grid.

Most of the numbers reproduced in the following tables are also available in punch-card form at the Smithsonian Astrophysical Observatory.

DESCRIPTION OF THE TABLES

A. The model atmospheres

Column 1 — τ_{5000}, optical depth at 5000 A;

Column 2 — temperatures in °K;

Column 3 — gas pressure in dynes cm^{-2};

Column 4 — number of electrons cm^{-3};

Column 5 — opacity per gram of stellar material at 5000 A;

Column 6 — density in g cm^{-3};

Column 7 — in the case of radiation pressure, log g_{eff} where

$$g_{eff} = g - \frac{\sigma}{c} \, \overline{\kappa} \, T_{eff}^4$$

and $\overline{\kappa}$ is the flux mean opacity;

— if radiation pressure is not included, the fraction of hydrogen that is ionized;

Column 8 — physical depth in km, reckoned from unit optical depth;

Column 9 — (given only for convective models) the fraction of the total flux carried by convection.

B. The monochromatic fluxes

In the work of Underhill, Mihalas, Gingerich, and others, there appears some precedent for tabulating the monochromatic flux, F_ν, which is defined so that

$$\int_0^\infty F_\nu \, d\nu = \frac{\sigma}{\pi} \, T_{eff}^4 \quad .$$

This definition has been adopted in these tables. The wavelengths have been given in microns.

In Figures 2 and 3, we have plotted the flux distribution as a function of frequency for 13 main-sequence models selected throughout the temperature range from 50,000° to 4000°.

In the calculations several different wavelength sets were used, depending on the effective temperature. In general, a certain amount of overlap in the wavelength sets has allowed most of the models to be tabulated together. The three hottest models provide an exception and are displayed in a group on the first page of the monochromatic fluxes. The next 12 pages show for the remaining 65 models monochromatic fluxes for all the wavelength points used in the frequency integration.

For blanketed models, an additional column gives the monochromatic blanketing coefficient η. Hence, the monochromatic flux refers to the continuum, but $F_\nu (1 - \eta)$ gives the net flux at a particular wavelength. The η's therefore allow broad-band filter predictions from the models. In the preparation of the blanketed models, the entire spectrum was divided into several zones, and at the division points calculations were made on both sides of the division (e.g., 4601^- and 4601^+). Since the monochromatic flux is always identical on both sides but the η values differ, discontinuities occur in the net fluxes at each division point.

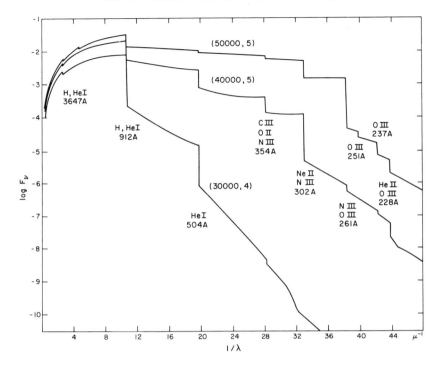

Figure 2.   Predicted flux distribution for three hot models.

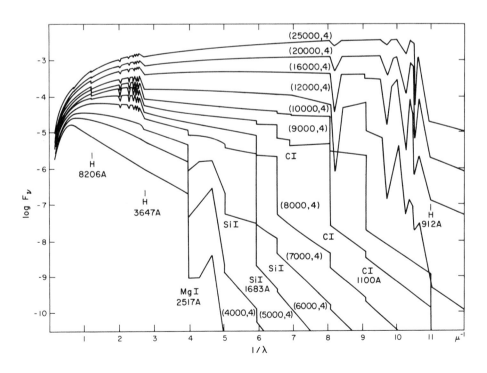

Figure 3.   Predicted continuum flux distributions for selected models in range
$25,000 \geq T_{eff} \geq 4000$.

C. The monochromatic limb darkening

The three-digit numbers in these tables are the fractional limb darkening relative to the center of the disk, with the leading decimal point omitted. Wavelengths in microns run horizontally, and the μ values are tabulated vertically.

D. Hγ profiles

The Hγ profiles for all the models have been calculated with a program supplied by Deane Peterson (1969) and operated in conjunction with the ATLAS program. Profiles are presented for both the Griem (1967) and the Edmonds, Schlüter, and Wells (E-S-W) (1967) theories. The $\Delta\lambda$, in angstroms from line center, run horizontally. The three-digit numbers give the relative flux, with the leading decimal point omitted. We note that for models $T_{eff} \geq 6000°$, the lines are normalized 100 A from line center, and for $T_{eff} \leq 5500°$ at 20 A from line center.

REFERENCES

Edmonds, F. N., Jr., Schlüter, H., and Wells, D. C., III. 1967, Mém.
    Roy. Astron. Soc., 71, 271.
Griem, H. R. 1967, Ap. J., 147, 1092.
Peterson, D. M. 1969, Smithsonian Ap. Obs. Spec. Rep. No. 293, 199 pp.

TABLES

50000-DEGREE, LOG G = 5.0

TAU 5000	TEMP	PRESSURE	ELECTRON DENSITY	OPACITY	DENSITY	LOG G EFF	HEIGHT (KM)
.0010	38470	2.234E+02	2.194E+13	4.751E-01	4.277E-11	4.9787	2763
.0016	38727	3.328E+02	3.247E+13	5.366E-01	6.329E-11	4.9775	2542
.0025	38898	4.842E+02	4.699E+13	6.210E-01	9.175E-11	4.9757	2331
.0040	39024	6.885E+02	6.662E+13	7.352E-01	1.300E-10	4.9732	2132
.0063	39153	9.586E+02	9.234E+13	8.824E-01	1.806E-10	4.9699	1944
.0100	39302	1.309E+03	1.256E+14	1.070E+00	2.458E-10	4.9656	1764
.0158	39487	1.761E+03	1.679E+14	1.305E+00	3.295E-10	4.9605	1590
.0251	39717	2.338E+03	2.217E+14	1.598E+00	4.352E-10	4.9542	1421
.0398	40006	3.075E+03	2.893E+14	1.958E+00	5.682E-10	4.9469	1254
.0631	40371	4.012E+03	3.739E+14	2.394E+00	7.351E-10	4.9386	1088
.1000	40838	5.203E+03	4.794E+14	2.921E+00	9.425E-10	4.9296	921
.1585	41428	6.718E+03	6.100E+14	3.548E+00	1.200E-09	4.9204	751
.2512	42177	8.661E+03	7.727E+14	4.292E+00	1.519E-09	4.9116	576
.3981	43116	1.117E+04	9.749E+14	5.165E+00	1.915E-09	4.9040	393
.6310	44275	1.443E+04	1.227E+15	6.179E+00	2.408E-09	4.8983	202
1.000	45680	1.872E+04	1.545E+15	7.356E+00	3.025E-09	4.8948	0
1.585	47344	2.440E+04	1.943E+15	8.702E+00	3.804E-09	4.8934	-215
2.512	49331	3.204E+04	2.451E+15	1.025E+01	4.789E-09	4.8942	-444
3.981	51588	4.235E+04	3.101E+15	1.206E+01	6.049E-09	4.8957	-688
6.310	54117	5.635E+04	3.933E+15	1.415E+01	7.671E-09	4.8977	-949
10.00	56970	7.538E+04	4.998E+15	1.656E+01	9.747E-09	4.9010	-1227
15.85	60103	1.013E+05	6.373E+15	1.938E+01	1.241E-08	4.9051	-1523
25.12	63604	1.369E+05	8.130E+15	2.264E+01	1.585E-08	4.9091	-1837
39.81	67537	1.857E+05	1.038E+16	2.635E+01	2.027E-08	4.9129	-2171
63.10	71839	2.526E+05	1.329E+16	3.073E+01	2.588E-08	4.9156	-2527
100.0	76656	3.443E+05	1.697E+16	3.560E+01	3.308E-08	4.9189	-2907
158.5	81677	4.702E+05	2.175E+16	4.149E+01	4.240E-08	4.9205	-3312
251.2	87398	6.423E+05	2.777E+16	4.794E+01	5.413E-08	4.9208	-3744
398.1	93713	8.787E+05	3.543E+16	5.538E+01	6.904E-08	4.9202	-4209
631.0	100905	1.205E+06	4.516E+16	6.377E+01	8.783E-08	4.9212	-4694

40000-DEGREE, LOG G = 5.0

TAU 5000	TEMP	PRESSURE	ELECTRON DENSITY	OPACITY	DENSITY	LOG G EFF	HEIGHT (KM)
.0010	31114	2.149E+02	2.552E+13	5.198E-01	5.209E-11	4.9896	2078
.0016	31181	3.161E+02	3.733E+13	5.995E-01	7.671E-11	4.9881	1914
.0025	31248	4.536E+02	5.328E+13	7.056E-01	1.102E-10	4.9861	1759
.0040	31332	6.365E+02	7.438E+13	8.446E-01	1.546E-10	4.9835	1613
.0063	31439	8.755E+02	1.018E+14	1.024E+00	2.123E-10	4.9802	1475
.0100	31583	1.184E+03	1.368E+14	1.249E+00	2.864E-10	4.9762	1343
.0158	31781	1.579E+03	1.813E+14	1.529E+00	3.799E-10	4.9715	1215
.0251	32053	2.086E+03	2.371E+14	1.875E+00	4.979E-10	4.9662	1089
.0398	32417	2.732E+03	3.072E+14	2.295E+00	6.447E-10	4.9605	964
.0631	32894	3.560E+03	3.946E+14	2.801E+00	8.275E-10	4.9546	839
.1000	33500	4.622E+03	5.033E+14	3.406E+00	1.054E-09	4.9488	712
.1585	34248	5.989E+03	6.387E+14	4.130E+00	1.334E-09	4.9434	581
.2512	35151	7.757E+03	8.076E+14	5.000E+00	1.680E-09	4.9385	445
.3981	36214	1.005E+04	1.019E+15	6.052E+00	2.108E-09	4.9343	303
.6310	37440	1.302E+04	1.282E+15	7.326E+00	2.631E-09	4.9309	155
1.000	38826	1.690E+04	1.611E+15	8.857E+00	3.277E-09	4.9285	0
1.585	40401	2.195E+04	2.021E+15	1.066E+01	4.072E-09	4.9273	-164
2.512	42214	2.864E+04	2.535E+15	1.274E+01	5.055E-09	4.9276	-339
3.981	44212	3.754E+04	3.186E+15	1.513E+01	6.301E-09	4.9280	-525
6.310	46455	4.950E+04	4.010E+15	1.782E+01	7.881E-09	4.9286	-726
10.00	49012	6.571E+04	5.053E+15	2.079E+01	9.900E-09	4.9307	-942
15.85	51798	8.788E+04	6.403E+15	2.423E+01	1.251E-08	4.9332	-1175
25.12	54954	1.183E+05	8.124E+15	2.808E+01	1.587E-08	4.9360	-1427
39.81	58466	1.602E+05	1.034E+16	3.250E+01	2.020E-08	4.9393	-1697
63.10	62263	2.180E+05	1.323E+16	3.776E+01	2.579E-08	4.9432	-1987
100.0	66284	2.975E+05	1.695E+16	4.396E+01	3.307E-08	4.9475	-2297
158.5	70425	4.061E+05	2.180E+16	5.159E+01	4.243E-08	4.9502	-2623
251.2	74965	5.540E+05	2.792E+16	6.018E+01	5.444E-08	4.9525	-2968
398.1	79852	7.559E+05	3.576E+16	7.022E+01	6.972E-08	4.9535	-3334
631.0	85362	1.033E+06	4.575E+16	8.157E+01	8.900E-08	4.9545	-3710

30000-DEGREE, LOG G = 4.0

TAU 5000	TEMP	PRESSURE	ELECTRON DENSITY	OPACITY	DENSITY	LOG G EFF	HEIGHT (KM)
.0010	18842	2.575E+01	4.954E+12	3.967E-01	1.051E-11	3.9714	15741
.0016	18939	3.881E+01	7.424E+12	4.365E-01	1.576E-11	3.9684	14655
.0025	19071	5.727E+01	1.088E+13	4.910E-01	2.312E-11	3.9647	13607
.0040	19245	8.281E+01	1.559E+13	5.635E-01	3.311E-11	3.9601	12599
.0063	19482	1.175E+02	2.183E+13	6.549E-01	4.643E-11	3.9551	11623
.0100	19792	1.639E+02	2.999E+13	7.677E-01	6.375E-11	3.9500	10669
.0158	20175	2.257E+02	4.050E+13	9.037E-01	8.614E-11	3.9452	9724
.0251	20627	3.080E+02	5.405E+13	1.067E+00	1.150E-10	3.9406	8778
.0398	21147	4.171E+02	7.143E+13	1.262E+00	1.519E-10	3.9360	7823
.0631	21738	5.619E+02	9.359E+13	1.492E+00	1.990E-10	3.9312	6850
.1000	22410	7.540E+02	1.218E+14	1.761E+00	2.591E-10	3.9258	5851
.1585	23183	1.009E+03	1.575E+14	2.067E+00	3.351E-10	3.9199	4815
.2512	24091	1.349E+03	2.027E+14	2.415E+00	4.313E-10	3.9134	3728
.3981	25164	1.806E+03	2.598E+14	2.800E+00	5.525E-10	3.9066	2575
.6310	26431	2.422E+03	3.319E+14	3.220E+00	7.056E-10	3.9001	1339
1.000	27909	3.264E+03	4.236E+14	3.685E+00	9.001E-10	3.8938	0
1.585	29614	4.413E+03	5.398E+14	4.199E+00	1.147E-09	3.8877	-1457
2.512	31574	5.986E+03	6.880E+14	4.795E+00	1.456E-09	3.8797	-3050
3.981	33856	8.120E+03	8.733E+14	5.514E+00	1.836E-09	3.8697	-4790
6.310	36440	1.095E+04	1.103E+15	6.481E+00	2.282E-09	3.8554	-6686
10.00	39366	1.462E+04	1.381E+15	7.666E+00	2.782E-09	3.8420	-8753
15.85	42497	1.952E+04	1.722E+15	8.858E+00	3.409E-09	3.8441	-11047
25.12	45761	2.637E+04	2.170E+15	1.010E+01	4.259E-09	3.8510	-13609
39.81	49239	3.610E+04	2.765E+15	1.154E+01	5.410E-09	3.8597	-16434
63.10	52986	4.985E+04	3.552E+15	1.322E+01	6.933E-09	3.8698	-19504
100.0	56748	6.920E+04	4.606E+15	1.538E+01	8.983E-09	3.8785	-22775
158.5	60835	9.606E+04	5.964E+15	1.784E+01	1.163E-08	3.8878	-26218
251.2	64977	1.332E+05	7.745E+15	2.091E+01	1.511E-08	3.8944	-29827
398.1	69415	1.842E+05	1.003E+16	2.447E+01	1.954E-08	3.8998	-33594
631.0	74226	2.543E+05	1.296E+16	2.862E+01	2.520E-08	3.9031	-37404

25000-DEGREE, LOG G = 4.0

TAU 5000	TEMP	PRESSURE	ELECTRON DENSITY	OPACITY	DENSITY	LOG G EFF	HEIGHT (KM)
.0010	15115	2.440E+01	5.833E+12	4.683E-01	1.245E-11	3.9824	11714
.0016	15319	3.561E+01	8.399E+12	5.302E-01	1.793E-11	3.9805	10936
.0025	15572	5.111E+01	1.186E+13	6.086E-01	2.533E-11	3.9787	10171
.0040	15859	7.230E+01	1.647E+13	7.078E-01	3.517E-11	3.9770	9423
.0063	16166	1.009E+02	2.256E+13	8.319E-01	4.812E-11	3.9753	8686
.0100	16478	1.391E+02	3.051E+13	9.866E-01	6.511E-11	3.9735	7960
.0158	16791	1.897E+02	4.083E+13	1.179E+00	8.713E-11	3.9713	7242
.0251	17109	2.562E+02	5.413E+13	1.416E+00	1.155E-10	3.9684	6528
.0398	17449	3.434E+02	7.111E+13	1.703E+00	1.518E-10	3.9648	5815
.0631	17838	4.572E+02	9.266E+13	2.043E+00	1.977E-10	3.9604	5097
.1000	18309	6.066E+02	1.198E+14	2.432E+00	2.554E-10	3.9555	4363
.1585	18893	8.039E+02	1.539E+14	2.865E+00	3.280E-10	3.9507	3601
.2512	19618	1.068E+03	1.969E+14	3.341E+00	4.198E-10	3.9461	2797
.3981	20505	1.426E+03	2.517E+14	3.856E+00	5.357E-10	3.9422	1937
.6310	21568	1.915E+03	3.215E+14	4.415E+00	6.836E-10	3.9390	1009
1.000	22810	2.591E+03	4.112E+14	5.029E+00	8.747E-10	3.9366	0
1.585	24227	3.526E+03	5.269E+14	5.710E+00	1.121E-09	3.9347	-1098
2.512	25838	4.829E+03	6.764E+14	6.474E+00	1.439E-09	3.9328	-2293
3.981	27663	6.644E+03	8.695E+14	7.326E+00	1.849E-09	3.9306	-3595
6.310	29745	9.174E+03	1.117E+15	8.270E+00	2.375E-09	3.9276	-5017
10.00	32160	1.270E+04	1.432E+15	9.316E+00	3.036E-09	3.9231	-6576
15.85	35016	1.758E+04	1.828E+15	1.058E+01	3.845E-09	3.9158	-8293
25.12	38449	2.415E+04	2.312E+15	1.231E+01	4.757E-09	3.9041	-10183
39.81	42547	3.284E+04	2.884E+15	1.425E+01	5.754E-09	3.8938	-12289
63.10	47319	4.495E+04	3.577E+15	1.560E+01	7.023E-09	3.8937	-14732
100.0	52814	6.292E+04	4.498E+15	1.671E+01	8.780E-09	3.8999	-17632
158.5	59030	8.990E+04	5.757E+15	1.805E+01	1.121E-08	3.9075	-20963

20000-DEGREE, LOG G = 4.0

TAU 5000	TEMP	PRESSURE	ELECTRON DENSITY	OPACITY	DENSITY	LOG G EFF	HEIGHT (KM)
.0010	13010	2.318E+01	6.274E+12	5.244E-01	1.409E-11	3.9917	8871
.0016	13220	3.331E+01	8.879E+12	6.073E-01	1.991E-11	3.9911	8257
.0025	13406	4.703E+01	1.236E+13	7.138E-01	2.774E-11	3.9904	7659
.0040	13561	6.535E+01	1.696E+13	8.524E-01	3.815E-11	3.9893	7080
.0063	13688	8.944E+01	2.297E+13	1.031E+00	5.177E-11	3.9878	6522
.0100	13812	1.207E+02	3.068E+13	1.254E+00	6.932E-11	3.9856	5981
.0158	13955	1.612E+02	4.052E+13	1.529E+00	9.173E-11	3.9830	5453
.0251	14133	2.135E+02	5.300E+13	1.861E+00	1.199E-10	3.9800	4930
.0398	14358	2.812E+02	6.878E+13	2.257E+00	1.553E-10	3.9766	4407
.0631	14652	3.694E+02	8.872E+13	2.719E+00	1.995E-10	3.9729	3875
.1000	15029	4.848E+02	1.139E+14	3.250E+00	2.545E-10	3.9693	3326
.1585	15507	6.371E+02	1.457E+14	3.850E+00	3.229E-10	3.9660	2752
.2512	16102	8.406E+02	1.862E+14	4.517E+00	4.080E-10	3.9633	2142
.3981	16827	1.116E+03	2.377E+14	5.242E+00	5.154E-10	3.9614	1486
.6310	17691	1.492E+03	3.038E+14	6.024E+00	6.531E-10	3.9604	775
1.000	18691	2.014E+03	3.888E+14	6.866E+00	8.324E-10	3.9602	0
1.585	19825	2.739E+03	4.994E+14	7.802E+00	1.065E-09	3.9604	-845
2.512	21094	3.752E+03	6.434E+14	8.859E+00	1.371E-09	3.9607	-1764
3.981	22508	5.168E+03	8.308E+14	1.006E+01	1.769E-09	3.9606	-2760
6.310	24092	7.144E+03	1.073E+15	1.140E+01	2.284E-09	3.9601	-3837
10.00	25893	9.908E+03	1.385E+15	1.286E+01	2.947E-09	3.9589	-5007
15.85	27976	1.379E+04	1.784E+15	1.439E+01	3.798E-09	3.9570	-6286
25.12	30432	1.928E+04	2.294E+15	1.599E+01	4.880E-09	3.9543	-7699
39.81	33363	2.708E+04	2.942E+15	1.768E+01	6.243E-09	3.9507	-9275
63.10	36884	3.808E+04	3.763E+15	1.988E+01	7.895E-09	3.9449	-11037
100.0	41169	5.306E+04	4.769E+15	2.314E+01	9.705E-09	3.9363	-12992
158.5	46178	7.334E+04	5.966E+15	2.626E+01	1.177E-08	3.9324	-15132

20000-DEGREE, LOG G = 3.0

TAU 5000	TEMP	PRESSURE	ELECTRON DENSITY	OPACITY	DENSITY	LOG G EFF	HEIGHT (KM)
.0010	12609	2.758E+00	7.818E+11	3.366E-01	1.706E-12	2.9475	123887
.0016	12820	4.260E+00	1.187E+12	3.509E-01	2.593E-12	2.9451	115863
.0025	13064	6.513E+00	1.782E+12	3.712E-01	3.887E-12	2.9429	107788
.0040	13316	9.847E+00	2.645E+12	3.991E-01	5.764E-12	2.9405	99738
.0063	13550	1.468E+01	3.877E+12	4.374E-01	8.442E-12	2.9376	91761
.0100	13747	2.154E+01	5.606E+12	4.893E-01	1.221E-11	2.9331	83912
.0158	13918	3.102E+01	7.966E+12	5.582E-01	1.738E-11	2.9261	76241
.0251	14102	4.384E+01	1.110E+13	6.464E-01	2.426E-11	2.9164	68742
.0398	14335	6.088E+01	1.517E+13	7.545E-01	3.314E-11	2.9048	61344
.0631	14639	8.334E+01	2.037E+13	8.822E-01	4.435E-11	2.8920	53927
.1000	15032	1.129E+02	2.694E+13	1.028E+00	5.837E-11	2.8789	46341
.1585	15537	1.521E+02	3.520E+13	1.187E+00	7.591E-11	2.8669	38422
.2512	16176	2.049E+02	4.566E+13	1.357E+00	9.801E-11	2.8572	29992
.3981	16965	2.772E+02	5.900E+13	1.535E+00	1.261E-10	2.8505	20876
.6310	17918	3.777E+02	7.622E+13	1.723E+00	1.625E-10	2.8467	10918
1.000	19039	5.190E+02	9.865E+13	1.927E+00	2.100E-10	2.8450	0
1.585	20325	7.191E+02	1.280E+14	2.150E+00	2.725E-10	2.8443	-11966
2.512	21783	1.003E+03	1.666E+14	2.401E+00	3.545E-10	2.8435	-25043
3.981	23424	1.404E+03	2.169E+14	2.683E+00	4.618E-10	2.8415	-39293
6.310	25271	1.970E+03	2.821E+14	2.993E+00	6.002E-10	2.8377	-54835
10.00	27377	2.766E+03	3.659E+14	3.326E+00	7.777E-10	2.8308	-71875
15.85	29815	3.882E+03	4.714E+14	3.672E+00	1.002E-09	2.8209	-90742
25.12	32689	5.431E+03	6.040E+14	4.091E+00	1.274E-09	2.8040	-111810
39.81	36103	7.489E+03	7.640E+14	4.701E+00	1.570E-09	2.7753	-135375
63.10	40113	1.012E+04	9.441E+14	5.387E+00	1.877E-09	2.7481	-162141
100.0	44738	1.379E+04	1.162E+15	5.795E+00	2.276E-09	2.7520	-193950
158.5	49972	1.942E+04	1.468E+15	6.192E+00	2.862E-09	2.7697	-231607

18000-DEGREE, LOG G = 4.0

TAU 5000	TEMP	PRESSURE	ELECTRON DENSITY	OPACITY	DENSITY	LOG G EFF	HEIGHT (KM)
.0010	12232	2.264E+01	6.426E+12	5.536E-01	1.484E-11	3.9944	7869
.0016	12362	3.221E+01	9.041E+12	6.509E-01	2.090E-11	3.9938	7320
.0025	12452	4.496E+01	1.252E+13	7.801E-01	2.898E-11	3.9929	6791
.0040	12522	6.167E+01	1.705E+13	9.468E-01	3.958E-11	3.9915	6287
.0063	12599	8.332E+01	2.287E+13	1.157E+00	5.319E-11	3.9897	5803
.0100	12697	1.112E+02	3.030E+13	1.416E+00	7.043E-11	3.9876	5333
.0158	12819	1.472E+02	3.969E+13	1.730E+00	9.243E-11	3.9852	4871
.0251	12978	1.937E+02	5.158E+13	2.105E+00	1.201E-10	3.9824	4411
.0398	13186	2.540E+02	6.661E+13	2.546E+00	1.550E-10	3.9794	3948
.0631	13458	3.328E+02	8.561E+13	3.056E+00	1.988E-10	3.9763	3474
.1000	13810	4.367E+02	1.097E+14	3.633E+00	2.537E-10	3.9734	2982
.1585	14256	5.748E+02	1.403E+14	4.282E+00	3.225E-10	3.9709	2466
.2512	14812	7.603E+02	1.796E+14	5.015E+00	4.085E-10	3.9688	1917
.3981	15489	1.011E+03	2.301E+14	5.845E+00	5.156E-10	3.9674	1329
.6310	16291	1.352E+03	2.952E+14	6.774E+00	6.501E-10	3.9666	692
1.000	17215	1.820E+03	3.789E+14	7.785E+00	8.223E-10	3.9666	0
1.585	18254	2.468E+03	4.869E+14	8.880E+00	1.046E-09	3.9672	-755
2.512	19412	3.372E+03	6.273E+14	1.009E+01	1.341E-09	3.9679	-1579
3.981	20693	4.636E+03	8.100E+14	1.146E+01	1.727E-09	3.9684	-2473
6.310	22116	6.403E+03	1.047E+15	1.301E+01	2.231E-09	3.9685	-3440
10.00	23715	8.874E+03	1.354E+15	1.470E+01	2.883E-09	3.9681	-4488
15.85	25545	1.234E+04	1.748E+15	1.651E+01	3.721E-09	3.9671	-5629
25.12	27678	1.723E+04	2.253E+15	1.842E+01	4.796E-09	3.9656	-6882
39.81	30198	2.419E+04	2.900E+15	2.038E+01	6.170E-09	3.9634	-8269
63.10	33192	3.414E+04	3.727E+15	2.244E+01	7.913E-09	3.9606	-9820
100.0	36800	4.827E+04	4.778E+15	2.498E+01	1.004E-08	3.9563	-11562
158.5	41146	6.783E+04	6.085E+15	2.889E+01	1.244E-08	3.9501	-13458

18000-DEGREE, LOG G = 3.0

TAU 5000	TEMP	PRESSURE	ELECTRON DENSITY	OPACITY	DENSITY	LOG G EFF	HEIGHT (KM)
.0010	11920	2.925E+00	8.622E+11	3.340E-01	1.945E-12	2.9657	104366
.0016	12133	4.498E+00	1.303E+12	3.514E-01	2.939E-12	2.9645	97284
.0025	12333	6.840E+00	1.950E+12	3.761E-01	4.394E-12	2.9632	90197
.0040	12492	1.026E+01	2.886E+12	4.105E-01	6.514E-12	2.9611	83215
.0063	12600	1.515E+01	4.215E+12	4.585E-01	9.546E-12	2.9574	76414
.0100	12688	2.191E+01	6.042E+12	5.235E-01	1.374E-11	2.9515	69845
.0158	12805	3.106E+01	8.472E+12	6.075E-01	1.933E-11	2.9439	63493
.0251	12967	4.324E+01	1.165E+13	7.124E-01	2.658E-11	2.9348	57292
.0398	13175	5.931E+01	1.573E+13	8.390E-01	3.587E-11	2.9244	51156
.0631	13451	8.041E+01	2.094E+13	9.880E-01	4.754E-11	2.9133	44985
.1000	13811	1.081E+02	2.752E+13	1.158E+00	6.201E-11	2.9022	38655
.1585	14273	1.448E+02	3.585E+13	1.349E+00	8.001E-11	2.8921	32040
.2512	14853	1.938E+02	4.642E+13	1.559E+00	1.023E-10	2.8840	25006
.3981	15565	2.604E+02	5.992E+13	1.784E+00	1.302E-10	2.8788	17408
.6310	16416	3.524E+02	7.732E+13	2.021E+00	1.662E-10	2.8768	9109
1.000	17411	4.816E+02	9.988E+13	2.270E+00	2.136E-10	2.8774	0
1.585	18545	6.646E+02	1.296E+14	2.545E+00	2.763E-10	2.8790	-9981
2.512	19822	9.246E+02	1.688E+14	2.855E+00	3.593E-10	2.8809	-20855
3.981	21246	1.292E+03	2.201E+14	3.207E+00	4.686E-10	2.8819	-32643
6.310	22834	1.813E+03	2.874E+14	3.606E+00	6.111E-10	2.8813	-45386
10.00	24622	2.543E+03	3.741E+14	4.037E+00	7.949E-10	2.8787	-59202
15.85	26669	3.571E+03	4.848E+14	4.488E+00	1.031E-09	2.8739	-74307
25.12	29059	5.024E+03	6.258E+14	4.956E+00	1.331E-09	2.8667	-91000
39.81	31884	7.068E+03	8.043E+14	5.465E+00	1.703E-09	2.8559	-109663
63.10	35245	9.878E+03	1.025E+15	6.180E+00	2.136E-09	2.8374	-130577
100.0	39248	1.356E+04	1.285E+15	7.199E+00	2.588E-09	2.8110	-153962
158.5	43891	1.849E+04	1.586E+15	7.947E+00	3.114E-09	2.8052	-180180

16000-DEGREE, LOG G = 4.0

TAU 5000	TEMP	PRESSURE	ELECTRON DENSITY	OPACITY	DENSITY	LOG G EFF	HEIGHT (KM)
.0010	11318	2.144E+01	6.533E+12	6.103E-01	1.528E-11	3.9960	7015
.0016	11354	3.004E+01	9.118E+12	7.343E-01	2.135E-11	3.9951	6536
.0025	11390	4.127E+01	1.248E+13	8.937E-01	2.926E-11	3.9940	6078
.0040	11440	5.583E+01	1.680E+13	1.094E+00	3.943E-11	3.9926	5641
.0063	11502	7.459E+01	2.230E+13	1.342E+00	5.244E-11	3.9909	5219
.0100	11586	9.872E+01	2.928E+13	1.644E+00	6.894E-11	3.9889	4807
.0158	11697	1.298E+02	3.810E+13	2.008E+00	8.987E-11	3.9867	4399
.0251	11844	1.700E+02	4.927E+13	2.440E+00	1.163E-10	3.9842	3991
.0398	12038	2.224E+02	6.340E+13	2.944E+00	1.496E-10	3.9815	3576
.0631	12292	2.910E+02	8.131E+13	3.521E+00	1.916E-10	3.9789	3151
.1000	12621	3.820E+02	1.040E+14	4.165E+00	2.449E-10	3.9766	2708
.1585	13039	5.038E+02	1.330E+14	4.874E+00	3.122E-10	3.9747	2240
.2512	13557	6.691E+02	1.703E+14	5.652E+00	3.980E-10	3.9735	1741
.3981	14186	8.954E+02	2.187E+14	6.521E+00	5.070E-10	3.9728	1205
.6310	14929	1.207E+03	2.818E+14	7.520E+00	6.454E-10	3.9726	626
1.000	15783	1.633E+03	3.644E+14	8.690E+00	8.183E-10	3.9727	0
1.585	16740	2.218E+03	4.713E+14	1.003E+01	1.038E-09	3.9731	-677
2.512	17800	3.025E+03	6.094E+14	1.151E+01	1.321E-09	3.9738	-1410
3.981	18964	4.146E+03	7.877E+14	1.313E+01	1.691E-09	3.9746	-2207
6.310	20245	5.709E+03	1.019E+15	1.491E+01	2.175E-09	3.9752	-3071
10.00	21667	7.897E+03	1.318E+15	1.689E+01	2.810E-09	3.9753	-4008
15.85	23272	1.096E+04	1.704E+15	1.903E+01	3.632E-09	3.9750	-5024
25.12	25120	1.529E+04	2.201E+15	2.132E+01	4.690E-09	3.9743	-6134
39.81	27275	2.141E+04	2.840E+15	2.370E+01	6.046E-09	3.9731	-7356
63.10	29809	3.014E+04	3.662E+15	2.618E+01	7.784E-09	3.9715	-8712
100.0	32821	4.267E+04	4.707E+15	2.870E+01	1.001E-08	3.9694	-10230
158.5	36421	6.066E+04	6.064E+15	3.179E+01	1.275E-08	3.9667	-11895

16000-DEGREE, LOG G = 3.0

TAU 5000	TEMP	PRESSURE	ELECTRON DENSITY	OPACITY	DENSITY	LOG G EFF	HEIGHT (KM)
.0010	11188	3.041E+00	9.423E+11	3.364E-01	2.181E-12	2.9789	88845
.0016	11326	4.643E+00	1.420E+12	3.582E-01	3.293E-12	2.9779	82607
.0025	11410	6.991E+00	2.120E+12	3.908E-01	4.926E-12	2.9760	76460
.0040	11451	1.033E+01	3.120E+12	4.377E-01	7.262E-12	2.9727	70524
.0063	11502	1.497E+01	4.493E+12	5.007E-01	1.048E-11	2.9679	64830
.0100	11580	2.124E+01	6.334E+12	5.834E-01	1.477E-11	2.9618	59351
.0158	11686	2.959E+01	8.736E+12	6.859E-01	2.041E-11	2.9546	54038
.0251	11830	4.061E+01	1.185E+13	8.114E-01	2.767E-11	2.9463	48823
.0398	12020	5.508E+01	1.581E+13	9.594E-01	3.694E-11	2.9372	43631
.0631	12274	7.411E+01	2.085E+13	1.130E+00	4.865E-11	2.9279	38374
.1000	12606	9.931E+01	2.724E+13	1.319E+00	6.338E-11	2.9193	32952
.1585	13031	1.330E+02	3.536E+13	1.528E+00	8.195E-11	2.9123	27260
.2512	13564	1.786E+02	4.583E+13	1.762E+00	1.052E-10	2.9070	21202
.3981	14214	2.407E+02	5.943E+13	2.028E+00	1.344E-10	2.9035	14694
.6310	14986	3.259E+02	7.708E+13	2.329E+00	1.710E-10	2.9017	7657
1.000	15879	4.436E+02	9.997E+13	2.658E+00	2.176E-10	2.9021	0
1.585	16890	6.083E+02	1.297E+14	3.005E+00	2.788E-10	2.9044	-8368
2.512	18020	8.413E+02	1.686E+14	3.386E+00	3.604E-10	2.9075	-17503
3.981	19271	1.172E+03	2.199E+14	3.814E+00	4.686E-10	2.9102	-27415
6.310	20651	1.638E+03	2.871E+14	4.299E+00	6.112E-10	2.9117	-38119
10.00	22188	2.296E+03	3.746E+14	4.840E+00	7.968E-10	2.9117	-49665
15.85	23923	3.221E+03	4.873E+14	5.422E+00	1.037E-09	2.9100	-62168
25.12	25922	4.528E+03	6.321E+14	6.036E+00	1.345E-09	2.9069	-75819
39.81	28254	6.378E+03	8.171E+14	6.666E+00	1.739E-09	2.9020	-90889
63.10	30998	9.006E+03	1.053E+15	7.333E+00	2.235E-09	2.8950	-107706
100.0	34257	1.270E+04	1.350E+15	8.141E+00	2.838E-09	2.8841	-126596
158.5	38160	1.773E+04	1.713E+15	9.401E+00	3.511E-09	2.8668	-147209

14000-DEGREE, LOG G = 4.0

TAU 5000	TEMP	PRESSURE	ELECTRON DENSITY	OPACITY	DENSITY	LOG G EFF	HEIGHT (KM)
.0010	10239	1.954E+01	6.541E+12	7.049E-01	1.548E-11	3.9967	6208
.0016	10262	2.696E+01	8.984E+12	8.558E-01	2.135E-11	3.9958	5797
.0025	10293	3.659E+01	1.213E+13	1.046E+00	2.894E-11	3.9948	5403
.0040	10334	4.904E+01	1.614E+13	1.281E+00	3.875E-11	3.9935	5025
.0063	10392	6.510E+01	2.124E+13	1.570E+00	5.128E-11	3.9919	4657
.0100	10470	8.578E+01	2.772E+13	1.920E+00	6.721E-11	3.9902	4296
.0158	10576	1.125E+02	3.591E+13	2.339E+00	8.744E-11	3.9882	3937
.0251	10715	1.471E+02	4.633E+13	2.838E+00	1.129E-10	3.9860	3576
.0398	10900	1.924E+02	5.958E+13	3.423E+00	1.451E-10	3.9837	3209
.0631	11140	2.518E+02	7.643E+13	4.096E+00	1.855E-10	3.9815	2831
.1000	11450	3.304E+02	9.790E+13	4.851E+00	2.362E-10	3.9795	2438
.1585	11840	4.358E+02	1.253E+14	5.676E+00	3.003E-10	3.9781	2021
.2512	12323	5.791E+02	1.606E+14	6.567E+00	3.821E-10	3.9772	1574
.3981	12905	7.763E+02	2.063E+14	7.524E+00	4.876E-10	3.9770	1092
.6310	13591	1.050E+03	2.661E+14	8.571E+00	6.238E-10	3.9774	568
1.000	14377	1.432E+03	3.445E+14	9.766E+00	8.012E-10	3.9780	0
1.585	15255	1.961E+03	4.481E+14	1.121E+01	1.027E-09	3.9787	-614
2.512	16226	2.691E+03	5.843E+14	1.295E+01	1.311E-09	3.9793	-1275
3.981	17287	3.693E+03	7.607E+14	1.497E+01	1.672E-09	3.9799	-1985
6.310	18443	5.073E+03	9.874E+14	1.718E+01	2.136E-09	3.9805	-2750
10.00	19712	6.991E+03	1.279E+15	1.954E+01	2.741E-09	3.9809	-3579
15.85	21122	9.670E+03	1.654E+15	2.209E+01	3.533E-09	3.9811	-4480
25.12	22720	1.344E+04	2.139E+15	2.484E+01	4.562E-09	3.9810	-5461
39.81	24555	1.877E+04	2.765E+15	2.778E+01	5.892E-09	3.9805	-6535
63.10	26682	2.634E+04	3.572E+15	3.086E+01	7.604E-09	3.9797	-7718
100.0	29177	3.714E+04	4.607E+15	3.402E+01	9.805E-09	3.9785	-9031
158.5	32130	5.271E+04	5.946E+15	3.737E+01	1.262E-08	3.9772	-10460

14000-DEGREE, LOG G = 3.0

TAU 5000	TEMP	PRESSURE	ELECTRON DENSITY	OPACITY	DENSITY	LOG G EFF	HEIGHT (KM)
.0010	10258	3.039E+00	1.021E+12	3.509E-01	2.391E-12	2.9869	76329
.0016	10276	4.581E+00	1.536E+12	3.837E-01	3.598E-12	2.9850	70936
.0025	10295	6.775E+00	2.265E+12	4.294E-01	5.317E-12	2.9823	65719
.0040	10328	9.823E+00	3.272E+12	4.917E-01	7.688E-12	2.9787	60722
.0063	10375	1.396E+01	4.624E+12	5.718E-01	1.089E-11	2.9741	55933
.0100	10444	1.949E+01	6.408E+12	6.764E-01	1.511E-11	2.9683	51319
.0158	10541	2.677E+01	8.706E+12	8.027E-01	2.059E-11	2.9616	46829
.0251	10674	3.631E+01	1.166E+13	9.547E-01	2.758E-11	2.9541	42398
.0398	10853	4.883E+01	1.543E+13	1.132E+00	3.647E-11	2.9462	37957
.0631	11090	6.532E+01	2.020E+13	1.333E+00	4.774E-11	2.9385	33432
.1000	11400	8.725E+01	2.628E+13	1.556E+00	6.197E-11	2.9318	28741
.1585	11795	1.168E+02	3.403E+13	1.794E+00	8.007E-11	2.9269	23792
.2512	12286	1.572E+02	4.406E+13	2.052E+00	1.033E-10	2.9239	18502
.3981	12880	2.132E+02	5.716E+13	2.333E+00	1.333E-10	2.9229	12804
.6310	13582	2.913E+02	7.440E+13	2.652E+00	1.720E-10	2.9233	6646
1.000	14392	4.001E+02	9.732E+13	3.034E+00	2.211E-10	2.9242	0
1.585	15303	5.505E+02	1.274E+14	3.485E+00	2.831E-10	2.9256	-7161
2.512	16316	7.593E+02	1.665E+14	3.989E+00	3.626E-10	2.9278	-14889
3.981	17428	1.051E+03	2.174E+14	4.534E+00	4.667E-10	2.9307	-23254
6.310	18642	1.463E+03	2.834E+14	5.125E+00	6.055E-10	2.9334	-32308
10.00	19976	2.043E+03	3.701E+14	5.792E+00	7.879E-10	2.9349	-42078
15.85	21459	2.859E+03	4.821E+14	6.517E+00	1.026E-09	2.9353	-52622
25.12	23140	4.010E+03	6.271E+14	7.302E+00	1.335E-09	2.9345	-64055
39.81	25072	5.639E+03	8.139E+14	8.137E+00	1.733E-09	2.9325	-76531
63.10	27311	7.951E+03	1.054E+15	9.001E+00	2.242E-09	2.9295	-90271
100.0	29935	1.125E+04	1.361E+15	9.898E+00	2.893E-09	2.9251	-105551
158.5	33039	1.595E+04	1.752E+15	1.090E+01	3.707E-09	2.9197	-122224

12000-DEGREE, LOG G = 4.0

TAU 5000	TEMP	PRESSURE	ELECTRON DENSITY	OPACITY	DENSITY	LOG G EFF	HEIGHT (KM)
.0010	9065	1.757E+01	6.325E+12	7.934E-01	1.640E-11	3.9973	5134
.0016	9086	2.421E+01	8.558E+12	9.509E-01	2.283E-11	3.9967	4789
.0025	9117	3.298E+01	1.142E+13	1.142E+00	3.140E-11	3.9959	4457
.0040	9160	4.453E+01	1.508E+13	1.371E+00	4.279E-11	3.9950	4138
.0063	9219	5.973E+01	1.977E+13	1.652E+00	5.773E-11	3.9939	3827
.0100	9298	7.964E+01	2.580E+13	1.996E+00	7.702E-11	3.9926	3523
.0158	9404	1.056E+02	3.357E+13	2.424E+00	1.015E-10	3.9912	3223
.0251	9542	1.392E+02	4.365E+13	2.965E+00	1.318E-10	3.9895	2924
.0398	9722	1.825E+02	5.669E+13	3.648E+00	1.684E-10	3.9876	2625
.0631	9954	2.379E+02	7.349E+13	4.496E+00	2.117E-10	3.9855	2322
.1000	10248	3.089E+02	9.491E+13	5.510E+00	2.623E-10	3.9835	2009
.1585	10615	4.009E+02	1.219E+14	6.647E+00	3.223E-10	3.9819	1678
.2512	11065	5.228E+02	1.560E+14	7.846E+00	3.959E-10	3.9808	1321
.3981	11606	6.884E+02	1.991E+14	9.055E+00	4.900E-10	3.9805	927
.6310	12239	9.182E+02	2.547E+14	1.027E+01	6.135E-10	3.9809	488
1.000	12964	1.241E+03	3.274E+14	1.156E+01	7.778E-10	3.9818	0
1.585	13772	1.697E+03	4.236E+14	1.300E+01	9.970E-10	3.9828	-540
2.512	14661	2.340E+03	5.515E+14	1.473E+01	1.285E-09	3.9838	-1130
3.981	15627	3.235E+03	7.214E+14	1.689E+01	1.654E-09	3.9846	-1766
6.310	16673	4.468E+03	9.444E+14	1.954E+01	2.118E-09	3.9852	-2449
10.00	17808	6.157E+03	1.232E+15	2.259E+01	2.704E-09	3.9855	-3180
15.85	19050	8.486E+03	1.599E+15	2.583E+01	3.459E-09	3.9858	-3968
25.12	20435	1.173E+04	2.070E+15	2.924E+01	4.439E-09	3.9860	-4825
39.81	21995	1.630E+04	2.678E+15	3.288E+01	5.717E-09	3.9860	-5762
63.10	23770	2.276E+04	3.464E+15	3.676E+01	7.378E-09	3.9858	-6788
100.0	25812	3.193E+04	4.476E+15	4.086E+01	9.530E-09	3.9853	-7919
158.5	28200	4.506E+04	5.786E+15	4.516E+01	1.230E-08	3.9847	-9136

12000-DEGREE, LOG G = 3.0

TAU 5000	TEMP	PRESSURE	ELECTRON DENSITY	OPACITY	DENSITY	LOG G EFF	HEIGHT (KM)
.0010	9051	2.933E+00	1.106E+12	3.815E-01	2.638E-12	2.9909	63729
.0016	9063	4.348E+00	1.630E+12	4.259E-01	3.920E-12	2.9889	59252
.0025	9085	6.323E+00	2.352E+12	4.851E-01	5.715E-12	2.9864	54946
.0040	9117	9.030E+00	3.325E+12	5.625E-01	8.180E-12	2.9831	50837
.0063	9164	1.267E+01	4.611E+12	6.608E-01	1.149E-11	2.9790	46903
.0100	9233	1.751E+01	6.286E+12	7.831E-01	1.583E-11	2.9742	43107
.0158	9330	2.387E+01	8.448E+12	9.330E-01	2.143E-11	2.9687	39401
.0251	9461	3.221E+01	1.123E+13	1.115E+00	2.855E-11	2.9625	35737
.0398	9637	4.313E+01	1.479E+13	1.332E+00	3.747E-11	2.9560	32063
.0631	9866	5.743E+01	1.935E+13	1.584E+00	4.849E-11	2.9496	28316
.1000	10161	7.629E+01	2.515E+13	1.867E+00	6.213E-11	2.9440	24423
.1585	10530	1.015E+02	3.256E+13	2.170E+00	7.920E-11	2.9399	20300
.2512	10984	1.359E+02	4.209E+13	2.490E+00	1.010E-10	2.9376	15860
.3981	11529	1.836E+02	5.448E+13	2.821E+00	1.294E-10	2.9373	11032
.6310	12170	2.508E+02	7.076E+13	3.173E+00	1.668E-10	2.9386	5756
1.000	12906	3.460E+02	9.237E+13	3.563E+00	2.164E-10	2.9408	0
1.585	13732	4.807E+02	1.212E+14	4.022E+00	2.813E-10	2.9434	-6241
2.512	14648	6.698E+02	1.598E+14	4.595E+00	3.644E-10	2.9457	-12953
3.981	15648	9.320E+02	2.107E+14	5.300E+00	4.691E-10	2.9477	-20127
6.310	16729	1.294E+03	2.767E+14	6.096E+00	6.029E-10	2.9497	-27798
10.00	17900	1.798E+03	3.617E+14	6.946E+00	7.776E-10	2.9517	-36048
15.85	19179	2.504E+03	4.715E+14	7.856E+00	1.008E-09	2.9531	-44957
25.12	20602	3.498E+03	6.139E+14	8.847E+00	1.309E-09	2.9538	-54600
39.81	22206	4.902E+03	7.985E+14	9.923E+00	1.701E-09	2.9537	-65063
63.10	24033	6.888E+03	1.038E+15	1.108E+01	2.206E-09	2.9527	-76474
100.0	26138	9.708E+03	1.344E+15	1.228E+01	2.861E-09	2.9510	-89008
158.5	28594	1.374E+04	1.741E+15	1.357E+01	3.697E-09	2.9488	-102493

12000-DEGREE, LOG G = 2.0

TAU 5000	TEMP	PRESSURE	ELECTRON DENSITY	OPACITY	DENSITY	LOG G EFF	HEIGHT (KM)
.0010	9137	3.041E-01	1.146E+11	2.941E-01	2.688E-13	1.9435	958064
.0016	9143	4.762E-01	1.794E+11	2.998E-01	4.207E-13	1.9408	899986
.0025	9149	7.405E-01	2.784E+11	3.084E-01	6.542E-13	1.9365	841986
.0040	9172	1.140E+00	4.274E+11	3.212E-01	1.005E-12	1.9301	784535
.0063	9197	1.733E+00	6.469E+11	3.396E-01	1.525E-12	1.9207	727704
.0100	9240	2.592E+00	9.619E+11	3.657E-01	2.274E-12	1.9077	671589
.0158	9308	3.801E+00	1.397E+12	3.997E-01	3.318E-12	1.8908	616023
.0251	9410	5.466E+00	1.986E+12	4.434E-01	4.722E-12	1.8697	560549
.0398	9557	7.712E+00	2.757E+12	4.962E-01	6.564E-12	1.8459	504491
.0631	9763	1.071E+01	3.750E+12	5.571E-01	8.917E-12	1.8213	446844
.1000	10040	1.473E+01	5.022E+12	6.243E-01	1.191E-11	1.7988	386405
.1585	10402	2.016E+01	6.647E+12	6.945E-01	1.572E-11	1.7820	321788
.2512	10860	2.770E+01	8.766E+12	7.678E-01	2.064E-11	1.7724	251616
.3981	11426	3.842E+01	1.158E+13	8.452E-01	2.715E-11	1.7700	174885
.6310	12103	5.387E+01	1.537E+13	9.300E-01	3.586E-11	1.7728	91026
1.000	12893	7.626E+01	2.055E+13	1.033E+00	4.738E-11	1.7776	0
1.585	13794	1.083E+02	2.758E+13	1.162E+00	6.228E-11	1.7816	-97945
2.512	14806	1.536E+02	3.699E+13	1.315E+00	8.114E-11	1.7863	-203038
3.981	15926	2.184E+02	4.934E+13	1.475E+00	1.063E-10	1.7944	-316217
6.310	17153	3.124E+02	6.579E+13	1.645E+00	1.405E-10	1.8037	-438144
10.00	18497	4.482E+02	8.769E+13	1.834E+00	1.867E-10	1.8105	-568876
15.85	19975	6.435E+02	1.166E+14	2.044E+00	2.481E-10	1.8132	-708703
25.12	21633	9.221E+02	1.542E+14	2.272E+00	3.284E-10	1.8127	-858759
39.81	23512	1.318E+03	2.029E+14	2.513E+00	4.318E-10	1.8086	-1021272
63.10	25664	1.880E+03	2.652E+14	2.764E+00	5.639E-10	1.8010	-1199430
100.0	28158	2.675E+03	3.438E+14	3.010E+00	7.318E-10	1.7894	-1397627
158.5	31079	3.798E+03	4.438E+14	3.293E+00	9.377E-10	1.7742	-1615107

11000-DEGREE, LOG G = 4.0

TAU 5000	TEMP	PRESSURE	ELECTRON DENSITY	OPACITY	DENSITY	LOG G EFF	HEIGHT (KM)
.0010	8433	1.835E+01	6.116E+12	7.092E-01	2.050E-11	3.9980	4134
.0016	8456	2.596E+01	8.218E+12	8.152E-01	2.979E-11	3.9977	3825
.0025	8488	3.643E+01	1.092E+13	9.406E-01	4.286E-11	3.9972	3529
.0040	8532	5.076E+01	1.443E+13	1.093E+00	6.093E-11	3.9967	3246
.0063	8592	7.018E+01	1.899E+13	1.285E+00	8.538E-11	3.9962	2974
.0100	8671	9.611E+01	2.499E+13	1.535E+00	1.175E-10	3.9954	2712
.0158	8775	1.300E+02	3.292E+13	1.874E+00	1.582E-10	3.9945	2459
.0251	8911	1.734E+02	4.353E+13	2.347E+00	2.071E-10	3.9934	2215
.0398	9087	2.274E+02	5.775E+13	3.018E+00	2.625E-10	3.9918	1979
.0631	9313	2.929E+02	7.666E+13	3.970E+00	3.212E-10	3.9898	1748
.1000	9600	3.709E+02	1.013E+14	5.277E+00	3.795E-10	3.9873	1518
.1585	9958	4.634E+02	1.320E+14	6.921E+00	4.359E-10	3.9846	1281
.2512	10396	5.769E+02	1.689E+14	8.725E+00	4.953E-10	3.9826	1026
.3981	10921	7.232E+02	2.124E+14	1.042E+01	5.680E-10	3.9816	738
.6310	11535	9.223E+02	2.652E+14	1.188E+01	6.673E-10	3.9819	400
1.000	12234	1.204E+03	3.325E+14	1.320E+01	8.083E-10	3.9829	0
1.585	13012	1.608E+03	4.216E+14	1.458E+01	1.006E-09	3.9842	-466
2.512	13866	2.188E+03	5.417E+14	1.623E+01	1.279E-09	3.9854	-996
3.981	14790	3.011E+03	7.031E+14	1.829E+01	1.640E-09	3.9865	-1583
6.310	15786	4.163E+03	9.185E+14	2.091E+01	2.108E-09	3.9872	-2221
10.00	16859	5.752E+03	1.202E+15	2.416E+01	2.698E-09	3.9876	-2907
15.85	18027	7.930E+03	1.566E+15	2.786E+01	3.444E-09	3.9879	-3644
25.12	19317	1.094E+04	2.033E+15	3.180E+01	4.399E-09	3.9881	-4440
39.81	20756	1.515E+04	2.632E+15	3.594E+01	5.641E-09	3.9882	-5309
63.10	22376	2.108E+04	3.403E+15	4.032E+01	7.267E-09	3.9882	-6260
100.0	24219	2.947E+04	4.402E+15	4.501E+01	9.377E-09	3.9880	-7305
158.5	26353	4.144E+04	5.690E+15	4.997E+01	1.211E-08	3.9876	-8425

11000-DEGREE, LOG G = 3.0

TAU 5000	TEMP	PRESSURE	ELECTRON DENSITY	OPACITY	DENSITY	LOG G EFF	HEIGHT (KM)
.0010	8400	2.906E+00	1.141E+12	3.870E-01	2.901E-12	2.9925	54989
.0016	8415	4.310E+00	1.660E+12	4.292E-01	4.357E-12	2.9909	50984
.0025	8438	6.291E+00	2.365E+12	4.836E-01	6.451E-12	2.9888	47148
.0040	8473	9.041E+00	3.310E+12	5.529E-01	9.393E-12	2.9863	43501
.0063	8523	1.280E+01	4.558E+12	6.407E-01	1.344E-11	2.9833	40023
.0100	8595	1.788E+01	6.201E+12	7.531E-01	1.884E-11	2.9798	36689
.0158	8694	2.459E+01	8.351E+12	8.980E-01	2.580E-11	2.9756	33469
.0251	8827	3.337E+01	1.117E+13	1.087E+00	3.446E-11	2.9706	30329
.0398	9001	4.464E+01	1.483E+13	1.332E+00	4.482E-11	2.9647	27232
.0631	9227	5.897E+01	1.956E+13	1.641E+00	5.683E-11	2.9583	24123
.1000	9514	7.717E+01	2.555E+13	2.007E+00	7.057E-11	2.9520	20927
.1585	9871	1.006E+02	3.305E+13	2.405E+00	8.671E-11	2.9468	17542
.2512	10307	1.318E+02	4.244E+13	2.807E+00	1.067E-10	2.9438	13852
.3981	10830	1.746E+02	5.438E+13	3.197E+00	1.326E-10	2.9431	9749
.6310	11442	2.347E+02	6.991E+13	3.581E+00	1.672E-10	2.9444	5146
1.000	12146	3.206E+02	9.049E+13	3.983E+00	2.140E-10	2.9468	0
1.585	12936	4.437E+02	1.181E+14	4.771E+00	2.771E-10	2.9498	-5694
2.512	13809	6.193E+02	1.552E+14	4.992E+00	3.605E-10	2.9528	-11907
3.981	14760	8.663E+02	2.049E+14	5.701E+00	4.681E-10	2.9553	-18590
6.310	15784	1.209E+03	2.707E+14	6.584E+00	6.035E-10	2.9571	-25715
10.00	16886	1.680E+03	3.555E+14	7.583E+00	7.761E-10	2.9587	-33316
15.85	18079	2.334E+03	4.646E+14	8.642E+00	1.000E-09	2.9601	-41481
25.12	19394	3.251E+03	6.052E+14	9.769E+00	1.294E-09	2.9611	-50312
39.81	20859	4.543E+03	7.875E+14	1.099E+01	1.679E-09	2.9617	-59891
63.10	22510	6.368E+03	1.023E+15	1.231E+01	2.180E-09	2.9615	-70310
100.0	24391	8.954E+03	1.328E+15	1.372E+01	2.829E-09	2.9606	-81693
158.5	26568	1.263E+04	1.723E+15	1.522E+01	3.658E-09	2.9593	-93858

11000-DEGREE, LOG G = 2.0

TAU 5000	TEMP	PRESSURE	ELECTRON DENSITY	OPACITY	DENSITY	LOG G EFF	HEIGHT (KM)
.0010	8452	3.146E-01	1.277E+11	2.964E-01	3.016E-13	1.9590	822460
.0016	8458	4.912E-01	1.988E+11	3.034E-01	4.715E-13	1.9562	771169
.0025	8468	7.611E-01	3.067E+11	3.137E-01	7.319E-13	1.9518	720098
.0040	8486	1.167E+00	4.671E+11	3.287E-01	1.124E-12	1.9454	669742
.0063	8515	1.765E+00	7.000E+11	3.502E-01	1.703E-12	1.9364	620228
.0100	8561	2.626E+00	1.029E+12	3.797E-01	2.535E-12	1.9243	571611
.0158	8632	3.832E+00	1.481E+12	4.190E-01	3.688E-12	1.9089	523758
.0251	8739	5.484E+00	2.085E+12	4.698E-01	5.230E-12	1.8903	476316
.0398	8891	7.699E+00	2.878E+12	5.332E-01	7.214E-12	1.8692	428691
.0631	9100	1.064E+01	3.900E+12	6.085E-01	9.704E-12	1.8472	380035
.1000	9376	1.451E+01	5.204E+12	6.925E-01	1.276E-11	1.8269	329193
.1585	9730	1.971E+01	6.865E+12	7.804E-01	1.660E-11	1.8116	274798
.2512	10172	2.687E+01	9.014E+12	8.694E-01	2.152E-11	1.8031	215522
.3981	10710	3.702E+01	1.184E+13	9.595E-01	2.804E-11	1.8014	150305
.6310	11348	5.167E+01	1.566E+13	1.054E+00	3.681E-11	1.8051	78518
1.000	12090	7.308E+01	2.085E+13	1.160E+00	4.874E-11	1.8119	0
1.585	12933	1.043E+02	2.797E+13	1.289E+00	6.470E-11	1.8192	-84910
2.512	13875	1.491E+02	3.768E+13	1.456E+00	8.534E-11	1.8249	-175664
3.981	14912	2.124E+02	5.066E+13	1.661E+00	1.116E-10	1.8302	-272121
6.310	16041	3.018E+02	6.766E+13	1.879E+00	1.459E-10	1.8373	-375045
10.00	17265	4.300E+02	9.000E+13	2.109E+00	1.921E-10	1.8447	-485368
15.85	18597	6.140E+02	1.194E+14	2.357E+00	2.545E-10	1.8501	-603548
25.12	20073	8.771E+02	1.581E+14	2.637E+00	3.366E-10	1.8523	-730071
39.81	21728	1.251E+03	2.085E+14	2.939E+00	4.433E-10	1.8521	-866103
63.10	23598	1.783E+03	2.735E+14	3.258E+00	5.818E-10	1.8492	-1013594
100.0	25740	2.537E+03	3.569E+14	3.591E+00	7.588E-10	1.8435	-1175197
158.5	28224	3.611E+03	4.637E+14	3.937E+00	9.843E-10	1.8362	-1349301

10000-DEGREE, LOG G = 4.5

TAU 5000	TEMP	PRESSURE	ELECTRON DENSITY	OPACITY	DENSITY	HYDROGEN IONIZATION	HEIGHT (KM)
.0010	7837	8.477E+01	1.209E+13	4.752E-01	1.415E-10	2.002E-01	-756
.0016	7860	1.220E+02	1.536E+13	5.192E-01	2.072E-10	1.737E-01	-688
.0025	7894	1.753E+02	1.964E+13	5.796E-01	3.015E-10	1.526E-01	-621
.0040	7932	2.505E+02	2.510E+13	6.542E-01	4.348E-10	1.352E-01	-555
.0063	7984	3.549E+02	3.233E+13	7.574E-01	6.184E-10	1.224E-01	-492
.0100	8051	4.955E+02	4.193E+13	8.996E-01	8.623E-10	1.138E-01	-432
.0158	8136	6.824E+02	5.491E+13	1.097E+00	1.180E-09	1.089E-01	-373
.0251	8253	9.183E+02	7.330E+13	1.400E+00	1.564E-09	1.097E-01	-319
.0398	8396	1.207E+03	9.876E+13	1.841E+00	2.012E-09	1.149E-01	-267
.0631	8593	1.543E+03	1.374E+14	2.591E+00	2.484E-09	1.295E-01	-220
.1000	8845	1.910E+03	1.947E+14	3.851E+00	2.924E-09	1.560E-01	-177
.1585	9166	2.292E+03	2.810E+14	6.084E+00	3.267E-09	2.016E-01	-138
.2512	9592	2.660E+03	4.121E+14	1.030E+01	3.409E-09	2.835E-01	-103
.3981	10111	3.004E+03	5.866E+14	1.733E+01	3.342E-09	4.118E-01	-71
.6310	10713	3.341E+03	7.802E+14	2.659E+01	3.158E-09	5.799E-01	-39
1.000	11405	3.715E+03	9.578E+14	3.487E+01	2.993E-09	7.510E-01	0
1.585	12184	4.209E+03	1.109E+15	3.906E+01	2.976E-09	8.746E-01	53
2.512	12962	4.942E+03	1.270E+15	4.068E+01	3.186E-09	9.356E-01	128
3.981	13823	6.066E+03	1.488E+15	4.192E+01	3.609E-09	9.668E-01	233
6.310	14758	7.783E+03	1.806E+15	4.410E+01	4.301E-09	9.820E-01	371
10.00	15658	1.030E+04	2.268E+15	4.852E+01	5.335E-09	9.887E-01	537
15.85	16702	1.392E+04	2.897E+15	5.371E+01	6.702E-09	9.930E-01	727
25.12	17962	1.910E+04	3.744E+15	5.923E+01	8.452E-09	9.958E-01	944
39.81	19105	2.641E+04	4.917E+15	6.814E+01	1.089E-08	9.970E-01	1184
63.10	20287	3.638E+04	6.424E+15	7.846E+01	1.403E-08	9.978E-01	1438

10000-DEGREE, LOG G = 4.0

TAU 5000	TEMP	PRESSURE	ELECTRON DENSITY	OPACITY	DENSITY	HYDROGEN IONIZATION	HEIGHT (KM)
.0010	7806	3.082E+01	6.259E+12	4.023E-01	4.770E-11	3.077E-01	-2683
.0016	7832	4.476E+01	8.133E+12	4.363E-01	7.102E-11	2.685E-01	-2445
.0025	7864	6.496E+01	1.057E+13	4.806E-01	1.052E-10	2.356E-01	-2212
.0040	7904	9.376E+01	1.376E+13	5.395E-01	1.541E-10	2.093E-01	-1988
.0063	7958	1.339E+02	1.797E+13	6.208E-01	2.218E-10	1.898E-01	-1772
.0100	8029	1.884E+02	2.361E+13	7.355E-01	3.124E-10	1.771E-01	-1566
.0158	8121	2.600E+02	3.129E+13	9.014E-01	4.283E-10	1.712E-01	-1371
.0251	8240	3.508E+02	4.190E+13	1.147E+00	5.691E-10	1.725E-01	-1188
.0398	8394	4.615E+02	5.694E+13	1.527E+00	7.287E-10	1.831E-01	-1017
.0631	8595	5.896E+02	7.876E+13	2.148E+00	8.927E-10	2.068E-01	-859
.1000	8853	7.301E+02	1.105E+14	3.189E+00	1.040E-09	2.492E-01	-714
.1585	9183	8.765E+02	1.560E+14	4.959E+00	1.143E-09	3.201E-01	-580
.2512	9596	1.024E+03	2.175E+14	7.837E+00	1.186E-09	4.305E-01	-454
.3981	10108	1.174E+03	2.917E+14	1.186E+01	1.173E-09	5.836E-01	-327
.6310	10713	1.340E+03	3.663E+14	1.584E+01	1.152E-09	7.462E-01	-184
1.000	11390	1.553E+03	4.356E+14	1.831E+01	1.180E-09	8.667E-01	0
1.585	12148	1.862E+03	5.104E+14	1.943E+01	1.281E-09	9.355E-01	251
2.512	12944	2.327E+03	6.099E+14	2.037E+01	1.479E-09	9.674E-01	590
3.981	13803	3.026E+03	7.505E+14	2.168E+01	1.789E-09	9.827E-01	1019
6.310	14716	4.054E+03	9.489E+14	2.364E+01	2.235E-09	9.901E-01	1532
10.00	15674	5.531E+03	1.224E+15	2.639E+01	2.845E-09	9.940E-01	2115
15.85	16711	7.611E+03	1.596E+15	2.977E+01	3.638E-09	9.962E-01	2759
25.12	17853	1.053E+04	2.091E+15	3.370E+01	4.659E-09	9.975E-01	3465
39.81	19066	1.460E+04	2.739E+15	3.831E+01	5.993E-09	9.983E-01	4232
63.10	20392	2.027E+04	3.578E+15	4.342E+01	7.737E-09	9.988E-01	5062

10000-DEGREE, LOG G = 4.0   (10X METALS)

TAU 5000	TEMP	PRESSURE	ELECTRON DENSITY	OPACITY	DENSITY	HYDROGEN IONIZATION	HEIGHT (KM)
.0010	7931	2.698E+01	6.351E+12	4.763E-01	3.953E-11	3.811E-01	-2878
.0016	7958	3.867E+01	8.277E+12	5.229E-01	5.819E-11	3.373E-01	-2636
.0025	7994	5.543E+01	1.079E+13	5.827E-01	8.523E-11	3.001E-01	-2399
.0040	8038	7.906E+01	1.407E+13	6.599E-01	1.236E-10	2.697E-01	-2170
.0063	8095	1.118E+02	1.839E+13	7.636E-01	1.764E-10	2.469E-01	-1950
.0100	8169	1.560E+02	2.416E+13	9.070E-01	2.467E-10	2.319E-01	-1739
.0158	8262	2.141E+02	3.193E+13	1.108E+00	3.367E-10	2.245E-01	-1538
.0251	8383	2.882E+02	4.259E+13	1.402E+00	4.462E-10	2.260E-01	-1348
.0398	8537	3.793E+02	5.741E+13	1.842E+00	5.715E-10	2.379E-01	-1169
.0631	8733	4.867E+02	7.838E+13	2.531E+00	7.030E-10	2.641E-01	-1000
.1000	8989	6.077E+02	1.084E+14	3.651E+00	8.242E-10	3.116E-01	-842
.1585	9314	7.378E+02	1.501E+14	5.461E+00	9.157E-10	3.889E-01	-693
.2512	9716	8.752E+02	2.044E+14	8.157E+00	9.686E-10	5.008E-01	-547
.3981	10217	1.024E+03	2.688E+14	1.164E+01	9.878E-10	6.461E-01	-396
.6310	10810	1.198E+03	3.352E+14	1.483E+01	1.010E-09	7.879E-01	-221
1.000	11471	1.428E+03	4.028E+14	1.691E+01	1.079E-09	8.868E-01	0
1.585	12220	1.760E+03	4.821E+14	1.814E+01	1.214E-09	9.436E-01	291
2.512	13014	2.254E+03	5.887E+14	1.936E+01	1.440E-09	9.707E-01	664
3.981	13876	2.984E+03	7.368E+14	2.092E+01	1.775E-09	9.841E-01	1120
6.310	14787	4.042E+03	9.422E+14	2.308E+01	2.243E-09	9.908E-01	1648
10.00	15746	5.548E+03	1.223E+15	2.593E+01	2.873E-09	9.943E-01	2239
15.85	16784	7.659E+03	1.600E+15	2.937E+01	3.686E-09	9.963E-01	2885
25.12	17923	1.061E+04	2.100E+15	3.335E+01	4.730E-09	9.976E-01	3589
39.81	19133	1.472E+04	2.754E+15	3.797E+01	6.093E-09	9.984E-01	4351
63.10	20466	2.045E+04	3.598E+15	4.302E+01	7.869E-09	9.989E-01	5175

10000-DEGREE, LOG G = 4.0   (1/10X METALS)

TAU 5000	TEMP	PRESSURE	ELECTRON DENSITY	OPACITY	DENSITY	HYDROGEN IONIZATION	HEIGHT (KM)
.0010	7762	3.245E+01	6.211E+12	3.766E-01	5.134E-11	2.834E-01	-2612
.0016	7786	4.737E+01	8.060E+12	4.067E-01	7.680E-11	2.458E-01	-2376
.0025	7815	6.910E+01	1.046E+13	4.459E-01	1.143E-10	2.143E-01	-2145
.0040	7854	1.002E+02	1.359E+13	4.990E-01	1.681E-10	1.893E-01	-1922
.0063	7907	1.436E+02	1.775E+13	5.738E-01	2.427E-10	1.712E-01	-1708
.0100	7977	2.026E+02	2.333E+13	6.798E-01	3.425E-10	1.594E-01	-1505
.0158	8070	2.799E+02	3.099E+13	8.368E-01	4.696E-10	1.544E-01	-1313
.0251	8191	3.774E+02	4.164E+13	1.071E+00	6.230E-10	1.564E-01	-1134
.0398	8348	4.954E+02	5.686E+13	1.440E+00	7.953E-10	1.674E-01	-967
.0631	8555	6.304E+02	7.915E+13	2.050E+00	9.695E-10	1.912E-01	-814
.1000	8819	7.766E+02	1.118E+14	3.085E+00	1.122E-09	2.334E-01	-674
.1585	9153	9.270E+02	1.587E+14	4.860E+00	1.226E-09	3.034E-01	-547
.2512	9573	1.076E+03	2.228E+14	7.804E+00	1.261E-09	4.141E-01	-427
.3981	10090	1.225E+03	3.001E+14	1.198E+01	1.236E-09	5.691E-01	-308
.6310	10698	1.389E+03	3.772E+14	1.617E+01	1.201E-09	7.361E-01	-173
1.000	11377	1.598E+03	4.470E+14	1.874E+01	1.216E-09	8.616E-01	0
1.585	12137	1.899E+03	5.206E+14	1.983E+01	1.307E-09	9.335E-01	240
2.512	12936	2.357E+03	6.178E+14	2.066E+01	1.497E-09	9.667E-01	567
3.981	13797	3.048E+03	7.562E+14	2.187E+01	1.801E-09	9.824E-01	988
6.310	14708	4.068E+03	9.527E+14	2.378E+01	2.242E-09	9.900E-01	1494
10.00	15668	5.539E+03	1.226E+15	2.647E+01	2.846E-09	9.939E-01	2074
15.85	16705	7.614E+03	1.597E+15	2.983E+01	3.636E-09	9.961E-01	2716
25.12	17849	1.053E+04	2.091E+15	3.374E+01	4.654E-09	9.975E-01	3422
39.81	19059	1.459E+04	2.739E+15	3.836E+01	5.985E-09	9.983E-01	4188
63.10	20390	2.026E+04	3.577E+15	4.344E+01	7.724E-09	9.988E-01	5019

10000-DEGREE, LOG G = 3.5

TAU 5000	TEMP	PRESSURE	ELECTRON DENSITY	OPACITY	DENSITY	LOG G EFF	HEIGHT (KM)
.0010	7774	1.383E+01	2.979E+12	3.480E-01	1.519E-11	3.4938	-10094
.0016	7796	1.595E+01	3.984E+12	3.706E-01	2.314E-11	3.4936	-9231
.0025	7828	2.348E+01	5.324E+12	4.033E-01	3.503E-11	3.4933	-8395
.0040	7866	3.434E+01	7.079E+12	4.461E-01	5.240E-11	3.4931	-7594
.0063	7921	4.967E+01	9.428E+12	5.082E-01	7.685E-11	3.4931	-6829
.0100	7993	7.058E+01	1.258E+13	5.975E-01	1.097E-10	3.4930	-6107
.0158	8083	9.839E+01	1.680E+13	7.230E-01	1.524E-10	3.4932	-5423
.0251	8211	1.337E+02	2.279E+13	9.244E-01	2.032E-10	3.4934	-4784
.0398	8366	1.766E+02	3.091E+13	1.219E+00	2.605E-10	3.4936	-4188
.0631	8575	2.264E+02	4.263E+13	1.707E+00	3.174E-10	3.4939	-3631
.1000	8839	2.811E+02	5.883E+13	2.481E+00	3.663E-10	3.4944	-3111
.1585	9169	3.398E+02	8.045E+13	3.668E+00	4.014E-10	3.4947	-2611
.2512	9593	4.016E+02	1.075E+14	5.343E+00	4.181E-10	3.4952	-2109
.3981	10102	4.710E+02	1.367E+14	7.104E+00	4.292E-10	3.4957	-1555
.6310	10684	5.580E+02	1.662E+14	8.403E+00	4.530E-10	3.4961	-880
1.000	11365	6.808E+02	1.991E+14	9.155E+00	5.014E-10	3.4969	0
1.585	12135	8.640E+02	2.415E+14	9.664E+00	5.856E-10	3.4972	1142
2.512	12923	1.139E+03	3.017E+14	1.041E+01	7.193E-10	3.4977	2565
3.981	13798	1.544E+03	3.850E+14	1.136E+01	9.084E-10	3.4981	4226
6.310	14762	2.132E+03	5.002E+14	1.256E+01	1.166E-09	3.4984	6116
10.00	15662	2.959E+03	6.596E+14	1.441E+01	1.514E-09	3.4989	8163
15.85	16784	4.118E+03	8.675E+14	1.619E+01	1.943E-09	3.4990	10372
25.12	18043	5.759E+03	1.142E+15	1.815E+01	2.500E-09	3.4992	12806
39.81	19227	8.069E+03	1.510E+15	2.091E+01	3.267E-09	3.4994	15439
63.10	20475	1.121E+04	1.976E+15	2.393E+01	4.249E-09	3.4994	18168

10000-DEGREE, LOG G = 3.0

TAU 5000	TEMP	PRESSURE	ELECTRON DENSITY	OPACITY	DENSITY	LOG G EFF	HEIGHT (KM)
.0010	7749	3.640E+00	1.267E+12	3.136E-01	4.562E-12	2.9839	-39195
.0016	7767	5.444E+00	1.765E+12	3.273E-01	7.073E-12	2.9835	-36003
.0025	7794	8.156E+00	2.446E+12	3.477E-01	1.096E-11	2.9830	-32905
.0040	7829	1.215E+01	3.362E+12	3.765E-01	1.683E-11	2.9822	-29940
.0063	7880	1.791E+01	4.597E+12	4.195E-01	2.534E-11	2.9808	-27125
.0100	7949	2.592E+01	6.263E+12	4.833E-01	3.707E-11	2.9787	-24474
.0158	8046	3.667E+01	8.540E+12	5.805E-01	5.226E-11	2.9789	-21986
.0251	8172	5.045E+01	1.165E+13	7.059E-01	7.059E-11	2.9792	-19658
.0398	8323	6.752E+01	1.581E+13	9.354E-01	9.171E-11	2.9797	-17464
.0631	8534	8.774E+01	2.168E+13	1.274E+00	1.127E-10	2.9807	-15385
.1000	8810	1.105E+02	2.955E+13	1.785E+00	1.309E-10	2.9822	-13396
.1585	9148	1.358E+02	3.936E+13	2.467E+00	1.456E-10	2.9832	-11404
.2512	9572	1.647E+02	5.079E+13	3.242E+00	1.578E-10	2.9848	-9275
.3981	10081	2.008E+02	6.334E+13	3.907E+00	1.728E-10	2.9864	-6806
.6310	10665	2.498E+02	7.767E+13	4.373E+00	1.965E-10	2.9878	-3772
1.000	11332	3.213E+02	9.602E+13	4.734E+00	2.336E-10	2.9900	0
1.585	12101	4.275E+02	1.209E+14	5.081E+00	2.884E-10	2.9910	4565
2.512	12928	5.850E+02	1.556E+14	5.543E+00	3.677E-10	2.9926	9921
3.981	13791	8.135E+02	2.038E+14	6.176E+00	4.772E-10	2.9940	15864
6.310	14819	1.142E+03	2.685E+14	6.836E+00	6.183E-10	2.9949	22417
10.00	15785	1.607E+03	3.587E+14	7.851E+00	8.085E-10	2.9963	29492
15.85	16877	2.254E+03	4.764E+14	8.931E+00	1.049E-09	2.9967	36994
25.12	18084	3.166E+03	6.294E+14	1.010E+01	1.364E-09	2.9974	45114
39.81	19318	4.435E+03	8.288E+14	1.151E+01	1.782E-09	2.9980	53676
63.10	21054	6.312E+03	1.085E+15	1.255E+01	2.321E-09	2.9983	63391

10000-DEGREE, LOG G = 2.0

TAU 5000	TEMP	PRESSURE	ELECTRON DENSITY	OPACITY	DENSITY	LOG G EFF	HEIGHT (KM)
.0010	7773	3.602E-01	1.550E+11	2.893E-01	3.860E-13	1.8906	-644264
.0016	7781	5.481E-01	2.322E+11	2.924E-01	5.938E-13	1.8863	-602555
.0025	7794	8.406E-01	3.485E+11	2.974E-01	9.240E-13	1.8794	-560429
.0040	7814	1.291E+00	5.203E+11	3.053E-01	1.445E-12	1.8683	-518555
.0063	7843	1.974E+00	7.682E+11	3.178E-01	2.253E-12	1.8500	-477458
.0100	7887	2.984E+00	1.117E+12	3.376E-01	3.466E-12	1.8194	-437484
.0158	7964	4.433E+00	1.606E+12	3.710E-01	5.182E-12	1.8107	-398818
.0251	8074	6.434E+00	2.275E+12	4.226E-01	7.467E-12	1.7964	-361536
.0398	8218	9.069E+00	3.171E+12	4.967E-01	1.030E-11	1.7876	-325325
.0631	8433	1.241E+01	4.366E+12	6.046E-01	1.344E-11	1.7916	-289618
.1000	8699	1.651E+01	5.859E+12	7.346E-01	1.685E-11	1.7978	-253268
.1585	9037	2.162E+01	7.688E+12	8.748E-01	2.060E-11	1.8093	-214610
.2512	9474	2.827E+01	9.905E+12	1.003E+00	2.502E-11	1.8268	-171454
.3981	9987	3.752E+01	1.271E+13	1.113E+00	3.098E-11	1.8444	-121930
.6310	10572	5.075E+01	1.640E+13	1.216E+00	3.923E-11	1.8602	-64904
1.000	11279	7.041E+01	2.145E+13	1.317E+00	5.077E-11	1.8841	0
1.585	12069	9.979E+01	2.850E+13	1.437E+00	6.703E-11	1.8951	72528
2.512	12913	1.424E+02	3.819E+13	1.588E+00	8.908E-11	1.9121	151170
3.981	13918	2.059E+02	5.174E+13	1.759E+00	1.183E-10	1.9268	235953
6.310	15005	2.970E+02	7.031E+13	1.968E+00	1.561E-10	1.9358	327767
10.00	15960	4.248E+02	9.544E+13	2.268E+00	2.079E-10	1.9497	423408
15.85	17176	6.124E+02	1.286E+14	2.552E+00	2.768E-10	1.9545	527618
25.12	17696	8.476E+02	1.730E+14	3.133E+00	3.716E-10	1.9620	626371
39.81	19494	1.194E+03	2.217E+14	3.250E+00	4.740E-10	1.9698	733896
63.10	21185	1.732E+03	2.960E+14	3.611E+00	6.322E-10	1.9758	856743

9500-DEGREE, LOG G = 4.0

TAU 5000	TEMP	PRESSURE	ELECTRON DENSITY	OPACITY	DENSITY	HYDROGEN IONIZATION	HEIGHT (KM)
.0010	7481	5.029E+01	5.977E+12	2.390E-01	9.120E-11	1.535E-01	-2201
.0016	7503	7.385E+01	7.642E+12	2.576E-01	1.359E-10	1.317E-01	-1990
.0025	7533	1.080E+02	9.816E+12	2.843E-01	2.008E-10	1.144E-01	-1785
.0040	7569	1.567E+02	1.261E+13	3.194E-01	2.933E-10	1.006E-01	-1585
.0063	7621	2.242E+02	1.637E+13	3.714E-01	4.200E-10	9.116E-02	-1394
.0100	7682	3.151E+02	2.125E+13	4.395E-01	5.889E-10	8.436E-02	-1212
.0158	7766	4.348E+02	2.804E+13	5.409E-01	8.060E-10	8.134E-02	-1039
.0251	7876	5.856E+02	3.760E+13	6.925E-01	1.070E-09	8.220E-02	-878
.0398	8007	7.699E+02	5.084E+13	9.113E-01	1.378E-09	8.627E-02	-727
.0631	8191	9.847E+02	7.131E+13	1.288E+00	1.707E-09	9.776E-02	-587
.1000	8441	1.215E+03	1.037E+14	1.973E+00	2.005E-09	1.211E-01	-463
.1585	8744	1.449E+03	1.521E+14	3.169E+00	2.238E-09	1.592E-01	-353
.2512	9166	1.667E+03	2.317E+14	5.673E+00	2.317E-09	2.344E-01	-258
.3981	9673	1.858E+03	3.414E+14	1.015E+01	2.242E-09	3.572E-01	-175
.6310	10279	2.033E+03	4.698E+14	1.679E+01	2.056E-09	5.361E-01	-93
1.000	10969	2.216E+03	5.837E+14	2.305E+01	1.878E-09	7.294E-01	0
1.585	11721	2.451E+03	6.681E+14	2.593E+01	1.808E-09	8.674E-01	128
2.512	12527	2.804E+03	7.472E+14	2.628E+01	1.867E-09	9.389E-01	322
3.981	13378	3.364E+03	8.547E+14	2.629E+01	2.065E-09	9.706E-01	607
6.310	14278	4.241E+03	1.018E+15	2.708E+01	2.420E-09	9.847E-01	1000
10.00	15241	5.560E+03	1.260E+15	2.887E+01	2.953E-09	9.914E-01	1492
15.85	16192	7.478E+03	1.607E+15	3.212E+01	3.712E-09	9.945E-01	2069
25.12	17308	1.020E+04	2.075E+15	3.575E+01	4.687E-09	9.966E-01	2719
39.81	18440	1.405E+04	2.710E+15	4.067E+01	5.998E-09	9.977E-01	3443
63.10	19806	1.948E+04	3.530E+15	4.550E+01	7.678E-09	9.985E-01	4243

9500-DEGREE, LOG G = 3.0

TAU 5000	TEMP	PRESSURE	ELECTRON DENSITY	OPACITY	DENSITY	LOG G EFF	HEIGHT (KM)
.0010	7395	4.953E+00	1.281E+12	2.084E-01	7.624E-12	2.9857	-30096
.0016	7418	7.737E+00	1.770E+12	2.098E-01	1.235E-11	2.9854	-27233
.0025	7449	1.205E+01	2.434E+12	2.173E-01	1.982E-11	2.9851	-24480
.0040	7487	1.856E+01	3.322E+12	2.315E-01	3.124E-11	2.9845	-21868
.0063	7541	2.801E+01	4.532E+12	2.574E-01	4.778E-11	2.9836	-19418
.0100	7603	4.124E+01	6.125E+12	2.942E-01	7.082E-11	2.9822	-17134
.0158	7694	5.906E+01	8.376E+12	3.563E-01	1.008E-10	2.9824	-15012
.0251	7813	8.166E+01	1.153E+13	4.539E-01	1.370E-10	2.9827	-13069
.0398	7960	1.091E+02	1.594E+13	6.043E-01	1.779E-10	2.9831	-11286
.0631	8168	1.402E+02	2.260E+13	8.735E-01	2.173E-10	2.9838	-9666
.1000	8443	1.727E+02	3.238E+13	1.350E+00	2.472E-10	2.9848	-8217
.1585	8761	2.055E+02	4.519E+13	2.080E+00	2.663E-10	2.9856	-6870
.2512	9181	2.378E+02	6.193E+13	3.232E+00	2.684E-10	2.9868	-5557
.3981	9680	2.714E+02	7.952E+13	4.512E+00	2.639E-10	2.9883	-4140
.6310	10262	3.118E+02	9.577E+13	5.424E+00	2.655E-10	2.9897	-2387
1.000	10931	3.689E+02	1.121E+14	5.813E+00	2.828E-10	2.9912	0
1.585	11676	4.569E+02	1.328E+14	5.994E+00	3.217E-10	2.9924	3283
2.512	12487	5.937E+02	1.629E+14	6.246E+00	3.876E-10	2.9936	7564
3.981	13350	8.009E+02	2.066E+14	6.736E+00	4.869E-10	2.9948	12763
6.310	14287	1.106E+03	2.682E+14	7.410E+00	6.250E-10	2.9958	18729
10.00	15282	1.543E+03	3.532E+14	8.305E+00	8.078E-10	2.9966	25297
15.85	16266	2.158E+03	4.694E+14	9.533E+00	1.050E-09	2.9972	32397
25.12	17492	3.017E+03	6.174E+14	1.064E+01	1.349E-09	2.9978	39992
39.81	20005	4.270E+03	7.717E+14	9.996E+00	1.654E-09	2.9981	48540
63.10	23073	6.516E+03	1.023E+15	9.897E+00	2.184E-09	2.9983	60718

9000-DEGREE, LOG G = 4.0

TAU 5000	TEMP	PRESSURE	ELECTRON DENSITY	OPACITY	DENSITY	HYDROGEN IONIZATION	HEIGHT (KM)
.0010	7178	8.879E+01	5.465E+12	1.396E-01	1.796E-10	7.108E-02	-1920
.0016	7201	1.287E+02	6.880E+12	1.535E-01	2.617E-10	6.139E-02	-1738
.0025	7227	1.856E+02	8.680E+12	1.720E-01	3.786E-10	5.350E-02	-1558
.0040	7262	2.651E+02	1.102E+13	1.972E-01	5.411E-10	4.749E-02	-1383
.0063	7305	3.740E+02	1.402E+13	2.302E-01	7.620E-10	4.289E-02	-1214
.0100	7356	5.206E+02	1.793E+13	2.735E-01	1.056E-09	3.955E-02	-1052
.0158	7424	7.134E+02	2.326E+13	3.341E-01	1.436E-09	3.771E-02	-896
.0251	7515	9.594E+02	3.075E+13	4.218E-01	1.909E-09	3.753E-02	-748
.0398	7635	1.262E+03	4.170E+13	5.548E-01	2.467E-09	3.939E-02	-609
.0631	7795	1.618E+03	5.837E+13	7.677E-01	3.085E-09	4.412E-02	-480
.1000	8007	2.012E+03	8.483E+13	1.133E+00	3.706E-09	5.344E-02	-364
.1585	8293	2.419E+03	1.294E+14	1.830E+00	4.234E-09	7.146E-02	-262
.2512	8687	2.793E+03	2.090E+14	3.348E+00	4.527E-09	1.081E-01	-177
.3981	9177	3.106E+03	3.392E+14	6.577E+00	4.510E-09	1.762E-01	-108
.6310	9787	3.353E+03	5.367E+14	1.337E+01	4.152E-09	3.032E-01	-52
1.000	10488	3.553E+03	7.655E+14	2.414E+01	3.604E-09	4.984E-01	0
1.585	11248	3.750E+03	9.487E+14	3.408E+01	3.131E-09	7.110E-01	59
2.512	12070	4.003E+03	1.058E+15	3.777E+01	2.872E-09	8.644E-01	144
3.981	12934	4.396E+03	1.135E+15	3.677E+01	2.833E-09	9.404E-01	283
6.310	13803	5.043E+03	1.243E+15	3.533E+01	2.998E-09	9.716E-01	505
10.00	14707	6.094E+03	1.422E+15	3.518E+01	3.374E-09	9.852E-01	836
15.85	15737	7.738E+03	1.701E+15	3.616E+01	3.974E-09	9.920E-01	1285
25.12	16780	1.020E+04	2.125E+15	3.913E+01	4.869E-09	9.951E-01	1844
39.81	17894	1.376E+04	2.718E+15	4.348E+01	6.092E-09	9.968E-01	2494
63.10	19121	1.880E+04	3.512E+15	4.871E+01	7.714E-09	9.979E-01	3228

9000-DEGREE, LOG G = 3.0

TAU 5000	TEMP	PRESSURE	ELECTRON DENSITY	OPACITY	DENSITY	LOG G EFF	HEIGHT (KM)
.0010	7077	8.804E+00	1.307E+12	1.134E-01	1.645E-11	2.9874	-23264
.0016	7102	1.395E+01	1.765E+12	1.139E-01	2.660E-11	2.9873	-20821
.0025	7131	2.192E+01	2.376E+12	1.184E-01	4.246E-11	2.9871	-18462
.0040	7169	3.382E+01	3.190E+12	1.278E-01	6.614E-11	2.9869	-16227
.0063	7217	5.093E+01	4.274E+12	1.434E-01	1.000E-10	2.9865	-14131
.0100	7279	7.460E+01	5.735E+12	1.674E-01	1.463E-10	2.9859	-12176
.0158	7361	1.060E+02	7.760E+12	2.043E-01	2.060E-10	2.9861	-10370
.0251	7466	1.457E+02	1.061E+13	2.607E-01	2.792E-10	2.9865	-8709
.0398	7604	1.937E+02	1.479E+13	3.518E-01	3.623E-10	2.9868	-7194
.0631	7789	2.478E+02	2.120E+13	5.099E-01	4.467E-10	2.9872	-5837
.1000	8029	3.045E+02	3.115E+13	7.960E-01	5.200E-10	2.9878	-4644
.1585	8345	3.591E+02	4.693E+13	1.354E+00	5.654E-10	2.9882	-3612
.2512	8745	4.074E+02	7.057E+13	2.433E+00	5.698E-10	2.9889	-2725
.3981	9228	4.488E+02	1.012E+14	4.261E+00	5.360E-10	2.9897	-1916
.6310	9816	4.864E+02	1.332E+14	6.586E+00	4.820E-10	2.9907	-1074
1.000	10484	5.286E+02	1.569E+14	8.065E+00	4.448E-10	2.9921	0
1.585	11250	5.896E+02	1.741E+14	8.239E+00	4.389E-10	2.9929	1617
2.512	12014	6.899E+02	1.950E+14	8.074E+00	4.719E-10	2.9941	4136
3.981	12861	8.550E+02	2.279E+14	8.052E+00	5.417E-10	2.9953	7739
6.310	13895	1.120E+03	2.782E+14	8.173E+00	6.528E-10	2.9960	12586
10.00	14757	1.520E+03	3.577E+14	9.086E+00	8.292E-10	2.9972	18396
15.85	15869	2.094E+03	4.642E+14	9.951E+00	1.050E-09	2.9975	24893
25.12	17142	2.936E+03	6.110E+14	1.099E+01	1.345E-09	2.9980	32377
39.81	18198	4.113E+03	8.118E+14	1.277E+01	1.763E-09	2.9984	40375
63.10	19420	5.733E+03	1.065E+15	1.454E+01	2.293E-09	2.9987	48779

9000-DEGREE, LOG G = 2.0

TAU 5000	TEMP	PRESSURE	ELECTRON DENSITY	OPACITY	DENSITY	LOG G EFF	HEIGHT (KM)
.0010	7011	4.866E-01	1.736E+11	1.871E-01	7.028E-13	1.9212	-382504
.0016	7025	8.069E-01	2.578E+11	1.687E-01	1.226E-12	1.9177	-347287
.0025	7041	1.373E+00	3.831E+11	1.517E-01	2.197E-12	1.9121	-312231
.0040	7068	2.359E+00	5.687E+11	1.399E-01	3.948E-12	1.9031	-278220
.0063	7108	4.015E+00	8.387E+11	1.348E-01	6.946E-12	1.8884	-246066
.0100	7165	6.641E+00	1.229E+12	1.385E-01	1.171E-11	1.8639	-216301
.0158	7248	1.050E+01	1.793E+12	1.538E-01	1.858E-11	1.8601	-189321
.0251	7361	1.574E+01	2.614E+12	1.854E-01	2.749E-11	1.8539	-165201
.0398	7515	2.222E+01	3.826E+12	2.439E-01	3.757E-11	1.8506	-143860
.0631	7719	2.952E+01	5.617E+12	3.493E-01	4.714E-11	1.8539	-125075
.1000	7979	3.700E+01	8.162E+12	5.305E-01	5.429E-11	1.8590	-108342
.1585	8304	4.418E+01	1.154E+13	8.240E-01	5.764E-11	1.8664	-92770
.2512	8714	5.088E+01	1.548E+13	1.226E+00	5.725E-11	1.8778	-77051
.3981	9195	5.780E+01	1.914E+13	1.580E+00	5.636E-11	1.8894	-58811
.6310	9779	6.660E+01	2.236E+13	1.750E+00	5.760E-11	1.9000	-34466
1.000	10432	8.033E+01	2.606E+13	1.794E+00	6.347E-11	1.9163	0
1.585	11202	1.029E+02	3.146E+13	1.818E+00	7.488E-11	1.9239	46940
2.512	11973	1.387E+02	3.986E+13	1.927E+00	9.407E-11	1.9357	106028
3.981	12890	1.941E+02	5.205E+13	2.067E+00	1.218E-10	1.9464	174485
6.310	13857	2.759E+02	6.938E+13	2.277E+00	1.598E-10	1.9539	250786
10.00	14945	3.968E+02	9.388E+13	2.544E+00	2.102E-10	1.9654	334403
15.85	15953	5.700E+02	1.278E+14	2.933E+00	2.798E-10	1.9714	422480
25.12	17320	8.157E+02	1.698E+14	3.216E+00	3.659E-10	1.9763	515288
39.81	18559	1.173E+03	2.285E+14	3.674E+00	4.895E-10	1.9801	615966
63.10	19952	1.667E+03	3.024E+14	4.129E+00	6.466E-10	1.9830	720139

8500-DEGREE, LOG G = 4.0

TAU 5000	TEMP	PRESSURE	ELECTRON DENSITY	OPACITY	DENSITY	HYDROGEN IONIZATION	HEIGHT (KM)	CONV/ TOTAL
.0010	6702	2.136E+02	4.030E+12	6.471E-02	4.843E-10	1.926E-02	-1588	.0000
.0016	6728	2.974E+02	4.997E+12	7.479E-02	6.729E-10	1.717E-02	-1441	.0000
.0025	6760	4.114E+02	6.235E+12	8.781E-02	9.278E-10	1.552E-02	-1298	.0000
.0040	6802	5.637E+02	7.868E+12	1.051E-01	1.265E-09	1.436E-02	-1158	.0000
.0063	6855	7.636E+02	1.003E+13	1.280E-01	1.701E-09	1.361E-02	-1022	.0000
.0100	6922	1.021E+03	1.301E+13	1.592E-01	2.253E-09	1.332E-02	-891	.0000
.0158	6996	1.348E+03	1.690E+13	1.992E-01	2.944E-09	1.325E-02	-765	.0000
.0251	7091	1.757E+03	2.250E+13	2.562E-01	3.783E-09	1.374E-02	-643	.0000
.0398	7216	2.250E+03	3.098E+13	3.421E-01	4.757E-09	1.506E-02	-526	.0000
.0631	7373	2.826E+03	4.391E+13	4.733E-01	5.834E-09	1.743E-02	-417	.0000
.1000	7566	3.474E+03	6.402E+13	6.805E-01	6.964E-09	2.134E-02	-316	.0000
.1585	7826	4.162E+03	9.917E+13	1.062E+00	8.013E-09	2.880E-02	-224	.0000
.2512	8187	4.821E+03	1.656E+14	1.871E+00	8.753E-09	4.414E-02	-146	.0000
.3981	8654	5.382E+03	2.891E+14	3.701E+00	9.000E-09	7.512E-02	-83	.0000
.6310	9263	5.804E+03	5.207E+14	8.356E+00	8.578E-09	1.422E-01	-35	.0000
1.000	9994	6.091E+03	8.851E+14	1.913E+01	7.536E-09	2.754E-01	0	.0003
1.585	10782	6.307E+03	1.290E+15	3.643E+01	6.293E-09	4.809E-01	31	.0051
2.512	11627	6.512E+03	1.591E+15	5.206E+01	5.267E-09	7.089E-01	67	.0018
3.981	12489	6.778E+03	1.730E+15	5.645E+01	4.700E-09	8.641E-01	121	.0000
6.310	13364	7.200E+03	1.799E+15	5.351E+01	4.493E-09	9.393E-01	213	.0000
10.00	14236	7.918E+03	1.891E+15	4.975E+01	4.565E-09	9.709E-01	373	.0000
15.85	15232	9.136E+03	2.061E+15	4.689E+01	4.877E-09	9.859E-01	631	.0000
25.12	16232	1.112E+04	2.375E+15	4.696E+01	5.528E-09	9.922E-01	1014	.0000
39.81	17293	1.418E+04	2.872E+15	4.942E+01	6.549E-09	9.952E-01	1521	.0000
63.10	18511	1.872E+04	3.588E+15	5.327E+01	7.984E-09	9.970E-01	2148	.0000

8500-DEGREE, LOG G = 3.0

TAU 5000	TEMP	PRESSURE	ELECTRON DENSITY	OPACITY	DENSITY	LOG G EFF	HEIGHT (KM)
.0010	6650	2.522E+01	1.218E+12	4.284E-02	5.606E-11	2.9865	-17502
.0016	6678	3.839E+01	1.594E+12	4.600E-02	8.550E-11	2.9867	-15612
.0025	6714	5.743E+01	2.092E+12	5.135E-02	1.278E-10	2.9869	-13800
.0040	6763	8.387E+01	2.765E+12	5.970E-02	1.859E-10	2.9872	-12093
.0063	6826	1.192E+02	3.687E+12	7.210E-02	2.622E-10	2.9876	-10498
.0100	6912	1.645E+02	5.022E+12	9.133E-02	3.573E-10	2.9884	-9023
.0158	6998	2.211E+02	6.712E+12	1.150E-01	4.744E-10	2.9887	-7648
.0251	7105	2.911E+02	9.144E+12	1.502E-01	6.142E-10	2.9892	-6352
.0398	7242	3.740E+02	1.280E+13	2.054E-01	7.713E-10	2.9896	-5147
.0631	7417	4.673E+02	1.846E+13	2.970E-01	9.350E-10	2.9898	-4045
.1000	7640	5.662E+02	2.756E+13	4.594E-01	1.087E-09	2.9902	-3059
.1585	7933	6.625E+02	4.301E+13	7.820E-01	1.200E-09	2.9904	-2208
.2512	8318	7.463E+02	6.981E+13	1.489E+00	1.238E-09	2.9907	-1507
.3981	8799	8.123E+02	1.131E+14	3.036E+00	1.186E-09	2.9911	-942
.6310	9394	8.607E+02	1.742E+14	6.153E+00	1.045E-09	2.9918	-470
1.000	10067	8.995E+02	2.335E+14	1.019E+01	8.836E-10	2.9929	0
1.585	10830	9.407E+02	2.700E+14	1.252E+01	7.670E-10	2.9937	615
2.512	11578	1.002E+03	2.869E+14	1.238E+01	7.266E-10	2.9948	1614
3.981	12419	1.111E+03	3.039E+14	1.144E+01	7.344E-10	2.9958	3312
6.310	13410	1.304E+03	3.339E+14	1.056E+01	7.909E-10	2.9965	6128
10.00	14226	1.625E+03	3.943E+14	1.084E+01	9.245E-10	2.9975	10179
15.85	15342	2.123E+03	4.828E+14	1.114E+01	1.110E-09	2.9978	15393
25.12	16552	2.890E+03	6.180E+14	1.198E+01	1.381E-09	2.9982	21926
39.81	17564	3.981E+03	8.098E+14	1.373E+01	1.777E-09	2.9986	29191
63.10	18740	5.501E+03	1.056E+15	1.552E+01	2.286E-09	2.9988	37036

8000-DEGREE, LOG G = 4.5

TAU 5000	TEMP	PRESSURE	ELECTRON DENSITY	OPACITY	DENSITY	HYDROGEN IONIZATION	HEIGHT (KM)	CONV/ TOTAL
.0010	6285	9.312E+02	3.904E+12	5.207E-02	2.283E-09	3.847E-03	-462	.0000
.0016	6307	1.256E+03	4.748E+12	6.168E-02	3.070E-09	3.471E-03	-423	.0000
.0025	6335	1.688E+03	5.838E+12	7.392E-02	4.109E-09	3.183E-03	-385	.0000
.0040	6374	2.256E+03	7.292E+12	8.991E-02	5.457E-09	2.989E-03	-347	.0000
.0063	6424	2.989E+03	9.256E+12	1.110E-01	7.176E-09	2.885E-03	-310	.0000
.0100	6482	3.925E+03	1.188E+13	1.383E-01	9.339E-09	2.846E-03	-274	.0000
.0158	6547	5.113E+03	1.533E+13	1.731E-01	1.205E-08	2.850E-03	-239	.0000
.0251	6628	6.605E+03	2.022E+13	2.209E-01	1.537E-08	2.952E-03	-204	.0000
.0398	6729	8.437E+03	2.739E+13	2.879E-01	1.933E-08	3.188E-03	-171	.0000
.0631	6852	1.064E+04	3.804E+13	3.836E-01	2.394E-08	3.590E-03	-139	.0000
.1000	7008	1.322E+04	5.488E+13	5.281E-01	2.906E-08	4.288E-03	-108	.0000
.1585	7209	1.613E+04	8.322E+13	7.609E-01	3.443E-08	5.521E-03	-79	.0000
.2512	7509	1.918E+04	1.412E+14	1.221E+00	3.919E-08	8.288E-03	-53	.0000
.3981	7892	2.206E+04	2.530E+14	2.107E+00	4.270E-08	1.372E-02	-30	.0000
.6310	8445	2.453E+04	5.141E+14	4.349E+00	4.382E-08	2.731E-02	-12	.0000
1.000	9249	2.622E+04	1.177E+15	1.163E+01	4.132E-08	6.657E-02	0	.0014
1.585	10157	2.714E+04	2.406E+15	3.183E+01	3.619E-08	1.558E-01	8	.2358
2.512	10706	2.782E+04	3.374E+15	5.385E+01	3.299E-08	2.398E-01	14	.5553
3.981	11127	2.854E+04	4.190E+15	7.633E+01	3.072E-08	3.200E-01	21	.6533
6.310	11516	2.937E+04	4.958E+15	1.004E+02	2.886E-08	4.031E-01	30	.7542
10.00	11932	3.039E+04	5.751E+15	1.276E+02	2.712E-08	4.977E-01	41	.8090
15.85	12421	3.169E+04	6.578E+15	1.572E+02	2.541E-08	6.075E-01	57	.7816
25.12	13087	3.338E+04	7.428E+15	1.859E+02	2.359E-08	7.389E-01	79	.5616
39.81	14484	3.578E+04	8.091E+15	1.908E+02	2.094E-08	9.064E-01	113	.1326
63.10	16552	4.014E+04	8.306E+15	1.544E+02	1.978E-08	9.782E-01	182	.0001

8000-DEGREE, LOG G = 4.0

TAU 5000	TEMP	PRESSURE	ELECTRON DENSITY	OPACITY	DENSITY	HYDROGEN IONIZATION	HEIGHT (KM)	CONV/ TOTAL
.0010	6278	4.077E+02	2.529E+12	3.650E-02	9.990E-10	5.745E-03	-1444	.0000
.0016	6301	5.547E+02	3.101E+12	4.309E-02	1.355E-09	5.186E-03	-1318	.0000
.0025	6332	7.504E+02	3.843E+12	5.157E-02	1.824E-09	4.765E-03	-1194	.0000
.0040	6373	1.008E+03	4.834E+12	6.273E-02	2.435E-09	4.487E-03	-1073	.0000
.0063	6425	1.340E+03	6.177E+12	7.753E-02	3.211E-09	4.346E-03	-955	.0000
.0100	6487	1.763E+03	7.986E+12	9.695E-02	4.185E-09	4.313E-03	-839	.0000
.0158	6556	2.298E+03	1.037E+13	1.218E-01	5.398E-09	4.347E-03	-727	.0000
.0251	6640	2.967E+03	1.376E+13	1.561E-01	6.881E-09	4.532E-03	-618	.0000
.0398	6745	3.784E+03	1.875E+13	2.048E-01	8.636E-09	4.933E-03	-512	.0000
.0631	6875	4.759E+03	2.627E+13	2.757E-01	1.065E-08	5.622E-03	-411	.0000
.1000	7038	5.886E+03	3.826E+13	3.851E-01	1.285E-08	6.814E-03	-315	.0000
.1585	7259	7.129E+03	5.940E+13	5.735E-01	1.506E-08	9.072E-03	-225	.0000
.2512	7573	8.386E+03	1.014E+14	9.495E-01	1.691E-08	1.388E-02	-147	.0000
.3981	7999	9.522E+03	1.882E+14	1.777E+00	1.801E-08	2.430E-02	-82	.0000
.6310	8584	1.041E+04	3.822E+14	3.991E+00	1.793E-08	4.976E-02	-33	.0000
1.000	9344	1.098E+04	7.958E+14	1.060E+01	1.648E-08	1.131E-01	0	.0006
1.585	10250	1.131E+04	1.509E+15	2.854E+01	1.385E-08	2.554E-01	22	.1570
2.512	10879	1.156E+04	2.058E+15	4.835E+01	1.203E-08	4.013E-01	40	.4109
3.981	11434	1.181E+04	2.486E+15	6.716E+01	1.067E-08	5.471E-01	63	.3838
6.310	12123	1.211E+04	2.855E+15	8.435E+01	9.360E-09	7.158E-01	93	.2608
10.00	13188	1.253E+04	3.068E+15	8.893E+01	8.146E-09	8.837E-01	141	.0083
15.85	14349	1.323E+04	3.111E+15	7.908E+01	7.619E-09	9.573E-01	231	.0000
25.12	15398	1.447E+04	3.219E+15	7.091E+01	7.667E-09	9.807E-01	395	.0000
39.81	16498	1.664E+04	3.490E+15	6.629E+01	8.146E-09	9.904E-01	669	.0000
63.10	17779	2.025E+04	3.997E+15	6.459E+01	9.082E-09	9.950E-01	1090	.0000

8000-DEGREE, LOG G = 4.0   (NO CONV)

TAU 5000	TEMP	PRESSURE	ELECTRON DENSITY	OPACITY	DENSITY	HYDROGEN IONIZATION	HEIGHT (KM)
.0010	6274	4.208E+02	2.553E+12	3.681E-02	1.032E-09	5.612E-03	-1434
.0016	6302	5.663E+02	3.135E+12	4.356E-02	1.383E-09	5.132E-03	-1312
.0025	6335	7.598E+02	3.888E+12	5.219E-02	1.846E-09	4.763E-03	-1192
.0040	6369	1.016E+03	4.814E+12	6.262E-02	2.457E-09	4.426E-03	-1072
.0063	6422	1.349E+03	6.158E+12	7.748E-02	3.235E-09	4.300E-03	-954
.0100	6484	1.772E+03	7.963E+12	9.690E-02	4.209E-09	4.276E-03	-840
.0158	6550	2.310E+03	1.028E+13	1.212E-01	5.431E-09	4.282E-03	-728
.0251	6637	2.980E+03	1.372E+13	1.560E-01	6.915E-09	4.497E-03	-619
.0398	6737	3.800E+03	1.854E+13	2.031E-01	8.682E-09	4.849E-03	-513
.0631	6869	4.783E+03	2.608E+13	2.744E-01	1.071E-08	5.549E-03	-412
.1000	7035	5.914E+03	3.812E+13	3.843E-01	1.292E-08	6.753E-03	-316
.1585	7260	7.156E+03	5.954E+13	5.753E-01	1.512E-08	9.060E-03	-227
.2512	7576	8.403E+03	1.019E+14	9.550E-01	1.694E-08	1.392E-02	-149
.3981	7975	9.548E+03	1.832E+14	1.725E+00	1.812E-08	2.350E-02	-84
.6310	8564	1.047E+04	3.755E+14	3.895E+00	1.811E-08	4.840E-02	-34
1.000	9305	1.106E+04	7.737E+14	1.013E+01	1.673E-08	1.083E-01	0
1.585	10127	1.143E+04	1.414E+15	2.539E+01	1.443E-08	2.297E-01	23
2.512	11063	1.167E+04	2.214E+15	5.477E+01	1.158E-08	4.485E-01	42
3.981	11971	1.188E+04	2.758E+15	8.083E+01	9.460E-09	6.842E-01	62
6.310	12835	1.215E+04	2.978E+15	8.854E+01	8.282E-09	8.440E-01	93
10.00	13732	1.257E+04	3.039E+15	8.356E+01	7.675E-09	9.290E-01	147
15.85	14664	1.332E+04	3.082E+15	7.515E+01	7.465E-09	9.673E-01	245
25.12	15614	1.461E+04	3.212E+15	6.871E+01	7.617E-09	9.836E-01	417
39.81	16661	1.683E+04	3.501E+15	6.511E+01	8.145E-09	9.913E-01	699
63.10	17621	2.037E+04	4.049E+15	6.667E+01	9.239E-09	9.945E-01	1108

8000-DEGREE, LOG G = 3.5

TAU 5000	TEMP	PRESSURE	ELECTRON DENSITY	OPACITY	DENSITY	LOG G EFF	HEIGHT (KM)	CONV/ TOTAL
.0010	6284	1.652E+02	1.620E+12	2.689E-02	4.032E-10	3.4972	-4609	.0000
.0016	6309	2.286E+02	2.008E+12	3.138E-02	5.561E-10	3.4973	-4188	.0000
.0025	6339	3.142E+02	2.501E+12	3.708E-02	7.613E-10	3.4974	-3773	.0000
.0040	6381	4.275E+02	3.176E+12	4.493E-02	1.029E-09	3.4975	-3370	.0000
.0063	6430	5.749E+02	4.056E+12	5.502E-02	1.374E-09	3.4977	-2980	.0000
.0100	6493	7.637E+02	5.284E+12	6.886E-02	1.808E-09	3.4979	-2602	.0000
.0158	6563	1.001E+03	6.893E+12	8.654E-02	2.345E-09	3.4979	-2238	.0000
.0251	6646	1.299E+03	9.152E+12	1.107E-01	3.004E-09	3.4980	-1884	.0000
.0398	6758	1.661E+03	1.263E+13	1.471E-01	3.774E-09	3.4981	-1545	.0000
.0631	6894	2.087E+03	1.787E+13	2.008E-01	4.644E-09	3.4981	-1224	.0000
.1000	7066	2.573E+03	2.628E+13	2.855E-01	5.574E-09	3.4982	-923	.0000
.1585	7307	3.092E+03	4.169E+13	4.424E-01	6.455E-09	3.4981	-649	.0000
.2512	7644	3.592E+03	7.247E+13	7.744E-01	7.113E-09	3.4980	-416	.0000
.3981	8078	4.020E+03	1.326E+14	1.519E+00	7.413E-09	3.4980	-229	.0000
.6310	8674	4.337E+03	2.627E+14	3.623E+00	7.171E-09	3.4980	-92	.0000
1.000	9408	4.530E+03	5.043E+14	9.453E+00	6.369E-09	3.4981	0	.0002
1.585	10219	4.648E+03	8.327E+14	2.180E+01	5.257E-09	3.4982	68	.0256
2.512	11049	4.737E+03	1.112E+15	3.661E+01	4.256E-09	3.4985	135	.0508
3.981	11919	4.835E+03	1.253E+15	4.386E+01	3.599E-09	3.4988	227	.0023
6.310	12823	4.988E+03	1.288E+15	4.194E+01	3.267E-09	3.4991	386	.0000
10.00	13729	5.262E+03	1.301E+15	3.735E+01	3.151E-09	3.4993	679	.0000
15.85	14655	5.761E+03	1.349E+15	3.367E+01	3.200E-09	3.4994	1205	.0000
25.12	15651	6.630E+03	1.466E+15	3.162E+01	3.422E-09	3.4995	2070	.0000
39.81	16698	8.066E+03	1.691E+15	3.157E+01	3.861E-09	3.4996	3358	.0000
63.10	17739	1.028E+04	2.052E+15	3.355E+01	4.584E-09	3.4997	5065	.0000

8000-DEGREE, LOG G = 3.0

TAU 5000	TEMP	PRESSURE	ELECTRON DENSITY	OPACITY	DENSITY	LOG G EFF	HEIGHT (KM)	CONV/ TOTAL
.0010	6258	6.456E+01	9.515E+11	2.007E-02	1.575E-10	2.9905	-14756	.0000
.0016	6286	9.176E+01	1.204E+12	2.293E-02	2.232E-10	2.9908	-13313	.0000
.0025	6320	1.290E+02	1.534E+12	2.684E-02	3.124E-10	2.9911	-11908	.0000
.0040	6364	1.788E+02	1.972E+12	3.214E-02	4.303E-10	2.9916	-10556	.0000
.0063	6416	2.440E+02	2.556E+12	3.923E-02	5.828E-10	2.9920	-9257	.0000
.0100	6484	3.276E+02	3.375E+12	4.917E-02	7.743E-10	2.9926	-8016	.0000
.0158	6555	4.332E+02	4.440E+12	6.178E-02	1.013E-09	2.9928	-6827	.0000
.0251	6645	5.644E+02	5.986E+12	7.989E-02	1.301E-09	2.9932	-5686	.0000
.0398	6763	7.222E+02	8.336E+12	1.072E-01	1.634E-09	2.9935	-4606	.0000
.0631	6912	9.046E+02	1.203E+13	1.500E-01	1.998E-09	2.9935	-3597	.0000
.1000	7101	1.106E+03	1.807E+13	2.211E-01	2.370E-09	2.9936	-2671	.0000
.1585	7360	1.313E+03	2.911E+13	3.585E-01	2.697E-09	2.9935	-1851	.0000
.2512	7717	1.502E+03	5.083E+13	6.648E-01	2.902E-09	2.9933	-1173	.0000
.3981	8167	1.654E+03	9.196E+13	1.386E+00	2.935E-09	2.9933	-649	.0000
.6310	8764	1.759E+03	1.730E+14	3.388E+00	2.735E-09	2.9935	-269	.0000
1.000	9480	1.822E+03	3.020E+14	8.302E+00	2.328E-09	2.9939	0	.0001
1.585	10251	1.864E+03	4.358E+14	1.590E+01	1.882E-09	2.9945	234	.0008
2.512	11086	1.902E+03	5.168E+14	2.119E+01	1.550E-09	2.9954	522	.0001
3.981	11944	1.957E+03	5.390E+14	2.098E+01	1.383E-09	2.9963	996	.0000
6.310	12823	2.058E+03	5.441E+14	1.862E+01	1.320E-09	2.9971	1875	.0000
10.00	13713	2.247E+03	5.620E+14	1.661E+01	1.335E-09	2.9977	3475	.0000
15.85	14627	2.587E+03	6.110E+14	1.559E+01	1.432E-09	2.9982	6146	.0000
25.12	15758	3.155E+03	6.995E+14	1.506E+01	1.604E-09	2.9985	10102	.0000
39.81	16798	4.071E+03	8.573E+14	1.598E+01	1.919E-09	2.9988	15586	.0000
63.10	18023	5.429E+03	1.078E+15	1.720E+01	2.358E-09	2.9991	22232	.0000

8000-DEGREE, LOG G = 2.0

TAU 5000	TEMP	PRESSURE	ELECTRON DENSITY	OPACITY	DENSITY	LOG G EFF	HEIGHT (KM)
.0010	6162	5.731E+00	2.259E+11	1.699E-02	1.390E-11	1.9026	-173392
.0016	6203	9.166E+00	3.129E+11	1.707E-02	2.218E-11	1.9021	-153946
.0025	6251	1.443E+01	4.360E+11	1.810E-02	3.477E-11	1.9013	-135087
.0040	6304	2.211E+01	6.025E+11	2.002E-02	5.296E-11	1.9018	-117252
.0063	6363	3.289E+01	8.295E+11	2.309E-02	7.817E-11	1.9046	-100542
.0100	6440	4.728E+01	1.157E+12	2.813E-02	1.111E-10	1.9091	-85121
.0158	6530	6.561E+01	1.616E+12	3.549E-02	1.519E-10	1.9110	-70992
.0251	6635	8.817E+01	2.275E+12	4.634E-02	2.006E-10	1.9141	-58028
.0398	6765	1.150E+02	3.267E+12	6.337E-02	2.559E-10	1.9164	-46117
.0631	6937	1.448E+02	4.888E+12	9.330E-02	3.123E-10	1.9169	-35479
.1000	7160	1.753E+02	7.627E+12	1.496E-01	3.625E-10	1.9177	-26239
.1585	7447	2.038E+02	1.241E+13	2.655E-01	3.966E-10	1.9188	-18513
.2512	7824	2.269E+02	2.103E+13	5.348E-01	4.035E-10	1.9205	-12384
.3981	8287	2.431E+02	3.521E+13	1.154E+00	3.786E-10	1.9237	-7648
.6310	8868	2.531E+02	5.566E+13	2.484E+00	3.225E-10	1.9292	-3803
1.000	9539	2.589E+02	7.416E+13	4.160E+00	2.614E-10	1.9377	0
1.585	10267	2.641E+02	8.213E+13	4.783E+00	2.225E-10	1.9436	5330
2.512	11053	2.738E+02	8.344E+13	4.380E+00	2.050E-10	1.9529	14845
3.981	11905	2.967E+02	8.533E+13	3.824E+00	2.033E-10	1.9623	32680
6.310	12812	3.436E+02	9.242E+13	3.488E+00	2.175E-10	1.9692	63386
10.00	13723	4.280E+02	1.081E+14	3.456E+00	2.515E-10	1.9756	109219
15.85	14864	5.647E+02	1.336E+14	3.531E+00	3.025E-10	1.9782	169517
25.12	16204	7.805E+02	1.723E+14	3.725E+00	3.773E-10	1.9823	245584
39.81	17260	1.099E+03	2.293E+14	4.262E+00	4.957E-10	1.9861	330842
63.10	18386	1.534E+03	3.014E+14	4.837E+00	6.472E-10	1.9882	418489

7500-DEGREE, LOG G = 4.0   (L/H=1.5)

TAU 5000	TEMP	PRESSURE	ELECTRON DENSITY	OPACITY	DENSITY	HYDROGEN IONIZATION	HEIGHT (KM)	CONV/ TOTAL
.0010	5898	6.847E+02	1.472E+12	2.276E-02	1.792E-09	1.790E-03	-1360	.0000
.0016	5921	9.189E+02	1.802E+12	2.715E-02	2.396E-09	1.633E-03	-1247	.0000
.0025	5949	1.229E+03	2.224E+12	3.270E-02	3.189E-09	1.510E-03	-1135	.0000
.0040	5986	1.633E+03	2.791E+12	3.995E-02	4.214E-09	1.431E-03	-1025	.0000
.0063	6033	2.154E+03	3.559E+12	4.949E-02	5.514E-09	1.394E-03	-918	.0000
.0100	6087	2.818E+03	4.580E+12	6.179E-02	7.148E-09	1.385E-03	-812	.0000
.0158	6148	3.657E+03	5.943E+12	7.769E-02	9.186E-09	1.401E-03	-709	.0000
.0251	6221	4.708E+03	7.848E+12	9.907E-02	1.169E-08	1.459E-03	-608	.0000
.0398	6312	6.000E+03	1.064E+13	1.291E-01	1.468E-08	1.585E-03	-510	.0000
.0631	6425	7.552E+03	1.489E+13	1.724E-01	1.815E-08	1.806E-03	-415	.0000
.1000	6562	9.372E+03	2.144E+13	2.358E-01	2.204E-08	2.160E-03	-324	.0000
.1585	6741	1.144E+04	3.264E+13	3.379E-01	2.618E-08	2.796E-03	-238	.0000
.2512	7002	1.363E+04	5.521E+13	5.312E-01	2.998E-08	4.180E-03	-160	.0000
.3981	7365	1.573E+04	1.038E+14	9.275E-01	3.280E-08	7.262E-03	-93	.0000
.6310	7846	1.752E+04	2.122E+14	1.816E+00	3.408E-08	1.442E-02	-40	.0000
1.000	8510	1.888E+04	4.820E+14	4.328E+00	3.328E-08	3.375E-02	0	.0001
1.585	9474	1.963E+04	1.217E+15	1.408E+01	2.944E-08	9.678E-02	24	.1099
2.512	10095	2.007E+04	1.940E+15	2.794E+01	2.661E-08	1.708E-01	40	.5179
3.981	10497	2.050E+04	2.500E+15	4.150E+01	2.486E-08	2.358E-01	56	.7406
6.310	10841	2.097E+04	3.019E+15	5.614E+01	2.347E-08	3.016E-01	76	.8060
10.00	11169	2.155E+04	3.533E+15	7.235E+01	2.229E-08	3.718E-01	101	.8820
15.85	11514	2.226E+04	4.073E+15	9.090E+01	2.121E-08	4.506E-01	134	.8643
25.12	11926	2.317E+04	4.678E+15	1.131E+02	2.006E-08	5.473E-01	178	.9016
39.81	12445	2.434E+04	5.328E+15	1.371E+02	1.887E-08	6.626E-01	238	.8446
63.10	13224	2.591E+04	5.985E+15	1.573E+02	1.752E-08	8.017E-01	324	.7893

7500-DEGREE, LOG G = 4.0   (L/H=2.5)

TAU 5000	TEMP	PRESSURE	ELECTRON DENSITY	OPACITY	DENSITY	HYDROGEN IONIZATION	HEIGHT (KM)	CONV/ TOTAL
.0010	5894	6.867E+02	1.463E+12	2.264E-02	1.798E-09	1.772E-03	-1363	.0000
.0016	5915	9.227E+02	1.783E+12	2.692E-02	2.408E-09	1.606E-03	-1250	.0000
.0025	5944	1.235E+03	2.206E+12	3.248E-02	3.207E-09	1.488E-03	-1138	.0000
.0040	5980	1.642E+03	2.765E+12	3.966E-02	4.240E-09	1.408E-03	-1028	.0000
.0063	6026	2.167E+03	3.520E+12	4.907E-02	5.553E-09	1.368E-03	-920	.0000
.0100	6081	2.835E+03	4.540E+12	6.138E-02	7.201E-09	1.362E-03	-815	.0000
.0158	6143	3.680E+03	5.897E+12	7.723E-02	9.252E-09	1.379E-03	-711	.0000
.0251	6215	4.737E+03	7.777E+12	9.838E-02	1.177E-08	1.434E-03	-610	.0000
.0398	6304	6.041E+03	1.051E+13	1.278E-01	1.480E-08	1.551E-03	-512	.0000
.0631	6414	7.612E+03	1.464E+13	1.701E-01	1.832E-08	1.757E-03	-417	.0000
.1000	6554	9.453E+03	2.121E+13	2.338E-01	2.226E-08	2.114E-03	-326	.0000
.1585	6743	1.152E+04	3.284E+13	3.398E-01	2.636E-08	2.795E-03	-241	.0000
.2512	7008	1.368E+04	5.591E+13	5.370E-01	3.008E-08	4.220E-03	-164	.0000
.3981	7257	1.587E+04	8.885E+13	8.064E-01	3.364E-08	6.043E-03	-95	.0000
.6310	7797	1.799E+04	2.018E+14	1.723E+00	3.524E-08	1.325E-02	-34	.0000
1.000	8718	1.916E+04	6.038E+14	5.635E+00	3.271E-08	4.309E-02	0	.0227
1.585	9407	1.981E+04	1.158E+15	1.305E+01	3.009E-08	9.004E-02	21	.4067
2.512	9790	2.037E+04	1.584E+15	2.027E+01	2.879E-08	1.289E-01	40	.6814
3.981	10091	2.097E+04	1.985E+15	2.810E+01	2.790E-08	1.667E-01	61	.7898
6.310	10361	2.169E+04	2.394E+15	3.700E+01	2.726E-08	2.058E-01	87	.8465
10.00	10620	2.256E+04	2.833E+15	4.743E+01	2.681E-08	2.478E-01	119	.9143
15.85	10882	2.365E+04	3.325E+15	5.995E+01	2.651E-08	2.941E-01	160	.9711
25.12	11157	2.502E+04	3.890E+15	7.512E+01	2.638E-08	3.459E-01	212	.9882
39.81	11447	2.676E+04	4.539E+15	9.322E+01	2.647E-08	4.024E-01	278	.5618
63.10	11744	2.902E+04	5.284E+15	1.140E+02	2.694E-08	4.602E-01	362	.9999

7500-DEGREE, LOG G = 4.0   (NO CONV)

TAU 5000	TEMP	PRESSURE	ELECTRON DENSITY	OPACITY	DENSITY	HYDROGEN IONIZATION	HEIGHT (KM)
.0010	5897	6.860E+02	1.472E+12	2.275E-02	1.796E-09	1.786E-03	-1362
.0016	5919	9.206E+02	1.796E+12	2.709E-02	2.401E-09	1.624E-03	-1249
.0025	5948	1.231E+03	2.220E+12	3.265E-02	3.196E-09	1.503E-03	-1138
.0040	5984	1.636E+03	2.784E+12	3.987E-02	4.222E-09	1.424E-03	-1028
.0063	6031	2.158E+03	3.547E+12	4.936E-02	5.527E-09	1.386E-03	-920
.0100	6085	2.823E+03	4.564E+12	6.163E-02	7.166E-09	1.376E-03	-815
.0158	6146	3.665E+03	5.927E+12	7.753E-02	9.208E-09	1.394E-03	-711
.0251	6219	4.718E+03	7.819E+12	9.878E-02	1.172E-08	1.449E-03	-610
.0398	6308	6.016E+03	1.058E+13	1.285E-01	1.472E-08	1.569E-03	-512
.0631	6420	7.576E+03	1.477E+13	1.713E-01	1.822E-08	1.783E-03	-417
.1000	6558	9.408E+03	2.133E+13	2.349E-01	2.214E-08	2.139E-03	-326
.1585	6746	1.147E+04	3.297E+13	3.409E-01	2.622E-08	2.821E-03	-240
.2512	6998	1.365E+04	5.489E+13	5.285E-01	3.004E-08	4.147E-03	-163
.3981	7337	1.578E+04	9.984E+13	8.954E-01	3.305E-08	6.927E-03	-95
.6310	7818	1.765E+04	2.054E+14	1.757E+00	3.449E-08	1.379E-02	-40
1.000	8486	1.903E+04	4.715E+14	4.209E+00	3.368E-08	3.262E-02	0
1.585	9327	1.987E+04	1.085E+15	1.190E+01	3.063E-08	8.290E-02	26
2.512	10299	2.033E+04	2.220E+15	3.434E+01	2.579E-08	2.017E-01	42
3.981	11293	2.060E+04	3.585E+15	7.741E+01	2.057E-08	4.090E-01	54
6.310	12257	2.084E+04	4.569E+15	1.198E+02	1.654E-08	6.482E-01	66
10.00	13207	2.112E+04	4.974E+15	1.350E+02	1.412E-08	8.270E-01	85
15.85	14108	2.156E+04	5.040E+15	1.282E+02	1.288E-08	9.178E-01	118
25.12	15099	2.233E+04	5.009E+15	1.134E+02	1.218E-08	9.629E-01	180
39.81	16085	2.372E+04	5.058E+15	1.008E+02	1.201E-08	9.817E-01	295
63.10	17154	2.616E+04	5.289E+15	9.196E+01	1.229E-08	9.905E-01	496

7500-DEGREE, LOG G = 3.0

TAU 5000	TEMP	PRESSURE	ELECTRON DENSITY	OPACITY	DENSITY	LOG G EFF	HEIGHT (KM)	CONV/ TOTAL
.0010	5896	1.243E+02	6.135E+11	1.145E-02	3.246E-10	2.9932	-13585	.0000
.0016	5920	1.713E+02	7.623E+11	1.341E-02	4.459E-10	2.9934	-12354	.0000
.0025	5952	2.343E+02	9.598E+11	1.603E-02	6.067E-10	2.9937	-11148	.0000
.0040	5992	3.170E+02	1.223E+12	1.950E-02	8.157E-10	2.9940	-9976	.0000
.0063	6041	4.239E+02	1.578E+12	2.409E-02	1.082E-09	2.9944	-8842	.0000
.0100	6100	5.599E+02	2.062E+12	3.019E-02	1.415E-09	2.9949	-7746	.0000
.0158	6165	7.313E+02	2.704E+12	3.805E-02	1.829E-09	2.9951	-6684	.0000
.0251	6245	9.450E+02	3.630E+12	4.900E-02	2.332E-09	2.9953	-5652	.0000
.0398	6345	1.204E+03	5.009E+12	6.474E-02	2.924E-09	2.9956	-4662	.0000
.0631	6467	1.511E+03	7.112E+12	8.783E-02	3.599E-09	2.9957	-3717	.0000
.1000	6620	1.863E+03	1.049E+13	1.236E-01	4.330E-09	2.9958	-2826	.0000
.1585	6823	2.249E+03	1.646E+13	1.850E-01	5.062E-09	2.9957	-2003	.0000
.2512	7110	2.636E+03	2.849E+13	3.083E-01	5.674E-09	2.9956	-1281	.0000
.3981	7517	2.980E+03	5.532E+13	6.004E-01	6.013E-09	2.9955	-693	.0000
.6310	8098	3.228E+03	1.211E+14	1.472E+00	5.907E-09	2.9953	-275	.0000
1.000	8767	3.381E+03	2.505E+14	3.899E+00	5.430E-09	2.9950	0	.0000
1.585	9666	3.461E+03	5.096E+14	1.189E+01	4.450E-09	2.9953	172	.0579
2.512	10363	3.507E+03	7.256E+14	2.212E+01	3.685E-09	2.9961	308	.2262
3.981	11044	3.549E+03	8.785E+14	3.124E+01	3.095E-09	2.9966	470	.1892
6.310	12033	3.602E+03	9.583E+14	3.475E+01	2.584E-09	2.9972	715	.0044
10.00	12946	3.693E+03	9.580E+14	3.129E+01	2.367E-09	2.9979	1171	.0000
15.85	14078	3.874E+03	9.423E+14	2.590E+01	2.244E-09	2.9982	2073	.0000
25.12	15463	4.242E+03	9.517E+14	2.124E+01	2.211E-09	2.9987	3883	.0000
39.81	16397	4.911E+03	1.051E+15	2.056E+01	2.390E-09	2.9991	6947	.0000
63.10	17555	6.018E+03	1.219E+15	2.053E+01	2.700E-09	2.9992	11513	.0000

7000-DEGREE, LOG G = 4.0

TAU 5000	TEMP	PRESSURE	ELECTRON DENSITY	OPACITY	DENSITY	HYDROGEN IONIZATION	HEIGHT (KM)	CONV/ TOTAL
.0010	5525	1.121E+03	7.948E+11	1.435E-02	3.137E-09	4.945E-04	-1260	.0000
.0016	5547	1.491E+03	9.760E+11	1.724E-02	4.156E-09	4.537E-04	-1158	.0000
.0025	5574	1.977E+03	1.211E+12	2.089E-02	5.484E-09	4.229E-04	-1056	.0000
.0040	5610	2.609E+03	1.529E+12	2.569E-02	7.188E-09	4.056E-04	-956	.0000
.0063	5654	3.417E+03	1.956E+12	3.193E-02	9.342E-09	3.991E-04	-858	.0000
.0100	5705	4.442E+03	2.529E+12	4.002E-02	1.204E-08	4.016E-04	-761	.0000
.0158	5764	5.734E+03	3.308E+12	5.061E-02	1.538E-08	4.138E-04	-667	.0000
.0251	5834	7.344E+03	4.401E+12	6.485E-02	1.946E-08	4.398E-04	-574	.0000
.0398	5920	9.314E+03	6.006E+12	8.473E-02	2.432E-08	4.880E-04	-484	.0000
.0631	6025	1.168E+04	8.433E+12	1.131E-01	2.996E-08	5.677E-04	-396	.0000
.1000	6150	1.446E+04	1.217E+13	1.543E-01	3.633E-08	6.909E-04	-312	.0000
.1585	6310	1.763E+04	1.848E+13	2.189E-01	4.316E-08	9.059E-04	-232	.0000
.2512	6530	2.105E+04	3.060E+13	3.330E-01	4.979E-08	1.337E-03	-159	.0000
.3981	6864	2.442E+04	5.918E+13	5.765E-01	5.489E-08	2.411E-03	-94	.0000
.6310	7290	2.736E+04	1.229E+14	1.074E+00	5.778E-08	4.850E-03	-42	.0000
1.000	7786	2.983E+04	2.573E+14	2.087E+00	5.871E-08	1.012E-02	0	.0000
1.585	8653	3.155E+04	7.315E+14	6.077E+00	5.483E-08	3.109E-02	30	.0164
2.512	9424	3.247E+04	1.534E+15	1.503E+01	5.001E-08	7.173E-02	48	.4303
3.981	9849	3.322E+04	2.174E+15	2.414E+01	4.752E-08	1.071E-01	63	.6675
6.310	10167	3.403E+04	2.756E+15	3.380E+01	4.588E-08	1.407E-01	80	.8502
10.00	10448	3.497E+04	3.346E+15	4.483E+01	4.461E-08	1.758E-01	101	.8791
15.85	10716	3.610E+04	3.977E+15	5.782E+01	4.362E-08	2.138E-01	127	.9002
25.12	10986	3.752E+04	4.680E+15	7.348E+01	4.282E-08	2.562E-01	159	.9290
39.81	11269	3.928E+04	5.483E+15	9.272E+01	4.221E-08	3.047E-01	201	.9474
63.10	11573	4.151E+04	6.415E+15	1.164E+02	4.178E-08	3.602E-01	254	.9606

7000-DEGREE, LOG G = 4.0   (NO CONV)

TAU 5000	TEMP	PRESSURE	ELECTRON DENSITY	OPACITY	DENSITY	HYDROGEN IONIZATION	HEIGHT (KM)
.0010	5529	1.154E+03	8.147E+11	1.467E-02	3.226E-09	4.928E-04	-1256
.0016	5550	1.518E+03	9.914E+11	1.747E-02	4.228E-09	4.530E-04	-1158
.0025	5581	1.997E+03	1.237E+12	2.126E-02	5.530E-09	4.296E-04	-1059
.0040	5619	2.618E+03	1.565E+12	2.617E-02	7.201E-09	4.162E-04	-961
.0063	5671	3.406E+03	2.031E+12	3.288E-02	9.284E-09	4.204E-04	-865
.0100	5710	4.410E+03	2.546E+12	4.021E-02	1.194E-08	4.087E-04	-770
.0158	5759	5.710E+03	3.264E+12	5.006E-02	1.533E-08	4.088E-04	-674
.0251	5834	7.331E+03	4.396E+12	6.478E-02	1.942E-08	4.401E-04	-580
.0398	5917	9.304E+03	5.959E+12	8.420E-02	2.431E-08	4.839E-04	-490
.0631	6021	1.169E+04	8.365E+12	1.124E-01	3.000E-08	5.616E-04	-402
.1000	6151	1.447E+04	1.219E+13	1.545E-01	3.635E-08	6.919E-04	-318
.1585	6314	1.763E+04	1.863E+13	2.204E-01	4.314E-08	9.147E-04	-238
.2512	6543	2.101E+04	3.128E+13	3.390E-01	4.960E-08	1.374E-03	-165
.3981	6824	2.442E+04	5.533E+13	5.453E-01	5.522E-08	2.234E-03	-100
.6310	7213	2.763E+04	1.100E+14	9.760E-01	5.901E-08	4.238E-03	-44
1.000	7765	3.025E+04	2.519E+14	2.044E+00	5.972E-08	9.730E-03	0
1.585	8440	3.213E+04	5.871E+14	4.742E+00	5.763E-08	2.370E-02	32
2.512	9584	3.318E+04	1.770E+15	1.807E+01	4.976E-08	8.323E-02	51
3.981	10600	3.366E+04	3.576E+15	5.125E+01	4.147E-08	2.022E-01	61
6.310	11562	3.395E+04	5.570E+15	1.084E+02	3.352E-08	3.898E-01	69
10.00	12594	3.421E+04	7.188E+15	1.720E+02	2.667E-08	6.325E-01	78
15.85	13583	3.452E+04	7.839E+15	1.957E+02	2.258E-08	8.149E-01	91
25.12	14688	3.501E+04	7.879E+15	1.826E+02	2.004E-08	9.219E-01	114
39.81	15605	3.587E+04	7.777E+15	1.626E+02	1.895E-08	9.610E-01	158
63.10	16358	3.737E+04	7.814E+15	1.490E+02	1.866E-08	9.767E-01	239

7000-DEGREE, LOG G = 3.0

TAU 5000	TEMP	PRESSURE	ELECTRON DENSITY	OPACITY	DENSITY	LOG G EFF	HEIGHT (KM)	CONV/ TOTAL
.0010	5547	2.138E+02	3.524E+11	7.051E-03	5.952E-10	2.9944	-12661	.0000
.0016	5570	2.895E+02	4.368E+11	8.380E-03	8.029E-10	2.9946	-11569	.0000
.0025	5599	3.898E+02	5.474E+11	1.010E-02	1.076E-09	2.9948	-10493	.0000
.0040	5637	5.207E+02	6.976E+11	1.237E-02	1.427E-09	2.9951	-9441	.0000
.0063	5684	6.886E+02	9.011E+11	1.537E-02	1.872E-09	2.9954	-8417	.0000
.0100	5739	9.014E+02	1.179E+12	1.933E-02	2.427E-09	2.9959	-7421	.0000
.0158	5800	1.169E+03	1.554E+12	2.449E-02	3.112E-09	2.9961	-6451	.0000
.0251	5878	1.500E+03	2.106E+12	3.174E-02	3.941E-09	2.9964	-5509	.0000
.0398	5974	1.899E+03	2.937E+12	4.215E-02	4.909E-09	2.9966	-4604	.0000
.0631	6088	2.370E+03	4.202E+12	5.715E-02	6.011E-09	2.9967	-3738	.0000
.1000	6227	2.914E+03	6.216E+12	7.970E-02	7.222E-09	2.9969	-2914	.0000
.1585	6406	3.518E+03	9.726E+12	1.166E-01	8.473E-09	2.9970	-2143	.0000
.2512	6645	4.150E+03	1.647E+13	1.834E-01	9.622E-09	2.9970	-1444	.0000
.3981	6964	4.758E+03	3.045E+13	3.154E-01	1.050E-08	2.9970	-839	.0000
.6310	7410	5.284E+03	6.368E+13	6.314E-01	1.089E-08	2.9968	-348	.0000
1.000	8038	5.657E+03	1.510E+14	1.587E+00	1.056E-08	2.9965	0	.0000
1.585	8916	5.862E+03	3.890E+14	5.376E+00	9.337E-09	2.9963	207	.0185
2.512	9661	5.956E+03	7.094E+14	1.343E+01	8.019E-09	2.9966	325	.4135
3.981	10094	6.030E+03	9.291E+14	2.099E+01	7.256E-09	2.9969	437	.7057
6.310	10460	6.107E+03	1.117E+15	2.881E+01	6.646E-09	2.9971	571	.7945
10.00	10823	6.195E+03	1.289E+15	3.706E+01	6.101E-09	2.9973	747	.8238
15.85	11269	6.300E+03	1.463E+15	4.615E+01	5.524E-09	2.9975	988	.7599
25.12	11987	6.435E+03	1.626E+15	5.422E+01	4.833E-09	2.9980	1342	.6767
39.81	14019	6.662E+03	1.615E+15	4.407E+01	3.902E-09	2.9986	2001	.0014
63.10	15767	7.235E+03	1.589E+15	3.367E+01	3.705E-09	2.9993	3686	.0000

7000-DEGREE, LOG G = 2.0

TAU 5000	TEMP	PRESSURE	ELECTRON DENSITY	OPACITY	DENSITY	LOG G EFF	HEIGHT (KM)	CONV/ TOTAL
.0010	5561	2.974E+01	1.330E+11	4.010E-03	8.241E-11	1.9325	-136943	.0000
.0016	5584	4.347E+01	1.711E+11	4.518E-03	1.200E-10	1.9345	-123215	.0000
.0025	5617	6.237E+01	2.233E+11	5.285E-03	1.712E-10	1.9375	-110092	.0000
.0040	5653	8.777E+01	2.905E+11	6.287E-03	2.395E-10	1.9410	-97605	.0000
.0063	5701	1.211E+02	3.848E+11	7.707E-03	3.276E-10	1.9450	-85757	.0000
.0100	5756	1.637E+02	5.115E+11	9.579E-03	4.386E-10	1.9512	-74553	.0000
.0158	5816	2.180E+02	6.799E+11	1.200E-02	5.782E-10	1.9535	-63795	.0000
.0251	5900	2.851E+02	9.444E+11	1.569E-02	7.454E-10	1.9571	-53588	.0000
.0398	5992	3.659E+02	1.316E+12	2.068E-02	9.417E-10	1.9602	-43950	.0000
.0631	6111	4.615E+02	1.913E+12	2.839E-02	1.164E-09	1.9616	-34830	.0000
.1000	6264	5.688E+02	2.915E+12	4.082E-02	1.398E-09	1.9640	-26409	.0000
.1585	6453	6.849E+02	4.634E+12	6.145E-02	1.631E-09	1.9639	-18711	.0000
.2512	6717	8.006E+02	8.113E+12	1.030E-01	1.826E-09	1.9639	-11984	.0000
.3981	7074	9.029E+02	1.559E+13	1.970E-01	1.941E-09	1.9633	-6507	.0000
.6310	7572	9.791E+02	3.357E+13	4.669E-01	1.928E-09	1.9616	-2495	.0000
1.000	8251	1.022E+03	7.757E+13	1.425E+00	1.751E-09	1.9590	0	.0000
1.585	8995	1.040E+03	1.530E+14	4.133E+00	1.462E-09	1.9615	1434	.0012
2.512	9782	1.048E+03	2.396E+14	9.179E+00	1.145E-09	1.9655	2565	.0937
3.981	10635	1.050E+03	2.935E+14	1.329E+01	9.004E-10	1.9698	3825	.0151
6.310	11445	1.053E+03	3.018E+14	1.318E+01	7.784E-10	1.9743	5910	.0001
10.00	12405	1.065E+03	2.918E+14	1.103E+01	7.048E-10	1.9814	10076	.0000
15.85	13435	1.102E+03	2.820E+14	8.938E+00	6.661E-10	1.9837	18832	.0000
25.12	14590	1.186E+03	2.825E+14	7.428E+00	6.546E-10	1.9873	36355	.0000
39.81	15556	1.358E+03	3.072E+14	6.986E+00	6.946E-10	1.9905	67178	.0000
63.10	16754	1.655E+03	3.529E+14	6.803E+00	7.745E-10	1.9921	114070	.0000

6500-DEGREE, LOG G = 4.0

TAU 5000	TEMP	PRESSURE	ELECTRON DENSITY	OPACITY	DENSITY	HYDROGEN IONIZATION	HEIGHT (KM)	CONV/ TOTAL
.0010	5186	1.703E+03	4.457E+11	9.688E-03	5.077E-09	1.242E-04	-1192	.0000
.0016	5219	2.242E+03	5.676E+11	1.195E-02	6.641E-09	1.196E-04	-1099	.0000
.0025	5230	2.947E+03	6.868E+11	1.423E-02	8.714E-09	1.054E-04	-1007	.0000
.0040	5260	3.873E+03	8.678E+11	1.751E-02	1.139E-08	1.002E-04	-914	.0000
.0063	5294	5.065E+03	1.099E+12	2.159E-02	1.479E-08	9.632E-05	-822	.0000
.0100	5346	6.580E+03	1.440E+12	2.729E-02	1.903E-08	9.984E-05	-732	.0000
.0158	5397	8.478E+03	1.870E+12	3.426E-02	2.429E-08	1.032E-04	-644	.0000
.0251	5457	1.087E+04	2.456E+12	4.340E-02	3.081E-08	1.098E-04	-557	.0000
.0398	5542	1.380E+04	3.373E+12	5.677E-02	3.852E-08	1.278E-04	-472	.0000
.0631	5632	1.737E+04	4.642E+12	7.439E-02	4.769E-08	1.501E-04	-389	.0000
.1000	5756	2.159E+04	6.771E+12	1.018E-01	5.799E-08	1.943E-04	-309	.0000
.1585	5893	2.646E+04	1.001E+13	1.409E-01	6.944E-08	2.571E-04	-232	.0000
.2512	6136	3.165E+04	1.793E+13	2.261E-01	7.973E-08	4.416E-04	-163	.0000
.3981	6333	3.702E+04	2.846E+13	3.303E-01	9.034E-08	6.490E-04	-100	.0000
.6310	6701	4.217E+04	5.923E+13	5.992E-01	9.720E-08	1.328E-03	-45	.0000
1.000	7157	4.663E+04	1.316E+14	1.159E+00	1.005E-07	2.954E-03	0	.0000
1.585	7749	5.007E+04	3.186E+14	2.488E+00	9.925E-08	7.380E-03	34	.0005
2.512	8600	5.233E+04	8.971E+14	6.792E+00	9.218E-08	2.264E-02	58	.1977
3.981	9141	5.386E+04	1.559E+15	1.258E+01	8.779E-08	4.145E-02	75	.4890
6.310	9491	5.537E+04	2.155E+15	1.860E+01	8.562E-08	5.882E-02	92	.7614
10.00	9779	5.704E+04	2.765E+15	2.550E+01	8.431E-08	7.673E-02	112	.8852
15.85	10047	5.901E+04	3.445E+15	3.394E+01	8.348E-08	9.660E-02	135	.8901
25.12	10297	6.138E+04	4.188E+15	4.397E+01	8.325E-08	1.178E-01	164	.8119
39.81	10547	6.431E+04	5.050E+15	5.644E+01	8.352E-08	1.417E-01	199	.8151
63.10	10800	6.794E+04	6.059E+15	7.201E+01	8.436E-08	1.683E-01	242	.9999

6500-DEGREE, LOG G = 3.0

TAU 5000	TEMP	PRESSURE	ELECTRON DENSITY	OPACITY	DENSITY	LOG G EFF	HEIGHT (KM)	CONV/ TOTAL
.0010	5186	3.561E+02	1.814E+11	4.378E-03	1.062E-09	2.9955	-11792	.0000
.0016	5213	4.769E+02	2.289E+11	5.285E-03	1.414E-09	2.9957	-10809	.0000
.0025	5235	6.362E+02	2.843E+11	6.334E-03	1.879E-09	2.9959	-9835	.0000
.0040	5268	8.449E+02	3.620E+11	7.759E-03	2.479E-09	2.9961	-8872	.0000
.0063	5311	1.113E+03	4.686E+11	9.645E-03	3.239E-09	2.9964	-7930	.0000
.0100	5363	1.451E+03	6.177E+11	1.218E-02	4.184E-09	2.9968	-7013	.0000
.0158	5419	1.876E+03	8.144E+11	1.540E-02	5.353E-09	2.9969	-6118	.0000
.0251	5493	2.401E+03	1.114E+12	2.004E-02	6.758E-09	2.9972	-5247	.0000
.0398	5580	3.035E+03	1.547E+12	2.642E-02	8.408E-09	2.9974	-4409	.0000
.0631	5683	3.790E+03	2.212E+12	3.563E-02	1.031E-08	2.9975	-3599	.0000
.1000	5818	4.657E+03	3.332E+12	5.007E-02	1.237E-08	2.9977	-2832	.0000
.1585	5975	5.630E+03	5.172E+12	7.208E-02	1.456E-08	2.9978	-2109	.0000
.2512	6229	6.628E+03	9.507E+12	1.189E-01	1.643E-08	2.9979	-1466	.0000
.3981	6447	7.640E+03	1.560E+13	1.797E-01	1.829E-08	2.9980	-883	.0000
.6310	6845	8.559E+03	3.365E+13	3.440E-01	1.927E-08	2.9979	-394	.0000
1.000	7312	9.325E+03	7.362E+13	6.913E-01	1.957E-08	2.9979	0	.0000
1.585	7974	9.861E+03	1.860E+14	1.742E+00	1.873E-08	2.9977	280	.0001
2.512	8895	1.014E+04	5.133E+14	5.944E+00	1.654E-08	2.9975	442	.2306
3.981	9402	1.031E+04	8.045E+14	1.120E+01	1.524E-08	2.9977	551	.5996
6.310	9728	1.047E+04	1.036E+15	1.637E+01	1.442E-08	2.9978	665	.7769
10.00	10003	1.064E+04	1.256E+15	2.206E+01	1.377E-08	2.9978	801	.8995
15.85	10265	1.084E+04	1.481E+15	2.865E+01	1.317E-08	2.9979	973	.9166
25.12	10535	1.108E+04	1.726E+15	3.663E+01	1.258E-08	2.9980	1193	.9999
39.81	10819	1.137E+04	1.988E+15	4.607E+01	1.201E-08	2.9981	1482	.9586
63.10	11145	1.172E+04	2.281E+15	5.762E+01	1.139E-08	2.9982	1866	.9539

6000-DEGREE, LOG G = 4.5

TAU 5000	TEMP	PRESSURE	ELECTRON DENSITY	OPACITY	DENSITY	HYDROGEN IONIZATION	HEIGHT (KM)	CONV/ TOTAL
.0010	4902	4.350E+03	5.121E+11	1.282E-02	1.372E-08	1.704E-05	-372	.0000
.0016	4916	5.645E+03	6.401E+11	1.573E-02	1.776E-08	1.499E-05	-346	.0000
.0025	4932	7.316E+03	8.018E+11	1.933E-02	2.294E-08	1.336E-05	-320	.0000
.0040	4951	9.469E+03	1.006E+12	2.379E-02	2.958E-08	1.212E-05	-294	.0000
.0063	4975	1.224E+04	1.267E+12	2.932E-02	3.805E-08	1.131E-05	-268	.0000
.0100	5005	1.580E+04	1.601E+12	3.615E-02	4.884E-08	1.087E-05	-242	.0000
.0158	5044	2.038E+04	2.037E+12	4.468E-02	6.250E-08	1.103E-05	-216	.0000
.0251	5096	2.623E+04	2.619E+12	5.542E-02	7.963E-08	1.203E-05	-189	.0000
.0398	5155	3.371E+04	3.382E+12	6.886E-02	1.012E-07	1.346E-05	-163	.0000
.0631	5230	4.321E+04	4.438E+12	8.624E-02	1.278E-07	1.626E-05	-137	.0000
.1000	5339	5.512E+04	6.053E+12	1.104E-01	1.597E-07	2.278E-05	-111	.0000
.1585	5471	6.967E+04	8.541E+12	1.449E-01	1.970E-07	3.430E-05	-85	.0000
.2512	5650	8.680E+04	1.299E+13	2.013E-01	2.377E-07	5.880E-05	-60	.0000
.3981	5885	1.056E+05	2.194E+13	3.041E-01	2.774E-07	1.133E-04	-37	.0000
.6310	6192	1.243E+05	4.199E+13	5.086E-01	3.105E-07	2.408E-04	-16	.0000
1.000	6590	1.411E+05	9.158E+13	9.459E-01	3.309E-07	5.657E-04	0	.0000
1.585	7106	1.548E+05	2.243E+14	1.940E+00	3.365E-07	1.467E-03	13	.0030
2.512	7772	1.648E+05	6.000E+14	4.367E+00	3.267E-07	4.189E-03	22	.1570
3.981	8315	1.724E+05	1.200E+15	8.001E+00	3.182E-07	8.707E-03	30	.4910
6.310	8722	1.797E+05	1.915E+15	1.236E+01	3.146E-07	1.412E-02	37	.7421
10.00	9055	1.875E+05	2.733E+15	1.754E+01	3.145E-07	2.022E-02	45	.8825
15.85	9350	1.965E+05	3.686E+15	2.385E+01	3.171E-07	2.708E-02	54	.9199
25.12	9622	2.071E+05	4.800E+15	3.157E+01	3.226E-07	3.472E-02	65	.9278
39.81	9887	2.198E+05	6.148E+15	4.137E+01	3.307E-07	4.342E-02	77	.9542
63.10	10151	2.353E+05	7.797E+15	5.391E+01	3.419E-07	5.330E-02	91	.9706

6000-DEGREE, LOG G = 4.0

TAU 5000	TEMP	PRESSURE	ELECTRON DENSITY	OPACITY	DENSITY	HYDROGEN IONIZATION	HEIGHT (KM)	CONV/ TOTAL
.0010	4900	2.186E+03	3.005E+11	7.781E-03	6.898E-09	2.869E-05	-1166	.0000
.0016	4914	2.862E+03	3.757E+11	9.505E-03	9.007E-09	2.525E-05	-1081	.0000
.0025	4931	3.738E+03	4.712E+11	1.165E-02	1.172E-08	2.250E-05	-996	.0000
.0040	4950	4.869E+03	5.926E+11	1.432E-02	1.521E-08	2.042E-05	-911	.0000
.0063	4974	6.324E+03	7.483E+11	1.765E-02	1.966E-08	1.902E-05	-828	.0000
.0100	5004	8.194E+03	9.485E+11	2.180E-02	2.533E-08	1.824E-05	-744	.0000
.0158	5044	1.059E+04	1.214E+12	2.703E-02	3.247E-08	1.852E-05	-661	.0000
.0251	5096	1.364E+04	1.571E+12	3.371E-02	4.140E-08	2.008E-05	-578	.0000
.0398	5155	1.752E+04	2.042E+12	4.213E-02	5.256E-08	2.232E-05	-495	.0000
.0631	5231	2.241E+04	2.711E+12	5.329E-02	6.624E-08	2.691E-05	-412	.0000
.1000	5339	2.845E+04	3.753E+12	6.926E-02	8.242E-08	3.689E-05	-331	.0000
.1585	5473	3.572E+04	5.412E+12	9.281E-02	1.009E-07	5.461E-05	-251	.0000
.2512	5652	4.407E+04	8.454E+12	1.323E-01	1.206E-07	9.124E-05	-176	.0000
.3981	5889	5.297E+04	1.471E+13	2.054E-01	1.391E-07	1.716E-04	-107	.0000
.6310	6197	6.164E+04	2.883E+13	3.514E-01	1.538E-07	3.580E-04	-48	.0000
1.000	6600	6.926E+04	6.417E+13	6.669E-01	1.622E-07	8.372E-04	0	.0000
1.585	7119	7.536E+04	1.585E+14	1.389E+00	1.634E-07	2.168E-03	37	.0005
2.512	7830	7.968E+04	4.485E+14	3.356E+00	1.564E-07	6.586E-03	64	.0856
3.981	8455	8.263E+04	9.671E+14	6.857E+00	1.491E-07	1.505E-02	84	.4954
6.310	8861	8.529E+04	1.509E+15	1.078E+01	1.456E-07	2.412E-02	102	.7649
10.00	9179	8.812E+04	2.085E+15	1.530E+01	1.440E-07	3.376E-02	121	.8724
15.85	9456	9.137E+04	2.723E+15	2.070E+01	1.436E-07	4.427E-02	144	.9437
25.12	9710	9.524E+04	3.441E+15	2.722E+01	1.443E-07	5.572E-02	171	.9294
39.81	9958	9.994E+04	4.286E+15	3.539E+01	1.461E-07	6.862E-02	203	.9560
63.10	10205	1.057E+05	5.297E+15	4.579E+01	1.488E-07	8.328E-02	242	.9724

6000-DEGREE, LOG G = 4.0   (NO BLKT)

TAU 5000	TEMP	PRESSURE	ELECTRON DENSITY	OPACITY	DENSITY	HYDROGEN IONIZATION	HEIGHT (KM)	CONV/ TOTAL
.0010	4855	2.231E+03	2.864E+11	7.636E-03	7.107E-09	2.205E-05	-1180	.0000
.0016	4864	2.920E+03	3.575E+11	9.336E-03	9.283E-09	1.882E-05	-1096	.0000
.0025	4871	3.812E+03	4.454E+11	1.143E-02	1.210E-08	1.585E-05	-1012	.0000
.0040	4878	4.965E+03	5.550E+11	1.403E-02	1.574E-08	1.334E-05	-929	.0000
.0063	4884	6.453E+03	6.910E+11	1.725E-02	2.043E-08	1.115E-05	-846	.0000
.0100	4889	8.371E+03	8.599E+11	2.121E-02	2.648E-08	9.305E-06	-764	.0000
.0158	4914	1.084E+04	1.086E+12	2.618E-02	3.411E-08	8.697E-06	-682	.0000
.0251	4945	1.400E+04	1.376E+12	3.233E-02	4.380E-08	8.511E-06	-600	.0000
.0398	4985	1.807E+04	1.752E+12	3.994E-02	5.607E-08	8.699E-06	-519	.0000
.0631	5046	2.327E+04	2.270E+12	4.960E-02	7.133E-08	1.006E-05	-437	.0000
.1000	5137	2.987E+04	3.022E+12	6.229E-02	8.997E-08	1.348E-05	-354	.0000
.1585	5261	3.812E+04	4.183E+12	8.007E-02	1.121E-07	2.086E-05	-272	.0000
.2512	5431	4.806E+04	6.212E+12	1.083E-01	1.369E-07	3.768E-05	-192	.0000
.3981	5660	5.921E+04	1.034E+13	1.603E-01	1.618E-07	7.811E-05	-118	.0000
.6310	5964	7.046E+04	1.998E+13	2.689E-01	1.827E-07	1.805E-04	-53	.0000
1.000	6360	8.042E+04	4.507E+13	5.124E-01	1.955E-07	4.576E-04	0	.0000
1.585	6876	8.826E+04	1.165E+14	1.096E+00	1.983E-07	1.282E-03	40	.0005
2.512	7585	9.369E+04	3.505E+14	2.717E+00	1.903E-07	4.197E-03	67	.0945
3.981	8207	9.732E+04	7.917E+14	5.588E+00	1.817E-07	1.007E-02	87	.4914
6.310	8626	1.006E+05	1.288E+15	8.908E+00	1.775E-07	1.684E-02	105	.7548
10.00	8955	1.040E+05	1.834E+15	1.279E+01	1.756E-07	2.431E-02	124	.8859
15.85	9241	1.078E+05	2.450E+15	1.744E+01	1.752E-07	3.260E-02	147	.9333
25.12	9502	1.124E+05	3.154E+15	2.310E+01	1.762E-07	4.178E-02	173	.9322
39.81	9755	1.179E+05	3.991E+15	3.024E+01	1.784E-07	5.226E-02	204	.9579
63.10	10006	1.246E+05	5.000E+15	3.937E+01	1.819E-07	6.426E-02	241	.9747

6000-DEGREE, LOG G = 4.0   (NO CONV)

TAU 5000	TEMP	PRESSURE	ELECTRON DENSITY	OPACITY	DENSITY	HYDROGEN IONIZATION	HEIGHT (KM)
.0010	4894	2.189E+03	2.979E+11	7.748E-03	6.917E-09	2.769E-05	-1164
.0016	4904	2.869E+03	3.711E+11	9.450E-03	9.047E-09	2.383E-05	-1079
.0025	4920	3.749E+03	4.657E+11	1.159E-02	1.179E-08	2.114E-05	-994
.0040	4939	4.885E+03	5.863E+11	1.427E-02	1.529E-08	1.918E-05	-909
.0063	4966	6.345E+03	7.426E+11	1.761E-02	1.976E-08	1.809E-05	-826
.0100	4999	8.218E+03	9.451E+11	2.178E-02	2.542E-08	1.771E-05	-742
.0158	5038	1.062E+04	1.209E+12	2.700E-02	3.258E-08	1.798E-05	-659
.0251	5093	1.367E+04	1.567E+12	3.369E-02	4.152E-08	1.967E-05	-576
.0398	5153	1.755E+04	2.041E+12	4.213E-02	5.267E-08	2.210E-05	-494
.0631	5233	2.243E+04	2.719E+12	5.339E-02	6.630E-08	2.708E-05	-411
.1000	5337	2.847E+04	3.744E+12	6.916E-02	8.251E-08	3.657E-05	-330
.1585	5474	3.574E+04	5.426E+12	9.299E-02	1.010E-07	5.486E-05	-250
.2512	5653	4.407E+04	8.482E+12	1.326E-01	1.206E-07	9.179E-05	-175
.3981	5892	5.295E+04	1.479E+13	2.064E-01	1.390E-07	1.732E-04	-107
.6310	6202	6.156E+04	2.912E+13	3.542E-01	1.534E-07	3.633E-04	-48
1.000	6601	6.914E+04	6.423E+13	6.673E-01	1.618E-07	8.399E-04	0
1.585	7109	7.528E+04	1.559E+14	1.370E+00	1.634E-07	2.130E-03	38
2.512	7794	7.972E+04	4.276E+14	3.218E+00	1.573E-07	6.240E-03	65
3.981	8645	8.258E+04	1.188E+15	8.437E+00	1.452E-07	1.902E-02	84
6.310	9706	8.416E+04	3.213E+15	2.620E+01	1.272E-07	5.904E-02	95
10.00	10787	8.499E+04	6.845E+15	7.493E+01	1.072E-07	1.496E-01	103
15.85	11939	8.548E+04	1.187E+16	1.829E+02	8.538E-08	3.262E-01	108
25.12	12848	8.587E+04	1.543E+16	2.920E+02	7.041E-08	5.144E-01	113
39.81	13901	8.630E+04	1.785E+16	3.775E+02	5.791E-08	7.232E-01	119
63.10	14943	8.690E+04	1.851E+16	3.866E+02	5.043E-08	8.610E-01	131

6000-DEGREE, LOG G = 4.0   (NO CONV OR BLKT)

TAU 5000	TEMP	PRESSURE	ELECTRON DENSITY	OPACITY	DENSITY	HYDROGEN IONIZATION	HEIGHT (KM)
.0010	4856	2.225E+03	2.859E+11	7.621E-03	7.086E-09	2.213E-05	-1181
.0016	4858	2.916E+03	3.544E+11	9.295E-03	9.282E-09	1.816E-05	-1096
.0025	4862	3.812E+03	4.407E+11	1.139E-02	1.213E-08	1.498E-05	-1012
.0040	4868	4.969E+03	5.495E+11	1.399E-02	1.579E-08	1.251E-05	-928
.0063	4877	6.460E+03	6.869E+11	1.723E-02	2.049E-08	1.065E-05	-846
.0100	4890	8.378E+03	8.610E+11	2.123E-02	2.650E-08	9.332E-06	-764
.0158	4911	1.084E+04	1.084E+12	2.618E-02	3.415E-08	8.563E-06	-682
.0251	4942	1.401E+04	1.373E+12	3.232E-02	4.385E-08	8.327E-06	-600
.0398	4986	1.807E+04	1.755E+12	3.997E-02	5.607E-08	8.777E-06	-519
.0631	5049	2.327E+04	2.277E+12	4.965E-02	7.129E-08	1.023E-05	-437
.1000	5138	2.987E+04	3.024E+12	6.231E-02	8.994E-08	1.355E-05	-354
.1585	5262	3.812E+04	4.186E+12	8.010E-02	1.121E-07	2.093E-05	-273
.2512	5431	4.805E+04	6.214E+12	1.083E-01	1.368E-07	3.773E-05	-193
.3981	5661	5.920E+04	1.034E+13	1.604E-01	1.618E-07	7.826E-05	-118
.6310	5964	7.045E+04	1.998E+13	2.688E-01	1.827E-07	1.805E-04	-53
1.000	6358	8.042E+04	4.488E+13	5.106E-01	1.956E-07	4.551E-04	0
1.585	6847	8.840E+04	1.111E+14	1.055E+00	1.994E-07	1.212E-03	40
2.512	7538	9.410E+04	3.292E+14	2.577E+00	1.923E-07	3.893E-03	69
3.981	8392	9.763E+04	9.811E+14	6.842E+00	1.778E-07	1.278E-02	88
6.310	9462	9.961E+04	2.862E+15	2.137E+01	1.567E-07	4.265E-02	100
10.00	10598	1.006E+05	6.710E+15	6.560E+01	1.325E-07	1.186E-01	107
15.85	11810	1.011E+05	1.260E+16	1.759E+02	1.055E-07	2.799E-01	111
25.12	12711	1.015E+05	1.692E+16	2.950E+02	8.741E-08	4.543E-01	115
39.81	13775	1.019E+05	2.032E+16	4.106E+02	7.108E-08	6.709E-01	120
63.10	14835	1.025E+05	2.153E+16	4.422E+02	6.088E-08	8.298E-01	129

6000-DEGREE, LOG G = 4.0   (10X METALS)

TAU 5000	TEMP	PRESSURE	ELECTRON DENSITY	OPACITY	DENSITY	HYDROGEN IONIZATION	HEIGHT (KM)
.0010	4914	7.236E+02	6.711E+11	2.574E-02	2.303E-09	1.415E-05	-1277
.0016	4918	9.269E+02	8.423E+11	3.176E-02	2.949E-09	1.153E-05	-1199
.0025	4922	1.188E+03	1.056E+12	3.910E-02	3.776E-09	9.500E-06	-1121
.0040	4929	1.525E+03	1.324E+12	4.802E-02	4.841E-09	7.929E-06	-1042
.0063	4939	1.961E+03	1.659E+12	5.881E-02	6.211E-09	6.749E-06	-963
.0100	4952	2.525E+03	2.081E+12	7.186E-02	7.976E-09	5.910E-06	-883
.0158	4972	3.257E+03	2.615E+12	8.762E-02	1.025E-08	5.371E-06	-802
.0251	5001	4.210E+03	3.298E+12	1.067E-01	1.317E-08	5.149E-06	-720
.0398	5042	5.451E+03	4.185E+12	1.298E-01	1.692E-08	5.306E-06	-638
.0631	5099	7.067E+03	5.353E+12	1.580E-01	2.169E-08	5.997E-06	-553
.1000	5180	9.173E+03	6.923E+12	1.921E-01	2.771E-08	7.717E-06	-468
.1585	5297	1.192E+04	9.089E+12	2.331E-01	3.521E-08	1.190E-05	-380
.2512	5431	1.552E+04	1.214E+13	2.814E-01	4.448E-08	2.243E-05	-290
.3981	5681	2.025E+04	1.668E+13	3.390E-01	5.576E-08	5.393E-05	-195
.6310	5984	2.639E+04	2.464E+13	4.239E-01	6.899E-08	1.611E-04	-96
1.000	6393	3.369E+04	4.361E+13	6.145E-01	8.239E-08	5.426E-04	0
1.585	6913	4.086E+04	9.691E+13	1.106E+00	9.231E-08	1.751E-03	82
2.512	7647	4.641E+04	2.803E+14	2.590E+00	9.440E-08	6.270E-03	141
3.981	8549	4.990E+04	8.406E+14	7.051E+00	8.956E-08	2.146E-02	179
6.310	9675	5.175E+04	2.428E+15	2.345E+01	7.848E-08	7.265E-02	200
10.00	10833	5.266E+04	5.321E+15	7.119E+01	6.460E-08	1.949E-01	213
15.85	11983	5.317E+04	8.737E+15	1.622E+02	5.059E-08	4.098E-01	222
25.12	12910	5.363E+04	1.083E+16	2.375E+02	4.164E-08	6.172E-01	232
39.81	13966	5.419E+04	1.190E+16	2.735E+02	3.504E-08	8.062E-01	247
63.10	14969	5.506E+04	1.205E+16	2.614E+02	3.154E-08	9.067E-01	273

6000-DEGREE, LOG G = 4.0 (1/10X METALS)

TAU 5000	TEMP	PRESSURE	ELECTRON DENSITY	OPACITY	DENSITY	HYDROGEN IONIZATION	HEIGHT (KM)
.0010	4825	3.198E+03	1.742E+11	4.753E-03	1.024E-08	2.917E-05	-1160
.0016	4827	4.331E+03	2.073E+11	5.554E-03	1.386E-08	2.491E-05	-1065
.0025	4830	5.863E+03	2.480E+11	6.531E-03	1.875E-08	2.135E-05	-971
.0040	4836	7.922E+03	2.987E+11	7.734E-03	2.531E-08	1.844E-05	-876
.0063	4845	1.066E+04	3.631E+11	9.239E-03	3.401E-08	1.616E-05	-783
.0100	4859	1.428E+04	4.469E+11	1.115E-02	4.542E-08	1.449E-05	-692
.0158	4881	1.900E+04	5.601E+11	1.366E-02	6.015E-08	1.347E-05	-602
.0251	4912	2.506E+04	7.152E+11	1.697E-02	7.885E-08	1.306E-05	-514
.0398	4958	3.271E+04	9.426E+11	2.158E-02	1.019E-07	1.354E-05	-429
.0631	5022	4.210E+04	1.288E+12	2.822E-02	1.296E-07	1.514E-05	-348
.1000	5111	5.326E+04	1.855E+12	3.839E-02	1.611E-07	1.866E-05	-270
.1585	5237	6.589E+04	2.891E+12	5.547E-02	1.945E-07	2.611E-05	-199
.2512	5406	7.929E+04	4.912E+12	8.571E-02	2.267E-07	4.132E-05	-136
.3981	5633	9.248E+04	9.310E+12	1.441E-01	2.537E-07	7.518E-05	-81
.6310	5934	1.044E+05	1.986E+13	2.655E-01	2.717E-07	1.578E-04	-36
1.000	6326	1.142E+05	4.728E+13	5.323E-01	2.788E-07	3.788E-04	0
1.585	6813	1.218E+05	1.200E+14	1.124E+00	2.758E-07	9.911E-04	27
2.512	7474	1.271E+05	3.467E+14	2.676E+00	2.620E-07	3.050E-03	47
3.981	8300	1.306E+05	1.021E+15	6.877E+00	2.409E-07	9.842E-03	61
6.310	9352	1.326E+05	3.011E+15	2.073E+01	2.126E-07	3.306E-02	70
10.00	10499	1.336E+05	7.356E+15	6.378E+01	1.809E-07	9.514E-02	75
15.85	11724	1.342E+05	1.448E+16	1.776E+02	1.459E-07	2.325E-01	78
25.12	12693	1.345E+05	2.056E+16	3.235E+02	1.199E-07	4.018E-01	81
39.81	13782	1.349E+05	2.557E+16	4.808E+02	9.668E-08	6.200E-01	84
63.10	14831	1.353E+05	2.767E+16	5.444E+02	8.198E-08	7.909E-01	89

6000-DEGREE, LOG G = 3.5

TAU 5000	TEMP	PRESSURE	ELECTRON DENSITY	OPACITY	DENSITY	LOG G EFF	HEIGHT (KM)	CONV/ TOTAL
.0010	4902	1.071E+03	1.803E+11	4.894E-03	3.377E-09	3.4992	-3647	.0000
.0016	4921	1.412E+03	2.265E+11	5.942E-03	4.436E-09	3.4992	-3369	.0000
.0025	4934	1.857E+03	2.817E+11	7.198E-03	5.820E-09	3.4992	-3093	.0000
.0040	4953	2.439E+03	3.537E+11	8.794E-03	7.614E-09	3.4992	-2818	.0000
.0063	4982	3.188E+03	4.503E+11	1.085E-02	9.897E-09	3.4993	-2545	.0000
.0100	5016	4.150E+03	5.759E+11	1.343E-02	1.279E-08	3.4993	-2276	.0000
.0158	5061	5.374E+03	7.455E+11	1.676E-02	1.642E-08	3.4993	-2010	.0000
.0251	5115	6.927E+03	9.742E+11	2.104E-02	2.094E-08	3.4994	-1746	.0000
.0398	5187	8.872E+03	1.303E+12	2.680E-02	2.645E-08	3.4994	-1485	.0000
.0631	5268	1.128E+04	1.762E+12	3.444E-02	3.312E-08	3.4995	-1228	.0000
.1000	5376	1.421E+04	2.488E+12	4.564E-02	4.089E-08	3.4995	-977	.0000
.1585	5503	1.767E+04	3.632E+12	6.209E-02	4.965E-08	3.4995	-735	.0000
.2512	5702	2.153E+04	6.062E+12	9.356E-02	5.839E-08	3.4995	-509	.0000
.3981	5906	2.551E+04	1.008E+13	1.410E-01	6.678E-08	3.4996	-307	.0000
.6310	6270	2.936E+04	2.233E+13	2.667E-01	7.238E-08	3.4996	-133	.0000
1.000	6686	3.247E+04	5.051E+13	5.163E-01	7.499E-08	3.4996	0	.0000
1.585	7216	3.494E+04	1.245E+14	1.093E+00	7.462E-08	3.4996	104	.0001
2.512	7962	3.665E+04	3.577E+14	2.813E+00	7.042E-08	3.4995	178	.0488
3.981	8628	3.770E+04	7.808E+14	6.257E+00	6.591E-08	3.4995	228	.5150
6.310	9018	3.861E+04	1.168E+15	9.910E+00	6.372E-08	3.4995	272	.7588
10.00	9315	3.957E+04	1.552E+15	1.399E+01	6.239E-08	3.4995	321	.8933
15.85	9572	4.068E+04	1.957E+15	1.876E+01	6.154E-08	3.4995	379	.8863
25.12	9811	4.200E+04	2.402E+15	2.447E+01	6.108E-08	3.4995	449	.9311
39.81	10046	4.361E+04	2.914E+15	3.159E+01	6.091E-08	3.4995	535	.9580
63.10	10284	4.557E+04	3.514E+15	4.057E+01	6.102E-08	3.4995	641	.9767

6000-DEGREE, LOG G = 3.0

TAU 5000	TEMP	PRESSURE	ELECTRON DENSITY	OPACITY	DENSITY	LOG G EFF	HEIGHT (KM)	CONV/ TOTAL
.0010	4894	5.016E+02	1.067E+11	3.156E-03	1.585E-09	2.9973	-11556	.0000
.0016	4913	6.697E+02	1.342E+11	3.793E-03	2.108E-09	2.9974	-10639	.0000
.0025	4925	8.917E+02	1.667E+11	4.546E-03	2.800E-09	2.9975	-9728	.0000
.0040	4946	1.184E+03	2.103E+11	5.527E-03	3.701E-09	2.9976	-8824	.0000
.0063	4975	1.562E+03	2.686E+11	6.792E-03	4.854E-09	2.9977	-7935	.0000
.0100	5010	2.047E+03	3.456E+11	8.407E-03	6.319E-09	2.9978	-7061	.0000
.0158	5055	2.666E+03	4.503E+11	1.051E-02	8.155E-09	2.9979	-6202	.0000
.0251	5110	3.447E+03	5.947E+11	1.326E-02	1.043E-08	2.9981	-5358	.0000
.0398	5182	4.419E+03	8.047E+11	1.703E-02	1.319E-08	2.9982	-4531	.0000
.0631	5265	5.612E+03	1.106E+12	2.214E-02	1.648E-08	2.9983	-3724	.0000
.1000	5375	7.041E+03	1.593E+12	2.982E-02	2.026E-08	2.9984	-2944	.0000
.1585	5507	8.698E+03	2.387E+12	4.143E-02	2.442E-08	2.9985	-2201	.0000
.2512	5711	1.050E+04	4.106E+12	6.404E-02	2.843E-08	2.9985	-1519	.0000
.3981	5915	1.234E+04	6.912E+12	9.762E-02	3.224E-08	2.9986	-912	.0000
.6310	6283	1.407E+04	1.562E+13	1.882E-01	3.460E-08	2.9986	-395	.0000
1.000	6701	1.546E+04	3.546E+13	3.685E-01	3.561E-08	2.9986	0	.0000
1.585	7250	1.653E+04	8.973E+13	8.143E-01	3.507E-08	2.9985	303	.0000
2.512	8049	1.723E+04	2.714E+14	2.344E+00	3.252E-08	2.9984	507	.0364
3.981	8719	1.761E+04	5.782E+14	5.506E+00	3.000E-08	2.9983	632	.4910
6.310	9096	1.793E+04	8.383E+14	8.806E+00	2.869E-08	2.9983	743	.7964
10.00	9374	1.826E+04	1.078E+15	1.233E+01	2.783E-08	2.9983	867	.8855
15.85	9623	1.865E+04	1.331E+15	1.651E+01	2.712E-08	2.9983	1015	.9285
25.12	9859	1.910E+04	1.606E+15	2.154E+01	2.653E-08	2.9984	1197	.9999
39.81	10089	1.964E+04	1.908E+15	2.764E+01	2.603E-08	2.9984	1424	.9157
63.10	10322	2.030E+04	2.249E+15	3.512E+01	2.562E-08	2.9985	1712	.9347

6000-DEGREE, LOG G = 2.0

TAU 5000	TEMP	PRESSURE	ELECTRON DENSITY	OPACITY	DENSITY	LOG G EFF	HEIGHT (KM)	CONV/ TOTAL
.0010	4916	8.882E+01	4.009E+10	1.541E-03	2.793E-10	1.9680	-121073	.0000
.0016	4930	1.241E+02	5.035E+10	1.775E-03	3.891E-10	1.9687	-110428	.0000
.0025	4949	1.721E+02	6.394E+10	2.082E-03	5.377E-10	1.9699	-99973	.0000
.0040	4970	2.366E+02	8.145E+10	2.472E-03	7.361E-10	1.9711	-89762	.0000
.0063	5002	3.218E+02	1.058E+11	2.999E-03	9.947E-10	1.9725	-79847	.0000
.0100	5041	4.320E+02	1.389E+11	3.690E-03	1.325E-09	1.9747	-70279	.0000
.0158	5087	5.733E+02	1.842E+11	4.602E-03	1.743E-09	1.9759	-61013	.0000
.0251	5150	7.503E+02	2.518E+11	5.893E-03	2.253E-09	1.9778	-52107	.0000
.0398	5220	9.681E+02	3.470E+11	7.622E-03	2.867E-09	1.9795	-43558	.0000
.0631	5312	1.231E+03	4.974E+11	1.018E-02	3.583E-09	1.9807	-35372	.0000
.1000	5425	1.538E+03	7.413E+11	1.406E-02	4.382E-09	1.9826	-27651	.0000
.1585	5573	1.882E+03	1.177E+12	2.043E-02	5.218E-09	1.9831	-20474	.0000
.2512	5764	2.245E+03	2.009E+12	3.151E-02	6.019E-09	1.9838	-14000	.0000
.3981	6020	2.604E+03	3.816E+12	5.318E-02	6.682E-09	1.9845	-8345	.0000
.6310	6343	2.926E+03	7.843E+12	9.668E-02	7.117E-09	1.9846	-3675	.0000
1.000	6794	3.189E+03	1.871E+13	2.052E-01	7.219E-09	1.9838	0	.0000
1.585	7418	3.366E+03	5.119E+13	5.397E-01	6.909E-09	1.9823	2525	.0000
2.512	8245	3.459E+03	1.486E+14	1.863E+00	6.170E-09	1.9813	3979	.0257
3.981	8891	3.499E+03	2.858E+14	4.648E+00	5.476E-09	1.9814	4777	.3737
6.310	9225	3.529E+03	3.793E+14	7.175E+00	5.106E-09	1.9822	5528	.7939
10.00	9505	3.557E+03	4.672E+14	1.003E+01	4.790E-09	1.9824	6395	.8625
15.85	9750	3.585E+03	5.483E+14	1.310E+01	4.515E-09	1.9828	7482	.8869
25.12	10001	3.611E+03	6.318E+14	1.671E+01	4.236E-09	1.9834	8904	.9156
39.81	10271	3.636E+03	7.181E+14	2.096E+01	3.941E-09	1.9841	10812	.9365
63.10	10582	3.653E+03	8.056E+14	2.584E+01	3.620E-09	1.9850	13443	.9479

5500-DEGREE, LOG G = 4.0

TAU 5000	TEMP	PRESSURE	ELECTRON DENSITY	OPACITY	DENSITY	HYDROGEN IONIZATION	HEIGHT (KM)	CONV/ TOTAL
.0010	4523	2.308E+03	2.337E+11	7.701E-03	8.083E-09	2.233E-06	-1114	.0000
.0016	4536	2.991E+03	2.930E+11	9.425E-03	1.044E-08	1.968E-06	-1040	.0000
.0025	4550	3.874E+03	3.674E+11	1.155E-02	1.349E-08	1.764E-06	-965	.0000
.0040	4567	5.017E+03	4.610E+11	1.415E-02	1.740E-08	1.609E-06	-891	.0000
.0063	4589	6.494E+03	5.797E+11	1.735E-02	2.242E-08	1.516E-06	-816	.0000
.0100	4615	8.403E+03	7.307E+11	2.128E-02	2.885E-08	1.472E-06	-742	.0000
.0158	4651	1.087E+04	9.273E+11	2.612E-02	3.703E-08	1.529E-06	-666	.0000
.0251	4698	1.405E+04	1.187E+12	3.213E-02	4.738E-08	1.709E-06	-591	.0000
.0398	4750	1.815E+04	1.523E+12	3.955E-02	6.053E-08	1.956E-06	-514	.0000
.0631	4819	2.341E+04	1.981E+12	4.886E-02	7.699E-08	2.470E-06	-437	.0000
.1000	4917	3.014E+04	2.643E+12	6.087E-02	9.714E-08	3.669E-06	-360	.0000
.1585	5038	3.866E+04	3.608E+12	7.673E-02	1.216E-07	6.016E-06	-282	.0000
.2512	5200	4.922E+04	5.183E+12	9.975E-02	1.500E-07	1.166E-05	-204	.0000
.3981	5416	6.170E+04	8.137E+12	1.385E-01	1.805E-07	2.636E-05	-128	.0000
.6310	5695	7.525E+04	1.457E+13	2.145E-01	2.093E-07	6.619E-05	-59	.0000
1.000	6056	8.818E+04	3.060E+13	3.804E-01	2.306E-07	1.805E-04	0	.0000
1.585	6512	9.910E+04	7.340E+13	7.557E-01	2.409E-07	5.188E-04	46	.0001
2.512	7103	1.074E+05	1.996E+14	1.678E+00	2.390E-07	1.633E-03	81	.0181
3.981	7732	1.132E+05	4.963E+14	3.568E+00	2.309E-07	4.533E-03	105	.2131
6.310	8193	1.182E+05	8.935E+14	5.987E+00	2.264E-07	8.612E-03	127	.4320
10.00	8550	1.231E+05	1.359E+15	8.858E+00	2.251E-07	1.344E-02	149	.5941
15.85	8854	1.287E+05	1.900E+15	1.231E+01	2.260E-07	1.897E-02	174	.7084
25.12	9128	1.351E+05	2.534E+15	1.654E+01	2.289E-07	2.523E-02	202	.5776
39.81	9372	1.428E+05	3.252E+15	2.152E+01	2.343E-07	3.186E-02	235	.6410
63.10	9714	1.519E+05	4.480E+15	3.071E+01	2.379E-07	4.362E-02	274	.9999

5500-DEGREE, LOG G = 3.0

TAU 5000	TEMP	PRESSURE	ELECTRON DENSITY	OPACITY	DENSITY	LOG G EFF	HEIGHT (KM)	CONV/ TOTAL
.0010	4512	6.176E+02	6.987E+10	2.739E-03	2.117E-09	2.9982	-11283	.0000
.0016	4528	8.101E+02	8.950E+10	3.336E-03	2.767E-09	2.9982	-10490	.0000
.0025	4539	1.060E+03	1.140E+11	4.085E-03	3.611E-09	2.9983	-9703	.0000
.0040	4557	1.382E+03	1.456E+11	5.011E-03	4.691E-09	2.9983	-8921	.0000
.0063	4582	1.799E+03	1.860E+11	6.154E-03	6.073E-09	2.9984	-8142	.0000
.0100	4613	2.337E+03	2.380E+11	7.559E-03	7.834E-09	2.9985	-7365	.0000
.0158	4653	3.031E+03	3.055E+11	9.291E-03	1.007E-08	2.9986	-6586	.0000
.0251	4702	3.925E+03	3.940E+11	1.143E-02	1.291E-08	2.9987	-5805	.0000
.0398	4765	5.076E+03	5.127E+11	1.408E-02	1.648E-08	2.9988	-5018	.0000
.0631	4840	6.554E+03	6.732E+11	1.743E-02	2.094E-08	2.9988	-4225	.0000
.1000	4937	8.437E+03	9.039E+11	2.180E-02	2.643E-08	2.9989	-3427	.0000
.1585	5059	1.080E+04	1.255E+12	2.787E-02	3.302E-08	2.9990	-2629	.0000
.2512	5240	1.364E+04	1.921E+12	3.818E-02	4.026E-08	2.9990	-1852	.0000
.3981	5426	1.685E+04	3.045E+12	5.449E-02	4.802E-08	2.9990	-1124	.0000
.6310	5747	2.008E+04	6.384E+12	9.687E-02	5.402E-08	2.9991	-493	.0000
1.000	6110	2.284E+04	1.430E+13	1.836E-01	5.777E-08	2.9991	0	.0000
1.585	6578	2.503E+04	3.650E+13	3.900E-01	5.877E-08	2.9991	375	.0000
2.512	7187	2.658E+04	1.038E+14	9.289E-01	5.698E-08	2.9990	643	.0018
3.981	7964	2.754E+04	3.102E+14	2.517E+00	5.281E-08	2.9989	816	.2943
6.310	8460	2.819E+04	5.610E+14	4.643E+00	5.033E-08	2.9989	944	.7058
10.00	8787	2.882E+04	8.004E+14	6.927E+00	4.901E-08	2.9989	1073	.8277
15.85	9060	2.951E+04	1.056E+15	9.635E+00	4.812E-08	2.9989	1219	.9857
25.12	9299	3.031E+04	1.328E+15	1.280E+01	4.758E-08	2.9989	1392	.9999
39.81	9530	3.127E+04	1.641E+15	1.676E+01	4.724E-08	2.9989	1601	.9519
63.10	9744	3.243E+04	1.983E+15	2.144E+01	4.723E-08	2.9989	1860	.8970

5000-DEGREE, LOG G = 4.0

TAU 5000	TEMP	PRESSURE	ELECTRON DENSITY	OPACITY	DENSITY	HYDROGEN IONIZATION	HEIGHT (KM)	CONV/ TOTAL
.0010	4071	2.294E+03	1.556E+11	7.546E-03	8.720E-09	5.935E-08	-1095	.0000
.0016	4084	2.994E+03	1.942E+11	9.155E-03	1.134E-08	5.421E-08	-1025	.0000
.0025	4100	3.907E+03	2.429E+11	1.113E-02	1.475E-08	5.084E-08	-954	.0000
.0040	4120	5.097E+03	3.045E+11	1.355E-02	1.915E-08	4.909E-08	-884	.0000
.0063	4145	6.644E+03	3.836E+11	1.652E-02	2.482E-08	4.949E-08	-813	.0000
.0100	4175	8.653E+03	4.853E+11	2.019E-02	3.210E-08	5.185E-08	-742	.0000
.0158	4211	1.126E+04	6.175E+11	2.473E-02	4.140E-08	5.702E-08	-671	.0000
.0251	4258	1.462E+04	7.948E+11	3.041E-02	5.318E-08	6.824E-08	-599	.0000
.0398	4308	1.895E+04	1.024E+12	3.739E-02	6.814E-08	8.296E-08	-528	.0000
.0631	4372	2.452E+04	1.337E+12	4.619E-02	8.689E-08	1.115E-07	-455	.0000
.1000	4463	3.164E+04	1.788E+12	5.742E-02	1.099E-07	1.786E-07	-383	.0000
.1585	4573	4.071E+04	2.422E+12	7.149E-02	1.379E-07	3.216E-07	-309	.0000
.2512	4721	5.225E+04	3.359E+12	8.925E-02	1.714E-07	7.134E-07	-235	.0000
.3981	4917	6.687E+04	4.791E+12	1.117E-01	2.106E-07	2.017E-06	-158	.0000
.6310	5173	8.528E+04	7.142E+12	1.424E-01	2.552E-07	7.137E-06	-79	.0000
1.000	5505	1.072E+05	1.213E+13	2.019E-01	3.013E-07	2.907E-05	0	.0000
1.585	5924	1.291E+05	2.650E+13	3.612E-01	3.372E-07	1.126E-04	69	.0000
2.512	6445	1.467E+05	7.245E+13	7.925E-01	3.518E-07	4.030E-04	119	.0051
3.981	7049	1.587E+05	2.079E+14	1.833E+00	3.477E-07	1.309E-03	153	.1927
6.310	7538	1.678E+05	4.411E+14	3.384E+00	3.433E-07	2.905E-03	180	.4789
10.00	7916	1.763E+05	7.486E+14	5.287E+00	3.429E-07	4.998E-03	205	.6986
15.85	8240	1.854E+05	1.140E+15	7.635E+00	3.456E-07	7.602E-03	231	.8157
25.12	8534	1.956E+05	1.634E+15	1.058E+01	3.510E-07	1.077E-02	261	.8846
39.81	8811	2.075E+05	2.258E+15	1.432E+01	3.593E-07	1.459E-02	294	.9276
63.10	9080	2.214E+05	3.048E+15	1.914E+01	3.706E-07	1.913E-02	332	.9555

5000-DEGREE, LOG G = 3.0

TAU 5000	TEMP	PRESSURE	ELECTRON DENSITY	OPACITY	DENSITY	LOG G EFF	HEIGHT (KM)	CONV/ TOTAL
.0010	4073	5.844E+02	5.190E+10	2.893E-03	2.219E-09	2.9988	-11109	.0000
.0016	4089	7.674E+02	6.593E+10	3.491E-03	2.903E-09	2.9988	-10390	.0000
.0025	4102	1.007E+03	8.336E+10	4.228E-03	3.798E-09	2.9989	-9669	.0000
.0040	4123	1.321E+03	1.059E+11	5.137E-03	4.955E-09	2.9989	-8949	.0000
.0063	4150	1.729E+03	1.350E+11	6.257E-03	6.446E-09	2.9989	-8229	.0000
.0100	4182	2.260E+03	1.725E+11	7.637E-03	8.361E-09	2.9990	-7508	.0000
.0158	4221	2.949E+03	2.214E+11	9.339E-03	1.081E-08	2.9991	-6786	.0000
.0251	4269	3.840E+03	2.860E+11	1.144E-02	1.392E-08	2.9991	-6061	.0000
.0398	4328	4.992E+03	3.721E+11	1.404E-02	1.785E-08	2.9992	-5333	.0000
.0631	4397	6.479E+03	4.872E+11	1.725E-02	2.280E-08	2.9992	-4598	.0000
.1000	4485	8.395E+03	6.461E+11	2.122E-02	2.896E-08	2.9993	-3854	.0000
.1585	4597	1.087E+04	8.676E+11	2.610E-02	3.657E-08	2.9993	-3098	.0000
.2512	4760	1.405E+04	1.203E+12	3.212E-02	4.565E-08	2.9993	-2321	.0000
.3981	4934	1.813E+04	1.686E+12	3.995E-02	5.684E-08	2.9994	-1522	.0000
.6310	5224	2.315E+04	2.750E+12	5.456E-02	6.854E-08	2.9994	-721	.0000
1.000	5549	2.849E+04	5.339E+12	8.885E-02	7.939E-08	2.9994	0	.0000
1.585	5971	3.315E+04	1.327E+13	1.803E-01	8.584E-08	2.9994	562	.0000
2.512	6497	3.654E+04	3.832E+13	4.186E-01	8.669E-08	2.9994	953	.0002
3.981	7153	3.876E+04	1.192E+14	1.059E+00	8.355E-08	2.9994	1213	.0457
6.310	7787	4.016E+04	2.993E+14	2.369E+00	7.911E-08	2.9993	1385	.3431
10.00	8184	4.135E+04	4.986E+14	3.848E+00	7.706E-08	2.9993	1538	.6688
15.85	8492	4.258E+04	7.183E+14	5.575E+00	7.601E-08	2.9993	1701	.8136
25.12	8756	4.397E+04	9.655E+14	7.649E+00	7.559E-08	2.9992	1886	.8882
39.81	8999	4.559E+04	1.251E+15	1.021E+01	7.567E-08	2.9992	2104	.9309
63.10	9233	4.752E+04	1.588E+15	1.343E+01	7.621E-08	2.9992	2364	.9570

5000-DEGREE, LOG G = 2.0

TAU 5000	TEMP	PRESSURE	ELECTRON DENSITY	OPACITY	DENSITY	LOG G EFF	HEIGHT (KM)	CONV/ TOTAL
.0010	4065	1.315E+02	1.412E+10	1.130E-03	5.001E-10	1.9873	-115125	.0000
.0016	4082	1.789E+02	1.877E+10	1.333E-03	6.779E-10	1.9875	-107004	.0000
.0025	4096	2.422E+02	2.475E+10	1.592E-03	9.145E-10	1.9878	-98994	.0000
.0040	4118	3.260E+02	3.251E+10	1.913E-03	1.224E-09	1.9882	-91107	.0000
.0063	4146	4.362E+02	4.257E+10	2.310E-03	1.627E-09	1.9887	-83328	.0000
.0100	4181	5.805E+02	5.568E+10	2.799E-03	2.147E-09	1.9896	-75633	.0000
.0158	4222	7.690E+02	7.273E+10	3.401E-03	2.817E-09	1.9900	-67995	.0000
.0251	4272	1.015E+03	9.522E+10	4.139E-03	3.673E-09	1.9908	-60385	.0000
.0398	4333	1.334E+03	1.250E+11	5.043E-03	4.762E-09	1.9915	-52765	.0000
.0631	4406	1.750E+03	1.647E+11	6.146E-03	6.142E-09	1.9920	-45104	.0000
.1000	4499	2.291E+03	2.190E+11	7.487E-03	7.876E-09	1.9927	-37353	.0000
.1585	4617	2.994E+03	2.957E+11	9.134E-03	1.003E-08	1.9929	-29471	.0000
.2512	4784	3.902E+03	4.172E+11	1.133E-02	1.261E-08	1.9932	-21421	.0000
.3981	4963	5.036E+03	6.181E+11	1.480E-02	1.569E-08	1.9934	-13388	.0000
.6310	5251	6.300E+03	1.160E+12	2.320E-02	1.855E-08	1.9935	-6006	.0000
1.000	5581	7.482E+03	2.565E+12	4.277E-02	2.073E-08	1.9936	0	.0000
1.585	6015	8.408E+03	6.954E+12	9.420E-02	2.160E-08	1.9936	4365	.0000
2.512	6560	9.040E+03	2.096E+13	2.309E-01	2.127E-08	1.9934	7304	.0000
3.981	7275	9.422E+03	7.003E+13	6.588E-01	1.988E-08	1.9927	9161	.0106
6.310	7977	9.630E+03	1.844E+14	1.734E+00	1.828E-08	1.9920	10271	.6189
10.00	8374	9.779E+03	2.956E+14	2.970E+00	1.743E-08	1.9918	11146	.6808
15.85	8633	9.931E+03	3.923E+14	4.203E+00	1.695E-08	1.9916	12096	.8350
25.12	8858	1.010E+04	4.942E+14	5.656E+00	1.658E-08	1.9916	13219	.9005
39.81	9068	1.030E+04	6.064E+14	7.427E+00	1.627E-08	1.9916	14588	.9376
63.10	9273	1.053E+04	7.330E+14	9.630E+00	1.599E-08	1.9917	16282	.9603

4500-DEGREE, LOG G = 4.0

TAU 5000	TEMP	PRESSURE	ELECTRON DENSITY	OPACITY	DENSITY	HYDROGEN IONIZATION	HEIGHT (KM)	CONV/ TOTAL
.0010	3627	2.697E+03	8.348E+10	6.070E-03	1.154E-08	8.151E-10	-955	.0000
.0016	3637	3.575E+03	1.026E+11	7.242E-03	1.526E-08	7.475E-10	-889	.0000
.0025	3649	4.738E+03	1.267E+11	8.676E-03	2.018E-08	7.024E-10	-823	.0000
.0040	3666	6.273E+03	1.581E+11	1.046E-02	2.662E-08	6.937E-10	-757	.0000
.0063	3687	8.282E+03	1.986E+11	1.268E-02	3.500E-08	7.126E-10	-691	.0000
.0100	3706	1.092E+04	2.473E+11	1.530E-02	4.597E-08	7.121E-10	-626	.0000
.0158	3733	1.436E+04	3.137E+11	1.866E-02	6.013E-08	7.732E-10	-561	.0000
.0251	3768	1.881E+04	4.040E+11	2.297E-02	7.814E-08	9.068E-10	-496	.0000
.0398	3809	2.453E+04	5.244E+11	2.840E-02	1.009E-07	1.116E-09	-432	.0000
.0631	3863	3.182E+04	6.962E+11	3.558E-02	1.292E-07	1.533E-09	-368	.0000
.1000	3940	4.094E+04	9.598E+11	4.556E-02	1.629E-07	2.533E-09	-305	.0000
.1585	4039	5.213E+04	1.363E+12	5.924E-02	2.021E-07	4.935E-09	-244	.0000
.2512	4171	6.565E+04	2.013E+12	7.843E-02	2.459E-07	1.208E-08	-183	.0000
.3981	4348	8.173E+04	3.084E+12	1.048E-01	2.928E-07	3.922E-08	-124	.0000
.6310	4584	1.009E+05	4.827E+12	1.384E-01	3.420E-07	1.748E-07	-63	.0000
1.000	4907	1.241E+05	7.706E+12	1.786E-01	3.922E-07	1.171E-06	0	.0000
1.585	5317	1.528E+05	1.281E+13	2.335E-01	4.449E-07	9.483E-06	68	.0000
2.512	5849	1.838E+05	2.905E+13	4.073E-01	4.860E-07	7.187E-05	135	.0016
3.981	6484	2.069E+05	9.339E+13	1.002E+00	4.933E-07	3.660E-04	182	.1926
6.310	6982	2.228E+05	2.226E+14	1.994E+00	4.931E-07	9.707E-04	214	.4612
10.00	7351	2.371E+05	4.030E+14	3.219E+00	4.980E-07	1.801E-03	243	.6340
15.85	7673	2.518E+05	6.521E+14	4.780E+00	5.062E-07	2.915E-03	272	.7700
25.12	7971	2.679E+05	9.907E+14	6.792E+00	5.177E-07	4.372E-03	304	.8522
39.81	8255	2.861E+05	1.446E+15	9.403E+00	5.329E-07	6.239E-03	338	.9052
63.10	8533	3.071E+05	2.053E+15	1.281E+01	5.522E-07	8.587E-03	377	.9408

4500-DEGREE, LOG G = 3.0

TAU 5000	TEMP	PRESSURE	ELECTRON DENSITY	OPACITY	DENSITY	LOG G EFF	HEIGHT (KM)	CONV/ TOTAL
.0010	3628	6.179E+02	3.055E+10	2.595E-03	2.636E-09	2.9995	-9933	.0000
.0016	3642	8.233E+02	3.846E+10	3.093E-03	3.499E-09	2.9995	-9259	.0000
.0025	3653	1.096E+03	4.802E+10	3.689E-03	4.647E-09	2.9995	-8584	.0000
.0040	3667	1.458E+03	6.018E+10	4.423E-03	6.159E-09	2.9995	-7910	.0000
.0063	3686	1.935E+03	7.600E+10	5.338E-03	8.132E-09	2.9995	-7238	.0000
.0100	3710	2.559E+03	9.649E+10	6.475E-03	1.069E-08	2.9996	-6571	.0000
.0158	3739	3.373E+03	1.234E+11	7.897E-03	1.398E-08	2.9996	-5908	.0000
.0251	3774	4.427E+03	1.593E+11	9.688E-03	1.819E-08	2.9996	-5249	.0000
.0398	3820	5.782E+03	2.088E+11	1.200E-02	2.348E-08	2.9996	-4595	.0000
.0631	3878	7.510E+03	2.782E+11	1.498E-02	3.005E-08	2.9996	-3947	.0000
.1000	3951	9.691E+03	3.783E+11	1.888E-02	3.805E-08	2.9996	-3304	.0000
.1585	4048	1.242E+04	5.278E+11	2.404E-02	4.759E-08	2.9996	-2665	.0000
.2512	4191	1.579E+04	7.732E+11	3.108E-02	5.839E-08	2.9996	-2028	.0000
.3981	4356	1.995E+04	1.125E+12	3.954E-02	7.093E-08	2.9996	-1383	.0000
.6310	4618	2.513E+04	1.733E+12	5.018E-02	8.424E-08	2.9996	-715	.0000
1.000	4942	3.169E+04	2.679E+12	6.241E-02	9.920E-08	2.9996	0	.0000
1.585	5362	3.966E+04	4.909E+12	8.946E-02	1.144E-07	2.9996	746	.0000
2.512	5892	4.690E+04	1.382E+13	1.936E-01	1.231E-07	2.9995	1353	.0000
3.981	6555	5.149E+04	5.083E+13	5.405E-01	1.214E-07	2.9995	1726	.0194
6.310	7205	5.417E+04	1.530E+14	1.325E+00	1.159E-07	2.9995	1952	.3704
10.00	7644	5.622E+04	2.934E+14	2.312E+00	1.131E-07	2.9995	2131	.6381
15.85	7972	5.825E+04	4.582E+14	3.453E+00	1.120E-07	2.9995	2312	.7439
25.12	8253	6.048E+04	6.555E+14	4.838E+00	1.119E-07	2.9995	2512	.8456
39.81	8509	6.304E+04	8.939E+14	6.562E+00	1.127E-07	2.9995	2742	.9042
63.10	8751	6.606E+04	1.184E+15	8.744E+00	1.142E-07	2.9995	3011	.9402

4000-DEGREE, LOG G = 4.0

TAU 5000	TEMP	PRESSURE	ELECTRON DENSITY	OPACITY	DENSITY	HYDROGEN IONIZATION	HEIGHT (KM)	CONV/ TOTAL
.0010	3236	3.804E+03	3.901E+10	4.233E-03	1.867E-08	7.653E-12	-797	.0000
.0016	3241	5.066E+03	4.830E+10	5.020E-03	2.501E-08	6.679E-12	-738	.0000
.0025	3249	6.749E+03	6.036E+10	5.987E-03	3.350E-08	6.075E-12	-681	.0000
.0040	3260	8.981E+03	7.580E+10	7.157E-03	4.486E-08	5.658E-12	-623	.0000
.0063	3272	1.194E+04	9.575E+10	8.577E-03	6.002E-08	5.444E-12	-566	.0000
.0100	3287	1.584E+04	1.215E+11	1.029E-02	8.019E-08	5.384E-12	-510	.0000
.0158	3309	2.099E+04	1.560E+11	1.243E-02	1.067E-07	5.776E-12	-455	.0000
.0251	3336	2.771E+04	2.020E+11	1.511E-02	1.410E-07	6.688E-12	-400	.0000
.0398	3360	3.653E+04	2.594E+11	1.820E-02	1.868E-07	7.338E-12	-346	.0000
.0631	3411	4.791E+04	3.488E+11	2.281E-02	2.417E-07	1.131E-11	-293	.0000
.1000	3464	6.226E+04	4.666E+11	2.853E-02	3.095E-07	1.758E-11	-240	.0000
.1585	3524	8.052E+04	6.293E+11	3.597E-02	3.929E-07	2.904E-11	-188	.0000
.2512	3654	1.023E+05	9.575E+11	5.028E-02	4.704E-07	9.848E-11	-138	.0000
.3981	3796	1.268E+05	1.477E+12	7.057E-02	5.496E-07	3.422E-10	-90	.0000
.6310	3993	1.538E+05	2.494E+12	1.046E-01	6.196E-07	1.687E-09	-44	.0000
1.000	4252	1.821E+05	4.517E+12	1.587E-01	6.764E-07	1.139E-08	0	.0000
1.585	4587	2.122E+05	8.250E+12	2.311E-01	7.224E-07	1.051E-07	43	.0039
2.512	5014	2.466E+05	1.429E+13	3.060E-01	7.634E-07	1.299E-06	89	.0200
3.981	5548	2.881E+05	2.612E+13	4.184E-01	8.042E-07	1.703E-05	142	.0765
6.310	6090	3.294E+05	6.173E+13	7.724E-01	8.369E-07	1.045E-04	192	.3599
10.00	6514	3.641E+05	1.334E+14	1.406E+00	8.644E-07	2.877E-04	233	.5519
15.85	6854	3.963E+05	2.447E+14	2.273E+00	8.939E-07	5.626E-04	270	.7333
25.12	7160	4.290E+05	4.114E+14	3.443E+00	9.259E-07	9.573E-04	306	.8296
39.81	7452	4.639E+05	6.561E+14	5.020E+00	9.615E-07	1.510E-03	343	.8904
63.10	7738	5.024E+05	1.010E+15	7.145E+00	1.002E-06	2.268E-03	382	.9316

4000-DEGREE, LOG G = 3.0

TAU 5000	TEMP	PRESSURE	ELECTRON DENSITY	OPACITY	DENSITY	LOG G EFF	HEIGHT (KM)	CONV/ TOTAL
.0010	3244	8.029E+02	1.390E+10	1.855E-03	3.850E-09	2.9998	-8602	.0000
.0016	3249	1.093E+03	1.729E+10	2.175E-03	5.244E-09	2.9998	-7958	.0000
.0025	3258	1.483E+03	2.173E+10	2.578E-03	7.107E-09	2.9998	-7323	.0000
.0040	3268	2.003E+03	2.722E+10	3.066E-03	9.599E-09	2.9998	-6695	.0000
.0063	3281	2.693E+03	3.442E+10	3.676E-03	1.289E-08	2.9998	-6077	.0000
.0100	3296	3.603E+03	4.356E+10	4.422E-03	1.723E-08	2.9998	-5469	.0000
.0158	3314	4.801E+03	5.531E+10	5.339E-03	2.292E-08	2.9998	-4868	.0000
.0251	3341	6.364E+03	7.157E+10	6.524E-03	3.024E-08	2.9998	-4276	.0000
.0398	3377	8.379E+03	9.418E+10	8.059E-03	3.948E-08	2.9998	-3695	.0000
.0631	3417	1.096E+04	1.241E+11	9.974E-03	5.115E-08	2.9998	-3122	.0000
.1000	3483	1.422E+04	1.731E+11	1.280E-02	6.499E-08	2.9998	-2560	.0000
.1585	3575	1.817E+04	2.523E+11	1.689E-02	8.060F-08	2.9998	-2016	.0000
.2512	3671	2.293E+04	3.674E+11	2.228E-02	9.866E-08	2.9998	-1484	.0000
.3981	3840	2.841E+04	6.163E+11	3.199E-02	1.160E-07	2.9998	-973	.0000
.6310	4026	3.448E+04	1.016E+12	4.490E-02	1.336E-07	2.9998	-486	.0000
1.000	4279	4.140E+04	1.739E+12	6.277E-02	1.503E-07	2.9998	0	.0000
1.585	4619	4.940E+04	2.932E+12	8.270E-02	1.657E-07	2.9998	506	.0000
2.512	5055	5.937E+04	4.857E+12	1.031E-01	1.818E-07	2.9998	1079	.0000
3.981	5612	7.107E+04	1.073E+13	1.648E-01	1.959E-07	2.9997	1698	.0005
6.310	6276	8.000E+04	3.851E+13	4.282E-01	1.971E-07	2.9997	2150	.1055
10.00	6825	8.555E+04	1.053E+14	9.447E-01	1.936E-07	2.9997	2435	.3775
15.85	7206	9.024E+04	1.986E+14	1.580E+00	1.932E-07	2.9997	2677	.6512
25.12	7524	9.495E+04	3.242E+14	2.376E+00	1.945E-07	2.9997	2920	.7884
39.81	7809	1.001E+05	4.893E+14	3.385E+00	1.971E-07	2.9997	3182	.8677
63.10	8075	1.058E+05	7.046E+14	4.681E+00	2.012E-07	2.9997	3473	.9170

4000-DEGREE, LOG G = 2.0

TAU 5000	TEMP	PRESSURE	ELECTRON DENSITY	OPACITY	DENSITY	LOG G EFF	HEIGHT (KM)	CONV/ TOTAL
.0010	3235	1.459E+02	4.481E+09	9.210E-04	6.982E-10	1.9978	-93284	.0000
.0016	3241	2.055E+02	5.700E+09	1.042E-03	9.819E-10	1.9978	-86124	.0000
.0025	3252	2.882E+02	7.322E+09	1.197E-03	1.373E-09	1.9977	-79035	.0000
.0040	3262	4.018E+02	9.350E+09	1.388E-03	1.910E-09	1.9977	-72056	.0000
.0063	3276	5.560E+02	1.202E+10	1.630E-03	2.633E-09	1.9978	-65211	.0000
.0100	3292	7.630E+02	1.543E+10	1.930E-03	3.600E-09	1.9978	-58514	.0000
.0158	3310	1.039E+03	1.983E+10	2.303E-03	4.882E-09	1.9978	-51951	.0000
.0251	3337	1.403E+03	2.600E+10	2.793E-03	6.543E-09	1.9979	-45538	.0000
.0398	3374	1.875E+03	3.469E+10	3.435E-03	8.654E-09	1.9979	-39291	.0000
.0631	3414	2.481E+03	4.627E+10	4.243E-03	1.133E-08	1.9979	-33186	.0000
.1000	3481	3.248E+03	6.571E+10	5.443E-03	1.453E-08	1.9979	-27239	.0000
.1585	3573	4.177E+03	9.728E+10	7.162E-03	1.819E-08	1.9979	-21537	.0000
.2512	3670	5.302E+03	1.431E+11	9.395E-03	2.246E-08	1.9980	-15988	.0000
.3981	3841	6.617E+03	2.382E+11	1.313E-02	2.673E-08	1.9979	-10642	.0000
.6310	4028	8.131E+03	3.791E+11	1.758E-02	3.128E-08	1.9979	-5414	.0000
1.000	4284	9.955E+03	6.126E+11	2.293E-02	3.597E-08	1.9978	0	.0000
1.585	4630	1.222E+04	9.790E+11	2.834E-02	4.084E-08	1.9977	5903	.0000
2.512	5072	1.517E+04	1.658E+12	3.559E-02	4.626E-08	1.9975	12670	.0000
3.981	5638	1.817E+04	4.821E+12	7.384E-02	4.984E-08	1.9973	18884	.0000
6.310	6320	2.002E+04	2.011E+13	2.231E-01	4.896E-08	1.9972	22611	.0067
10.00	7028	2.099E+04	7.173E+13	6.300E-01	4.602F-08	1.9971	24636	.3922
15.85	7511	2.163E+04	1.505E+14	1.210E+00	4.421E-08	1.9970	26072	.8575
25.12	7813	2.224E+04	2.295E+14	1.801E+00	4.353E-08	1.9969	27479	.7805
39.81	8066	2.291E+04	3.202E+14	2.509E+00	4.325E-08	1.9969	29054	.8706
63.10	8296	2.369E+04	4.270E+14	3.389E+00	4.324E-08	1.9968	30884	.9205

M O N O C H R O M A T I C   F L U X E S

WAVE	(50000,5)	(40000,5)	(30000,4)	WAVE	(50000,5)	(40000,5)	(30000,4)
.0167	1.22E-08	2.76E-11	7.72E-19	.0649	1.22E-02	3.68E-03	4.32E-05
.0173	2.42E-08	6.82E-11	3.63E-18	.0694	1.27E-02	4.01E-03	5.99E-05
.0180-	4.75E-08	1.68E-10	1.70E-17	.0745	1.31E-02	4.38E-03	8.44E-05
.0180+	5.04E-08	1.68E-10	1.70E-17	.0805	1.35E-02	4.79E-03	1.20E-04
.0186	9.03E-08	3.69E-10	6.48E-17	.0875	1.39E-02	5.22E-03	1.72E-04
.0193-	1.61E-07	8.08E-10	2.47E-16	.0912-	1.40E-02	5.43E-03	2.04E-04
.0193+	1.64E-07	8.09E-10	2.47E-16	.0912+	3.13E-02	2.07E-02	7.51E-03
.0203-	3.59E-07	2.35E-09	1.52E-15	.1005	2.91E-02	1.95E-02	7.40E-03
.0203+	3.64E-07	2.35E-09	1.52E-15	.1119	2.65E-02	1.80E-02	7.12E-03
.0211	6.30E-07	4.95E-09	5.43E-15	.1262	2.35E-02	1.62E-02	6.66E-03
.0219	1.09E-06	1.04E-08	1.94E-14	.1448	2.02E-02	1.42E-02	6.02E-03
.0228-	1.88E-06	2.18E-08	6.83E-14	.1698	1.67E-02	1.19E-02	5.21E-03
.0228+	4.34E-06	5.21E-08	7.27E-14	.2051-	1.29E-02	9.39E-03	4.25E-03
.0237-	6.81E-06	9.73E-08	2.25E-13	.2051+	1.34E-02	9.68E-03	4.28E-03
.0237+	1.54E-05	1.17E-07	2.26E-13	.2601	9.42E-03	6.94E-03	3.19E-03
.0251-	2.39E-05	2.89E-07	1.20E-12	.3122	7.08E-03	5.27E-03	2.50E-03
.0251+	3.21E-05	2.96E-07	1.20E-12	.3422-	6.11E-03	4.57E-03	2.21E-03
.0261-	4.16E-05	5.26E-07	3.48E-12	.3422+	6.11E-03	4.59E-03	2.23E-03
.0261+	1.40E-03	7.81E-07	3.53E-12	.3647-	5.50E-03	4.14E-03	2.04E-03
.0280	1.42E-03	1.88E-06	2.11E-11	.3647+	5.74E-03	4.47E-03	2.47E-03
.0302-	1.42E-03	4.55E-06	1.24E-10	.3680	5.66E-03	4.41E-03	2.44E-03
.0302+	4.84E-03	1.19E-04	1.36E-10	.5699-	2.63E-03	2.07E-03	1.19E-03
.0326	5.37E-03	1.21E-04	6.93E-10	.5699+	2.64E-03	2.08E-03	1.19E-03
.0354-	5.94E-03	1.38E-04	3.47E-09	.6634	2.01E-03	1.58E-03	9.17E-04
.0354+	6.83E-03	3.71E-04	4.47E-09	.7437	1.63E-03	1.28E-03	7.50E-04
.0384	7.48E-03	3.95E-04	1.78E-08	.7846	1.47E-03	1.16E-03	6.82E-04
.0419-	8.17E-03	4.57E-04	7.00E-08	.8193	1.36E-03	1.07E-03	6.32E-04
.0419+	8.17E-03	4.61E-04	7.09E-08	.8206-	1.36E-03	1.07E-03	6.30E-04
.0460	8.90E-03	5.86E-04	2.65E-07	.8206+	1.37E-03	1.08E-03	6.52E-04
.0504-	9.61E-03	7.82E-04	8.41E-07	1.1169	7.66E-04	6.08E-04	3.70E-04
.0504+	1.02E-02	2.64E-03	1.45E-05	1.4588-	4.60E-04	3.65E-04	2.24E-04
.0537	1.08E-02	2.87E-03	1.81E-05	1.4588+	4.61E-04	3.67E-04	2.26E-04
.0575-	1.13E-02	3.14E-03	2.41E-05	1.8463	2.93E-04	2.33E-04	1.44E-04
.0575+	1.13E-02	3.14E-03	2.43E-05	2.2794-	1.95E-04	1.55E-04	9.56E-05
.0610	1.17E-02	3.39E-03	3.19E-05	2.2794+	1.95E-04	1.55E-04	9.59E-05

MONOCHROMATIC FLUX IS IN ERGS/CM**2/SEC/STER/HZ (WITHOUT PI).

KEY TO THE DISCONTINUITIES:

.0180- N IV, O IV
.0193- C IV, NE III
.0203- N IV, O IV, NE III
.0228- HE II, O III
.0237- O III
.0251- O III
.0261- N III, O III
.0302- NE II, N III
.0354- C III, O II, N III
.0419- N II
.0504- HE I
.0575- NE I
.0912- H I, HE II
.2051- HE II
.3422- HE I
.3647- H I, HE II
.5699- HE II
.8206- H I, HE II
1.4588- H I, HE I, HE II
2.2794- H I, HE I, HE II

M O N O C H R O M A T I C   F L U X E S

MODEL	.0504	.0545	.0593	.0649	.0701	.0804
(25000,4)	1.15E-07	2.67E-07	6.32E-07	1.51E-06	2.97E-06	8.64E-06
(20000,4)	1.34E-09	4.66E-09	1.61E-08	5.47E-08	1.40E-07	6.09E-07
(20000,3)	1.73E-09	5.86E-09	1.97E-08	6.57E-08	1.66E-07	7.05E-07
(18000,4)	1.80E-10	7.35E-10	2.96E-09	1.17E-08	3.36E-08	1.75E-07
(18000,3)	2.38E-10	9.58E-10	3.79E-09	1.48E-08	4.20E-08	2.15E-07
(16000,4)	1.81E-11	8.76E-11	4.17E-10	1.94E-09	6.33E-09	4.01E-08
(16000,3)	2.34E-11	1.12E-10	5.30E-10	2.46E-09	7.97E-09	5.02E-08
(14000,4)	8.71E-13	4.91E-12	2.69E-11	1.44E-10	5.22E-10	4.01E-09
(14000,3)	1.38E-12	8.16E-12	4.73E-11	2.69E-10	1.02E-09	8.25E-09
(12000,4)	2.19E-15	1.50E-14	1.04E-13	7.62E-13	3.76E-12	5.36E-11
(12000,3)	1.99E-14	1.43E-13	1.00E-12	6.87E-12	3.05E-11	3.30E-10
(12000,2)	4.69E-14	3.58E-13	2.69E-12	1.98E-11	9.12E-11	1.00E-09
(11000,4)	7.48E-17	6.38E-16	5.74E-15	5.66E-14	3.61E-13	7.65E-12
(11000,3)	2.01E-16	1.63E-15	1.32E-14	1.11E-13	6.11E-13	1.03E-11
(11000,2)	3.15E-15	2.84E-14	2.49E-13	2.13E-12	1.10E-11	1.47E-10

MODEL	.0912-	.0912+	.0940B	.0950B	.0955B	.0965B
(25000,4)	1.91E-05	1.91E-05	1.26E-03	5.80E-05	3.12E-03	3.54E-03
(20000,4)	1.86E-06	1.86E-06	8.11E-05	6.79E-06	5.83E-04	8.85E-04
(20000,3)	2.11E-06	2.11E-06	4.27E-04	9.77E-06	8.70E-04	9.53E-04
(18000,4)	6.17E-07	6.17E-07	1.52E-05	2.42E-06	1.83E-04	3.56E-04
(18000,3)	7.45E-07	7.45E-07	1.14E-04	3.20E-06	4.10E-04	4.92E-04
(16000,4)	1.64E-07	1.64E-07	2.21E-06	7.34E-07	3.13E-05	8.92E-05
(16000,3)	2.07E-07	2.07E-07	1.48E-06	8.93E-07	1.34E-04	1.94E-04
(14000,4)	1.98E-08	1.98E-08	2.97E-07	1.70E-07	1.95E-06	8.43E-06
(14000,3)	4.11E-08	4.11E-08	8.20E-07	1.93E-07	1.86E-05	4.07E-05
(12000,4)	4.99E-10	4.99E-10	2.96E-08	1.97E-08	6.42E-08	1.32E-07
(12000,3)	2.20E-09	2.20E-09	4.07E-08	2.52E-08	2.73E-07	1.26E-06
(12000,2)	6.46E-09	6.46E-09	2.53E-07	3.27E-08	4.42E-06	7.03E-06
(11000,4)	9.66E-11	9.66E-11	7.17E-09	4.00E-09	1.19E-08	1.61E-08
(11000,3)	1.12E-10	1.12E-10	8.34E-09	6.55E-09	1.75E-08	3.65E-08
(11000,2)	1.10E-09	1.10E-09	1.62E-08	8.09E-09	2.25E-07	5.47E-07

MODEL	.0975B	.1000B	.1015B	.1030B	.1050B	.1100-
(25000,4)	1.56E-03	3.86E-03	3.71E-03	2.71E-03	3.80E-03	3.73E-03
(20000,4)	1.22E-04	1.28E-03	1.12E-03	4.25E-04	1.30E-03	1.32E-03
(20000,3)	5.28E-04	1.04E-03	1.03E-03	8.50E-04	1.08E-03	1.11E-03
(18000,4)	2.45E-05	6.81E-04	5.39E-04	1.19E-04	7.12E-04	7.43E-04
(18000,3)	1.61E-04	5.82E-04	5.64E-04	3.75E-04	6.12E-04	6.40E-04
(16000,4)	3.62E-06	2.88E-04	1.87E-04	1.86E-05	3.18E-04	3.50E-04
(16000,3)	2.48E-05	2.71E-04	2.49E-04	1.06E-04	2.92E-04	3.12E-04
(14000,4)	4.83E-07	7.45E-05	3.14E-05	1.66E-06	9.18E-05	1.14E-04
(14000,3)	1.50E-06	8.51E-05	6.84E-05	1.13E-05	9.59E-05	1.07E-04
(12000,4)	5.01E-08	4.11E-06	7.21E-07	1.28E-07	6.69E-06	1.13E-05
(12000,3)	7.08E-08	7.99E-06	4.25E-06	2.53E-07	1.00E-05	1.26E-05
(12000,2)	5.20E-07	1.11E-05	1.04E-05	3.69E-06	1.33E-05	1.57E-05
(11000,4)	1.25E-08	9.01E-08	3.88E-08	3.08E-08	1.95E-07	4.50E-07
(11000,3)	1.48E-08	4.56E-07	1.66E-07	3.94E-08	6.61E-07	9.65E-07
(11000,2)	3.02E-08	1.28E-06	1.06E-06	1.71E-07	1.62E-06	2.03E-06

MONOCHROMATIC FLUX IS IN ERGS/CM**2/SEC/STER/HZ (WITHOUT PI).

MONOCHROMATIC FLUXES

MODEL	.1100+	.1190B	.1220B	.1239-	.1239+
(25000,4)	3.74E-03	3.56E-03	2.69E-03	3.47E-03	3.47E-03
(20000,4)	1.37E-03	1.32E-03	5.35E-04	1.31E-03	1.32E-03
(20000,3)	1.13E-03	1.17E-03	9.76E-04	1.18E-03	1.19E-03
(18000,4)	8.22E-04	7.96E-04	1.80E-04	7.89E-04	8.06E-04
(18000,3)	6.78E-04	7.10E-04	4.80E-04	7.24E-04	7.32E-04
(16000,4)	4.46E-04	4.27E-04	3.91E-05	4.26E-04	4.45E-04
(16000,3)	3.70E-04	3.92E-04	1.64E-04	4.03E-04	4.16E-04
(14000,4)	2.07E-04	1.87E-04	5.90E-06	1.86E-04	2.05E-04
(14000,3)	1.74E-04	1.85E-04	2.59E-05	1.91E-04	2.08E-04
(12000,4)	6.86E-05	4.82E-05	7.71E-07	4.60E-05	5.94E-05
(12000,3)	6.07E-05	6.09E-05	1.40E-06	6.28E-05	7.96E-05
(12000,2)	4.68E-05	5.58E-05	1.49E-05	6.09E-05	7.21E-05
(11000,4)	2.62E-05	1.19E-05	2.41E-07	1.05E-05	1.78E-05
(11000,3)	2.66E-05	2.37E-05	3.15E-07	2.39E-05	3.81E-05
(11000,2)	2.21E-05	2.61E-05	1.62E-06	2.86E-05	4.06E-05

MODEL	.0912+	.1100-		.1100+		.1239-		.1239+	
(10000,4.5)	1.05E-09	1.83E-08		1.06E-06		1.39E-06		2.51E-05	
(10000,4)	1.04E-09	1.81E-08		2.09E-06		2.73E-06		2.56E-05	
(10000,4,10X)	1.62E-09	2.59E-08		2.86E-06		2.88E-06		1.32E-05	
(10000,4,.1X)	8.94E-10	1.60E-08		1.71E-06		2.31E-06		2.80E-05	
(10000,3.5)	1.05E-09	1.81E-08		3.41E-06		4.41E-06		2.57E-05	
(10000,3)	1.11E-09	1.87E-08		4.74E-06		6.30E-06		2.51E-05	
(10000,2)	3.10E-09	2.96E-08		5.93E-06		8.91E-06		2.06E-05	
(9500,4)	3.58E-10	7.51E-09		2.59E-08		7.69E-08		1.36E-05	
(9500,3)	3.28E-10	6.98E-09		3.25E-07		5.19E-07		1.48E-05	
(9000,4)	1.35E-10	3.34E-09		4.45E-09		2.35E-08		4.54E-06	
(9000,3)	1.07E-10	2.77E-09		4.19E-09		2.19E-08		6.29E-06	
(9000,2)	1.02E-10	2.66E-09		1.95E-08		4.99E-08		6.90E-06	
(8500,4)	2.52E-11	8.36E-10	0	1.00E-09	0	6.27E-09	0	4.79E-07	9
(8500,3)	2.22E-11	7.53E-10	0	9.26E-10	1	5.85E-09	1	1.15E-06	9
(8000,4.5)	5.36E-12	2.32E-10	0	2.57E-10	0	1.87E-09	0	3.92E-09	2
(8000,4)	5.05E-12	2.21E-10	0	2.44E-10	0	1.78E-09	0	4.41E-09	4
(8000,4) *	4.80E-12	2.11E-10	0	2.24E-10	0	1.65E-09	0	4.50E-09	4
(8000,3.5)	5.06E-12	2.21E-10	0	2.45E-10	0	1.79E-09	0	5.54E-09	5
(8000,3)	4.30E-12	1.93E-10	0	2.18E-10	0	1.62E-09	0	7.02E-09	7
(8000,2)	2.54E-12	1.25E-10	0	1.69E-10	0	1.30E-09	0	2.58E-08	10
(7500,4)	9.84E-13	5.69E-11	0	6.11E-11	0	5.23E-10	0	7.61E-10	1
(7500,4) +	9.89E-13	5.71E-11	0	6.03E-11	0	5.17E-10	0	7.46E-10	1
(7500,4) *	9.64E-13	5.59E-11	0	6.02E-11	0	5.17E-10	0	7.54E-10	1
(7500,3)	9.22E-13	5.39E-11	0	5.73E-11	0	4.95E-10	0	7.48E-10	1
(7000,4)	1.54E-13	1.22E-11	0	1.29E-11	0	1.31E-10	0	1.63E-10	0
(7000,4) *	1.57E-13	1.24E-11	0	1.20E-11	0	1.23E-10	0	1.56E-10	0
(7000,3)	1.67E-13	1.31E-11	0	1.36E-11	0	1.38E-10	0	1.75E-10	0
(7000,2)	2.03E-13	1.54E-11	0	1.52E-11	0	1.52E-10	0	1.87E-10	0
(6500,4)	2.25E-14	2.48E-12	0	2.60E-12	0	3.17E-11	0	3.60E-11	0
(6500,3)	2.18E-14	2.42E-12	0	2.49E-12	0	3.06E-11	0	3.41E-11	0
(6000,4.5)	3.91E-15	5.83E-13	0	6.11E-13	0	8.77E-12	0	9.23E-12	1
(6000,4)	4.04E-15	5.98E-13	0	6.22E-13	0	8.90E-12	0	9.49E-12	0
(6000,4) *	4.55E-15	6.60E-13	0	6.58E-13	0	9.37E-12	0	9.49E-12	0
(6000,4) **	3.27E-15	5.02E-13		5.22E-13		7.62E-12		8.02E-12	
(6000,4) * **	3.90E-15	5.81E-13		5.78E-13		8.36E-12		8.36E-12	
(6000,4,10X)	5.52E-15	7.75E-13		7.75E-13		1.08E-11		1.08E-11	
(6000,4,1/10X)	3.23E-15	4.97E-13		4.93E-13		7.27E-12		7.07E-12	
(6000,3.5)	4.09E-15	6.04E-13	0	6.13E-13	0	8.81E-12	0	9.48E-12	1
(6000,3)	3.81E-15	5.70E-13	0	5.77E-13	0	8.34E-12	0	9.04E-12	1
(6000,2)	5.13E-15	7.29E-13	0	7.21E-13	0	1.02E-11	0	1.01E-11	1
(5500,4)	2.54E-16	6.03E-14	0	6.23E-14	0	1.15E-12	0	1.17E-12	2
(5500,3)	2.44E-16	5.85E-14	0	5.91E-14	0	1.10E-12	0	1.15E-12	0
(5000,4)	5.00E-18	2.33E-15	0	2.37E-15	0	6.28E-14	0	6.29E-14	0
(5000,3)	5.12E-18	2.38E-15	0	2.39E-15	0	6.37E-14	0	6.38E-14	0
(5000,2)	4.55E-18	2.15E-15	0	2.16E-15	0	5.83E-14	0	5.92E-14	6
(4500,4)	4.42E-20	4.62E-17	0	4.64E-17	0	1.92E-15	0	1.92E-15	0
(4500,3)	4.31E-20	4.52E-17	0	4.51E-17	0	1.88E-15	0	1.88E-15	0
(4000,4)	2.67E-22	6.69E-19	0	6.66E-19	0	4.47E-17	0	4.47E-17	0
(4000,3)	6.09E-22	1.33E-18	0	1.33E-18	0	8.23E-17	0	8.23E-17	0
(4000,2)	5.22E-22	1.17E-18	0	1.17E-18	0	7.35E-17	0	7.35E-17	0

MONOCHROMATIC FLUXES

MODEL	.1444-		.1444+		.1527-		.1527+		.1623-	
(25000,4)	3.13E-03		3.13E-03		3.00E-03		3.00E-03		2.85E-03	
(20000,4)	1.27E-03		1.27E-03		1.24E-03		1.24E-03		1.21E-03	
(20000,3)	1.22E-03		1.22E-03		1.21E-03		1.21E-03		1.20E-03	
(18000,4)	8.16E-04		8.17E-04		8.06E-04		8.06E-04		7.92E-04	
(18000,3)	7.74E-04		7.74E-04		7.79E-04		7.79E-04		7.81E-04	
(16000,4)	4.86E-04		4.87E-04		4.87E-04		4.87E-04		4.85E-04	
(16000,3)	4.58E-04		4.59E-04		4.67E-04		4.67E-04		4.74E-04	
(14000,4)	2.60E-04		2.61E-04		2.64E-04		2.65E-04		2.68E-04	
(14000,3)	2.44E-04		2.45E-04		2.52E-04		2.52E-04		2.59E-04	
(12000,4)	1.16E-04		1.18E-04		1.21E-04		1.23E-04		1.25E-04	
(12000,3)	1.08E-04		1.09E-04		1.14E-04		1.15E-04		1.19E-04	
(12000,2)	9.69E-05		9.75E-05		1.06E-04		1.06E-04		1.14E-04	
(11000,4)	6.87E-05		7.02E-05		7.27E-05		7.56E-05		7.76E-05	
(11000,3)	6.37E-05		6.49E-05		6.79E-05		6.96E-05		7.25E-05	
(11000,2)	5.76E-05		5.83E-05		6.34E-05		6.42E-05		6.94E-05	
(10000,4.5)	2.73E-05		3.00E-05		3.15E-05		3.82E-05		3.98E-05	
(10000,4)	2.86E-05		3.08E-05		3.23E-05		3.71E-05		3.86E-05	
(10000,4,10X)	1.18E-05		1.24E-05		1.36E-05		3.41E-05		3.51E-05	
(10000,4,.1X)	3.26E-05		3.54E-05		3.69E-05		3.75E-05		3.92E-05	
(10000,3.5)	2.94E-05		3.10E-05		3.26E-05		3.61E-05		3.77E-05	
(10000,3)	2.96E-05		3.09E-05		3.26E-05		3.53E-05		3.71E-05	
(10000,2)	2.80E-05		2.87E-05		3.16E-05		3.31E-05		3.61E-05	
(9500,4)	1.45E-05		1.72E-05		1.81E-05		2.57E-05		2.67E-05	
(9500,3)	1.74E-05		1.92E-05		2.03E-05		2.42E-05		2.54E-05	
(9000,4)	4.25E-06		6.18E-06		6.63E-06		1.61E-05		1.67E-05	
(9000,3)	6.91E-06		8.86E-06		9.52E-06		1.53E-05		1.61E-05	
(9000,2)	9.24E-06		1.05E-05		1.15E-05		1.44E-05		1.56E-05	
(8500,4)	2.74E-07	6	7.04E-07	8	8.21E-07	7	8.78E-06	9	8.90E-06	8
(8500,3)	9.97E-07	8	1.94E-06	9	2.13E-06	8	8.85E-06	9	9.23E-06	8
(8000,4.5)	2.51E-08	1	2.65E-08	1	5.15E-08	1	1.77E-06	11	1.85E-06	10
(8000,4)	2.53E-08	1	2.75E-08	2	5.31E-08	1	2.23E-06	11	2.25E-06	10
(8000,4) *	2.48E-08	1	2.71E-08	2	5.22E-08	1	2.71E-06	11	2.55E-06	11
(8000,3.5)	2.63E-08	1	3.05E-08	2	5.75E-08	2	2.86E-06	11	2.77E-06	11
(8000,3)	2.65E-08	2	3.47E-08	3	6.27E-08	3	3.01E-06	11	3.01E-06	11
(8000,2)	4.23E-08	5	1.13E-07	9	1.63E-07	8	3.29E-06	11	3.50E-06	10
(7500,4)	7.03E-09	0	7.21E-09	0	1.51E-08	0	2.37E-07	15	2.06E-07	12
(7500,4) +	6.92E-09	0	7.10E-09	0	1.49E-08	0	1.88E-07	14	2.00E-07	12
(7500,4) *	6.98E-09	0	7.16E-09	0	1.50E-08	0	2.26E-07	15	1.91E-07	12
(7500,3)	7.08E-09	1	7.45E-09	1	1.57E-08	1	3.67E-07	17	3.54E-07	16
(7000,4)	1.93E-09	0	1.94E-09	0	4.39E-09	0	1.02E-08	7	1.89E-08	5
(7000,4) *	1.83E-09	0	1.85E-09	0	4.19E-09	0	1.06E-08	9	1.89E-08	5
(7000,3)	2.09E-09	0	2.13E-09	0	4.78E-09	0	1.33E-08	11	2.29E-08	7
(7000,2)	2.29E-09	0	2.37E-09	0	5.32E-09	0	1.76E-08	18	2.78E-08	12
(6500,4)	5.38E-10	0	5.39E-10	0	1.31E-09	0	1.99E-09	3	4.53E-09	2
(6500,3)	5.36E-10	0	5.39E-10	0	1.31E-09	0	2.04E-09	2	4.67E-09	2
(6000,4.5)	1.70E-10	0	1.70E-10	0	4.38E-10	0	5.41E-10	2	1.37E-09	2
(6000,4)	1.74E-10	0	1.74E-10	0	4.49E-10	0	5.38E-10	1	1.36E-09	0
(6000,4) *	1.76E-10	0	1.76E-10	0	4.54E-10	0	5.19E-10	0	1.32E-09	0
(6000,4) **	1.51E-10		1.51E-10		3.93E-10		4.27E-10		1.11E-09	
(6000,4) * **	1.59E-10		1.59E-10		4.11E-10		4.20E-10		1.10E-09	
(6000,4,10X)	1.97E-10		1.97E-10		5.04E-10		5.12E-10		1.32E-09	
(6000,4,1/10X)	1.40E-10		1.40E-10		3.67E-10		4.81E-10		1.17E-09	
(6000,3.5)	1.74E-10	0	1.74E-10	0	4.49E-10	0	5.41E-10	0	1.38E-09	0
(6000,3)	1.68E-10	0	1.68E-10	0	4.34E-10	0	5.18E-10	0	1.33E-09	0
(6000,2)	1.90E-10	0	1.90E-10	1	4.88E-10	1	5.59E-10	1	1.45E-09	1
(5500,4)	2.90E-11	0	2.90E-11	1	8.24E-11	0	9.47E-11	2	2.68E-10	2
(5500,3)	2.85E-11	0	2.85E-11	0	8.10E-11	0	9.09E-11	0	2.59E-10	0
(5000,4)	2.37E-12	0	2.37E-12	0	7.71E-12	0	8.49E-12	5	2.78E-11	2
(5000,3)	2.40E-12	0	2.40E-12	0	7.81E-12	0	8.55E-12	1	2.80E-11	1
(5000,2)	2.25E-12	0	2.25E-12	0	7.33E-12	0	8.14E-12	1	2.68E-11	1
(4500,4)	1.19E-13	0	1.19E-13	0	4.57E-13	0	4.90E-13	2	1.90E-12	2
(4500,3)	1.17E-13	0	1.17E-13	0	4.50E-13	0	4.70E-13	4	1.83E-12	4
(4000,4)	4.74E-15	0	4.74E-15	0	2.17E-14	0	2.13E-14	2	1.00E-13	1
(4000,3)	7.99E-15	0	7.99E-15	0	3.55E-14	0	3.26E-14	12	1.51E-13	13
(4000,2)	7.25E-15	0	7.25E-15	0	3.23E-14	0	3.04E-14	17	1.42E-13	15

M O N O C H R O M A T I C   F L U X E S

MODEL	.1623+	.1683-	.1683+	.1988-	.1988+
(25000,4)	2.85E-03	2.77E-03	2.77E-03	2.38E-03	2.38E-03
(20000,4)	1.21E-03	1.19E-03	1.19E-03	1.09E-03	1.09E-03
(20000,3)	1.20E-03	1.19E-03	1.19E-03	1.12E-03	1.12E-03
(18000,4)	7.92E-04	7.84E-04	7.84E-04	7.39E-04	7.39E-04
(18000,3)	7.81E-04	7.80E-04	7.80E-04	7.58E-04	7.58E-04
(16000,4)	4.85E-04	4.83E-04	4.83E-04	4.71E-04	4.71E-04
(16000,3)	4.74E-04	4.76E-04	4.76E-04	4.79E-04	4.79E-04
(14000,4)	2.68E-04	2.69E-04	2.69E-04	·2.72E-04	2.72E-04
(14000,3)	2.59E-04	2.62E-04	2.62E-04	2.74E-04	2.74E-04
(12000,4)	1.25E-04	1.27E-04	1.27E-04	1.34E-04	1.34E-04
(12000,3)	1.19E-04	1.21E-04	1.22E-04	1.32E-04	1.32E-04
(12000,2)	1.14E-04	1.19E-04	1.19E-04	1.38E-04	1.38E-04
(11000,4)	7.76E-05	7.88E-05	7.95E-05	8.54E-05	8.55E-05
(11000,3)	7.25E-05	7.42E-05	7.46E-05	8.22E-05	8.22E-05
(11000,2)	6.94E-05	7.24E-05	7.26E-05	8.55E-05	8.55E-05
(10000,4.5)	3.98E-05	4.07E-05	4.23E-05	4.77E-05	4.78E-05
(10000,4)	3.87E-05	3.96E-05	4.07E-05	4.58E-05	4.59E-05
(10000,4,10X)	3.51E-05	3.57E-05	4.36E-05	4.95E-05	4.99E-05
(10000,4,.1X)	3.92E-05	4.01E-05	4.03E-05	4.52E-05	4.52E-05
(10000,3.5)	3.78E-05	3.87E-05	3.95E-05	4.45E-05	4.46E-05
(10000,3)	3.71E-05	3.81E-05	3.87E-05	4.39E-05	4.39E-05
(10000,2)	3.61E-05	3.78E-05	3.81E-05	4.59E-05	4.59E-05
(9500,4)	2.68E-05	2.73E-05	2.93E-05	3.34E-05	3.35E-05
(9500,3)	2.54E-05	2.61E-05	2.70E-05	3.10E-05	3.10E-05
(9000,4)	1.68E-05	1.71E-05	2.03E-05	2.38E-05	2.40E-05
(9000,3)	1.61E-05	1.66E-05	1.81E-05	2.10E-05	2.11E-05
(9000,2)	1.56E-05	1.63E-05	1.69E-05	2.06E-05	2.06E-05
(8500,4)	8.98E-06 8	9.04E-06 8	1.43E-05 8	1.74E-05 8	1.77E-05 8
(8500,3)	9.28E-06 8	9.45E-06 8	1.22E-05 8	1.48E-05 7	1.49E-05 7
(8000,4.5)	1.91E-06 10	1.93E-06 10	8.15E-06 11	1.24E-05 10	1.31E-05 10
(8000,4)	2.31E-06 10	2.32E-06 10	8.14E-06 11	1.15E-05 10	1.20E-05 10
(8000,4) *	2.62E-06 11	2.51E-06 11	9.07E-06 11	1.16E-05 10	1.21E-05 10
(8000,3.5)	2.83E-06 11	2.76E-06 11	8.09E-06 11	1.04E-05 10	1.08E-05 10
(8000,3)	3.06E-06 11	3.03E-06 10	7.30E-06 11	9.41E-06 10	9.68E-06 10
(8000,2)	3.54E-06 10	3.64E-06 10	5.99E-06 10	7.88E-06 9	8.01E-06 9
(7500,4)	2.23E-07 12	2.02E-07 10	4.01E-06 15	6.61E-06 14	7.36E-06 14
(7500,4) +	2.13E-07 13	2.00E-07 10	3.52E-06 15	6.38E-06 14	7.13E-06 14
(7500,4) *	2.03E-07 12	1.96E-07 10	4.22E-06 15	5.83E-06 14	6.49E-06 14
(7500,3)	3.72E-07 16	3.48E-07 14	3.70E-06 17	5.58E-06 16	6.04E-06 16
(7000,4)	1.92E-08 5	2.78E-08 4	1.85E-06 28	3.32E-06 26	4.18E-06 27
(7000,4) *	1.93E-08 5	2.79E-08 4	2.46E-06 28	3.46E-06 26	4.38E-06 27
(7000,3)	2.34E-08 7	3.30E-08 6	1.59E-06 33	2.90E-06 30	3.47E-06 31
(7000,2)	2.86E-08 12	3.89E-08 9	1.16E-06 36	2.19E-06 31	2.50E-06 32
(6500,4)	4.56E-09 2	7.26E-09 2	5.69E-07 43	1.09E-06 40	1.75E-06 42
(6500,3)	4.71E-09 2	7.49E-09 2	3.20E-07 48	8.95E-07 45	1.30E-06 46
(6000,4.5)	1.37E-09 2	2.31E-09 2	1.13E-07 59	2.03E-07 50	4.92E-07 55
(6000,4)	1.37E-09 0	2.31E-09 0	9.92E-08 60	1.86E-07 50	4.67E-07 55
(6000,4) *	1.33E-09 0	2.25E-09 0	1.16E-07 63	1.83E-07 49	4.99E-07 56
(6000,4) **	1.11E-09	1.91E-09	2.81E-08	5.31E-08	2.01E-07
(6000,4) * **	1.10E-09	1.89E-09	3.18E-08	5.42E-08	2.07E-07
(6000,4,10X)	1.32E-09	2.26E-09	2.82E-09	2.26E-08	2.47E-08
(6000,4,1/10X)	1.18E-09	1.98E-09	3.46E-07	8.81E-07	1.09E-06
(6000,3.5)	1.38E-09 0	2.33E-09 0	9.90E-08 71	1.97E-07 55	4.94E-07 62
(6000,3)	1.33E-09 0	2.26E-09 0	3.41E-08 68	1.33E-07 52	3.29E-07 61
(6000,2)	1.45E-09 1	2.47E-09 1	3.42E-08 69	1.48E-07 55	3.09E-07 65
(5500,4)	2.68E-10 2	4.82E-10 2	3.12E-09 47	1.24E-08 19	3.13E-08 52
(5500,3)	2.60E-10 0	4.69E-10 0	1.98E-09 43	1.14E-08 18	2.97E-08 51
(5000,4)	2.79E-11 2	5.44E-11 2	1.20E-10 17	1.36E-09 10	1.50E-09 12
(5000,3)	2.81E-11 1	5.49E-11 1	1.07E-10 9	1.33E-09 6	1.57E-09 11
(5000,2)	2.68E-11 1	5.26E-11 1	9.00E-11 0	1.27E-09 4	1.56E-09 8
(4500,4)	1.92E-12 1	4.13E-12 2	5.42E-12 2	1.12E-10 1	1.15E-10 1
(4500,3)	1.84E-12 3	3.96E-12 3	5.38E-12 2	1.13E-10 1	1.20E-10 1
(4000,4)	1.01E-13 1	2.41E-13 1	2.57E-13 2	8.88E-12 2	1.19E-11 2
(4000,3)	1.47E-13 6	3.51E-13 6	3.04E-13 2	1.05E-11 2	1.39E-11 2
(4000,2)	1.39E-13 15	3.32E-13 13	2.78E-13 2	9.74E-12 2	1.18E-11 2

MONOCHROMATIC   FLUXES

MODEL	.2138		.2312		.2517−		.2517+		.2807	
(25000,4)	2.22E−03		2.05E−03		1.88E−03		1.88E−03		1.67E−03	
(20000,4)	1.04E−03		9.91E−04		9.38E−04		9.38E−04		8.71E−04	
(20000,3)	1.09E−03		1.04E−03		9.87E−04		9.87E−04		9.17E−04	
(18000,4)	7.18E−04		6.95E−04		6.68E−04		6.68E−04		6.33E−04	
(18000,3)	7.42E−04		7.22E−04		6.97E−04		6.97E−04		6.62E−04	
(16000,4)	4.65E−04		4.57E−04		4.48E−04		4.48E−04		4.35E−04	
(16000,3)	4.77E−04		4.72E−04		4.64E−04		4.64E−04		4.52E−04	
(14000,4)	2.73E−04		2.74E−04		2.75E−04		2.75E−04		2.76E−04	
(14000,3)	2.77E−04		2.80E−04		2.82E−04		2.82E−04		2.83E−04	
(12000,4)	1.38E−04		1.42E−04		1.46E−04		1.46E−04		1.53E−04	
(12000,3)	1.36E−04		1.41E−04		1.46E−04		1.46E−04		1.53E−04	
(12000,2)	1.45E−04		1.52E−04		1.59E−04		1.59E−04		1.67E−04	
(11000,4)	8.86E−05		9.24E−05		9.72E−05		9.72E−05		1.04E−04	
(11000,3)	8.58E−05		9.00E−05		9.51E−05		9.52E−05		1.02E−04	
(11000,2)	9.07E−05		9.63E−05		1.02E−04		1.02E−04		1.10E−04	
(10000,4.5)	5.07E−05		5.42E−05		5.87E−05		5.89E−05		6.56E−05	
(10000,4)	4.87E−05		5.22E−05		5.67E−05		5.67E−05		6.35E−05	
(10000,4,10X)	5.31E−05		5.70E−05		6.19E−05		6.20E−05		6.94E−05	
(10000,4,.1X)	4.79E−05		5.13E−05		5.56E−05		5.56E−05		6.22E−05	
(10000,3.5)	4.73E−05		5.08E−05		5.52E−05		5.53E−05		6.19E−05	
(10000,3)	4.67E−05		5.02E−05		5.46E−05		5.46E−05		6.12E−05	
(10000,2)	4.94E−05		5.34E−05		5.82E−05		5.82E−05		6.51E−05	
(9500,4)	3.59E−05		3.88E−05		4.26E−05		4.28E−05		4.88E−05	
(9500,3)	3.33E−05		3.62E−05		3.99E−05		4.00E−05		4.58E−05	
(9000,4)	2.62E−05		2.88E−05		3.21E−05		3.24E−05		3.78E−05	
(9000,3)	2.30E−05		2.54E−05		2.86E−05		2.87E−05		3.39E−05	
(9000,2)	2.25E−05		2.49E−05		2.80E−05		2.81E−05		3.31E−05	
(8500,4)	1.99E−05	7	2.21E−05	6	2.48E−05	6	2.54E−05	6	3.02E−05	5
(8500,3)	1.67E−05	7	1.87E−05	6	2.13E−05	5	2.16E−05	5	2.60E−05	4
(8000,4.5)	1.62E−05	10	1.85E−05	9	2.04E−05	8	2.21E−05	8	2.64E−05	7
(8000,4)	1.45E−05	10	1.65E−05	9	1.85E−05	8	1.96E−05	8	2.37E−05	7
(8000,4) *	1.44E−05	10	1.62E−05	9	1.81E−05	8	1.93E−05	8	2.34E−05	7
(8000,3.5)	1.27E−05	9	1.45E−05	9	1.64E−05	8	1.73E−05	8	2.11E−05	7
(8000,3)	1.13E−05	9	1.29E−05	8	1.49E−05	7	1.54E−05	7	1.91E−05	6
(8000,2)	9.27E−06	9	1.08E−05	8	1.27E−05	7	1.29E−05	7	1.63E−05	5
(7500,4)	9.98E−06	14	1.16E−05	13	1.28E−05	12	1.48E−05	12	1.85E−05	11
(7500,4) +	9.97E−06	14	1.18E−05	13	1.30E−05	12	1.52E−05	12	1.88E−05	11
(7500,4) *	8.57E−06	14	9.93E−06	13	1.10E−05	12	1.27E−05	12	1.62E−05	11
(7500,3)	7.78E−06	15	9.16E−06	14	1.05E−05	12	1.15E−05	13	1.47E−05	11
(7000,4)	6.79E−06	26	8.01E−06	25	8.32E−06	23	1.12E−05	24	1.45E−05	22
(7000,4) *	6.83E−06	26	7.83E−06	25	8.02E−06	23	1.08E−05	24	1.40E−05	22
(7000,3)	5.27E−06	30	6.36E−06	28	7.06E−06	25	8.68E−06	26	1.15E−05	23
(7000,2)	3.68E−06	30	4.65E−06	28	5.60E−06	24	6.40E−06	25	8.81E−06	22
(6500,4)	4.02E−06	43	4.71E−06	40	4.30E−06	36	7.65E−06	40	1.05E−05	37
(6500,3)	2.81E−06	46	3.59E−06	44	3.79E−06	38	5.81E−06	42	8.32E−06	39
(6000,4.5)	2.39E−06	59	2.53E−06	56	1.66E−06	45	5.46E−06	56	7.78E−06	53
(6000,4)	2.22E−06	59	2.38E−06	55	1.61E−06	44	5.07E−06	56	7.24E−06	52
(6000,4) *	2.35E−06	59	2.43E−06	55	1.60E−06	44	5.11E−06	55	7.22E−06	52
(6000,4) **	1.68E−06		1.63E−06		8.78E−07		4.23E−06		6.10E−06	
(6000,4) * **	1.81E−06		1.67E−06		8.68E−07		4.28E−06		6.07E−06	
(6000,4,10X)	6.81E−07		2.29E−07		2.66E−07		4.55E−06		5.95E−06	
(6000,4,1/10X)	1.97E−06		2.66E−06		3.23E−06		4.01E−06		5.86E−06	
(6000,3.5)	2.10E−06	65	2.33E−06	61	1.70E−06	49	4.78E−06	61	6.88E−06	57
(6000,3)	1.41E−06	63	1.77E−06	59	1.45E−06	48	3.75E−06	59	5.75E−06	55
(6000,2)	1.18E−06	68	1.52E−06	62	1.46E−06	50	3.03E−06	61	4.72E−06	55
(5500,4)	9.65E−07	75	7.95E−07	69	3.14E−07	42	2.98E−06	73	4.47E−06	69
(5500,3)	7.57E−07	77	6.92E−07	70	3.22E−07	43	2.47E−06	73	3.85E−06	69
(5000,4)	2.78E−07	92	1.14E−07	81	4.02E−08	25	1.56E−06	91	2.45E−06	88
(5000,3)	2.02E−07	91	9.31E−08	77	4.00E−08	22	1.30E−06	90	2.14E−06	87
(5000,2)	1.00E−07	83	6.66E−08	66	4.05E−08	21	8.75E−07	87	1.63E−06	84
(4500,4)	4.00E−08	98	5.73E−09	66	5.52E−09	15	6.80E−07	99	1.14E−06	98
(4500,3)	1.63E−08	93	3.12E−09	40	4.86E−09	9	6.35E−07	98	1.08E−06	97
(4000,4)	3.72E−09	97	8.63E−10	71	9.19E−10	24	2.00E−07	99	3.50E−07	99
(4000,3)	6.36E−10	84	3.61E−10	37	7.26E−10	7	1.83E−07	99	3.36E−07	98
(4000,2)	1.28E−10	32	2.22E−10	11	6.25E−10	2	9.86E−08	98	2.30E−07	97

M O N O C H R O M A T I C   F L U X E S

MODEL	.3172		.3647-		.3862B		.3897B		.3930B	
(25000,4)	1.46E-03		1.26E-03		1.55E-03		1.38E-03		1.52E-03	
(20000,4)	7.98E-04		7.20E-04		1.01E-03		8.67E-04		1.00E-03	
(20000,3)	8.38E-04		7.56E-04		1.01E-03		9.58E-04		9.92E-04	
(18000,4)	5.93E-04		5.47E-04		8.31E-04		6.96E-04		8.26E-04	
(18000,3)	6.19E-04		5.72E-04		8.30E-04		7.77E-04		8.16E-04	
(16000,4)	4.19E-04		3.98E-04		6.66E-04		5.41E-04		6.66E-04	
(16000,3)	4.35E-04		4.13E-04		6.66E-04		6.13E-04		6.57E-04	
(14000,4)	2.75E-04		2.72E-04		5.16E-04		4.03E-04		5.20E-04	
(14000,3)	2.82E-04		2.78E-04		5.16E-04		4.64E-04		5.11E-04	
(12000,4)	1.60E-04		1.66E-04		3.79E-04		2.78E-04		3.86E-04	
(12000,3)	1.60E-04		1.67E-04		3.79E-04		3.29E-04		3.78E-04	
(12000,2)	1.74E-04		1.79E-04		3.52E-04		3.40E-04		3.49E-04	
(11000,4)	1.12E-04		1.21E-04		3.13E-04		2.17E-04		3.22E-04	
(11000,3)	1.11E-04		1.20E-04		3.16E-04		2.66E-04		3.15E-04	
(11000,2)	1.18E-04		1.27E-04		2.93E-04		2.81E-04		2.91E-04	
(10000,4.5)	7.41E-05		8.40E-05		2.15E-04		1.31E-04		2.42E-04	
(10000,4)	7.21E-05		8.21E-05		2.35E-04		1.48E-04		2.51E-04	
(10000,4,10X)	7.84E-05		8.88E-05		2.43E-04		1.57E-04		2.58E-04	
(10000,4,.1X)	7.05E-05		8.02E-05		2.34E-04		1.46E-04		2.50E-04	
(10000,3.5)	7.04E-05		8.04E-05		2.48E-04		1.73E-04		2.55E-04	
(10000,3)	6.96E-05		7.95E-05		2.52E-04		2.00E-04		2.54E-04	
(10000,2)	7.34E-05		8.29E-05		2.38E-04		2.25E-04		2.37E-04	
(9500,4)	5.66E-05		6.61E-05		1.97E-04		1.21E-04		2.16E-04	
(9500,3)	5.35E-05		6.29E-05		2.21E-04		1.64E-04		2.24E-04	
(9000,4)	4.50E-05		5.39E-05		1.59E-04		9.94E-05		1.78E-04	
(9000,3)	4.08E-05		4.96E-05		1.87E-04		1.29E-04		1.93E-04	
(9000,2)	3.98E-05		4.83E-05		1.87E-04		1.68E-04		1.87E-04	
(8500,4)	3.64E-05	4	4.41E-05	2	1.27E-04	6	8.24E-05	4	1.44E-04	6
(8500,3)	3.20E-05	3	3.96E-05	2	1.55E-04	6	1.02E-04	5	1.63E-04	6
(8000,4.5)	3.19E-05	6	3.86E-05	4	8.73E-05	7	6.19E-05	6	9.77E-05	7
(8000,4)	2.90E-05	6	3.57E-05	4	9.51E-05	8	6.58E-05	6	1.06E-04	8
(8000,4) *	2.88E-05	6	3.56E-05	4	9.42E-05	8	6.55E-05	6	1.07E-04	8
(8000,3.5)	2.63E-05	5	3.29E-05	3	1.04E-04	8	7.05E-05	6	1.16E-04	8
(8000,3)	2.41E-05	5	3.06E-05	3	1.15E-04	8	7.71E-05	7	1.25E-04	8
(8000,2)	2.10E-05	4	2.74E-05	2	1.32E-04	8	1.00E-04	7	1.35E-04	8
(7500,4)	2.32E-05	9	2.91E-05	7	6.75E-05	10	5.10E-05	9	7.40E-05	11
(7500,4) +	2.33E-05	9	2.89E-05	7	6.71E-05	10	5.16E-05	9	7.25E-05	11
(7500,4) *	2.08E-05	9	2.70E-05	7	6.00E-05	10	4.63E-05	9	6.69E-05	11
(7500,3)	1.90E-05	9	2.46E-05	6	8.12E-05	13	5.88E-05	11	8.85E-05	13
(7000,4)	1.88E-05	19	2.42E-05	15	4.80E-05	19	3.95E-05	17	5.19E-05	20
(7000,4) *	1.82E-05	19	2.37E-05	15	4.72E-05	19	3.89E-05	17	5.21E-05	20
(7000,3)	1.53E-05	20	2.04E-05	14	5.58E-05	24	4.40E-05	21	6.03E-05	24
(7000,2)	1.22E-05	17	1.70E-05	12	6.64E-05	27	5.15E-05	24	7.05E-05	27

MODEL	.3172		.3647-		.3716B		.3760B		.3786B	
(6500,4)	1.41E-05	33	1.89E-05	26	2.23E-05	28	2.70E-05	30	2.88E-05	30
(6500,3)	1.18E-05	34	1.64E-05	26	2.31E-05	31	2.98E-05	34	3.22E-05	35
(6000,4.5)	1.08E-05	48	1.48E-05	38	1.63E-05	39	1.83E-05	40	1.91E-05	40
(6000,4)	1.02E-05	47	1.40E-05	37	1.60E-05	38	1.84E-05	40	1.93E-05	40
(6000,4) *	1.01E-05	47	1.39E-05	37	1.59E-05	38	1.83E-05	40	1.91E-05	40
(6000,4) **	8.62E-06		1.19E-05		1.34E-05		1.51E-05		1.57E-05	
(6000,4) * **	8.53E-06		1.18E-05		1.32E-05		1.50E-05		1.56E-05	
(6000,4,10X)	7.73E-06		1.00E-05		1.08E-05		1.19E-05		1.22E-05	
(6000,4,1/10X)	8.43E-06		1.19E-05		1.34E-05		1.52E-05		1.58E-05	
(6000,3.5)	9.73E-06	51	1.35E-05	40	1.62E-05	42	1.92E-05	45	2.02E-05	46
(6000,3)	8.52E-06	49	1.23E-05	38	1.61E-05	42	1.94E-05	45	2.06E-05	46
(6000,2)	7.19E-06	48	1.07E-05	36	1.73E-05	47	2.18E-05	51	2.32E-05	52
(5500,4)	6.52E-06	64	9.31E-06	52	1.03E-05	53	1.13E-05	54	1.16E-05	54
(5500,3)	5.81E-06	64	8.55E-06	52	1.04E-05	54	1.18E-05	56	1.23E-05	56
(5000,4)	3.71E-06	84	5.50E-06	75	5.90E-06	74	6.35E-06	74	6.49E-06	74
(5000,3)	3.38E-06	83	5.16E-06	73	5.75E-06	74	6.28E-06	75	6.44E-06	74
(5000,2)	2.82E-06	81	4.60E-06	71	5.35E-06	74	5.94E-06	75	6.11E-06	75
(4500,4)	1.82E-06	95	2.79E-06	90	2.99E-06	89	3.19E-06	89	3.25E-06	89
(4500,3)	1.74E-06	95	2.69E-06	89	2.86E-06	89	3.09E-06	89	3.15E-06	89
(4000,4)	5.92E-07	97	9.88E-07	93	1.03E-06	92	1.10E-06	92	1.12E-06	92
(4000,3)	5.81E-07	97	9.82E-07	92	1.01E-06	92	1.08E-06	92	1.11E-06	91
(4000,2)	4.65E-07	96	8.65E-07	91	8.31E-07	91	9.08E-07	91	9.36E-07	91

MONOCHROMATIC FLUXES

MODEL	.3978B	.4019B	.4019B	.4062B	.4110B
(25000,4)	1.35E-03	1.47E-03		1.45E-03	1.28E-03
(20000,4)	8.49E-04	9.76E-04		9.61E-04	8.12E-04
(20000,3)	9.35E-04	9.66E-04		9.53E-04	8.96E-04
(18000,4)	6.82E-04	8.08E-04		7.96E-04	6.54E-04
(18000,3)	7.59E-04	7.96E-04		7.86E-04	7.29E-04
(16000,4)	5.31E-04	6.54E-04		6.44E-04	5.11E-04
(16000,3)	6.00E-04	6.42E-04		6.34E-04	5.77E-04
(14000,4)	3.97E-04	5.13E-04		5.05E-04	3.83E-04
(14000,3)	4.55E-04	5.01E-04		4.96E-04	4.39E-04
(12000,4)	2.75E-04	3.83E-04		3.78E-04	2.66E-04
(12000,3)	3.24E-04	3.72E-04		3.68E-04	3.13E-04
(12000,2)	3.36E-04	3.44E-04		3.41E-04	3.28E-04
(11000,4)	2.15E-04	3.21E-04		3.17E-04	2.09E-04
(11000,3)	2.62E-04	3.11E-04		3.08E-04	2.54E-04
(11000,2)	2.78E-04	2.88E-04	.4019B	2.86E-04	2.72E-04
(10000,4.5)	1.30E-04	2.55E-04	2.55E-04	2.44E-04 0	1.31E-04
(10000,4)	1.46E-04	2.57E-04	2.57E-04	2.49E-04 0	1.47E-04
(10000,4,10X)	1.56E-04	2.63E-04	2.63E-04	2.56E-04 0	1.56E-04
(10000,4,.1X)	1.45E-04	2.55E-04	2.55E-04	2.48E-04 0	1.45E-04
(10000,3.5)	1.70E-04	2.56E-04	2.56E-04	2.51E-04 0	1.68E-04
(10000,3)	1.96E-04	2.52E-04	2.52E-04	2.49E-04 0	1.92E-04
(10000,2)	2.22E-04	2.35E-04	2.35E-04	2.33E-04 0	2.18E-04
(9500,4)	1.20E-04	2.24E-04	2.24E-04	2.16E-04 0	1.21E-04
(9500,3)	1.61E-04	2.23E-04	2.23E-04	2.20E-04 0	1.59E-04
(9000,4)	9.91E-05	1.88E-04	1.88E-04	1.80E-04 0	1.01E-04
(9000,3)	1.27E-04	1.94E-04	1.94E-04	1.90E-04 0	1.26E-04
(9000,2)	1.65E-04	1.85E-04	1.85E-04	1.84E-04 0	1.63E-04
(8500,4)	8.27E-05 4	1.55E-04 6	1.55E-04 5	1.48E-04 5	8.45E-05 3
(8500,3)	1.01E-04 5	1.65E-04 6	1.65E-04 5	1.62E-04 5	1.02E-04 4
(8000,4.5)	6.26E-05 6	1.06E-04 8	1.06E-04 6	1.02E-04 6	6.45E-05 4
(8000,4)	6.65E-05 6	1.14E-04 8	1.14E-04 7	1.10E-04 7	6.83E-05 5
(8000,4) *	6.61E-05 6	1.17E-04 8	1.17E-04 7	1.11E-04 7	6.79E-05 5
(8000,3.5)	7.09E-05 6	1.23E-04 8	1.23E-04 7	1.18E-04 7	7.24E-05 5
(8000,3)	7.72E-05 6	1.29E-04 8	1.29E-04 7	1.25E-04 7	7.83E-05 5
(8000,2)	9.86E-05 7	1.34E-04 8	1.34E-04 7	1.33E-04 7	9.79E-05 6
(7500,4)	5.19E-05 9	7.82E-05 11	7.82E-05 7	7.66E-05 6	5.36E-05 5
(7500,4) +	5.23E-05 9	7.58E-05 11	7.58E-05 6	7.49E-05 6	5.40E-05 5
(7500,4) *	4.72E-05 9	7.25E-05 11	7.60E-05 7	7.30E-05 6	5.08E-05 5
(7500,3)	5.92E-05 10	9.24E-05 13	9.24E-05 11	9.02E-05 11	6.07E-05 8
(7000,4)	4.04E-05 17	5.44E-05 19	5.44E-05 10	5.40E-05 10	4.21E-05 9
(7000,4) *	3.98E-05 17	5.59E-05 20	5.59E-05 10	5.45E-05 10	4.15E-05 8
(7000,3)	4.48E-05 21	6.27E-05 24	6.27E-05 16	6.21E-05 16	4.63E-05 13
(7000,2)	5.19E-05 24	7.20E-05 27	7.20E-05 22	7.13E-05 22	5.30E-05 18

MODEL	.3930B	.4019B	.4019+	.4450-	.4450+
(6500,4)	3.40E-05 31	3.55E-05 25	3.55E-05 16	3.97E-05 14	3.97E-05 10
(6500,3)	3.87E-05 36	4.03E-05 35	4.04E-05 23	4.41E-05 21	4.41E-05 11
(6000,4.5)	2.18E-05 40	2.27E-05 39	2.27E-05 25	2.65E-05 23	2.65E-05 10
(6000,4)	2.22E-05 40	2.31E-05 40	2.32E-05 25	2.69E-05 23	2.69E-05 10
(6000,4) *	2.23E-05 41	2.34E-05 40	2.34E-05 26	2.70E-05 23	2.70E-05 10
(6000,4) **	1.81E-05	1.89E-05	1.89E-05	2.23E-05	2.23E-05
(6000,4) * **	1.82E-05	1.91E-05	1.91E-05	2.24E-05	2.24E-05
(6000,4,10X)	1.66E-05	1.77E-05	1.77E-05	2.19E-05	2.19E-05
(6000,4,1/10X)	1.79E-05	1.87E-05	1.87E-05	2.19E-05	2.19E-05
(6000,3.5)	2.35E-05 46	2.45E-05 45	2.45E-05 35	2.83E-05 31	2.83E-05 13
(6000,3)	2.40E-05 46	2.50E-05 46	2.50E-05 35	2.88E-05 32	2.88E-05 13
(6000,2)	2.70E-05 52	2.80E-05 51	2.80E-05 45	3.16E-05 41	3.16E-05 18
(5500,4)	1.34E-05 53	1.40E-05 52	1.40E-05 35	1.71E-05 32	1.71E-05 13
(5500,3)	1.42E-05 55	1.49E-05 54	1.49E-05 45	1.80E-05 41	1.80E-05 18
(5000,4)	7.48E-06 73	7.93E-06 72	7.93E-06 46	1.00E-05 42	1.00E-05 19
(5000,3)	7.59E-06 74	8.09E-06 73	8.09E-06 55	1.03E-05 50	1.03E-05 25
(5000,2)	7.40E-06 74	7.95E-06 73	7.95E-06 58	1.04E-05 54	1.04E-05 32
(4500,4)	3.74E-06 88	3.98E-06 87	3.98E-06 54	5.14E-06 49	5.14E-06 24
(4500,3)	3.74E-06 88	4.01E-06 87	4.01E-06 57	5.26E-06 52	5.26E-06 33
(4000,4)	1.30E-06 90	1.40E-06 89	1.40E-06 60	1.94E-06 54	1.94E-06 31
(4000,3)	1.32E-06 90	1.43E-06 89	1.43E-06 63	2.02E-06 57	2.02E-06 42
(4000,2)	1.18E-06 90	1.30E-06 89	1.30E-06 66	1.94E-06 60	1.94E-06 48

M O N O C H R O M A T I C   F L U X E S

MODEL	.4142B		.4281B		.4348B		.4400B		.4601-	
(25000,4)	1.41E-03		1.34E-03		1.18E-03		1.29E-03			
(20000,4)	9.36E-04		8.97E-04		7.53E-04		8.64E-04			
(20000,3)	9.29E-04		8.91E-04		8.31E-04		8.60E-04			
(18000,4)	7.76E-04		7.46E-04		6.09E-04		7.20E-04			
(18000,3)	7.68E-04		7.38E-04		6.79E-04		7.14E-04			
(16000,4)	6.30E-04		6.07E-04		4.79E-04		5.87E-04			
(16000,3)	6.21E-04		5.99E-04		5.40E-04		5.80E-04			
(14000,4)	4.95E-04		4.79E-04		3.61E-04		4.65E-04			
(14000,3)	4.86E-04		4.70E-04		4.13E-04		4.57E-04			
(12000,4)	3.71E-04		3.61E-04		2.53E-04		3.51E-04			
(12000,3)	3.61E-04		3.51E-04		2.96E-04		3.42E-04			
(12000,2)	3.37E-04		3.29E-04		3.15E-04		3.23E-04			
(11000,4)	3.11E-04		3.04E-04		2.00E-04		2.96E-04			
(11000,3)	3.03E-04		2.95E-04		2.41E-04		2.88E-04			
(11000,2)	2.82E-04		2.77E-04		2.61E-04		2.71E-04			
(10000,4.5)	2.39E-04		2.40E-04		1.30E-04		2.34E-04		2.29E-04	
(10000,4)	2.45E-04		2.42E-04		1.43E-04		2.36E-04		2.29E-04	
(10000,4,10X)	2.52E-04		2.48E-04		1.52E-04		2.42E-04		2.35E-04	
(10000,4,.1X)	2.44E-04		2.41E-04		1.41E-04		2.35E-04		2.28E-04	
(10000,3.5)	2.47E-04		2.42E-04		1.61E-04		2.36E-04		2.28E-04	
(10000,3)	2.45E-04		2.39E-04		1.82E-04		2.34E-04		2.25E-04	
(10000,2)	2.30E-04		2.26E-04		2.09E-04		2.22E-04		2.16E-04	
(9500,4)	2.13E-04		2.11E-04		1.20E-04		2.07E-04		2.02E-04	
(9500,3)	2.17E-04		2.12E-04		1.51E-04		2.08E-04		2.01E-04	
(9000,4)	1.78E-04		1.78E-04		1.00E-04		1.75E-04		1.72E-04	
(9000,3)	1.88E-04		1.84E-04		1.22E-04		1.80E-04		1.75E-04	
(9000,2)	1.82E-04		1.79E-04		1.55E-04		1.76E-04		1.70E-04	
(8500,4)	1.46E-04	5	1.48E-04	5	8.51E-05	3	1.46E-04	5	1.45E-04	4
(8500,3)	1.59E-04	5	1.58E-04	5	9.99E-05	3	1.55E-04	5	1.51E-04	5
(8000,4.5)	1.02E-04	6	1.05E-04	6	6.64E-05	4	1.05E-04	6	1.07E-04	6
(8000,4)	1.09E-04	6	1.12E-04	6	6.98E-05	4	1.11E-04	6	1.12E-04	6
(8000,4) *	1.10E-04	7	1.13E-04	6	6.94E-05	4	1.12E-04	6	1.13E-04	6
(8000,3.5)	1.17E-04	7	1.19E-04	6	7.34E-05	4	1.17E-04	6	1.17E-04	6
(8000,3)	1.24E-04	7	1.24E-04	7	7.84E-05	5	1.22E-04	6	1.20E-04	6
(8000,2)	1.31E-04	7	1.30E-04	7	9.49E-05	6	1.28E-04	7	1.24E-04	6
(7500,4)	7.70E-05	6	7.92E-05	6	5.57E-05	5	7.97E-05	6	8.13E-05	6
(7500,4) +	7.55E-05	6	7.76E-05	6	5.58E-05	5	7.83E-05	6	7.99E-05	6
(7500,4) *	7.31E-05	6	7.56E-05	6	5.30E-05	5	7.58E-05	6	7.75E-05	6
(7500,3)	8.99E-05	10	9.09E-05	10	6.19E-05	8	9.05E-05	10	9.06E-05	9
(7000,4)	5.47E-05	10	5.65E-05	10	4.44E-05	8	5.75E-05	9	5.92E-05	9
(7000,4) *	5.49E-05	10	5.70E-05	10	4.39E-05	8	5.76E-05	9	5.94E-05	9
(7000,3)	6.25E-05	15	6.40E-05	15	4.82E-05	12	6.46E-05	15	6.57E-05	14
(7000,2)	7.12E-05	21	7.17E-05	21	5.42E-05	17	7.16E-05	20	7.17E-05	19

MONOCHROMATIC FLUX IS IN ERGS/CM**2/SEC/STER/HZ (WITHOUT PI).

KEY TO THE MODELS:

   *   NO CONVECTION
  ** NO BLANKETING
  +   L/H = 2.5

THE (6000,4,10X) AND (6000,4,.1X) MODELS HAVE NEITHER CONVECTION NOR BLANKETING.

MONOCHROMATIC FLUXES

MODEL	.4601+	.4821B	.4869B	.4901B	.5450-
(25000,4)	1.20E-03	1.11E-03	9.86E-04	1.09E-03	
(20000,4)	8.13E-04	7.58E-04	6.42E-04	7.40E-04	
(20000,3)	8.11E-04	7.60E-04	7.10E-04	7.43E-04	
(18000,4)	6.80E-04	6.35E-04	5.24E-04	6.21E-04	
(18000,3)	6.75E-04	6.35E-04	5.84E-04	6.22E-04	
(16000,4)	5.56E-04	5.21E-04	4.16E-04	5.10E-04	
(16000,3)	5.51E-04	5.20E-04	4.69E-04	5.09E-04	
(14000,4)	4.42E-04	4.15E-04	3.18E-04	4.07E-04	
(14000,3)	4.35E-04	4.12E-04	3.62E-04	4.04E-04	
(12000,4)	3.36E-04	3.16E-04	2.27E-04	3.10E-04	
(12000,3)	3.27E-04	3.11E-04	2.62E-04	3.06E-04	
(12000,2)	3.12E-04	2.99E-04	2.85E-04	2.95E-04	
(11000,4)	2.85E-04	2.67E-04	1.81E-04	2.62E-04	
(11000,3)	2.76E-04	2.63E-04	2.14E-04	2.59E-04	
(11000,2)	2.62E-04	2.53E-04	2.37E-04	2.49E-04	
(10000,4.5)	2.29E-04	2.09E-04	1.28E-04	2.05E-04	1.95E-04
(10000,4)	2.29E-04	2.13E-04	1.38E-04	2.10E-04	1.95E-04
(10000,4,10X)	2.35E-04	2.20E-04	1.45E-04	2.16E-04	2.00E-04
(10000,4,.1X)	2.28E-04	2.12E-04	1.36E-04	2.08E-04	1.94E-04
(10000,3.5)	2.28E-04	2.15E-04	1.51E-04	2.12E-04	1.94E-04
(10000,3)	2.25E-04	2.15E-04	1.67E-04	2.11E-04	1.92E-04
(10000,2)	2.16E-04	2.08E-04	1.93E-04	2.05E-04	1.88E-04
(9500,4)	2.02E-04	1.87E-04	1.17E-04	1.84E-04	1.73E-04
(9500,3)	2.01E-04	1.91E-04	1.41E-04	1.88E-04	1.72E-04
(9000,4)	1.72E-04	1.59E-04	1.01E-04	1.57E-04	1.51E-04
(9000,3)	1.75E-04	1.66E-04	1.16E-04	1.64E-04	1.51E-04
(9000,2)	1.70E-04	1.64E-04	1.43E-04	1.62E-04	1.50E-04
(8500,4)	1.45E-04 4	1.34E-04 4	8.70E-05 2	1.32E-04 4	1.31E-04 3
(8500,3)	1.51E-04 4	1.43E-04 4	9.79E-05 2	1.41E-04 4	1.33E-04 3
(8000,4.5)	1.07E-04 5	1.01E-04 5	7.01E-05 3	1.00E-04 5	1.03E-04 4
(8000,4)	1.12E-04 5	1.06E-04 5	7.28E-05 3	1.05E-04 5	1.06E-04 4
(8000,4) *	1.13E-04 5	1.05E-04 5	7.25E-05 3	1.04E-04 5	1.05E-04 4
(8000,3.5)	1.17E-04 5	1.10E-04 5	7.56E-05 3	1.09E-04 5	1.08E-04 4
(8000,3)	1.20E-04 6	1.14E-04 5	7.94E-05 4	1.13E-04 5	1.09E-04 4
(8000,2)	1.24E-04 6	1.20E-04 6	9.14E-05 5	1.19E-04 6	1.11E-04 5
(7500,4)	8.13E-05 5	7.91E-05 5	5.96E-05 4	7.90E-05 5	8.15E-05 4
(7500,4) +	7.99E-05 5	7.87E-05 5	5.93E-05 4	7.87E-05 5	8.11E-05 4
(7500,4) *	7.79E-05 5	7.49E-05 5	5.74E-05 4	7.48E-05 5	7.78E-05 4
(7500,3)	9.06E-05 7	8.75E-05 7	6.47E-05 5	8.70E-05 7	8.68E-05 6
(7000,4)	5.92E-05 7	5.93E-05 7	4.87E-05 5	5.96E-05 6	6.27E-05 6
(7000,4) *	5.94E-05 7	5.85E-05 6	4.82E-05 5	5.87E-05 6	6.21E-05 6
(7000,3)	6.57E-05 9	6.52E-05 8	5.18E-05 7	6.53E-05 8	6.72E-05 7
(7000,2)	7.17E-05 10	7.06E-05 10	5.66E-05 8	7.04E-05 10	7.04E-05 8

MODEL	.4879	.5400-
(6500,4)	4.29E-05 8	4.59E-05 7
(6500,3)	4.69E-05 10	4.95E-05 9
(6000,4.5)	2.96E-05 9	3.29E-05 8
(6000,4)	3.00E-05 9	3.32E-05 8
(6000,4) *	3.00E-05 9	3.32E-05 8
(6000,4) **	2.52E-05	2.83E-05
(6000,4) * **	2.52E-05	2.82E-05
(6000,4,10X)	2.51E-05	2.98E-05
(6000,4,1/10X)	2.46E-05	2.75E-05
(6000,3.5)	3.14E-05 12	3.46E-05 10
(6000,3)	3.18E-05 12	3.50E-05 11
(6000,2)	3.44E-05 16	3.73E-05 15
(5500,4)	1.97E-05 12	2.26E-05 11
(5500,3)	2.06E-05 17	2.36E-05 15
(5000,4)	1.19E-05 18	1.42E-05 16
(5000,3)	1.24E-05 22	1.48E-05 20
(5000,2)	1.26E-05 29	1.52E-05 27
(4500,4)	6.28E-06 22	7.75E-06 19
(4500,3)	6.47E-06 30	8.03E-06 26
(4000,4)	2.54E-06 27	3.34E-06 24
(4000,3)	2.65E-06 38	3.51E-06 33
(4000,2)	2.61E-06 43	3.54E-06 39

MONOCHROMATIC FLUXES

MODEL	.5450		.6137		.7022		.8206-		.8206+	
(25000,4)	9.20E-04		7.59E-04		6.07E-04		4.67E-04		4.91E-04	
(20000,4)	6.37E-04		5.34E-04		4.35E-04		3.41E-04		3.65E-04	
(20000,3)	6.41E-04		5.39E-04		4.40E-04		3.45E-04		3.66E-04	
(18000,4)	5.39E-04		4.55E-04		3.73E-04		2.95E-04		3.20E-04	
(18000,3)	5.40E-04		4.58E-04		3.77E-04		2.98E-04		3.20E-04	
(16000,4)	4.47E-04		3.80E-04		3.15E-04		2.51E-04		2.77E-04	
(16000,3)	4.46E-04		3.81E-04		3.17E-04		2.53E-04		2.76E-04	
(14000,4)	3.60E-04		3.10E-04		2.59E-04		2.09E-04		2.36E-04	
(14000,3)	3.58E-04		3.09E-04		2.59E-04		2.09E-04		2.34E-04	
(12000,4)	2.78E-04		2.42E-04		2.05E-04		1.67E-04		1.95E-04	
(12000,3)	2.74E-04		2.39E-04		2.03E-04		1.66E-04		1.92E-04	
(12000,2)	2.67E-04		2.36E-04		2.03E-04		1.67E-04		1.86E-04	
(11000,4)	2.38E-04		2.08E-04		1.77E-04		1.46E-04		1.74E-04	
(11000,3)	2.33E-04		2.05E-04		1.75E-04		1.44E-04		1.71E-04	
(11000,2)	2.27E-04		2.02E-04		1.74E-04		1.45E-04		1.65E-04	
(10000,4.5)	1.95E-04		1.72E-04		1.49E-04		1.25E-04		1.50E-04	
(10000,4)	1.95E-04		1.71E-04		1.48E-04		1.24E-04		1.50E-04	
(10000,4,10X)	2.00E-04		1.77E-04		1.53E-04		1.27E-04		1.54E-04	
(10000,4,.1X)	1.94E-04		1.70E-04		1.47E-04		1.23E-04		1.50E-04	
(10000,3.5)	1.94E-04		1.70E-04		1.47E-04		1.23E-04		1.50E-04	
(10000,3)	1.92E-04		1.69E-04		1.46E-04		1.22E-04		1.49E-04	
(10000,2)	1.88E-04		1.68E-04		1.46E-04		1.22E-04		1.45E-04	
(9500,4)	1.73E-04		1.54E-04		1.34E-04		1.13E-04		1.38E-04	
(9500,3)	1.72E-04		1.52E-04		1.32E-04		1.11E-04		1.38E-04	
(9000,4)	1.51E-04		1.36E-04		1.20E-04		1.03E-04		1.24E-04	
(9000,3)	1.51E-04		1.35E-04		1.19E-04		1.01E-04		1.26E-04	
(9000,2)	1.50E-04		1.34E-04		1.17E-04		9.96E-05		1.24E-04	
(8500,4)	1.31E-04	2	1.20E-04	2	1.07E-04	2	9.34E-05	1	1.11E-04	0
(8500,3)	1.33E-04	2	1.20E-04	2	1.07E-04	2	9.19E-05	1	1.14E-04	0
(8000,4.5)	1.03E-04	3	9.81E-05	3	9.14E-05	2	8.22E-05	2	9.07E-05	0
(8000,4)	1.06E-04	3	9.98E-05	3	9.21E-05	2	8.22E-05	2	9.33E-05	0
(8000,4) *	1.05E-04	3	9.91E-05	3	9.16E-05	2	8.19E-05	2	9.29E-05	0
(8000,3.5)	1.08E-04	3	1.00E-04	3	9.20E-05	2	8.16E-05	2	9.54E-05	0
(8000,3)	1.09E-04	3	1.01E-04	3	9.20E-05	2	8.10E-05	2	9.76E-05	0
(8000,2)	1.11E-04	4	1.02E-04	3	9.14E-05	2	7.96E-05	2	1.01E-04	0
(7500,4)	8.15E-05	2	7.97E-05	2	7.62E-05	1	7.04E-05	1	7.61E-05	0
(7500,4) +	8.11E-05	2	7.96E-05	2	7.62E-05	1	7.04E-05	1	7.59E-05	0
(7500,4) *	7.96E-05	2	7.79E-05	2	7.47E-05	1	6.93E-05	1	7.58E-05	0
(7500,3)	8.68E-05	3	8.29E-05	3	7.78E-05	2	7.06E-05	2	8.09E-05	0
(7000,4)	6.27E-05	2	6.34E-05	2	6.27E-05	1	5.99E-05	1	6.25E-05	0
(7000,4) *	6.21E-05	2	6.26E-05	2	6.20E-05	1	5.93E-05	1	6.19E-05	0
(7000,3)	6.72E-05	5	6.65E-05	4	6.47E-05	3	6.08E-05	3	6.62E-05	0
(7000,2)	7.04E-05	6	6.85E-05	5	6.54E-05	4	6.06E-05	3	6.99E-05	0

MODEL	.5400+		.6095		.6994		.8206-		.8206+	
(6500,4)	4.59E-05	4	4.83E-05	3	4.97E-05	2	4.92E-05	2	5.01E-05	0
(6500,3)	4.95E-05	7	5.13E-05	6	5.20E-05	5	5.08E-05	4	5.30E-05	0
(6000,4.5)	3.29E-05	3	3.59E-05	3	3.83E-05	2	3.93E-05	2	3.95E-05	0
(6000,4)	3.32E-05	3	3.61E-05	3	3.84E-05	2	3.94E-05	2	3.97E-05	0
(6000,4) *	3.32E-05	3	3.61E-05	3	3.84E-05	2	3.94E-05	2	3.97E-05	0
(6000,4) **	2.83E-05		3.12E-05		3.37E-05		3.52E-05		3.53E-05	
(6000,4) * **	2.82E-05		3.11E-05		3.37E-05		3.51E-05		3.53E-05	
(6000,4,10X)	2.98E-05		3.32E-05		3.68E-05		3.81E-05		3.84E-05	
(6000,4,1/10X)	2.75E-05		3.04E-05		3.28E-05		3.44E-05		3.45E-05	
(6000,3.5)	3.46E-05	5	3.74E-05	4	3.96E-05	4	4.04E-05	3	4.09E-05	0
(6000,3)	3.50E-05	5	3.77E-05	4	3.98E-05	4	4.05E-05	3	4.12E-05	0
(6000,2)	3.73E-05	8	3.97E-05	6	4.14E-05	5	4.17E-05	5	4.33E-05	0
(5500,4)	2.26E-05	4	2.56E-05	4	2.83E-05	3	3.02E-05	3	3.03E-05	0
(5500,3)	2.36E-05	7	2.65E-05	6	2.92E-05	5	3.10E-05	4	3.11E-05	0
(5000,4)	1.42E-05	6	1.67E-05	5	1.94E-05	4	2.17E-05	4	2.17E-05	0
(5000,3)	1.48E-05	10	1.74E-05	9	2.01E-05	8	2.24E-05	7	2.24E-05	0
(5000,2)	1.52E-05	14	1.79E-05	12	2.07E-05	10	2.30E-05	9	2.31E-05	0
(4500,4)	7.75E-06	10	9.52E-06	8	1.16E-05	7	1.38E-05	6	1.38E-05	1
(4500,3)	8.03E-06	16	9.86E-06	13	1.21E-05	11	1.43E-05	9	1.43E-05	1
(4000,4)	3.34E-06	12	4.45E-06	10	5.91E-06	8	7.67E-06	7	7.67E-06	1
(4000,3)	3.51E-06	21	4.68E-06	17	6.22E-06	14	8.03E-06	12	8.03E-06	1
(4000,2)	3.54E-06	25	4.76E-06	21	6.38E-06	17	8.23E-06	14	8.24E-06	2

MONOCHROMATIC FLUXES

MODEL	1.0504		1.4588-		1.4588+		1.7791		2.2794-	
(25000,4)	3.17E-04		1.74E-04		1.77E-04		1.22E-04		7.62E-05	
(20000,4)	2.40E-04		1.35E-04		1.38E-04		9.60E-05		6.07E-05	
(20000,3)	2.42E-04		1.37E-04		1.39E-04		9.68E-05		6.13E-05	
(18000,4)	2.13E-04		1.21E-04		1.23E-04		8.62E-05		5.48E-05	
(18000,3)	2.14E-04		1.22E-04		1.24E-04		8.69E-05		5.53E-05	
(16000,4)	1.86E-04		1.06E-04		1.09E-04		7.67E-05		4.90E-05	
(16000,3)	1.86E-04		1.07E-04		1.10E-04		7.71E-05		4.93E-05	
(14000,4)	1.59E-04		9.24E-05		9.54E-05		6.74E-05		4.32E-05	
(14000,3)	1.59E-04		9.26E-05		9.54E-05		6.75E-05		4.33E-05	
(12000,4)	1.33E-04		7.84E-05		8.17E-05		5.80E-05		3.74E-05	
(12000,3)	1.32E-04		7.79E-05		8.10E-05		5.76E-05		3.72E-05	
(12000,2)	1.31E-04		7.81E-05		8.05E-05		5.76E-05		3.74E-05	
(11000,4)	1.20E-04		7.11E-05		7.46E-05		5.31E-05		3.43E-05	
(11000,3)	1.19E-04		7.04E-05		7.38E-05		5.26E-05		3.41E-05	
(11000,2)	1.17E-04		7.03E-05		7.30E-05		5.24E-05		3.41E-05	
(10000,4.5)	1.06E-04		6.38E-05		6.72E-05		4.81E-05		3.13E-05	
(10000,4)	1.05E-04		6.34E-05		6.69E-05		4.79E-05		3.11E-05	
(10000,4,10X)	1.08E-04		6.47E-05		6.81E-05		4.87E-05		3.17E-05	
(10000,4,.1X)	1.05E-04		6.31E-05		6.66E-05		4.76E-05		3.10E-05	
(10000,3.5)	1.05E-04		6.30E-05		6.65E-05		4.76E-05		3.10E-05	
(10000,3)	1.04E-04		6.27E-05		6.62E-05		4.74E-05		3.09E-05	
(10000,2)	1.03E-04		6.25E-05		6.55E-05		4.71E-05		3.08E-05	
(9500,4)	9.80E-05		5.96E-05		6.31E-05		4.53E-05		2.95E-05	
(9500,3)	9.71E-05		5.88E-05		6.24E-05		4.47E-05		2.92E-05	
(9000,4)	9.02E-05		5.61E-05		5.93E-05		4.29E-05		2.80E-05	
(9000,3)	8.99E-05		5.52E-05		5.87E-05		4.23E-05		2.77E-05	
(9000,2)	8.88E-05		5.44E-05		5.78E-05		4.17E-05		2.73E-05	
(8500,4)	8.31E-05	0	5.33E-05	0	5.63E-05	0	4.10E-05	0	2.68E-05	0
(8500,3)	8.33E-05	0	5.21E-05	0	5.56E-05	0	4.02E-05	0	2.63E-05	0
(8000,4.5)	7.30E-05	0	5.03E-05	0	5.25E-05	0	3.88E-05	0	2.54E-05	0
(8000,4)	7.38E-05	0	4.99E-05	0	5.24E-05	0	3.86E-05	0	2.53E-05	0
(8000,4) *	7.35E-05	0	4.98E-05	0	5.22E-05	0	3.85E-05	0	2.52E-05	0
(8000,3.5)	7.41E-05	0	4.93E-05	0	5.20E-05	0	3.82E-05	0	2.50E-05	0
(8000,3)	7.45E-05	0	4.87E-05	0	5.17E-05	0	3.78E-05	0	2.48E-05	0
(8000,2)	7.45E-05	0	4.74E-05	0	5.08E-05	0	3.70E-05	0	2.43E-05	0
(7500,4)	6.38E-05	0	4.62E-05	0	4.79E-05	0	3.60E-05	0	2.37E-05	0
(7500,4) +	6.38E-05	0	4.64E-05	0	4.82E-05	0	3.62E-05	0	2.37E-05	0
(7500,4) *	6.35E-05	0	4.61E-05	0	4.78E-05	0	3.60E-05	0	2.36E-05	0
(7500,3)	6.54E-05	0	4.55E-05	0	4.79E-05	0	3.56E-05	0	2.34E-05	0
(7000,4)	5.51E-05	0	4.27E-05	0	4.38E-05	0	3.37E-05	0	2.22E-05	0
(7000,4) *	5.46E-05	0	4.24E-05	0	4.35E-05	0	3.35E-05	0	2.21E-05	0
(7000,3)	5.67E-05	0	4.22E-05	0	4.39E-05	0	3.34E-05	0	2.21E-05	0
(7000,2)	5.79E-05	0	4.13E-05	0	4.36E-05	0	3.28E-05	0	2.18E-05	0
(6500,4)	4.64E-05	0	3.86E-05	0	3.92E-05	0	3.10E-05	0	2.06E-05	0
(6500,3)	4.81E-05	0	3.87E-05	0	3.97E-05	0	3.10E-05	0	2.06E-05	0
(6000,4.5)	3.83E-05	0	3.45E-05	0	3.47E-05	0	2.84E-05	0	1.89E-05	0
(6000,4)	3.85E-05	0	3.45E-05	0	3.47E-05	0	2.83E-05	0	1.89E-05	0
(6000,4) *	3.85E-05	0	3.44E-05	0	3.47E-05	0	2.83E-05	0	1.89E-05	0
(6000,4) **	3.50E-05		3.26E-05		3.28E-05		2.72E-05		1.81E-05	
(6000,4) * **	3.50E-05		3.25E-05		3.28E-05		2.71E-05		1.81E-05	
(6000,4,10X)	3.73E-05		3.26E-05		3.28E-05		2.65E-05		1.79E-05	
(6000,4,1/10X)	3.44E-05		3.23E-05		3.25E-05		2.71E-05		1.81E-05	
(6000,3.5)	3.94E-05	0	3.48E-05	0	3.52E-05	0	2.86E-05	0	1.90E-05	0
(6000,3)	3.96E-05	0	3.47E-05	0	3.52E-05	0	2.85E-05	0	1.90E-05	0
(6000,2)	4.09E-05	0	3.47E-05	0	3.57E-05	0	2.84E-05	0	1.90E-05	0
(5500,4)	3.09E-05	0	3.00E-05	0	3.01E-05	0	2.56E-05	0	1.72E-05	0
(5500,3)	3.16E-05	0	3.03E-05	0	3.06E-05	0	2.57E-05	0	1.73E-05	0
(5000,4)	2.37E-05	0	2.55E-05	0	2.55E-05	0	2.28E-05	0	1.54E-05	0
(5000,3)	2.43E-05	0	2.58E-05	0	2.59E-05	0	2.30E-05	0	1.55E-05	0
(5000,2)	2.49E-05	0	2.60E-05	1	2.62E-05	1	2.30E-05	1	1.56E-05	0
(4500,4)	1.65E-05	1	2.08E-05	1	2.08E-05	1	2.03E-05	1	1.37E-05	1
(4500,3)	1.69E-05	1	2.11E-05	1	2.12E-05	1	2.05E-05	2	1.38E-05	1
(4000,4)	1.03E-05	1	1.54E-05	1	1.54E-05	1	1.71E-05	1	1.17E-05	1
(4000,3)	1.07E-05	1	1.58E-05	2	1.58E-05	2	1.73E-05	3	1.18E-05	2
(4000,2)	1.09E-05	1	1.60E-05	3	1.60E-05	3	1.74E-05	3	1.19E-05	2

M O N O C H R O M A T I C   F L U X E S

MODEL	2.2794+		2.6904		3.2823−		3.2823+		6.5647	
(25000,4)	7.66E−05		5.57E−05		3.78E−05		3.79E−05		9.67E−06	
(20000,4)	6.11E−05		4.47E−05		3.07E−05		3.08E−05		8.06E−06	
(20000,3)	6.16E−05		4.51E−05		3.10E−05		3.10E−05		8.12E−06	
(18000,4)	5.52E−05		4.05E−05		2.79E−05		2.79E−05		7.39E−06	
(18000,3)	5.56E−05		4.09E−05		2.81E−05		2.82E−05		7.45E−06	
(16000,4)	4.94E−05		3.64E−05		2.51E−05		2.52E−05		6.71E−06	
(16000,3)	4.97E−05		3.66E−05		2.53E−05		2.53E−05		6.76E−06	
(14000,4)	4.37E−05		3.22E−05		2.23E−05		2.24E−05		6.01E−06	
(14000,3)	4.38E−05		3.23E−05		2.24E−05		2.25E−05		6.04E−06	
(12000,4)	3.79E−05		2.80E−05		1.94E−05		1.95E−05		5.28E−06	
(12000,3)	3.77E−05		2.79E−05		1.94E−05		1.95E−05		5.27E−06	
(12000,2)	3.78E−05		2.81E−05		1.95E−05		1.96E−05		5.33E−06	
(11000,4)	3.49E−05		2.58E−05		1.79E−05		1.81E−05		4.89E−06	
(11000,3)	3.46E−05		2.57E−05		1.78E−05		1.80E−05		4.88E−06	
(11000,2)	3.45E−05		2.57E−05		1.79E−05		1.80E−05		4.90E−06	
(10000,4.5)	3.18E−05		2.36E−05		1.65E−05		1.66E−05		4.55E−06	
(10000,4)	3.17E−05		2.35E−05		1.64E−05		1.65E−05		4.52E−06	
(10000,4,10X)	3.22E−05		2.40E−05		1.67E−05		1.68E−05		4.60E−06	
(10000,4,.1X)	3.15E−05		2.34E−05		1.63E−05		1.64E−05		4.50E−06	
(10000,3.5)	3.15E−05		2.34E−05		1.63E−05		1.64E−05		4.51E−06	
(10000,3)	3.14E−05		2.34E−05		1.64E−05		1.64E−05		4.57E−06	
(10000,2)	3.13E−05		2.33E−05		1.63E−05		1.64E−05		4.54E−06	
(9500,4)	3.01E−05		2.24E−05		1.56E−05		1.57E−05		4.29E−06	
(9500,3)	2.98E−05		2.21E−05		1.54E−05		1.56E−05		4.26E−06	
(9000,4)	2.86E−05		2.13E−05		1.49E−05		1.50E−05		4.13E−06	
(9000,3)	2.82E−05		2.11E−05		1.47E−05		1.49E−05		4.12E−06	
(9000,2)	2.79E−05		2.08E−05		1.46E−05		1.47E−05		4.09E−06	
(8500,4)	2.73E−05	0	2.03E−05	0	1.42E−05	0	1.43E−05	0	3.91E−06	0
(8500,3)	2.69E−05	0	2.01E−05	0	1.40E−05	0	1.41E−05	0	3.89E−06	0
(8000,4.5)	2.57E−05	0	1.92E−05	0	1.34E−05	0	1.34E−05	0	3.67E−06	0
(8000,4)	2.57E−05	0	1.91E−05	0	1.34E−05	0	1.34E−05	0	3.67E−06	0
(8000,4) *	2.57E−05	0	1.91E−05	0	1.34E−05	0	1.34E−05	0	3.67E−06	0
(8000,3.5)	2.55E−05	0	1.90E−05	0	1.33E−05	0	1.34E−05	0	3.67E−06	0
(8000,3)	2.54E−05	0	1.89E−05	0	1.32E−05	0	1.33E−05	0	3.65E−06	0
(8000,2)	2.49E−05	0	1.86E−05	0	1.30E−05	0	1.32E−05	0	3.62E−06	0
(7500,4)	2.40E−05	0	1.79E−05	0	1.25E−05	0	1.26E−05	0	3.44E−06	0
(7500,4) +	2.40E−05	0	1.79E−05	0	1.25E−05	0	1.25E−05	0	3.44E−06	0
(7500,4) *	2.39E−05	0	1.79E−05	0	1.25E−05	0	1.26E−05	0	3.44E−06	0
(7500,3)	2.39E−05	0	1.78E−05	0	1.25E−05	0	1.26E−05	0	3.45E−06	0
(7000,4)	2.24E−05	0	1.68E−05	0	1.17E−05	0	1.18E−05	0	3.23E−06	0
(7000,4) *	2.23E−05	0	1.67E−05	0	1.17E−05	0	1.17E−05	0	3.23E−06	0
(7000,3)	2.24E−05	0	1.68E−05	0	1.18E−05	0	1.18E−05	0	3.25E−06	0
(7000,2)	2.22E−05	0	1.66E−05	0	1.17E−05	0	1.18E−05	0	3.25E−06	0
(6500,4)	2.07E−05	0	1.55E−05	0	1.09E−05	0	1.09E−05	0	3.02E−06	0
(6500,3)	2.08E−05	0	1.57E−05	0	1.10E−05	0	1.10E−05	0	3.04E−06	0
(6000,4.5)	1.89E−05	0	1.43E−05	0	1.00E−05	0	1.00E−05	0	2.79E−06	0
(6000,4)	1.89E−05	0	1.43E−05	0	1.00E−05	0	1.01E−05	0	2.79E−06	0
(6000,4) *	1.89E−05	0	1.43E−05	0	1.00E−05	0	1.01E−05	0	2.79E−06	0
(6000,4) **	1.82E−05		1.37E−05		9.64E−06		9.64E−06		2.69E−06	
(6000,4) * **	1.82E−05		1.37E−05		9.64E−06		9.64E−06		2.69E−06	
(6000,4,10X)	1.79E−05		1.36E−05		9.61E−06		9.61E−06		2.71E−06	
(6000,4,1/10X)	1.81E−05		1.37E−05		9.62E−06		9.62E−06		2.68E−06	
(6000,3.5)	1.91E−05	0	1.44E−05	0	1.01E−05	0	1.01E−05	0	2.81E−06	0
(6000,3)	1.91E−05	0	1.44E−05	0	1.01E−05	0	1.01E−05	0	2.81E−06	0
(6000,2)	1.92E−05	0	1.45E−05	0	1.02E−05	0	1.02E−05	0	2.84E−06	0
(5500,4)	1.72E−05	0	1.30E−05	0	9.16E−06	0	9.17E−06	0	2.56E−06	0
(5500,3)	1.73E−05	0	1.31E−05	0	9.24E−06	0	9.24E−06	0	2.57E−06	0
(5000,4)	1.54E−05	0	1.17E−05	0	8.27E−06	0	8.27E−06	0	2.32E−06	0
(5000,3)	1.56E−05	0	1.18E−05	0	8.34E−06	0	8.34E−06	0	2.34E−06	0
(5000,2)	1.56E−05	0	1.19E−05	0	8.40E−06	0	8.40E−06	0	2.35E−06	0
(4500,4)	1.37E−05	1	1.04E−05	1	7.34E−06	0	7.34E−06	0	2.05E−06	0
(4500,3)	1.38E−05	1	1.05E−05	1	7.40E−06	1	7.40E−06	1	2.07E−06	0
(4000,4)	1.17E−05	1	8.93E−06	1	6.36E−06	1	6.36E−06	1	1.80E−06	0
(4000,3)	1.18E−05	2	9.02E−06	1	6.42E−06	1	6.42E−06	1	1.82E−06	0
(4000,2)	1.19E−05	2	9.07E−06	2	6.46E−06	1	6.46E−06	1	1.83E−06	0

L I M B   D A R K E N I N G

50000-DEGREE, LOG G = 5.0

MU	.20	.30	.36	.40	.45	.50	.55	.60	.70	.80	1.00	1.20	1.60	2.20	MU
.90	979	985	987	987	988	989	990	991	992	993	994	995	996	997	.90
.80	956	969	974	973	975	978	980	982	984	986	988	990	993	995	.80
.70	931	952	960	958	962	966	969	971	976	979	982	985	989	992	.70
.60	904	932	944	941	946	953	957	960	966	970	975	979	984	989	.60
.50	873	911	926	922	929	937	943	947	955	961	967	973	980	985	.50
.40	839	886	906	900	909	920	927	932	943	950	958	965	974	981	.40
.35	819	872	894	887	897	910	918	924	935	944	953	961	971	979	.35
.30	797	856	881	873	884	898	908	914	927	937	947	956	967	976	.30
.25	773	839	867	857	870	886	896	904	918	929	940	950	963	973	.25
.20	746	818	850	838	853	870	883	891	907	920	932	944	958	970	.20
.15	713	794	829	815	832	852	866	875	894	908	922	935	951	965	.15
.10	673	762	802	785	804	827	843	854	875	892	908	923	942	958	.10

40000-DEGREE, LOG G = 5.0

MU	.20	.30	.36	.40	.45	.50	.55	.60	.70	.80	1.00	1.20	1.60	2.20	MU
.90	976	983	985	985	986	987	988	989	990	991	992	993	995	996	.90
.80	951	964	970	969	971	974	976	978	980	982	985	987	990	992	.80
.70	924	944	953	952	955	960	963	965	969	973	976	980	984	989	.70
.60	894	922	934	932	937	944	948	951	957	962	967	972	978	984	.60
.50	860	897	913	910	917	925	931	936	944	950	956	963	971	980	.50
.40	821	869	889	885	893	904	912	917	928	936	944	953	964	974	.40
.35	798	853	876	870	880	892	901	907	919	928	937	947	959	971	.35
.30	774	835	861	853	865	879	888	895	909	919	930	941	955	968	.30
.25	747	815	844	835	847	863	874	882	898	910	921	934	949	964	.25
.20	716	792	824	813	827	845	858	867	885	898	911	926	943	960	.20
.15	679	764	801	787	803	824	838	849	869	885	899	916	936	955	.15
.10	634	730	771	753	772	796	812	825	849	867	884	903	926	948	.10

30000-DEGREE, LOG G = 4.0

MU	.20	.30	.36	.40	.45	.50	.55	.60	.70	.80	1.00	1.20	1.60	2.20	MU
.90	958	971	975	975	977	979	981	982	984	986	987	989	991	993	.90
.80	915	940	949	948	952	957	960	963	968	971	974	978	981	985	.80
.70	868	906	922	919	925	933	938	942	949	955	960	965	971	977	.70
.60	817	871	891	887	895	906	913	919	929	937	944	951	960	968	.60
.50	762	831	858	851	862	876	885	894	907	917	926	936	947	958	.50
.40	701	787	820	810	824	841	854	864	881	894	905	918	932	946	.40
.35	668	762	799	787	802	822	836	847	866	880	893	907	923	939	.35
.30	633	736	776	761	779	801	816	829	850	866	879	895	913	931	.30
.25	595	707	751	733	752	777	793	808	831	849	864	882	902	922	.25
.20	553	674	723	701	722	749	768	784	810	830	846	867	889	912	.20
.15	508	637	689	664	687	716	737	755	784	806	824	847	873	899	.15
.10	457	592	648	617	642	675	698	718	751	776	795	822	851	883	.10

25000-DEGREE, LOG G = 4.0

MU	.20	.30	.36	.40	.45	.50	.55	.60	.70	.80	1.00	1.20	1.60	2.20	MU
.90	951	967	973	973	975	977	979	981	983	985	987	989	991	993	.90
.80	899	933	946	944	948	953	957	961	966	970	973	977	982	986	.80
.70	845	897	916	912	919	928	933	939	947	954	959	965	972	979	.70
.60	787	858	884	878	887	899	907	915	926	935	942	951	961	970	.60
.50	725	816	849	839	852	867	878	888	903	915	924	936	948	961	.50
.40	658	769	811	796	811	831	845	857	877	892	903	919	934	950	.40
.35	621	744	790	771	789	811	827	840	863	880	892	909	926	943	.35
.30	583	716	767	745	764	789	806	822	847	865	879	898	917	936	.30
.25	543	687	743	715	737	764	784	801	829	850	864	885	907	928	.25
.20	499	655	715	682	706	736	758	777	808	832	847	871	895	918	.20
.15	452	618	684	644	670	704	728	750	784	810	827	853	880	905	.15
.10	401	575	646	597	626	664	691	715	753	782	801	830	859	888	.10

LIMB DARKENING

20000-DEGREE, LOG G = 4.0

MU	.20	.30	.36	.40	.45	.50	.55	.60	.70	.80	1.00	1.20	1.60	2.20	MU
.90	944	964	972	971	973	976	978	980	982	985	986	988	991	993	.90
.80	885	927	942	940	944	950	954	958	964	969	972	977	981	987	.80
.70	824	888	912	906	913	922	929	935	944	952	957	964	971	980	.70
.60	760	847	879	869	879	892	901	909	923	933	940	950	960	972	.60
.50	691	802	843	828	841	858	871	881	899	913	921	935	948	963	.50
.40	617	753	805	781	798	820	836	850	872	890	900	917	934	954	.40
.35	578	727	784	755	774	799	816	832	857	877	888	907	926	948	.35
.30	537	699	762	727	748	775	795	813	841	863	875	897	918	942	.30
.25	494	670	739	695	719	750	772	791	823	848	860	885	909	935	.25
.20	449	638	713	661	687	721	746	768	803	831	844	872	898	928	.20
.15	401	603	685	620	650	688	716	740	780	811	825	856	885	918	.15
.10	349	563	652	573	605	648	679	707	752	786	801	836	869	906	.10

20000-DEGREE, LOG G = 3.0

MU	.20	.30	.36	.40	.45	.50	.55	.60	.70	.80	1.00	1.20	1.60	2.20	MU
.90	940	961	969	966	969	972	975	977	980	982	984	987	989	992	.90
.80	879	921	936	931	936	943	948	952	959	964	967	973	978	984	.80
.70	816	878	902	892	901	911	919	925	936	944	949	958	966	975	.70
.60	749	833	866	851	862	877	887	896	911	922	929	941	953	966	.60
.50	680	785	827	804	819	838	852	864	883	898	907	922	937	955	.50
.40	606	733	784	753	771	795	812	827	852	871	881	901	920	943	.40
.35	568	705	761	724	744	770	789	806	834	855	866	888	910	936	.35
.30	529	675	736	693	715	744	765	784	815	838	850	875	899	927	.30
.25	489	644	710	660	684	715	738	759	793	820	832	860	887	918	.25
.20	447	610	681	623	649	683	708	731	769	798	811	842	872	907	.20
.15	405	573	649	582	609	646	673	698	740	772	785	820	854	894	.15
.10	361	531	611	534	562	601	631	658	704	740	752	792	829	874	.10

18000-DEGREE, LOG G = 4.0

MU	.20	.30	.36	.40	.45	.50	.55	.60	.70	.80	1.00	1.20	1.60	2.20	MU
.90	940	962	971	969	972	975	977	979	982	984	986	988	990	993	.90
.80	878	923	940	937	942	948	952	956	963	968	971	976	981	986	.80
.70	812	882	909	902	909	919	926	932	942	950	955	962	970	979	.70
.60	744	839	875	863	874	887	897	905	920	931	937	948	958	971	.60
.50	672	792	839	820	834	852	865	876	895	909	917	932	946	962	.50
.40	595	742	801	772	789	812	828	843	867	886	895	914	931	952	.40
.35	554	715	780	744	764	790	808	825	852	873	883	904	923	947	.35
.30	512	687	758	715	737	765	786	805	835	859	869	893	915	941	.30
.25	468	657	735	682	707	739	762	783	817	843	854	881	905	935	.25
.20	421	625	710	646	673	709	735	758	796	826	838	867	895	927	.20
.15	373	590	683	605	635	675	704	730	773	806	818	852	882	918	.15
.10	321	551	652	555	589	634	667	696	745	782	794	832	866	907	.10

18000-DEGREE, LOG G = 3.0

MU	.20	.30	.36	.40	.45	.50	.55	.60	.70	.80	1.00	1.20	1.60	2.20	MU
.90	938	960	969	966	969	972	974	977	980	982	984	987	989	992	.90
.80	874	919	936	930	936	943	948	952	959	964	967	973	978	984	.80
.70	807	876	902	891	900	911	918	925	936	944	949	958	966	976	.70
.60	738	830	866	849	861	876	886	896	911	923	929	941	953	966	.60
.50	666	782	827	803	818	837	851	863	883	899	907	922	938	956	.50
.40	590	729	785	750	769	793	811	826	852	872	881	901	920	944	.40
.35	551	701	763	721	742	769	789	806	835	857	866	889	911	937	.35
.30	510	672	739	690	713	743	764	784	816	840	850	876	900	929	.30
.25	469	641	714	656	681	714	738	759	794	822	832	861	888	921	.25
.20	426	608	686	619	645	681	708	731	771	801	812	844	874	911	.20
.15	383	571	656	577	605	644	673	699	743	777	787	823	856	898	.15
.10	339	530	620	528	558	600	631	660	708	746	755	796	833	881	.10

LIMB DARKENING

16000-DEGREE, LOG G = 4.0

MU	.20	.30	.36	.40	.45	.50	.55	.60	.70	.80	1.00	1.20	1.60	2.20	MU
.90	934	960	969	968	970	973	975	977	981	983	985	987	990	993	.90
.80	867	918	938	933	938	945	950	954	961	966	969	974	979	985	.80
.70	797	874	905	896	904	914	922	928	939	947	952	960	968	978	.70
.60	723	828	871	855	866	881	891	900	915	927	933	945	956	969	.60
.50	646	780	835	810	824	843	857	869	889	905	913	928	942	960	.50
.40	565	727	796	759	777	801	819	834	860	880	889	909	927	950	.40
.35	523	700	775	730	751	778	797	815	844	867	876	898	919	944	.35
.30	479	671	753	699	722	752	774	794	827	852	862	887	910	938	.30
.25	434	641	730	665	691	724	749	771	808	836	846	874	900	932	.25
.20	387	609	706	627	656	693	721	745	786	818	829	860	889	924	.20
.15	338	574	679	584	615	658	689	716	762	798	808	844	876	915	.15
.10	288	536	649	533	568	616	651	682	734	774	784	824	860	905	.10

16000-DEGREE, LOG G = 3.0

MU	.20	.30	.36	.40	.45	.50	.55	.60	.70	.80	1.00	1.20	1.60	2.20	MU
.90	933	959	968	965	968	971	974	976	979	982	984	986	989	992	.90
.80	865	916	935	928	934	941	946	951	958	964	967	972	977	984	.80
.70	795	871	901	889	897	908	916	923	935	943	948	957	965	975	.70
.60	721	824	865	845	857	872	883	893	909	922	927	940	951	966	.60
.50	645	774	826	797	813	833	847	860	881	897	904	921	936	955	.50
.40	566	720	785	743	763	788	806	822	849	870	878	899	919	943	.40
.35	526	692	763	714	735	763	783	801	831	855	863	887	909	936	.35
.30	484	662	740	682	705	736	759	779	812	838	847	874	898	929	.30
.25	441	631	715	647	672	706	731	754	791	820	829	859	886	920	.25
.20	397	598	689	608	636	673	701	726	767	800	808	842	872	911	.20
.15	353	562	660	565	595	636	666	693	739	776	784	822	855	899	.15
.10	309	523	626	516	547	591	624	654	705	746	753	796	833	883	.10

14000-DEGREE, LOG G = 4.0

MU	.20	.30	.36	.40	.45	.50	.55	.60	.70	.80	1.00	1.20	1.60	2.20	MU
.90	927	956	968	965	968	971	974	976	979	982	984	986	989	992	.90
.80	852	911	935	929	934	941	946	950	958	964	967	972	978	984	.80
.70	775	864	901	889	897	908	916	923	935	944	949	957	966	976	.70
.60	695	816	866	845	857	872	883	893	909	923	929	941	952	967	.60
.50	613	764	828	797	812	832	847	860	882	899	906	923	938	957	.50
.40	527	710	788	743	762	787	806	823	851	873	881	902	922	946	.40
.35	483	682	767	712	734	762	783	802	833	858	867	891	912	940	.35
.30	438	652	745	680	703	735	759	780	815	843	852	879	903	933	.30
.25	392	622	722	644	670	706	732	755	795	826	835	865	892	926	.25
.20	345	589	698	603	633	673	702	728	772	807	817	850	880	918	.20
.15	298	555	672	558	591	635	668	698	747	786	795	833	866	909	.15
.10	249	517	642	504	541	591	628	661	717	761	769	812	849	897	.10

14000-DEGREE, LOG G = 3.0

MU	.20	.30	.36	.40	.45	.50	.55	.60	.70	.80	1.00	1.20	1.60	2.20	MU
.90	927	955	967	963	966	970	972	975	978	981	983	986	988	992	.90
.80	852	910	933	925	930	938	943	948	956	962	965	971	976	983	.80
.70	775	862	898	883	892	903	912	919	931	941	945	955	963	974	.70
.60	696	812	861	837	850	866	877	888	905	918	924	937	949	964	.60
.50	615	760	823	787	803	824	839	853	875	893	900	917	933	953	.50
.40	531	705	781	731	751	777	796	814	842	865	872	895	914	940	.40
.35	488	676	759	700	722	751	773	792	824	849	857	882	904	933	.35
.30	445	646	736	666	690	723	747	768	804	832	840	868	893	926	.30
.25	401	615	712	630	656	692	719	742	782	814	822	853	880	917	.25
.20	357	582	687	590	619	658	687	714	758	793	800	836	866	907	.20
.15	313	547	659	546	576	619	652	681	730	769	775	815	849	895	.15
.10	269	508	627	495	527	574	609	641	696	740	744	789	827	880	.10

L I M B   D A R K E N I N G

12000-DEGREE, LOG G = 4.0

MU	.20	.30	.36	.40	.45	.50	.55	.60	.70	.80	1.00	1.20	1.60	2.20	MU
.90	916	951	966	962	965	968	971	973	977	981	982	985	988	991	.90
.80	830	902	930	922	928	935	941	945	954	960	964	970	975	982	.80
.70	744	851	894	879	887	899	908	915	928	939	944	953	962	973	.70
.60	656	798	857	831	843	860	872	882	900	915	922	935	947	963	.60
.50	566	744	817	779	795	816	832	846	870	889	898	915	931	951	.50
.40	476	687	776	720	740	767	787	805	836	860	870	892	913	938	.40
.35	430	657	754	687	710	740	763	783	817	845	854	880	902	931	.35
.30	385	627	732	652	677	711	736	759	797	827	838	866	891	924	.30
.25	339	595	708	613	641	678	707	732	775	809	819	851	879	916	.25
.20	294	562	683	570	601	643	674	703	750	788	798	834	865	906	.20
.15	248	528	656	521	555	602	638	669	723	765	774	815	850	896	.15
.10	204	490	626	465	502	555	594	630	690	738	745	792	830	883	.10

12000-DEGREE, LOG G = 3.0

MU	.20	.30	.36	.40	.45	.50	.55	.60	.70	.80	1.00	1.20	1.60	2.20	MU
.90	915	951	965	960	963	967	970	972	977	980	982	984	987	991	.90
.80	830	900	929	919	925	933	938	943	952	959	962	968	974	981	.80
.70	744	849	892	874	883	895	904	912	926	936	941	951	960	971	.70
.60	656	796	854	825	837	855	867	878	897	912	918	932	944	960	.60
.50	568	740	814	771	787	810	826	841	865	885	892	910	927	948	.50
.40	479	683	771	710	731	759	780	798	830	854	863	886	907	934	.40
.35	435	653	749	677	700	731	754	775	810	838	846	873	896	926	.35
.30	390	622	726	642	667	701	727	750	789	820	828	858	883	918	.30
.25	346	590	702	603	631	669	697	722	766	800	808	841	870	909	.25
.20	303	557	676	561	591	632	664	692	740	778	785	823	854	898	.20
.15	260	522	649	514	546	592	626	657	710	753	759	801	836	885	.15
.10	219	485	618	461	495	544	581	616	675	723	726	774	813	870	.10

12000-DEGREE, LOG G = 2.0

MU	.20	.30	.36	.40	.45	.50	.55	.60	.70	.80	1.00	1.20	1.60	2.20	MU
.90	916	948	962	956	959	963	967	969	974	977	979	982	985	989	.90
.80	833	896	923	910	916	925	931	937	946	954	956	963	969	978	.80
.70	749	842	883	860	870	884	893	902	916	928	931	942	952	966	.70
.60	664	786	842	807	821	839	852	864	884	900	904	920	933	952	.60
.50	581	729	798	750	767	790	807	822	848	869	873	894	912	937	.50
.40	497	669	752	688	708	736	757	776	808	834	838	864	887	919	.40
.35	456	638	728	655	677	707	729	750	786	814	818	848	872	909	.35
.30	416	606	703	620	643	675	700	722	761	793	796	829	857	897	.30
.25	376	573	676	583	607	641	668	692	735	770	772	809	839	884	.25
.20	338	539	648	543	568	604	632	659	705	743	744	785	818	869	.20
.15	301	503	617	501	526	563	593	621	671	713	711	756	792	850	.15
.10	268	465	582	455	479	517	547	576	629	675	670	719	759	825	.10

11000-DEGREE, LOG G = 4.0

MU	.20	.30	.36	.40	.45	.50	.55	.60	.70	.80	1.00	1.20	1.60	2.20	MU
.90	908	948	964	960	963	966	969	972	976	979	981	984	987	991	.90
.80	815	895	927	917	923	931	937	942	950	958	962	968	974	981	.80
.70	722	841	889	871	880	892	901	909	923	934	940	950	959	971	.70
.60	628	786	850	821	833	850	863	874	894	909	917	931	944	959	.60
.50	535	729	809	765	782	804	821	836	861	882	891	909	926	947	.50
.40	442	670	767	703	724	752	773	792	825	851	862	885	906	933	.40
.35	396	640	744	669	692	723	747	768	805	834	845	872	895	926	.35
.30	350	609	721	631	657	692	719	743	783	816	827	857	883	917	.30
.25	305	577	697	591	619	658	688	715	760	796	807	841	870	908	.25
.20	261	544	672	546	577	621	654	683	734	774	785	823	855	898	.20
.15	218	509	645	495	530	578	615	648	705	750	760	802	838	887	.15
.10	176	471	615	436	475	529	570	607	670	721	729	777	817	873	.10

L I M B   D A R K E N I N G

11000-DEGREE, LOG G = 3.0

MU	.20	.30	.36	.40	.45	.50	.55	.60	.70	.80	1.00	1.20	1.60	2.20	MU
.90	907	947	963	958	961	965	968	971	975	979	981	984	987	990	.90
.80	814	894	926	914	921	929	935	940	949	956	960	967	973	980	.80
.70	721	840	888	867	876	889	899	907	921	933	938	948	958	969	.70
.60	628	785	849	815	829	846	859	871	891	907	914	928	941	958	.60
.50	536	727	807	759	776	799	816	832	858	878	887	906	923	944	.50
.40	444	668	764	696	717	746	768	787	820	847	856	880	901	930	.40
.35	399	638	742	661	685	717	741	763	800	829	838	866	890	921	.35
.30	355	607	718	624	650	686	712	736	777	810	820	850	877	912	.30
.25	311	575	694	584	613	652	681	708	753	789	799	833	862	902	.25
.20	269	541	668	541	571	614	646	676	726	766	775	813	846	891	.20
.15	228	507	641	492	525	571	607	639	695	740	747	791	827	878	.15
.10	189	469	610	437	472	522	561	597	659	709	713	763	802	861	.10

11000-DEGREE, LOG G = 2.0

MU	.20	.30	.36	.40	.45	.50	.55	.60	.70	.80	1.00	1.20	1.60	2.20	MU
.90	908	945	961	954	958	962	965	968	973	977	978	982	985	989	.90
.80	818	890	921	907	913	922	929	935	944	952	955	962	969	977	.80
.70	727	834	880	856	866	880	890	899	914	926	930	941	951	964	.70
.60	637	776	838	815	834	847	860	880	897	902	918	931	950		.60
.50	548	717	794	743	760	783	801	817	844	865	870	891	909	935	.50
.40	462	656	748	679	699	728	749	769	802	830	835	861	884	916	.40
.35	419	624	724	644	666	697	721	743	780	810	815	845	869	906	.35
.30	378	592	699	608	632	665	691	714	755	788	792	826	853	894	.30
.25	338	559	673	570	595	630	658	683	728	765	768	805	835	881	.25
.20	299	525	645	529	555	592	622	649	698	738	740	781	814	866	.20
.15	263	490	615	486	512	550	582	611	663	707	707	752	789	847	.15
.10	231	452	581	439	464	503	535	566	622	670	665	716	756	822	.10

10000-DEGREE, LOG G = 4.5

MU	.20	.30	.36	.40	.45	.50	.55	.60	.70	.80	1.00	1.20	1.60	2.20	MU
.90	900	946	964	957	960	964	967	970	974	978	980	984	987	991	.90
.80	800	890	926	910	918	925	932	938	947	955	959	966	973	980	.80
.70	700	834	887	860	872	883	893	903	918	931	936	948	957	969	.70
.60	602	777	848	805	822	837	852	865	886	904	911	927	941	958	.60
.50	505	720	807	744	766	787	806	823	852	875	884	904	922	945	.50
.40	409	660	765	676	704	730	755	777	813	842	852	879	901	930	.40
.35	363	630	743	639	670	699	727	752	792	825	835	864	890	922	.35
.30	318	599	720	598	633	666	697	724	770	806	816	849	877	913	.30
.25	274	568	696	554	593	630	665	695	745	786	795	832	863	904	.25
.20	232	535	671	506	549	590	629	662	719	763	772	814	848	894	.20
.15	192	501	645	452	501	546	589	626	689	738	746	793	830	882	.15
.10	154	465	616	391	445	496	543	585	654	709	715	767	809	868	.10

10000-DEGREE, LOG G = 4.0

MU	.20	.30	.36	.40	.45	.50	.55	.60	.70	.80	1.00	1.20	1.60	2.20	MU
.90	899	946	964	957	960	964	967	970	974	978	980	984	987	990	.90
.80	797	891	927	911	918	925	932	938	947	955	959	966	973	980	.80
.70	698	836	889	861	873	883	894	903	918	931	937	947	957	969	.70
.60	599	779	850	807	823	838	853	865	887	904	912	927	941	958	.60
.50	502	722	810	747	768	788	807	824	852	875	884	904	922	945	.50
.40	408	663	768	680	707	732	757	777	814	843	853	878	901	930	.40
.35	362	633	746	643	673	701	729	752	793	825	835	864	890	922	.35
.30	317	603	723	603	636	668	699	725	770	806	816	849	877	913	.30
.25	274	572	700	559	597	632	666	695	746	786	795	832	863	904	.25
.20	232	539	675	511	553	593	630	663	719	763	772	813	847	893	.20
.15	193	506	649	458	505	548	590	627	689	738	745	792	829	881	.15
.10	156	469	620	398	449	497	544	585	654	708	714	766	808	866	.10

L I M B   D A R K E N I N G

10000-DEGREE, LOG G = 4.0   (10X METALS)

MU	.20	.30	.36	.40	.45	.50	.55	.60	.70	.80	1.00	1.20	1.60	2.20	MU
.90	902	948	965	958	961	965	968	971	975	979	981	984	987	991	.90
.80	803	895	929	913	920	927	934	940	949	957	961	967	974	981	.80
.70	706	842	892	864	876	887	897	906	921	933	939	949	959	971	.70
.60	609	787	854	811	827	843	857	870	891	907	915	929	943	959	.60
.50	514	731	815	752	774	794	813	830	857	879	888	907	925	947	.50
.40	420	674	774	686	714	740	764	785	820	848	858	883	905	933	.40
.35	375	644	753	650	681	710	737	760	800	831	841	869	894	925	.35
.30	330	614	730	611	646	678	708	734	778	813	822	854	881	917	.30
.25	287	583	707	568	607	642	676	705	754	793	802	838	868	907	.25
.20	245	551	682	522	564	604	642	674	728	771	780	820	853	897	.20
.15	205	517	656	470	517	561	603	638	699	746	754	799	836	885	.15
.10	167	480	626	410	462	511	557	597	664	717	723	774	814	871	.10

10000-DEGREE, LOG G = 4.0   (1/10X METALS)

MU	.20	.30	.36	.40	.45	.50	.55	.60	.70	.80	1.00	1.20	1.60	2.20	MU
.90	898	945	963	957	960	964	967	969	974	978	980	983	987	990	.90
.80	796	889	925	911	918	925	931	937	947	955	959	966	972	980	.80
.70	696	833	886	860	872	883	893	902	917	930	936	947	957	969	.70
.60	596	775	846	806	822	837	851	864	885	902	911	926	940	957	.60
.50	499	717	805	746	767	786	805	822	850	873	883	903	921	943	.50
.40	404	657	762	678	705	730	754	775	811	840	851	877	900	928	.40
.35	358	627	740	640	671	699	726	749	790	822	833	862	888	920	.35
.30	313	596	717	600	634	665	696	722	767	803	814	846	875	911	.30
.25	270	564	693	557	594	629	663	692	742	782	793	829	861	901	.25
.20	228	531	668	508	550	589	626	659	715	759	769	810	845	890	.20
.15	188	497	642	455	501	544	586	622	684	733	742	788	826	878	.15
.10	151	461	612	393	444	493	539	579	648	703	709	762	804	863	.10

10000-DEGREE, LOG G = 3.5

MU	.20	.30	.36	.40	.45	.50	.55	.60	.70	.80	1.00	1.20	1.60	2.20	MU
.90	898	945	964	957	960	964	967	970	974	978	980	984	987	990	.90
.80	796	890	927	911	918	925	931	937	947	955	959	966	972	980	.80
.70	695	834	889	861	872	883	893	902	917	930	936	947	957	969	.70
.60	597	778	850	806	822	837	852	864	886	903	911	926	940	957	.60
.50	500	721	810	747	768	788	806	823	851	874	883	903	921	944	.50
.40	405	662	768	680	706	731	755	776	812	841	851	877	900	929	.40
.35	360	632	746	643	673	700	727	751	791	823	833	863	888	921	.35
.30	316	602	723	603	636	667	697	723	768	804	814	847	875	912	.30
.25	273	570	700	561	597	631	665	694	743	783	793	829	860	902	.25
.20	232	538	675	513	554	592	629	661	716	760	769	810	844	891	.20
.15	193	505	649	461	505	547	588	624	685	735	742	788	826	878	.15
.10	157	469	620	401	450	496	541	581	649	704	708	761	803	862	.10

10000-DEGREE, LOG G = 3.0

MU	.20	.30	.36	.40	.45	.50	.55	.60	.70	.80	1.00	1.20	1.60	2.20	MU
.90	897	945	963	956	960	963	967	969	974	978	980	983	986	990	.90
.80	795	889	925	909	917	924	931	937	946	954	958	965	972	979	.80
.70	695	833	887	859	871	882	892	901	916	929	935	946	956	968	.70
.60	596	776	848	804	821	836	851	863	884	902	909	924	938	956	.60
.50	499	718	807	744	766	786	805	821	849	871	881	901	919	942	.50
.40	405	659	765	677	705	729	753	774	809	838	848	874	896	926	.40
.35	360	629	743	641	671	698	725	748	788	820	830	859	884	918	.35
.30	317	598	720	602	635	665	695	720	764	800	810	842	870	908	.30
.25	274	567	697	560	595	629	662	690	739	779	788	824	855	898	.25
.20	234	535	672	513	553	590	625	657	711	755	763	804	838	886	.20
.15	196	501	646	462	505	545	585	619	679	728	734	781	818	872	.15
.10	161	465	617	405	450	494	537	575	642	696	700	752	794	856	.10

LIMB DARKENING

10000-DEGREE, LOG G = 2.0

MU	.20	.30	.36	.40	.45	.50	.55	.60	.70	.80	1.00	1.20	1.60	2.20	MU
.90	899	943	961	953	957	961	965	967	972	976	978	982	985	989	.90
.80	799	884	921	904	912	920	927	933	943	951	954	962	968	977	.80
.70	700	826	880	851	864	876	886	895	911	924	929	940	950	964	.70
.60	603	767	838	795	812	828	842	855	877	894	900	916	930	949	.60
.50	510	707	794	734	756	776	794	811	839	861	868	889	908	933	.50
.40	420	645	749	668	694	718	741	762	796	825	832	859	882	915	.40
.35	377	614	726	632	660	687	712	734	773	805	811	842	867	904	.35
.30	335	583	702	595	625	653	681	705	748	783	789	823	851	893	.30
.25	295	550	677	555	587	617	647	674	720	759	764	802	833	880	.25
.20	257	517	650	513	546	578	610	639	690	732	736	778	812	864	.20
.15	222	483	622	468	501	535	569	600	655	702	703	750	787	846	.15
.10	191	447	591	419	452	487	522	555	614	665	662	714	754	822	.10

9500-DEGREE, LOG G = 4.0

MU	.20	.30	.36	.40	.45	.50	.55	.60	.70	.80	1.00	1.20	1.60	2.20	MU
.90	893	943	962	954	958	961	965	968	973	977	979	983	986	990	.90
.80	786	886	924	905	912	920	927	934	944	953	957	964	971	979	.80
.70	681	828	885	851	864	876	887	897	913	927	933	945	955	968	.70
.60	579	770	845	793	810	827	843	857	880	899	906	923	938	955	.60
.50	479	711	804	729	752	774	795	813	844	868	877	899	918	941	.50
.40	384	651	761	658	687	715	741	764	803	834	844	872	896	926	.40
.35	339	621	739	618	651	682	712	738	781	816	826	857	884	918	.35
.30	295	590	716	576	613	648	681	709	758	796	806	841	870	908	.30
.25	253	558	692	531	572	610	647	679	733	775	784	823	856	898	.25
.20	213	526	667	481	527	569	610	645	705	752	760	804	839	887	.20
.15	175	492	641	427	477	524	569	608	674	726	732	781	820	874	.15
.10	141	456	612	366	421	473	522	565	638	696	700	755	798	859	.10

9500-DEGREE, LOG G = 3.0

MU	.20	.30	.36	.40	.45	.50	.55	.60	.70	.80	1.00	1.20	1.60	2.20	MU
.90	891	943	962	954	958	962	965	968	973	977	979	983	986	990	.90
.80	783	886	924	905	913	920	927	933	944	952	957	964	971	979	.80
.70	676	828	885	853	864	876	887	896	913	926	932	944	954	967	.70
.60	574	769	845	795	812	828	843	856	879	897	905	921	936	954	.60
.50	475	710	803	733	754	775	795	812	842	866	876	897	916	939	.50
.40	381	650	760	663	690	716	741	763	801	831	842	869	893	923	.40
.35	336	620	738	625	655	684	712	736	779	813	823	853	880	914	.35
.30	293	589	715	584	618	650	680	708	755	792	802	837	866	904	.30
.25	253	558	692	540	577	612	646	676	728	770	780	818	850	893	.25
.20	214	525	667	492	533	571	609	642	700	746	754	797	832	881	.20
.15	178	492	640	440	483	526	567	604	667	718	725	773	812	866	.15
.10	146	456	611	380	428	474	519	559	629	685	689	743	786	849	.10

9000-DEGREE, LOG G = 4.0

MU	.20	.30	.36	.40	.45	.50	.55	.60	.70	.80	1.00	1.20	1.60	2.20	MU
.90	886	941	961	949	954	958	962	966	971	976	978	982	986	989	.90
.80	775	882	922	895	905	913	922	929	941	950	954	963	970	978	.80
.70	666	823	881	836	852	866	879	890	909	923	929	942	953	966	.70
.60	561	763	840	774	795	813	832	848	874	894	901	919	935	953	.60
.50	460	703	797	704	732	757	781	802	836	862	870	894	915	938	.50
.40	365	641	754	628	663	694	725	751	794	827	835	866	892	922	.40
.35	320	610	731	586	625	661	694	723	771	808	816	850	879	913	.35
.30	277	579	707	542	585	625	662	694	747	788	796	834	865	904	.30
.25	237	547	683	495	542	586	627	663	721	766	773	815	849	893	.25
.20	198	514	658	445	497	545	590	629	693	742	748	795	832	881	.20
.15	163	480	631	391	447	500	549	591	662	715	720	772	813	868	.15
.10	131	444	601	332	393	450	503	549	626	685	688	745	789	852	.10

L I M B   D A R K E N I N G

9000-DEGREE, LOG G = 3.0

MU	.20	.30	.36	.40	.45	.50	.55	.60	.70	.80	1.00	1.20	1.60	2.20	MU
.90	883	942	962	951	955	959	963	966	971	976	978	982	985	989	.90
.80	770	884	923	898	907	915	923	930	941	950	955	963	970	978	.80
.70	661	825	884	842	855	868	880	891	908	923	929	942	953	966	.70
.60	554	766	844	781	800	817	834	849	873	893	901	918	934	952	.60
.50	453	707	802	714	739	762	784	802	835	861	870	893	913	937	.50
.40	359	647	759	641	671	700	727	752	793	826	835	864	889	920	.40
.35	315	617	737	601	635	666	697	724	770	806	815	848	876	911	.35
.30	274	585	714	558	595	630	665	694	745	785	794	831	861	901	.30
.25	235	554	690	512	553	592	629	662	718	763	771	812	845	889	.25
.20	199	522	665	463	507	550	591	627	689	738	745	790	827	877	.20
.15	166	488	638	409	457	504	549	588	656	710	715	766	806	862	.15
.10	137	452	609	349	402	452	501	544	618	677	679	735	779	844	.10

9000-DEGREE, LOG G = 2.0

MU	.20	.30	.36	.40	.45	.50	.55	.60	.70	.80	1.00	1.20	1.60	2.20	MU
.90	881	940	961	950	954	959	962	965	970	975	977	981	984	988	.90
.80	767	880	921	898	906	914	921	928	939	948	952	960	967	976	.80
.70	659	819	880	842	855	866	878	888	905	919	925	938	949	962	.70
.60	553	759	838	782	799	815	831	845	868	888	895	913	928	948	.60
.50	453	699	796	717	739	760	780	798	828	854	863	885	905	931	.50
.40	362	637	751	646	673	699	723	746	784	816	825	854	879	912	.40
.35	319	606	728	608	637	665	692	717	760	795	804	836	864	901	.35
.30	279	575	704	568	599	629	659	686	734	773	781	817	847	889	.30
.25	241	543	679	525	558	591	624	654	705	748	756	796	829	875	.25
.20	207	510	653	479	515	550	585	617	674	721	727	771	807	860	.20
.15	177	476	625	430	467	505	542	577	639	690	693	742	781	841	.15
.10	151	440	594	378	415	454	494	530	597	653	652	706	749	817	.10

8500-DEGREE, LOG G = 4.0

MU	.20	.30	.36	.40	.45	.50	.55	.60	.70	.80	1.00	1.20	1.60	2.20	MU
.90	881	938	957	943	949	954	959	963	969	974	976	981	985	989	.90
.80	765	876	913	882	894	904	915	923	937	947	950	960	968	977	.80
.70	653	812	869	817	835	852	867	881	902	918	923	938	951	963	.70
.60	545	748	822	747	772	795	817	835	864	886	892	913	931	949	.60
.50	442	683	774	672	704	734	762	786	824	852	859	887	909	933	.50
.40	347	617	723	589	630	668	702	731	779	814	822	856	885	915	.40
.35	302	583	696	546	591	632	670	702	755	794	802	839	871	905	.35
.30	260	548	668	500	549	594	636	671	729	772	779	821	856	894	.30
.25	219	512	639	451	505	554	600	639	701	748	756	801	839	882	.25
.20	182	475	607	401	459	512	561	603	671	722	729	779	820	868	.20
.15	149	435	573	348	410	467	519	564	638	693	700	754	798	852	.15
.10	119	391	534	293	357	417	473	521	599	659	665	724	771	832	.10

8500-DEGREE, LOG G = 3.0

MU	.20	.30	.36	.40	.45	.50	.55	.60	.70	.80	1.00	1.20	1.60	2.20	MU
.90	878	940	960	947	951	956	960	964	970	975	977	981	985	989	.90
.80	762	879	918	890	900	909	918	926	938	948	952	961	969	977	.80
.70	648	817	875	829	845	859	873	884	903	919	925	939	951	964	.70
.60	539	755	832	764	785	805	824	840	866	888	895	914	931	949	.60
.50	437	693	786	692	720	746	770	791	826	854	862	888	909	933	.50
.40	344	629	738	614	648	681	711	737	782	816	825	858	885	915	.40
.35	300	597	713	571	610	645	679	708	758	796	805	841	871	905	.35
.30	260	563	687	527	568	608	645	677	732	774	782	822	855	894	.30
.25	222	529	659	479	524	567	608	644	704	750	758	802	838	882	.25
.20	187	494	630	427	477	524	569	608	673	724	730	780	819	868	.20
.15	156	456	597	372	426	477	525	568	639	694	699	753	796	851	.15
.10	129	414	560	313	370	424	476	522	598	658	662	721	768	830	.10

LIMB DARKENING

8000-DEGREE, LOG G = 4.5

MU	.20	.30	.36	.40	.45	.50	.55	.60	.70	.80	1.00	1.20	1.60	2.20	MU
.90	899	931	950	945	949	952	956	959	965	971	973	978	983	987	.90
.80	794	860	899	885	893	901	908	916	929	939	944	954	964	973	.80
.70	686	788	846	819	832	844	857	869	889	906	913	929	944	958	.70
.60	577	716	793	747	765	783	801	818	847	870	879	901	922	942	.60
.50	467	643	736	667	692	717	741	764	802	831	842	870	897	924	.50
.40	360	569	677	579	612	645	677	705	752	790	801	836	869	903	.40
.35	308	532	646	532	570	607	642	673	726	767	779	817	854	892	.35
.30	257	494	614	484	527	568	606	641	699	744	756	797	837	879	.30
.25	211	454	580	434	481	527	569	607	670	719	732	776	818	866	.25
.20	171	414	543	382	435	485	531	572	640	691	706	753	797	850	.20
.15	135	371	504	331	388	442	491	534	606	661	677	726	773	832	.15
.10	106	325	460	281	341	397	448	493	569	627	644	695	744	811	.10

8000-DEGREE, LOG G = 4.0

MU	.20	.30	.36	.40	.45	.50	.55	.60	.70	.80	1.00	1.20	1.60	2.20	MU
.90	888	932	952	941	946	951	956	959	966	971	973	979	983	988	.90
.80	775	864	904	877	889	899	908	915	930	940	945	955	965	974	.80
.70	661	796	853	809	827	842	856	869	891	908	914	930	946	960	.70
.60	551	726	801	735	760	781	801	819	849	873	880	903	924	944	.60
.50	445	656	747	655	686	714	741	765	805	835	843	873	900	926	.50
.40	340	584	690	567	607	643	676	706	756	794	803	840	873	906	.40
.35	292	547	660	520	565	604	642	674	730	772	781	821	858	894	.35
.30	247	510	629	473	521	564	606	642	702	748	757	801	841	882	.30
.25	206	471	596	423	475	523	568	608	673	723	732	780	823	868	.25
.20	169	431	561	371	428	480	529	572	642	695	705	756	802	853	.20
.15	137	389	523	319	380	436	488	533	608	664	675	729	778	835	.15
.10	111	343	480	268	331	389	443	490	569	629	641	698	748	813	.10

8000-DEGREE, LOG G = 4.0   (NO CONV)

MU	.20	.30	.36	.40	.45	.50	.55	.60	.70	.80	1.00	1.20	1.60	2.20	MU
.90	874	934	953	933	940	947	953	958	966	972	973	979	984	988	.90
.80	751	867	905	862	877	891	903	914	930	942	945	956	966	974	.80
.70	633	799	855	788	811	832	851	867	891	910	914	931	947	960	.70
.60	522	730	803	709	740	769	796	817	850	875	881	905	926	944	.60
.50	415	661	749	625	666	703	736	763	806	838	844	875	902	926	.50
.40	320	588	692	538	587	632	672	706	758	797	804	842	875	906	.40
.35	276	552	662	492	546	594	638	675	732	775	783	824	860	895	.35
.30	235	514	631	445	503	556	603	643	705	751	759	804	843	883	.30
.25	197	476	598	398	460	516	567	609	676	726	735	783	825	869	.25
.20	162	435	563	350	415	475	528	573	645	698	707	759	804	854	.20
.15	132	392	524	303	370	431	487	534	611	668	678	732	780	836	.15
.10	107	346	481	255	322	385	442	492	572	632	644	700	750	814	.10

8000-DEGREE, LOG G = 3.5

MU	.20	.30	.36	.40	.45	.50	.55	.60	.70	.80	1.00	1.20	1.60	2.20	MU
.90	873	935	954	936	943	949	954	959	967	972	974	979	984	988	.90
.80	751	869	907	868	882	894	906	916	931	942	946	957	966	975	.80
.70	633	802	859	797	818	837	855	870	893	911	915	932	947	961	.70
.60	521	735	810	720	749	776	800	820	853	877	882	906	926	945	.60
.50	417	667	758	638	676	710	741	767	809	840	846	877	903	928	.50
.40	322	597	704	551	597	639	677	709	761	799	806	844	876	908	.40
.35	279	561	675	505	555	601	643	678	735	777	784	826	861	897	.35
.30	239	525	645	458	512	562	607	646	708	754	761	806	845	885	.30
.25	202	487	614	409	467	521	570	611	678	728	735	785	827	871	.25
.20	168	448	580	359	421	478	530	575	647	701	707	761	806	856	.20
.15	140	407	544	309	373	433	488	535	612	670	677	734	782	838	.15
.10	116	363	502	257	323	384	442	491	572	634	641	702	752	816	.10

LIMB DARKENING

8000-DEGREE, LOG G = 3.0

MU	.20	.30	.36	.40	.45	.50	.55	.60	.70	.80	1.00	1.20	1.60	2.20	MU
.90	869	936	956	938	944	950	955	960	967	973	974	979	984	988	.90
.80	743	871	911	873	885	897	908	917	932	943	947	957	967	975	.80
.70	625	806	864	803	823	840	857	871	895	912	917	933	948	961	.70
.60	514	740	816	729	755	780	803	823	855	879	884	907	927	946	.60
.50	412	673	766	649	683	715	745	770	811	842	848	878	904	928	.50
.40	320	605	713	563	605	644	681	712	764	802	808	846	877	909	.40
.35	278	571	686	517	563	607	647	682	738	780	786	828	862	898	.35
.30	239	535	657	469	520	567	611	649	710	756	762	808	846	885	.30
.25	204	498	626	420	475	526	573	614	681	731	736	786	828	872	.25
.20	173	460	593	369	427	482	533	577	649	703	708	762	807	857	.20
.15	146	420	558	316	377	435	489	536	613	672	676	735	782	838	.15
.10	125	376	518	263	325	385	441	490	572	635	639	701	752	816	.10

8000-DEGREE, LOG G = 2.0

MU	.20	.30	.36	.40	.45	.50	.55	.60	.70	.80	1.00	1.20	1.60	2.20	MU
.90	865	937	958	942	947	952	957	961	967	973	975	979	984	988	.90
.80	737	873	915	881	891	902	911	919	933	944	948	957	966	975	.80
.70	618	810	871	816	832	847	862	875	896	913	918	933	946	960	.70
.60	508	747	825	746	768	789	809	827	856	879	886	907	925	944	.60
.50	408	683	778	671	699	726	752	775	813	843	850	878	901	927	.50
.40	320	617	728	590	624	658	690	718	765	802	810	845	874	906	.40
.35	281	584	701	546	584	621	656	687	739	780	787	826	858	895	.35
.30	245	550	674	500	541	582	620	654	712	756	763	806	841	882	.30
.25	213	515	645	452	497	540	582	619	682	730	736	783	821	867	.25
.20	185	478	614	402	449	496	541	581	649	701	706	758	799	851	.20
.15	162	440	580	349	399	448	496	539	612	669	672	728	773	831	.15
.10	145	397	540	294	345	396	446	491	569	630	630	692	739	805	.10

7500-DEGREE, LOG G = 4.0   (L/H=1.5)

MU	.20	.30	.36	.40	.45	.50	.55	.60	.70	.80	1.00	1.20	1.60	2.20	MU
.90	888	929	948	939	943	948	952	956	963	969	971	977	982	986	.90
.80	774	856	895	873	882	892	901	909	924	936	940	951	963	972	.80
.70	659	782	840	801	816	831	846	860	882	900	907	924	942	956	.70
.60	544	708	783	723	745	766	788	806	838	863	872	895	918	938	.60
.50	432	633	723	638	668	697	725	750	791	822	833	862	892·	919	.50
.40	325	556	660	548	587	624	658	689	740	778	791	826	863	897	.40
.35	276	517	627	501	544	585	623	657	713	755	769	807	846	885	.35
.30	229	478	592	453	501	545	587	624	685	730	745	786	828	871	.30
.25	189	436	556	404	456	505	550	590	655	704	720	763	808	856	.25
.20	155	394	517	355	411	464	512	555	624	676	694	739	786	839	.20
.15	127	349	476	308	367	422	473	517	590	645	665	711	760	820	.15
.10	106	302	430	262	322	378	430	476	552	609	632	679	728	796	.10

7500-DEGREE, LOG G = 4.0   (L/H=2.5)

MU	.20	.30	.36	.40	.45	.50	.55	.60	.70	.80	1.00	1.20	1.60	2.20	MU
.90	902	927	944	946	949	952	955	958	964	968	972	977	982	985	.90
.80	800	851	886	886	893	900	907	913	925	934	941	951	963	969	.80
.70	693	773	829	820	831	843	853	864	882	898	908	923	941	952	.70
.60	583	694	770	747	763	780	795	810	837	859	872	893	917	934	.60
.50	470	614	710	666	688	710	732	753	788	817	832	859	889	913	.50
.40	358	536	648	575	605	635	664	690	735	772	789	821	857	891	.40
.35	304	496	616	527	561	596	628	657	708	748	766	801	840	879	.35
.30	252	457	583	477	517	555	590	623	680	724	742	780	821	865	.30
.25	205	417	549	426	470	512	553	589	650	697	717	758	800	851	.25
.20	165	378	513	373	423	471	514	553	619	670	691	733	777	835	.20
.15	132	338	472	323	376	427	474	516	586	640	662	706	752	816	.15
.10	112	293	426	274	331	384	433	477	550	606	631	676	721	793	.10

L I M B   D A R K E N I N G

7500-DEGREE, LOG G = 4.0   (NO CONV)

MU	.20	.30	.36	.40	.45	.50	.55	.60	.70	.80	1.00	1.20	1.60	2.20	MU
.90	863	930	949	921	932	940	948	954	963	969	971	977	983	987	.90
.80	731	859	896	840	861	879	893	905	924	936	940	952	964	972	.80
.70	606	787	841	756	788	814	836	855	882	902	907	925	943	956	.70
.60	490	714	785	671	712	747	777	802	839	865	872	896	920	939	.60
.50	384	639	725	583	634	678	715	746	792	825	834	864	894	920	.50
.40	289	563	663	494	554	606	650	687	742	782	793	829	865	898	.40
.35	246	523	630	450	514	569	617	656	716	759	771	810	849	886	.35
.30	207	483	596	406	473	532	582	623	688	734	748	789	831	872	.30
.25	172	442	560	363	432	493	546	590	659	708	723	767	811	858	.25
.20	142	399	521	320	391	454	509	555	627	680	697	742	789	841	.20
.15	117	355	480	279	350	414	470	518	594	649	668	715	763	822	.15
.10	098	307	433	238	307	371	428	477	556	614	635	683	731	798	.10

7500-DEGREE, LOG G = 3.0

MU	.20	.30	.36	.40	.45	.50	.55	.60	.70	.80	1.00	1.20	1.60	2.20	MU
.90	868	932	952	930	939	946	951	956	964	970	972	977	983	987	.90
.80	741	863	902	857	875	888	900	909	926	938	942	953	964	973	.80
.70	623	794	851	780	806	827	845	860	885	904	909	927	944	958	.70
.60	508	724	798	700	734	761	786	808	842	867	873	898	922	941	.60
.50	400	653	744	615	657	692	724	750	795	828	834	867	897	922	.50
.40	307	580	686	525	575	618	656	689	744	785	792	832	869	901	.40
.35	265	543	656	478	532	578	620	657	717	762	768	812	852	889	.35
.30	229	505	625	431	487	537	583	623	688	737	744	791	835	875	.30
.25	196	466	591	381	441	496	545	588	658	710	718	769	815	861	.25
.20	168	426	556	332	395	453	506	551	626	681	689	743	793	844	.20
.15	146	385	518	283	349	409	463	511	590	650	658	715	766	825	.15
.10	132	339	473	237	301	362	418	468	550	613	623	682	735	802	.10

7000-DEGREE, LOG G = 4.0

MU	.20	.30	.36	.40	.45	.50	.55	.60	.70	.80	1.00	1.20	1.60	2.20	MU
.90	877	926	944	932	938	943	949	953	961	967	970	975	981	985	.90
.80	753	850	886	860	872	884	895	904	920	931	938	948	961	970	.80
.70	629	773	826	783	802	820	837	852	876	894	904	919	939	952	.70
.60	509	694	764	701	728	754	777	798	830	854	867	888	914	934	.60
.50	394	614	698	615	651	684	714	739	781	811	828	854	886	913	.50
.40	286	532	629	526	570	610	646	677	728	765	785	816	855	888	.40
.35	238	490	592	480	528	571	611	645	700	741	762	796	837	875	.35
.30	196	448	553	433	485	533	576	613	672	715	739	774	818	860	.30
.25	161	403	513	387	443	494	539	579	642	688	714	750	796	844	.25
.20	133	358	470	342	401	454	502	543	609	658	687	725	772	825	.20
.15	112	311	424	298	358	413	462	505	574	626	658	696	743	803	.15
.10	098	261	374	254	314	369	419	463	535	590	624	663	709	777	.10

7000-DEGREE, LOG G = 4.0   (NO CONV)

MU	.20	.30	.36	.40	.45	.50	.55	.60	.70	.80	1.00	1.20	1.60	2.20	MU
.90	847	925	945	914	923	935	944	951	960	966	969	975	981	986	.90
.80	703	849	889	824	846	869	886	899	918	931	937	948	961	970	.80
.70	568	775	830	732	769	802	827	846	874	894	902	919	939	954	.70
.60	444	699	768	644	692	734	766	791	829	856	866	888	914	935	.60
.50	332	620	703	557	614	662	701	734	782	815	828	856	887	915	.50
.40	240	539	635	470	534	591	638	676	731	770	788	819	856	891	.40
.35	203	497	599	427	496	557	606	645	705	746	766	799	839	878	.35
.30	170	455	561	387	459	521	572	613	677	721	743	778	820	863	.30
.25	140	411	522	348	420	484	537	581	647	695	719	755	799	847	.25
.20	117	366	480	309	382	447	501	546	616	666	692	730	775	829	.20
.15	099	320	434	271	343	407	462	509	582	634	664	702	747	808	.15
.10	088	270	384	233	302	365	421	468	543	598	631	670	714	782	.10

L I M B   D A R K E N I N G

7000-DEGREE, LOG G = 3.0

MU	.20	.30	.36	.40	.45	.50	.55	.60	.70	.80	1.00	1.20	1.60	2.20	MU
.90	866	929	949	933	938	943	949	953	962	968	970	976	982	986	.90
.80	738	858	897	860	872	883	894	904	921	933	938	950	962	972	.80
.70	616	786	843	783	801	819	836	853	878	897	904	922	941	955	.70
.60	492	714	787	699	725	751	776	797	832	859	866	891	917	938	.60
.50	381	640	728	611	646	681	711	738	784	818	826	858	891	918	.50
.40	289	563	665	519	563	604	642	676	732	773	784	821	861	895	.40
.35	250	525	632	472	519	564	607	644	705	749	761	801	844	883	.35
.30	216	485	597	423	475	525	571	611	676	724	737	780	825	869	.30
.25	187	443	560	374	431	485	534	577	646	697	711	757	805	853	.25
.20	165	401	520	328	388	444	496	541	613	668	683	731	781	835	.20
.15	148	356	477	284	345	403	456	502	578	635	653	702	754	815	.15
.10	139	308	428	241	301	358	412	459	538	597	618	669	720	789	.10

7000-DEGREE, LOG G = 2.0

MU	.20	.30	.36	.40	.45	.50	.55	.60	.70	.80	1.00	1.20	1.60	2.20	MU
.90	846	933	953	924	934	941	948	954	963	970	970	977	983	987	.90
.80	715	865	905	845	865	879	895	905	924	937	938	952	964	972	.80
.70	598	797	856	764	791	815	837	853	883	902	905	925	943	957	.70
.60	484	729	804	679	715	748	776	800	838	865	868	896	920	940	.60
.50	387	659	750	589	636	675	712	743	790	825	829	863	894	920	.50
.40	308	588	692	498	551	600	645	680	739	781	784	827	864	898	.40
.35	275	551	662	450	509	562	608	647	711	757	761	807	848	886	.35
.30	248	514	630	403	466	521	571	613	682	732	735	785	829	872	.30
.25	225	475	596	358	421	480	533	578	651	705	708	762	809	857	.25
.20	208	435	559	312	377	437	493	540	618	675	679	736	785	839	.20
.15	195	393	519	268	333	394	451	500	582	642	647	706	758	818	.15
.10	189	348	473	226	289	349	407	457	540	603	610	671	723	791	.10

6500-DEGREE, LOG G = 4.0

MU	.20	.30	.36	.40	.45	.50	.55	.60	.70	.80	1.00	1.20	1.60	2.20	MU
.90	851	922	938	926	934	941	946	952	960	965	970	974	981	985	.90
.80	705	842	875	850	865	879	891	902	918	929	938	946	959	968	.80
.70	568	760	810	769	793	815	834	849	873	890	903	917	936	949	.70
.60	441	675	742	686	720	749	773	793	825	849	866	884	910	929	.60
.50	317	590	671	602	642	677	708	735	777	806	828	849	881	907	.50
.40	220	504	596	513	562	606	644	676	725	760	787	811	848	882	.40
.35	182	460	557	469	523	570	610	644	697	735	765	791	829	867	.35
.30	149	416	516	427	483	533	576	612	669	709	742	769	809	852	.30
.25	123	371	474	384	443	495	540	579	638	682	717	745	786	834	.25
.20	105	325	429	342	403	456	503	543	607	653	691	719	761	815	.20
.15	092	278	382	300	361	415	464	506	571	620	661	691	731	792	.15
.10	085	231	330	257	316	370	419	462	532	582	627	657	696	765	.10

6500-DEGREE, LOG G = 3.0

MU	.20	.30	.36	.40	.45	.50	.55	.60	.70	.80	1.00	1.20	1.60	2.20	MU
.90	850	926	946	925	933	940	946	952	960	966	969	975	981	985	.90
.80	707	853	889	847	862	877	890	901	918	930	937	948	961	970	.80
.70	567	778	830	764	788	811	832	848	873	892	901	918	939	953	.70
.60	436	702	769	678	713	744	769	791	826	853	864	886	914	934	.60
.50	332	623	704	592	634	671	704	732	778	810	825	852	886	913	.50
.40	249	542	634	500	550	597	638	672	725	764	783	814	854	889	.40
.35	215	500	597	454	510	560	603	640	697	739	760	793	837	875	.35
.30	188	457	559	411	470	522	568	607	668	712	736	771	817	860	.30
.25	167	413	518	368	429	483	531	572	637	684	710	747	795	843	.25
.20	149	368	474	325	387	443	493	536	604	654	682	721	770	823	.20
.15	139	322	426	284	345	402	453	497	567	619	652	691	741	801	.15
.10	134	273	374	242	300	356	407	452	525	580	615	655	705	772	.10

LIMB DARKENING

6000-DEGREE, LOG G = 4.5

MU	.20	.30	.36	.40	.45	.50	.55	.60	.70	.80	1.00	1.20	1.60	2.20	MU
.90	806	916	929	924	932	939	945	950	958	963	969	973	980	983	.90
.80	629	829	856	845	862	876	888	898	913	924	936	943	958	965	.80
.70	464	740	780	764	789	811	828	844	867	884	901	912	933	945	.70
.60	315	648	701	681	714	742	766	787	818	841	864	878	906	923	.60
.50	203	553	619	595	636	672	702	727	767	795	824	842	875	899	.50
.40	131	457	534	507	556	599	635	666	713	747	782	802	839	871	.40
.35	106	409	491	464	516	562	601	634	685	721	760	781	819	855	.35
.30	087	361	447	421	476	524	565	601	656	695	736	758	797	838	.30
.25	074	314	401	378	435	485	529	566	625	667	711	734	772	819	.25
.20	065	268	355	335	394	446	491	530	592	637	684	708	745	798	.20
.15	058	222	308	292	351	404	451	492	557	604	655	679	713	774	.15
.10	054	179	259	249	307	360	408	451	519	568	622	646	675	744	.10

6000-DEGREE, LOG G = 4.0

MU	.20	.30	.36	.40	.45	.50	.55	.60	.70	.80	1.00	1.20	1.60	2.20	MU
.90	804	915	930	921	930	937	944	949	957	963	968	972	979	983	.90
.80	618	827	858	839	857	873	885	896	912	923	934	943	957	965	.80
.70	445	737	784	756	783	805	824	840	865	883	899	911	932	945	.70
.60	304	646	707	670	705	736	761	783	816	840	862	877	905	923	.60
.50	204	553	627	583	626	664	696	723	765	794	822	840	873	899	.50
.40	137	460	545	495	547	592	630	662	711	746	780	801	838	871	.40
.35	114	414	503	452	507	555	595	630	683	721	757	779	817	856	.35
.30	097	368	459	410	467	517	560	597	654	694	734	757	796	839	.30
.25	085	322	415	368	427	479	524	563	623	666	709	732	771	820	.25
.20	076	277	369	326	386	440	486	527	590	636	682	706	744	799	.20
.15	070	233	322	285	345	399	447	489	555	603	652	677	712	775	.15
.10	067	190	273	243	301	355	404	447	516	567	619	644	674	745	.10

6000-DEGREE, LOG G = 4.0 (NO BLKT)

MU	.20	.30	.36	.40	.45	.50	.55	.60	.70	.80	1.00	1.20	1.60	2.20	MU
.90	737	909	924	916	926	934	940	946	954	960	966	971	979	982	.90
.80	503	815	845	831	849	865	879	890	907	919	931	939	955	963	.80
.70	337	720	764	743	771	795	815	831	857	876	894	906	929	942	.70
.60	226	623	682	654	691	722	748	771	806	831	855	870	900	919	.60
.50	154	525	597	564	609	648	681	710	753	784	815	832	867	893	.50
.40	112	429	510	475	528	574	613	646	698	734	771	791	829	863	.40
.35	098	381	467	432	487	536	578	614	669	709	748	769	808	847	.35
.30	088	335	422	389	447	498	543	581	640	682	725	746	785	829	.30
.25	081	290	378	347	407	460	507	547	609	654	700	722	759	810	.25
.20	076	246	333	307	367	422	470	512	578	625	673	695	731	788	.20
.15	072	205	288	267	327	382	432	475	544	594	645	667	698	762	.15
.10	071	167	243	228	287	342	392	437	508	560	615	636	658	732	.10

6000-DEGREE, LOG G = 4.0 (NO CONV)

MU	.20	.30	.36	.40	.45	.50	.55	.60	.70	.80	1.00	1.20	1.60	2.20	MU
.90	768	913	931	908	923	933	941	947	957	963	968	972	979	983	.90
.80	571	824	860	818	846	866	881	894	912	924	934	942	956	965	.80
.70	408	735	786	730	769	797	820	839	865	883	899	911	931	946	.70
.60	278	644	710	643	692	729	758	782	817	840	862	877	903	924	.60
.50	185	553	631	558	615	658	694	723	766	795	823	841	872	900	.50
.40	124	461	548	475	537	587	628	661	712	747	780	801	837	872	.40
.35	104	415	506	434	498	550	594	629	684	721	758	780	817	857	.35
.30	089	369	462	393	459	513	559	596	654	695	734	757	795	840	.30
.25	078	323	417	353	420	475	523	562	623	666	709	733	771	821	.25
.20	070	278	371	313	380	436	485	526	590	636	682	706	744	800	.20
.15	065	233	323	273	338	395	445	488	555	603	652	677	712	775	.15
.10	062	191	274	232	295	352	402	446	516	566	618	643	674	746	.10

L I M B D A R K E N I N G

6000-DEGREE, LOG G = 4.0 (1/10X METALS)

MU	.20	.30	.36	.40	.45	.50	.55	.60	.70	.80	1.00	1.20	1.60	2.20	MU
.90	819	902	923	906	919	930	938	945	954	960	966	970	977	982	.90
.80	658	805	844	812	838	859	875	888	907	919	931	939	953	963	.80
.70	515	707	763	721	758	788	812	831	858	877	895	906	926	942	.70
.60	393	610	680	631	679	716	747	771	808	832	857	870	897	919	.60
.50	295	513	595	544	599	644	681	711	755	786	816	833	864	893	.50
.40	222	418	509	459	520	571	613	648	700	737	773	792	826	864	.40
.35	195	372	465	418	481	534	579	616	672	711	751	770	805	847	.35
.30	176	327	420	377	442	497	543	583	642	684	727	747	782	830	.30
.25	162	284	375	337	402	459	507	548	612	656	702	722	757	810	.25
.20	155	243	330	298	363	420	470	513	580	627	676	696	728	788	.20
.15	153	204	286	259	323	381	432	476	546	596	647	668	695	762	.15
.10	154	172	242	222	283	340	392	438	510	562	616	636	655	731	.10

6000-DEGREE, LOG G = 4.0 (10X METALS)

MU	.20	.30	.36	.40	.45	.50	.55	.60	.70	.80	1.00	1.20	1.60	2.20	MU
.90	969	901	919	899	916	927	936	943	953	960	966	970	977	982	.90
.80	940	802	837	801	833	854	872	885	905	918	930	939	952	963	.80
.70	913	703	753	707	751	782	807	826	855	875	893	906	925	943	.70
.60	886	604	669	617	671	709	741	766	804	829	853	871	896	920	.60
.50	861	507	586	531	592	636	674	704	749	781	811	833	863	895	.50
.40	836	411	502	449	514	563	606	641	693	730	767	792	827	867	.40
.35	824	364	461	410	477	527	572	609	664	704	744	770	806	851	.35
.30	811	319	420	372	440	491	537	576	634	676	719	746	784	834	.30
.25	799	275	379	335	403	454	502	542	602	647	694	722	760	816	.25
.20	786	232	339	300	366	418	466	507	570	617	666	695	733	795	.20
.15	772	192	300	265	329	380	429	472	535	585	637	666	702	772	.15
.10	758	156	262	232	293	343	391	434	498	550	605	634	666	745	.10

6000-DEGREE, LOG G = 4.0 (NO CONV OR BLKT)

MU	.20	.30	.36	.40	.45	.50	.55	.60	.70	.80	1.00	1.20	1.60	2.20	MU
.90	711	905	924	904	919	929	938	944	954	961	966	970	977	982	.90
.80	482	809	846	810	837	858	875	888	907	920	931	939	953	963	.80
.70	321	713	766	717	756	787	811	830	858	877	895	906	926	942	.70
.60	212	617	684	628	677	715	746	770	807	832	856	871	897	919	.60
.50	144	521	600	541	597	642	679	710	755	786	816	833	864	894	.50
.40	104	426	514	455	518	569	612	647	700	737	773	792	827	865	.40
.35	092	380	470	415	479	533	578	615	672	711	750	770	806	848	.35
.30	083	334	426	374	440	496	543	582	642	684	726	747	783	831	.30
.25	076	289	381	335	401	458	507	548	612	656	701	723	758	811	.25
.20	071	246	336	296	362	420	470	513	580	627	675	697	730	789	.20
.15	067	205	291	258	323	381	432	477	546	596	647	668	697	764	.15
.10	066	167	246	220	283	340	393	438	510	562	616	637	658	734	.10

6000-DEGREE, LOG G = 3.5

MU	.20	.30	.36	.40	.45	.50	.55	.60	.70	.80	1.00	1.20	1.60	2.20	MU
.90	802	916	934	919	929	936	943	948	956	962	968	972	980	984	.90
.80	613	830	865	836	855	870	883	894	911	923	933	942	957	966	.80
.70	447	743	794	752	779	802	821	838	864	882	897	911	933	946	.70
.60	316	655	720	664	700	731	757	779	814	839	859	876	905	925	.60
.50	219	565	643	575	620	658	691	719	763	793	819	839	875	901	.50
.40	155	475	562	486	539	585	624	657	708	744	776	799	839	873	.40
.35	132	429	520	443	499	548	589	624	679	718	753	777	819	857	.35
.30	115	383	477	401	459	510	553	591	649	691	729	754	797	840	.30
.25	104	338	433	358	418	471	517	557	618	663	703	730	773	821	.25
.20	095	293	387	317	378	432	480	521	586	633	676	703	746	800	.20
.15	090	249	340	277	337	391	440	482	550	599	646	674	714	776	.15
.10	087	208	290	235	293	348	397	441	511	563	613	640	675	746	.10

LIMB DARKENING

6000-DEGREE, LOG G = 3.0

MU	.20	.30	.36	.40	.45	.50	.55	.60	.70	.80	1.00	1.20	1.60	2.20	MU
.90	789	917	936	917	927	935	942	948	956	962	967	972	980	984	.90
.80	594	833	870	832	852	868	882	893	910	923	932	942	957	966	.80
.70	441	748	802	746	775	799	818	835	862	882	895	910	933	947	.70
.60	322	662	730	657	694	726	753	776	813	839	857	876	906	926	.60
.50	233	575	655	566	612	652	687	716	761	793	817	839	875	902	.50
.40	175	487	576	476	531	579	619	653	706	744	773	799	840	875	.40
.35	154	442	535	433	491	541	584	621	677	718	750	777	820	859	.35
.30	141	397	493	391	451	503	548	587	647	691	726	754	799	842	.30
.25	129	353	450	350	410	465	512	553	616	662	700	729	775	824	.25
.20	121	309	405	309	370	426	474	517	583	632	673	702	747	802	.20
.15	116	267	358	270	330	385	435	478	547	598	642	672	715	778	.15
.10	113	228	309	230	287	342	392	436	508	561	608	638	676	748	.10

6000-DEGREE, LOG G = 2.0

MU	.20	.30	.36	.40	.45	.50	.55	.60	.70	.80	1.00	1.20	1.60	2.20	MU
.90	802	922	942	915	926	935	942	947	957	964	967	973	980	985	.90
.80	634	844	882	828	851	868	880	892	912	926	932	944	959	968	.80
.70	495	764	820	742	773	796	817	835	865	887	895	913	936	950	.70
.60	389	683	755	652	690	723	751	776	817	845	857	879	910	930	.60
.50	315	602	687	558	606	648	685	716	765	800	816	843	880	907	.50
.40	265	519	615	466	523	573	616	653	711	752	772	803	847	882	.40
.35	246	478	577	423	483	536	581	620	682	726	748	782	827	867	.35
.30	231	437	538	381	443	498	546	587	652	699	724	759	807	851	.30
.25	218	396	497	340	403	459	509	552	621	670	698	734	783	833	.25
.20	208	356	454	302	364	420	471	516	587	639	670	707	757	812	.20
.15	199	317	409	265	324	380	432	477	551	605	639	677	726	788	.15
.10	192	282	360	230	284	338	389	435	510	566	604	642	687	758	.10

5500-DEGREE, LOG G = 4.0

MU	.20	.30	.36	.40	.45	.50	.55	.60	.70	.80	1.00	1.20	1.60	2.20	MU
.90	713	905	920	915	925	933	940	945	954	960	966	970	978	982	.90
.80	500	809	838	829	848	864	877	889	906	918	931	939	955	963	.80
.70	357	711	754	741	770	793	813	830	856	875	894	905	929	942	.70
.60	265	611	667	652	689	721	747	770	805	829	855	870	899	918	.60
.50	207	512	580	562	608	647	680	708	752	782	814	831	866	892	.50
.40	170	414	491	474	527	572	611	644	695	731	770	789	828	862	.40
.35	156	366	446	431	486	534	576	611	666	704	747	767	807	845	.35
.30	145	319	400	388	446	496	539	577	635	676	722	743	784	827	.30
.25	134	274	355	346	405	456	502	541	603	647	696	718	758	807	.25
.20	125	231	309	304	363	416	463	504	569	616	668	690	728	784	.20
.15	116	190	264	262	321	375	423	465	533	583	637	660	694	757	.15
.10	108	153	218	221	278	331	380	423	494	545	603	626	653	725	.10

5500-DEGREE, LOG G = 3.0

MU	.20	.30	.36	.40	.45	.50	.55	.60	.70	.80	1.00	1.20	1.60	2.20	MU
.90	727	906	925	911	921	930	937	943	953	960	966	970	978	982	.90
.80	528	811	848	820	840	857	873	886	905	918	930	938	954	963	.80
.70	388	716	769	727	758	785	808	827	855	874	892	904	928	942	.70
.60	302	621	688	634	676	712	742	766	803	829	853	868	899	919	.60
.50	242	527	604	545	595	638	674	704	749	781	811	830	866	894	.50
.40	201	434	518	458	514	563	604	639	692	729	766	787	828	864	.40
.35	186	389	474	416	474	525	568	605	662	702	742	765	807	847	.35
.30	172	344	429	374	434	486	532	571	631	674	717	740	784	829	.30
.25	160	301	384	333	393	447	494	535	598	644	691	715	758	809	.25
.20	150	261	338	293	353	407	456	498	564	612	662	687	729	785	.20
.15	140	224	293	254	312	366	415	459	527	578	631	656	694	759	.15
.10	132	192	248	216	270	323	372	416	487	540	596	621	653	726	.10

LIMB DARKENING

5000-DEGREE, LOG G = 4.0

MU	.20	.30	.36	.40	.45	.50	.55	.60	.70	.80	1.00	1.20	1.60	2.20	MU
.90	940	891	906	906	916	925	933	939	949	955	963	967	977	981	.90
.80	886	782	811	811	831	849	864	876	896	909	925	933	952	960	.80
.70	836	672	714	715	745	771	794	812	842	862	885	896	924	937	.70
.60	788	563	617	620	659	693	722	747	785	812	842	857	892	912	.60
.50	741	457	520	527	574	614	650	680	728	760	798	815	857	883	.50
.40	695	355	425	436	489	536	576	611	666	706	750	770	816	851	.40
.35	671	307	379	392	447	495	539	576	635	677	725	746	792	832	.35
.30	647	262	333	348	405	455	501	540	602	647	699	720	767	812	.30
.25	622	219	288	306	363	415	462	503	569	616	671	693	738	790	.25
.20	596	179	245	265	321	374	422	465	533	583	641	664	706	764	.20
.15	569	144	204	224	280	332	381	425	495	548	609	632	669	734	.15
.10	540	114	164	185	237	289	337	382	454	508	572	595	623	698	.10

5000-DEGREE, LOG G = 3.0

MU	.20	.30	.36	.40	.45	.50	.55	.60	.70	.80	1.00	1.20	1.60	2.20	MU
.90	934	890	910	899	913	924	932	939	949	956	963	967	976	981	.90
.80	874	782	818	800	827	847	863	876	896	910	925	933	951	960	.80
.70	820	675	725	704	740	769	793	813	842	862	884	896	922	938	.70
.60	768	570	632	609	655	692	722	747	786	813	842	857	891	912	.60
.50	718	469	539	517	570	613	650	680	727	760	797	815	855	884	.50
.40	669	373	446	428	486	534	576	611	666	705	749	770	814	852	.40
.35	644	328	400	385	444	494	538	576	634	676	724	745	791	834	.35
.30	620	285	355	343	402	454	500	539	601	646	697	720	766	814	.30
.25	594	245	310	302	361	413	461	502	567	615	669	692	738	791	.25
.20	569	209	267	263	320	373	421	463	531	581	639	663	706	766	.20
.15	541	178	226	225	279	331	379	422	493	545	606	630	669	736	.15
.10	512	153	187	187	238	287	335	378	451	505	569	593	623	699	.10

5000-DEGREE, LOG G = 2.0

MU	.20	.30	.36	.40	.45	.50	.55	.60	.70	.80	1.00	1.20	1.60	2.20	MU
.90	931	893	915	896	910	922	931	938	949	956	963	967	977	981	.90
.80	868	787	828	793	821	844	861	876	896	910	924	933	951	961	.80
.70	809	684	740	693	734	766	791	812	842	863	884	896	923	939	.70
.60	754	585	651	598	648	688	720	746	786	813	841	857	892	914	.60
.50	701	491	561	507	563	609	648	679	727	761	796	815	857	886	.50
.40	648	403	472	421	480	531	574	610	666	706	748	770	817	854	.40
.35	622	363	428	379	439	491	536	574	634	676	722	746	794	836	.35
.30	595	326	385	340	399	452	498	538	601	646	695	720	769	816	.30
.25	569	292	342	302	359	412	459	500	566	614	667	692	741	794	.25
.20	541	263	301	265	320	371	419	461	530	580	636	662	709	769	.20
.15	512	239	263	231	281	331	378	420	491	543	602	629	672	739	.15
.10	481	222	228	200	242	288	333	376	447	501	564	590	626	702	.10

4500-DEGREE, LOG G = 4.0

MU	.20	.30	.36	.40	.45	.50	.55	.60	.70	.80	1.00	1.20	1.60	2.20	MU
.90	985	869	882	886	897	907	916	924	937	946	955	960	974	978	.90
.80	968	739	765	772	794	814	832	848	873	890	910	918	946	955	.80
.70	951	611	649	661	692	722	749	772	809	834	862	875	914	928	.70
.60	934	489	535	552	593	632	667	696	743	776	813	829	879	899	.60
.50	915	373	428	450	499	544	584	620	677	717	762	781	838	866	.50
.40	896	270	330	355	408	459	505	545	609	656	709	729	791	828	.40
.35	886	225	283	311	365	417	465	507	575	624	681	702	764	807	.35
.30	875	183	241	269	323	376	425	469	540	592	653	674	734	783	.30
.25	864	148	201	229	283	336	386	432	505	560	623	645	701	757	.25
.20	851	117	164	192	245	297	347	393	469	525	592	614	664	728	.20
.15	837	093	132	158	207	258	308	355	431	490	559	580	621	693	.15
.10	820	077	103	126	172	220	269	314	391	451	522	543	568	651	.10

L I M B   D A R K E N I N G

4500-DEGREE, LOG G = 3.0

MU	.20	.30	.36	.40	.45	.50	.55	.60	.70	.80	1.00	1.20	1.60	2.20	MU
.90	984	864	883	881	894	906	916	924	936	945	955	960	973	978	.90
.80	967	732	767	764	790	812	831	847	872	889	909	918	944	954	.80
.70	949	605	653	652	688	720	747	771	807	832	861	874	912	928	.70
.60	931	486	544	545	590	630	665	695	741	774	811	828	875	899	.60
.50	911	377	439	444	496	542	583	618	674	713	759	779	834	866	.50
.40	890	282	343	352	406	456	502	542	605	652	705	727	787	828	.40
.35	878	240	297	309	363	414	462	503	570	620	677	700	760	807	.35
.30	866	203	255	268	322	373	422	465	535	587	648	671	731	783	.30
.25	852	171	216	230	282	333	382	427	500	554	618	641	699	757	.25
.20	838	145	180	194	244	294	343	389	463	520	586	609	662	728	.20
.15	821	126	149	162	208	256	304	350	425	483	552	576	618	693	.15
.10	800	115	123	133	173	218	265	310	385	444	515	538	566	651	.10

4000-DEGREE, LOG G = 4.0

MU	.20	.30	.36	.40	.45	.50	.55	.60	.70	.80	1.00	1.20	1.60	2.20	MU
.90	995	842	860	869	883	896	908	917	931	941	952	957	973	977	.90
.80	989	691	724	743	770	795	817	835	863	882	904	912	942	953	.80
.70	983	549	596	623	661	697	728	754	794	822	854	866	908	925	.70
.60	976	420	477	510	558	602	640	674	724	761	802	818	870	895	.60
.50	969	307	369	406	460	510	555	595	655	699	749	767	826	860	.50
.40	960	213	274	313	370	424	473	517	585	636	694	714	776	820	.40
.35	956	175	231	270	327	382	433	478	550	604	665	686	747	797	.35
.30	952	142	193	230	286	341	393	440	515	571	636	657	715	772	.30
.25	947	115	158	193	248	302	354	402	479	538	606	627	680	744	.25
.20	944	093	128	160	211	264	316	364	443	504	575	596	639	712	.20
.15	941	078	102	130	177	228	279	326	406	469	542	562	592	674	.15
.10	941	071	081	104	146	193	242	288	369	432	507	526	535	628	.10

4000-DEGREE, LOG G = 3.0

MU	.20	.30	.36	.40	.45	.50	.55	.60	.70	.80	1.00	1.20	1.60	2.20	MU
.90	000	831	857	865	882	896	907	917	932	942	953	957	971	976	.90
.80	000	674	721	737	768	794	817	835	863	883	904	913	938	951	.80
.70	000	533	595	618	660	696	728	754	795	823	854	866	903	923	.70
.60	000	410	479	507	558	602	641	675	726	763	803	819	863	892	.60
.50	000	306	375	406	462	512	557	596	657	701	751	768	818	857	.50
.40	000	224	284	316	373	426	475	518	588	638	696	715	767	817	.40
.35	000	191	243	275	332	385	435	480	553	606	667	688	738	794	.35
.30	000	164	207	237	292	345	396	442	517	573	638	659	706	769	.30
.25	000	142	175	202	254	307	357	404	481	539	607	629	671	742	.25
.20	000	127	147	171	218	269	319	365	445	504	575	597	632	710	.20
.15	000	117	124	142	185	232	281	327	407	468	541	562	586	673	.15
.10	000	115	107	119	154	197	243	287	366	428	503	525	530	627	.10

4000-DEGREE, LOG G = 2.0

MU	.20	.30	.36	.40	.45	.50	.55	.60	.70	.80	1.00	1.20	1.60	2.20	MU
.90	998	833	860	864	881	895	907	917	931	941	952	957	970	976	.90
.80	997	681	728	736	768	794	816	834	862	881	903	912	937	951	.80
.70	995	546	604	618	660	695	726	752	793	821	853	866	901	923	.70
.60	994	431	492	509	558	601	639	672	724	760	801	817	862	892	.60
.50	994	340	392	411	463	511	555	593	654	698	748	766	817	857	.50
.40	995	270	306	324	376	426	473	516	584	634	692	713	766	817	.40
.35	996	242	268	286	336	385	433	477	549	602	664	685	737	795	.35
.30	999	221	235	250	297	346	395	439	513	569	634	656	705	770	.30
.25	000	204	207	218	261	308	356	401	477	534	602	626	670	742	.25
.20	000	193	182	189	227	271	318	363	440	499	570	593	631	710	.20
.15	000	186	164	165	196	236	281	325	402	462	535	558	585	673	.15
.10	000	184	152	146	169	203	244	286	362	423	497	520	530	628	.10

H GAMMA (E-S-W)

MODEL	0	.5	1	2	4	6	8	12	16	20	25	30	35	40	70A	WIDTH
50000,5	819	828	848	875	918	947	965	983	991	994	997	998	999	999	000	1.95
40000,5	744	754	778	816	876	917	943	971	984	990	994	996	997	998	000	3.03
30000,4	528	588	671	757	862	917	948	976	987	992	995	997	998		000	3.51
25000,4	449	554	626	704	814	881	922	962	979	987	992	995	997	998	000	4.62
20000,4	422	532	578	647	760	837	887	943	968	980	988	992	995	996	999	5.95
20000,3	436	571	659	767	890	943	967	986	993	996	997	998	999	999	000	3.01
18000,4	415	510	549	615	729	811	867	930	960	975	985	990	993	995	999	6.75
18000,3	434	549	626	733	865	927	957	981	990	994	996	998	998	999	000	3.50
16000,4	406	479	514	579	694	781	842	915	950	969	981	988	992	994	999	7.74
16000,3	424	517	586	693	834	907	943	975	987	992	995	997	998	998	000	4.13
14000,4	379	438	472	536	651	743	812	895	938	961	976	984	989	992	999	9.00
14000,3	405	474	538	644	795	879	925	966	982	989	994	996	997	998	000	4.97
12000,4	296	380	413	475	590	688	764	864	917	947	967	979	985	990	998	10.92
12000,3	356	412	471	577	740	840	898	952	974	984	991	994	996	997	999	6.20
12000,2	399	501	624	781	918	962	979	992	996	997	998	999	999	999	000	2.71
11000,4	257	343	374	433	543	640	721	833	897	933	959	973	981	987	998	12.63
11000,3	305	372	428	530	698	808	875	941	968	981	988	992	995	996	999	7.15
11000,2	355	451	567	732	892	948	971	988	994	996	998	999	999	999	000	3.28
10000,4.5	289	314	337	381	458	528	590	696	779	839	891	924	946	961	993	19.83
10000,4	288	316	345	398	494	578	653	772	850	899	936	958	971	979	996	15.57
10000,4,10X	301	332	363	418	515	600	676	792	865	910	943	963	974	982	997	14.52
10000,.1X	283	309	337	391	487	572	647	766	846	896	934	956	970	978	996	15.86
10000,3.5	282	321	358	425	547	653	739	852	913	945	967	978	985	990	998	11.83
10000,3	284	333	380	469	628	749	831	917	954	972	983	989	993	995	999	8.76
10000,2	313	396	496	663	852	927	959	983	991	995	997	998	999	999	000	4.10
9500,4	271	300	328	381	472	552	622	736	817	873	917	944	961	972	995	17.53
9500,3	262	308	352	435	579	698	787	891	939	962	978	985	990	993	999	10.22
9000,4	262	294	324	378	471	548	614	718	796	852	900	931	951	964	994	18.83
9000,3	243	292	335	413	542	649	737	852	914	946	967	979	986	990	998	11.95
9000,2	247	323	399	535	747	863	920	966	982	989	994	996	997	998	999	6.00
8500,4	226	275	318	384	483	559	621	719	790	842	889	921	943	958	992	19.42
8500,3	208	271	322	402	526	623	703	817	886	927	954	970	980	985	997	13.54
8000,4.5	232	304	359	438	544	617	675	762	824	869	907	932	950	963	993	16.99
8000,4	215	289	344	422	532	609	669	760	825	869	909	935	952	965	993	17.02
8000,4 *	211	285	339	418	525	600	659	747	808	852	893	921	942	956	992	18.33
8000,3.5	204	277	332	413	526	606	670	765	830	877	916	941	958	969	994	16.60
8000,3	189	270	328	414	535	624	694	796	863	906	940	960	972	980	996	14.57
8000,2	162	275	346	453	610	727	810	905	948	968	981	988	991	994	999	9.52
7500,4	218	331	403	499	616	691	747	827	878	913	941	958	970	978	996	12.90
7500,4 +	222	338	412	504	621	706	768	852	902	932	955	969	978	984	997	11.59
7500,4 *	213	324	395	488	600	673	727	800	850	884	916	937	953	965	993	15.14
7500,3	188	300	369	466	589	672	737	825	881	916	945	962	973	981	996	13.10
7000,4	210	385	478	588	705	772	820	884	921	945	964	975	982	987	998	9.41
7000,4 *	205	382	473	576	688	753	797	855	892	917	940	956	968	975	995	11.54
7000,3	188	359	447	553	674	752	807	880	922	947	966	977	984	988	998	9.79
7000,2	173	332	420	530	667	753	813	888	931	955	972	981	987	991	998	9.49

KEY TO THE MODELS:

* NO CONVECTION
+ L/H = 2.5

H GAMMA (GRIEM)

MODEL	0	.5	1	2	4	6	8	12	16	20	25	30	35	40	70A	WIDTH
50000,5	819	828	846	870	915	943	960	979	988	993	996	997	998	999	000	2.13
40000,5	744	754	775	807	867	907	934	965	980	988	993	995	997	998	000	3.34
30000,4	528	583	657	739	842	901	936	970	983	990	994	996	997	998	000	3.94
25000,4	449	547	614	687	793	862	906	953	974	984	990	994	996	997	999	5.15
20000,4	422	527	569	631	737	814	868	930	960	975	985	990	993	995	999	6.60
20000,3	436	567	643	742	867	928	957	982	991	994	997	998	999	999	000	3.41
18000,4	415	505	540	600	705	787	846	916	951	970	982	988	992	994	999	7.48
18000,3	434	545	611	708	840	909	945	976	988	993	996	997	998	999	000	3.95
16000,4	406	475	506	564	670	755	819	899	940	962	977	985	990	993	999	8.54
16000,3	424	513	571	666	806	886	929	968	983	990	994	996	997	998	000	4.66
14000,4	379	435	464	520	627	716	786	876	926	953	971	981	987	991	999	9.87
14000,3	405	469	523	617	764	855	908	958	977	987	992	995	997	997	000	5.57
12000,4	296	377	405	460	566	659	736	842	903	937	962	975	983	988	998	11.90
12000,3	356	406	456	550	706	811	876	941	968	981	989	993	995	996	999	6.90
12000,2	399	491	596	746	896	951	973	989	995	997	998	999	999	999	000	3.07
11000,4	257	339	366	419	519	612	692	809	880	922	952	968	978	984	998	13.67
11000,3	305	366	413	504	663	777	851	928	961	976	986	991	994	996	999	7.92
11000,2	355	439	538	694	865	934	963	985	992	996	997	998	999	999	000	3.72
10000,4.5	289	311	331	370	443	509	569	674	758	821	878	915	939	955	996	20.91
10000,4	288	312	337	385	473	553	626	746	830	885	927	951	966	976	998	16.65
10000,4,10X	301	328	356	405	494	576	649	767	846	897	935	957	970	979	998	15.57
10000,4,.1X	283	305	330	378	467	547	620	741	826	882	925	950	965	975	998	16.96
10000,3.5	282	317	348	408	520	621	708	829	897	935	961	975	983	988	999	12.83
10000,3	284	327	368	446	593	714	801	900	945	966	980	987	991	994	999	9.64
10000,2	313	385	470	623	819	908	948	979	989	994	996	998	998	999	000	4.62
9500,4	271	296	321	368	454	530	598	711	797	857	906	936	955	968	997	18.60
9500,3	262	302	341	414	547	663	755	870	926	955	973	983	988	992	999	11.16
9000,4	262	290	317	366	452	527	591	697	777	837	888	922	944	959	996	19.82
9000,3	243	286	324	393	515	618	705	829	898	937	962	975	983	988	999	12.94
9000,2	247	314	379	499	704	832	900	958	978	987	993	995	997	998	000	6.69
8500,4	226	268	307	369	464	539	601	699	773	828	878	912	936	953	995	20.32
8500,3	208	264	309	383	500	595	674	793	869	915	947	966	976	983	998	14.54
8000,4.5	232	295	346	422	525	599	658	747	811	858	899	926	945	959	996	17.61
8000,4	215	280	331	405	512	589	650	742	809	858	900	928	947	961	996	17.79
8000,4 *	210	277	327	401	505	580	639	729	793	840	883	914	936	952	995	19.08
8000,3.5	204	269	319	395	504	585	649	746	815	864	906	934	952	965	997	17.48
8000,3	189	262	314	394	512	600	670	775	847	894	932	954	968	977	998	15.51
8000,2	162	265	328	428	578	693	781	887	937	962	978	986	990	993	999	10.38
7500,4	218	321	387	480	597	672	730	812	868	905	935	954	967	976	998	13.52
7500,4 +	222	327	395	487	599	685	750	838	891	924	950	965	975	983	998	12.20
7500,4 *	212	314	379	469	581	655	709	786	838	875	909	932	949	961	996	15.76
7500,3	188	289	353	444	567	651	715	807	867	907	938	958	970	979	998	13.90
7000,4	210	371	458	568	687	756	807	873	913	940	960	973	981	987	999	9.91
7000,4 *	204	368	454	557	671	738	783	844	884	911	935	952	964	973	997	12.02
7000,3	188	344	427	531	654	732	789	866	912	941	962	974	982	987	999	10.42
7000,2	173	317	399	506	641	730	792	873	920	948	968	979	985	990	999	10.21

KEY TO THE MODELS:

*   NO CONVECTION
+   L/H = 2.5

H GAMMA (GRIEM)

MODEL	0	.5	1	1.5	2	3	4	5	6	8	10	12	16	20	30A	WIDTH
6500,4	216	459	564	631	679	747	792	826	851	889	915	934	958	972	988	6.20
6500,3	187	427	528	594	642	710	758	794	823	866	897	920	950	967	986	7.12
6000,4.5	248	554	674	742	787	845	882	907	926	950	965	974	985	991	996	3.51
6000,4	237	552	669	735	780	838	874	900	919	944	960	970	982	989	996	3.74
6000,4 **	265	590	706	770	811	864	898	920	937	958	970	978	988	992	997	3.10
6000,4 *	233	547	663	729	773	829	865	890	908	933	949	960	974	982	992	4.32
6000,4 ***	263	587	701	764	804	855	887	909	925	946	959	968	980	986	994	3.66
6000,4,10X	316	633	743	799	835	88C	907	926	939	956	968	975	985	990	996	2.95
6000,4,.1X	255	582	701	765	806	858	891	912	927	948	961	970	981	987	994	3.58
6000,3.5	218	532	650	717	763	822	860	887	907	934	952	964	979	987	995	4.15
6000,3	207	525	640	707	753	812	851	879	900	929	948	961	977	986	994	4.36
6000,2	193	501	612	678	725	788	831	862	886	921	943	958	976	985	994	4.73

MODEL	0.0	.1	.2	.3	.4	.5	.6	.8	1.0	1.2	1.6	2.0	3.0	5.0	10A	WIDTH
5500,4	256	280	383	536	616	657	688	736	772	799	840	868	913	955	988	1.96
5500,3	213	231	313	488	602	645	676	723	758	785	826	855	901	946	984	2.20
5000,4	350	386	523	666	734	771	798	838	866	886	915	934	962	983	996	1.10
5000,3	278	316	445	635	729	767	794	833	860	881	909	928	956	980	995	1.22
5000,2	226	257	371	581	718	759	786	824	852	872	902	921	950	976	994	1.35
4500,4	508	538	667	774	825	855	875	905	924	938	957	969	984	994	999	.63
4500,3	416	463	605	757	824	853	872	901	920	934	953	965	980	992	998	.71
4000,4	765	781	867	923	947	961	969	980	986	990	994	996	998	999	000	.18
4000,3	668	708	824	912	943	956	965	976	983	987	992	995	998	999	000	.23
4000,2	596	646	775	899	945	958	966	977	983	987	992	994	998	999	000	.26

H GAMMA (E-S-W)

MODEL	0	.5	1	1.5	2	3	4	5	6	8	10	12	16	20	30A	WIDTH
6500,4	216	478	585	652	699	764	807	839	863	899	923	940	962	974	989	5.83
6500,3	187	446	550	617	663	730	775	810	838	879	908	929	955	970	987	6.65
6000,4.5	248	575	695	760	803	858	893	917	934	956	968	977	987	992	997	3.27
6000,4	237	572	689	754	797	852	886	909	927	950	964	973	984	990	996	3.48
6000,4 **	265	610	726	787	827	877	908	928	943	962	974	981	989	993	997	2.88
6000,4 *	233	567	683	747	790	843	877	899	916	939	954	964	976	984	992	4.06
6000,4 ***	263	607	721	781	819	868	897	917	931	950	963	971	982	987	994	3.43
6000,4,10X	316	656	761	815	849	891	916	933	945	961	971	978	986	991	996	2.74
6000,4,.1X	255	601	720	782	822	870	900	920	934	953	964	973	983	988	994	3.36
6000,3.5	218	553	672	738	780	837	872	897	916	942	957	969	981	988	995	3.86
6000,3	207	546	663	728	771	828	865	891	910	937	954	966	980	987	995	4.04
6000,2	193	523	635	700	745	807	847	877	899	931	950	964	979	987	995	4.35

MODEL	0.0	.1	.2	.3	.4	.5	.6	.8	1.0	1.2	1.6	2.0	3.0	5.0	10A	WIDTH
5500,4	256	280	385	548	635	677	708	755	790	816	855	881	922	961	990	1.81
5500,3	213	231	313	495	621	665	697	744	777	804	842	869	912	953	986	2.03
5000,4	350	389	528	680	752	788	815	853	879	898	925	942	967	986	997	1.01
5000,3	278	316	446	643	747	785	811	848	874	894	920	937	962	983	996	1.12
5000,2	226	257	372	585	736	778	804	841	867	887	913	931	957	980	995	1.24
4500,4	508	543	674	788	840	868	887	915	933	946	963	973	986	995	999	.58
4500,3	416	464	606	765	838	866	885	912	930	942	959	970	983	994	999	.66
4000,4	765	785	873	931	954	966	974	983	988	991	995	997	999	000	000	.17
4000,3	668	710	826	918	950	962	970	980	985	989	993	996	998	999	000	.21
4000,2	596	646	776	903	951	964	971	980	986	989	993	996	998	999	000	.24

KEY TO THE MODELS:

* NO CONVECTION
** NO BLANKETING